Adhesion Mo

T0222052

Series on

Modern Insights into Disease from Molecules to Man

Series Editor
Victor R. Preedy
Professor of Nutritional Biochemistry
School of Biomedical & Health Sciences
King's College London
and
Professor of Clinical Biochemistry
King's College Hospital
UK

Books in this Series

Under publication

- Adhesion Molecules
- Apoptosis

Under preparation

- Adipokines
- Cytokines

Further books in the series

- Micro RNAs
- Stem Cells
- Molecular Motors
- Telomeres
- Alcohol and Genes
- Oxidative Stress

Adhesion Molecules

Editor

Victor R. Preedy
Professor of Nutritional Biochemistry
School of Biomedical & Health Sciences
King's College London
and
Professor of Clinical Biochemistry
King's College Hospital
UK

CRC Press
Taylor & Francis Group
Boca Raton London New York

CRC Press is an imprint of the
Taylor & Francis Group, an **informa** business

Science Publishers
Enfield, New Hampshire

CRC Press
Taylor & Francis Group
6000 Broken Sound Parkway NW, Suite 300
Boca Raton, FL 33487-2742

First issued in paperback 2017

ISBN-13: 978-1-57808-671-9 (hbk)
ISBN-13: 978-1-138-11789-1 (pbk)

This book contains information obtained from authentic and highly regarded sources. While all reasonable efforts have been made to publish reliable data and information, neither the author[s] nor the publisher can accept any legal responsibility or liability for any errors or omissions that may be made. The publishers wish to make clear that any views or opinions expressed in this book by individual editors, authors or contributors are personal to them and do not necessarily reflect the views/opinions of the publishers. The information or guidance contained in this book is intended for use by medical, scientific or healthcare professionals and is provided strictly as a supplement to the medical or other professional's own judgement, their knowledge of the patient's medical history, relevant manufacturer's instructions and the appropriate best practice guidelines. Because of the rapid advances in medical science, any information or advice on dosages, procedures or diagnoses should be independently verified. The reader is strongly urged to consult the relevant national drug formulary and the drug companies' and device or material manufacturers' printed instructions, and their websites, before administering or utilizing any of the drugs, devices or materials mentioned in this book. This book does not indicate whether a particular treatment is appropriate or suitable for a particular individual. Ultimately it is the sole responsibility of the medical professional to make his or her own professional judgements, so as to advise and treat patients appropriately. The authors and publishers have also attempted to trace the copyright holders of all material reproduced in this publication and apologize to copyright holders if permission to publish in this form has not been obtained. If any copyright material has not been acknowledged please write and let us know so we may rectify in any future reprint.

Cover Illustration designed by Deirdre R. Coombe is a composite image compiled from cartoons of adhesion molecules appearing in figures in Chapter One. The publications inspiring these cartoons are cited in Chapter One.

Library of Congress Cataloging-in-Publication Data
Adhesion molecules / editor, Victor R. Preedy.
     p.; cm.
  Includes bibliographical references and index.
  ISBN 978-1-57808-671-9 (hardcover)
 1. Cell adhesion molecules. I. Preedy, Victor R.
  [DNLM: 1. Cell Adhesion Molecules. 2. Cell Adhesion. QU 55.7 A234 2010]
  QP552.C42A337 2010
  541'.22--dc22

                                                          2009054197

Visit the Taylor & Francis Web site at
http://www.taylorandfrancis.com

and the CRC Press Web site at
http://www.crcpress.com

Preface

The adhesion molecules are a family of proteins located on the cell surface and are involved in cell-to-cell or cell-extracellular matrix binding. They have a role in maintaining the cellular integrity of tissues and the functional viability of organs. In many diseases the normative features of adhesion molecules are perturbed. Thus, understanding how adhesion molecules behave in health and disease is a key step in the advancement of science and health in general. This book **Adhesion Molecules** is designed to impart important information on adhesion molecules. It is multidisciplinary and covers all scientific levels.

Adhesion Molecules has four major sections, covering (1) General aspects, (2) Cellular, metabolic and organ specific diseases, (3) The immune system and cancer and finally (4) Cardiovascular and neuronal aspects of adhesion molecules. The simplistic nature of the main section headings, however, should not detract from the fact that individual chapters are highly detailed. For example, in the first part of the book, chapters cover the structure and classification of adhesion molecules, immunohistochemical localization, junctional and inflammatory adhesion molecules and molecular imaging. The rest of the book is equally detailed, covering adhesion molecules in relation to signalling pathways, gene expression, arginine, stem cells, neutrophil migration, leukocytes decompression sickness, ischemia reperfusion injury, diabetes, obesity, metabolic syndrome, weight loss, hypoxia, kidney disease, smoking, osteoprotegerin, tumors, bioinformatics, high-throughput technologies, chemotherapy, hematopoiesis, lipoproteins, fatty acids, atrial fibrillation and heart disease, brain and dementia. Effectively, the book **Adhesion Molecules** has wide coverage from the cell to the whole body.

There are three distinguishing features of **Adhesion Molecules** that sets it apart from other scientific texts. As well as an Abstract, each chapter has one or more Key Facts or Key Features which expands areas of interest for the novice. There are also at least five definitions and explanations of key terms or words used in each chapter. Finally, each chapter is summarised with at least five bullets points. The book is well-illustrated with numerous figures and tables and the entire book has an excellent index. As a consequence of these features the readership is designed to be broad, from the novice to leading expert.

Contributors of **Adhesion Molecules** are all either international or national experts, leading authorities or are carrying out ground breaking and innovative work on their subject. The book is an essential book for research scientists, pathologists, molecular biologists, biochemists or cellular biochemists, clinicians, health care professionals, general practitioners as well as those interested in disease in general.

Adhesion Molecules is part of the **Modern Insights into Disease from Molecules to Man** series.

Professor Victor R. Preedy

Contents

Feature, Structure and Classification of Adhesion Molecules: An Overview

Deirdre R. Coombe[1] **and Danielle E. Dye**[2]
[1]Molecular Immunology, School of Biomedical Sciences,
Curtin University of Technology, Level 3, MRF Building, Rear, 50 Murray Street,
Perth WA 6000, E-mail: D.Coombe@curtin.edu.au
[2]Department of Health, Western Australia, 3rd Floor, C Block, 189 Royal St.,
East Perth WA 6004, E-mail: ddye06@gmail.com

ABSTRACT

Cell adhesion molecules allow cells to communicate with each other and with their environment. The interactions between cell adhesion receptors and their ligands orchestrate the assembly of cells into tissues, organs and systems, and lead to the formation of multicellular organisms. Numerous different adhesive events, which may be synergistic or antagonistic, are required for organizing a tissue or a cell behavior. It is the balance between these events that determines the structure of a tissue, or whether cells stay in tight association with other cells or migrate around the body.

There are four main families of cell adhesion receptors: the immunoglobulin superfamily, integrins, cadherins and selectins. The cadherins mediate strong cell-cell adhesion and play a fundamental role in morphogenesis and development, while integrins are critical to cell-matrix interactions and cell migration through the extracellular matrix. Immunoglobulin superfamily members contribute to and modulate cell-cell interactions. These three large families contribute to adhesive interactions between many cell types in different tissues. The selectins have three family members and they perform a very specific role critical for leukocyte migration from the vasculature.

Key terms are defined at the end of the chapter.

INTRODUCTION

Cell adhesion molecules (CAMs) mediate vital cell behaviors: cell-to-cell and cell-to-matrix adhesion. These interactions are largely responsible for the precise, distinctive organization required for a collection of cells to form a tissue, for tissues to form organs and for organs to form a system. Cell adhesion is the primary way in which cells communicate with each other and is important in development and morphogenesis, cell migration, and the regulation of gene expression, cell division and cell death (Alberts *et al.* 2002).

Cell adhesion is the result of multiple different adhesion molecule–ligand binding events and involves cross-talk and cooperation between many different adhesion molecules, each binding its ligands with different affinities (Chen and Gumbiner 2006). Cell adhesion events range from very stable to transient. For example, the interactions between muscle cells are strong and stable, whereas the adhesion events that enable cells to move through blood vessel walls are weaker and transitory. There is considerable redundancy between cell adhesion molecules, increasing the complexity of cell-cell and cell-matrix interactions.

All CAMs are integral membrane proteins comprising extracellular, transmembrane and cytoplasmic domains. These extracellular domains extend from the cell and bind ligands on other cells or within the extracellular matrix (ECM), while cytoplasmic domains interact with cytoskeletal proteins to provide an intracellular anchor. Binding can occur between adhesion molecules of the same type (homophilic binding), between adhesion molecules of a different type (heterophilic binding), or via an intermediary 'linker' that binds other adhesion molecules.

There are four main types of adhesion molecules: immunoglobulin superfamily members, integrins, cadherins and the selectins.

IMMUNOGLOBULIN SUPERFAMILY (IgSF)

The IgSF is one of the largest and most diverse protein families in the genome with over 750 members. Proteins of the IgSF possess one or more immunoglobulin (Ig)-like domains, which are domains homologous to the basic structural unit of immunoglobulin (antibody) molecules. Most IgSF members are cell surface proteins, and many are CAMs (Aricescu and Jones 2007).

The Immunoglobulin (Ig) Domains

Immunoglobulin domains consist of 70 to 110 amino acids and are classified into two subtypes: the variable (V) and constant (C) domains (Barclay 2003). Both V and C subtypes share a characteristic sandwich structure, consisting of two sheets of anti-parallel β strands stabilized by a disulfide bridge, producing a compact structure relatively insensitive to proteolytic cleavage. The V subtype contains the antigen-binding properties and consists of nine β strands, while the C subtype mediates effector functions and contains seven β strands (Barclay 2003).

As these domains were first described in immunoglobulins, it was thought they had developed to allow antibody recognition and were specific to the immune system. However, as increasing numbers of proteins were found with similar domains, it became clear that this arrangement is ancient in evolutionary terms and mediates a wide array of interactions outside the immune system (Barclay 2003). These sequences are called IgSF domains or Ig-like domains, distinguishing them from the domains of immunoglobulins (Barclay 2003).

The IgSF domains are classified as V, C1, C2 or I (intermediate), according to sequence patterns and length. C1 corresponds to the C domain of immunoglobulins, whereas C2 is a variant found in most other IgSF proteins. The I domain is structurally similar to the V domain, but also contains features characteristic of C domains (Harpaz and Chothia 1994). IgSF domains contain relatively few highly conserved residues, and recognizing these domains can be difficult. Characteristic features of IgSF domains include alternating hydrophobic residues in the β strands and the conserved cysteine residues that form disulfide bonds between the two β-sheets.

Structure of IgSF Cell Adhesion Molecules

Most IgSF members are type I membrane proteins, consisting of an amino N-terminal extracellular region, a single transmembrane domain and a carboxyl C-terminal cytoplasmic domain (Barclay 2003). The extracellular domains of IgSF CAMs contain a number of Ig-like domains and many also contain one or more fibronectin type III (FNIII) domains (Fig. 1). The FNIII domain was first described in the ECM protein fibronectin and consists of overlapping β-sheets containing seven anti-parallel β-strands. This topology is similar to the immunoglobulin C domain, although the FNIII domain is not stabilized by a disulfide bridge (Harpaz and Chothia 1994).

In CAMs with multiple Ig-like domains, one or more V-type domain(s) are usually located closest to the N-terminus, followed by one or more C2-type domain(s). When present, the FNIII domains are adjacent to the membrane (Aricescu and Jones 2007). The transmembrane domains of IgSF CAMs are small, while the intracellular domains vary in length, with many containing signaling motifs and/or regions that interact with cytoskeletal or adaptor elements (Barclay 2003). Thus, interactions of IgSF CAMs at the cell surface can lead to signaling within the cell (outside-in signaling) and inside-out signaling.

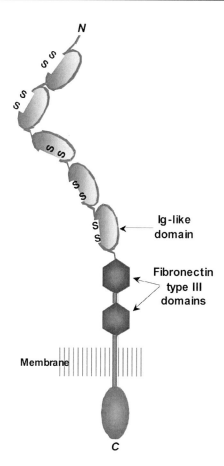

Fig. 1 Neural cell adhesion molecule (NCAM). The IgSF member NCAM has a large extracellular domain containing five Ig-like domains stabilized by disulfide bonds and two fibronectin type III repeats, and short transmembrane and intracellular domains.

Interactions of IgSF Cell Adhesion Molecules

IgSF CAMs bind both homophilic and heterophilic ligands. These interactions are often mediated through the N-terminal Ig-like domains, which commonly bind other Ig-like domains, but may interact with integrins and carbohydrates (Barclay 2003).

Unlike the high-affinity binding that occurs between antibody and antigen, interactions between IgSF CAMs and their ligands are weak (Barclay 2003). However, crystal structures of the extracellular domains from several IgSF CAMs suggest these molecules form ordered clusters at cell-cell contacts

(Aricescu and Jones 2007), with functional receptor-ligand interactions resulting from the cumulative effect of many specific, weak binding events. These clusters require interactions outside of the primary ligand binding site in both *cis* (between molecules on the same cell) and *trans* (between molecules on another cell surface) as shown in Fig. 2 (Aricescu and Jones 2007). IgSF proteins form *trans* interactions with molecules such as integrins (Fig. 2). For example, the interaction between intercellular adhesion molecule (ICAM-1) and specific integrins probably involves a heterophilic adhesion 'zipper' (Luo *et al.* 2007). Other IgSF proteins engage in homophilic binding events and a double zipper-like adhesion complex appears to stabilize NCAM adhesion.

INTEGRINS

The integrins are a large family of CAMs involved in cell-cell adhesion and cell interactions with the ECM proteins laminin, collagen, fibronectin and vitronectin. The extracellular domains of integrins bind to ECM proteins, while their cytoplasmic domains engage the cell's cytoskeleton. Integrins mediate communication between the extra- and intracellular environments; integrin-cytoskeleton interactions affect the binding affinity and avidity of integrins for ECM ligands (inside-out signaling), whereas ECM-integrin interactions lead to changes in the shape and composition of the cell architecture (outside-in signaling) (Shimaoka *et al.* 2002). All integrin-ligand interactions depend on the presence of divalent cations such as Ca^{2+}, Mg^{2+} and Mn^{2+}.

Integrin Structure: The α and β Subunits

Integrins are heterodimeric transmembrane glycoproteins, consisting of non-covalently bound α and β subunits. Eighteen α and eight β subunits are described in humans, assembling into 24 different integrins (Arnaout *et al.* 2007). The α and β chains share no homology, being distinct polypeptides with specific domain structures. The extracellular domains from both subunits contribute to the ligand-binding site of the heterodimer (Takada *et al.* 2007).

The extracellular domain of the α subunit contains seven repeats of approximately 60 residues, which fold into a seven-bladed propeller structure with β-sheets arranged around a central axis. A subset of integrin α chains have an insertion domain (αA) containing a cation binding site, located between repeats two and three of the propeller. C-terminal to the propeller is an Ig-like Thigh domain, followed by two β-sandwich modules, Calf-1 and Calf-2, and a small transmembrane domain (Fig. 3A) (Humphries *et al.* 2003, Arnaout *et al.* 2007). The intracellular domains of the α subunits show little homology, except for a conserved motif proximal to the transmembrane region that associates with the cytoplasmic tail of the β chain (Arnaout *et al.* 2007).

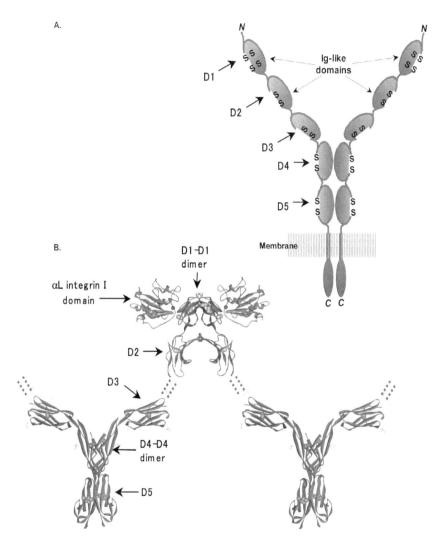

Fig. 2 Dimerization structure of intercellular adhesion molecule (ICAM-1). (A) Schematic of *cis*-dimerization of two ICAM-1 molecules (orange and blue), interactions of Ig-like domains four and five (D4, D5) are shown. (B) Ribbon diagram of two ICAM-1 dimers. The interaction of Ig-like domains D1 and D2 of one molecule from each dimer with its neighbor forms the binding site for the αA domain of αL integrin (green) (PDB code: 1MQ8) (Shimaoka *et al.* 2003). An approximate alignment of the D1-D2 structures with the dimer of D3-D5 (PDB code: 1P53) (Yang *et al.* 2004) is shown. Yellow sulfur molecules indicate the position of disulfide bonds. Divalent cations (Mg^{++}) are in pink. Figure adapted from Yang *et al.* (2004).

Color image of this figure appears in the color plate section at the end of the book.

The extracellular region of integrin β subunits contains a hybrid domain, a plexin-semaphorin-integrin (PSI) domain, four epidermal growth factor (EGF)-like repeats and a β tail domain (βTD) (Fig. 3A). The hybrid domain is an Ig-like domain comprising an αA-like motif (the βA domain) inserted between two β-sheets. The βA domain contains a metal-binding motif and binds divalent cations when the α subunit does not contain an αA domain (Takada *et al.* 2007). The PSI domain forms a two-stranded anti-parallel β-sheet flanked by two short helices and contributes to integrin activation. The cytoplasmic tails of integrin β chains are short and highly conserved and contain one or two phosphorylation motifs. The tails recruit proteins such as talin, which bind actin filaments, thus connecting integrins to the actin cytoskeleton. Integrin-cytoskeletal interactions are essential for all integrin-mediated functions (Takada *et al.* 2007).

The Integrin Heterodimer

The extracellular structure of integrin heterodimers consists of a 'head' and 'legs'. The head, which mediates integrin-ligand interactions, is the major point of contact between α and β subunits. Contacts are formed by interactions of the propeller (α chain) with the βA domain (β chain) (Arnaout *et al.* 2007). The legs are formed by the Thigh and Calf domains of the α subunit and the PSI, EGF and βTD domains of the β subunit (Arnaout 2002). The 'knee' of the α subunit lies at the junction of the Thigh and Calf-1 domains, whereas the corresponding region of the β chain lies in the PSI–EGF-1/2 region. The knees allows integrins to adopt a bent or upright conformation, which is critical to integrin activation (Fig. 3) (Arnaout 2002). Additional contacts between α and β subunits exist between Calf-1 and EGF3; Calf-2 and EGF4; Calf-2 and βTD; and the α and β cytoplasmic tails (Arnaout 2002).

Integrin-Ligand Binding

Integrins may bind divalent cations via two different domains, αA and βA. The αA domain is found in the α chains: α1, α2, α10, α11, αM, αX, αL and αD, and mediates cation binding when present. If the αA domain is absent, cation binding occurs via the βA domain found in all β chains (Arnaout *et al.* 2007).

The ligand-binding pocket is formed by the interface between blades 2 and 3 of the propeller domain of the α chain and the βA domain of the β chain. Its orientation alters slightly depending on whether or not the α chain contains the αA domain. Integrins are classified according to the presence or absence of the αA domain and/or their ligands (Fig. 4). Four αA-containing α chains (α1, α2, α10 and α11) combine with β1 to give a distinct laminin/collagen binding subfamily. The non αA-containing integrins (α3β1, α6β1, α7β1 and α6β4) are selective laminin receptors and bind to a different site on laminin than the αA-containing integrins

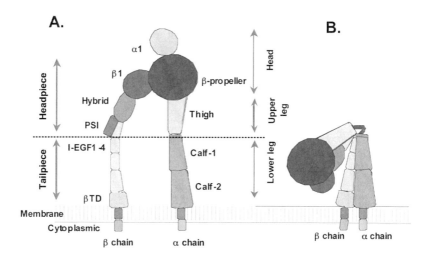

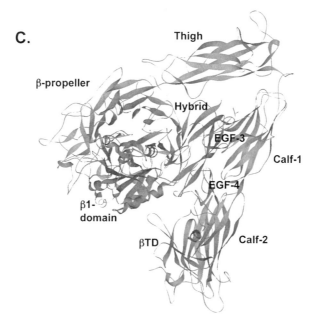

Fig. 3   Integrin organization. (A) Schematic of activated αXβ2-integrin. (B) The integrin in an inactive conformation. (A) and (B) were adapted from Luo *et al.* (2007). (C) Ribbon diagram showing the interaction between the extracellular segments of αV (blue) and β3 (orange) subunits (PDB code: 1JV2) (Xiong *et al.* 2001).

Color image of this figure appears in the color plate section at the end of the book.

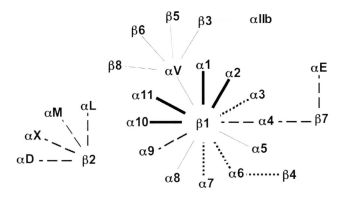

Fig. 4 The integrin family. Lines indicate heterodimeric α–β pairings identified on mammalian cells. The lines joining α–β pairs indicate the ligands they bind: bold line, laminin/collagen binding; dotted line, selective laminin binding; dashed line, LDV or LEV motif binding; and standard line, RGD binding integrins.

(Humphries *et al.* 2006). The αV-integrins, and α5β1, α8β1 and αIIbβ3, recognize the RGD (arginine, glycine, aspartic acid) motif, which is important for integrin binding to fibronectin, vitronectin and collagen. Although many RGD-binding integrins interact with the same ligands, they bind with different affinities, reflecting the preciseness of the fit of the RGD motif with the ligand binding pockets created by different α-β chain combinations (Humphries *et al.* 2006). Integrins α4β1, α4β7, α9β1 and αEβ7 recognize an acidic motif; LDV (leucine, aspartic acid, valine) found in fibronectin and some Ig-SF members. The β2-integrins (αDβ2, αLβ2, αMβ2, αXβ2) bind a similar LEV motif, where the aspartic acid (D) is replaced by glutamate (E) (Humphries *et al.* 2006).

Integrin Activation

Integrins display three major activation states: inactive (low affinity), active (high affinity) and ligand occupied (Askari *et al.* 2009). In the switchblade model of integrin activation, an inactive integrin is bent over (Fig. 3), causing the ligand-binding site to be buried. Integrins are active when upright, fully exposing the ligand-binding pocket (Shimaoka *et al.* 2002, Askari *et al.* 2009).

When inactive, integrins bind ligands with low affinity. This low-affinity interaction stimulates intracellular signals that activate the cytoskeletal protein talin, which binds to the cytoplasmic tail of the β subunit. This disrupts the inhibitory association between the α and β chains, allows the cytoplasmic and transmembrane regions of the two chains to separate, and leads to the extracellular domain changing from a bent to an extended form to allow high-affinity ligand binding (Askari *et al.* 2009).

Although integrins bind their ligands with high affinity following inside-out activation, stable binding requires their cytoplasmic domains to be anchored to the cytoskeleton. Integrin-ligand binding triggers integrins to cluster on the cell surface and leads to the recruitment of enzymes and adaptor molecules that form adhesion complexes linking integrins to the cytoskeleton (Arnaout *et al.* 2007).

CADHERINS

The cadherin superfamily mediates Ca^{2+}-dependent cell-cell adhesion. Cadherins play critical roles in tissue organization and morphogenesis, involving cell recognition and sorting, coordinated cell movement and the formation and maintenance of cell and tissue patterning (Ivanov *et al.* 2001). Cadherins are defined by the presence of cadherin repeats in their extracellular domains. More than 100 cadherins have been identified and are classified into subgroups: classical type I cadherins, type II cadherins, desmosomal cadherins, protocadherins, and other cadherin-related molecules (Ivanov *et al.* 2001, Morishita and Yagi 2007).

Basic Structure

Most cadherins comprise an N-terminal extracellular domain containing five or six cadherin repeats, a short transmembrane region and a C-terminal intracellular domain (Ivanov *et al.* 2001). The cadherin repeat unit is about 110 amino acids and consists of seven strands arranged in a compact β-barrel structure similar to the Ig-like repeat (Fig. 5C) (Van Roy and Berx 2008). These repeats are linked by highly conserved Ca^{2+}-binding motifs that provide stability to the extracellular domain (Ivanov *et al.* 2001). The cadherin domains are numbered, with the repeat closest to the N-terminus designated extracellular cadherin (EC) 1 (Fig. 5). The C-terminal intracellular domain binds to a variety of molecules that link cadherins to the cytoskeleton.

Cadherin Subgroups

Classical (Type I) Cadherins

The classical cadherins were the first to be described and are named after the tissue in which they were identified. The extracellular domain of classical cadherins consists of an N-terminal pre-domain followed by five EC domains, with EC5 characterized by four conserved cysteines not present in the other repeats (Patel *et al.* 2003). EC1 contains a conserved tryptophan at position two of the mature protein (W2) and a hydrophobic pocket capable of accommodating the W2 of a neighboring EC1 (Fig. 5C).

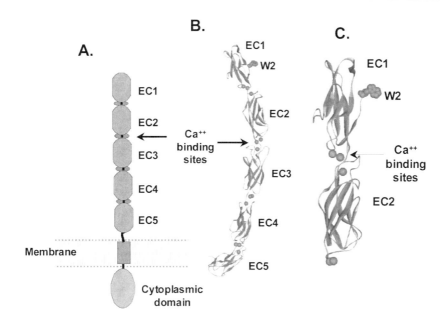

Fig. 5 Type 1 cadherin. (A) Schematic of E-cadherin, a classical cadherin with five extracellular cadherin domains (EC1-EC5) (blue) stabilized by Ca^{2+} at domain interfaces (green). (B) Ribbon diagram of the C-cadherin ectodomain (blue) with bound calcium molecules (green) from *Xenopus laevis* (PDB code:1L3W) (Boggon *et al.* 2002). (C) Ribbon diagram (represented according to secondary structure) of human E-cadherin domains EC1 and EC2 showing the Ig-like fold, the conserved tryptophan residue (W2) (orange) and the bound Ca^{2+} (PDB code: 2O72) (Parisini *et al.* 2007).

Color image of this figure appears in the color plate section at the end of the book.

Following cleavage of the pre-domain, the N-terminal residues (including W2) form an adhesion arm that interacts with the acceptor pocket in the body of EC1 and a salt bridge also forms between the N-terminal aspartic acid (D1) and a conserved glutamic acid (E89) near the acceptor pocket (Van Roy and Berx 2008). If this interaction occurs within a molecule, the molecule is closed and cannot dimerize. The closed form is in equilibrium with an open form that can interact with neighboring molecules. This 'strand swapping' probably occurs first between molecules on the same cell (*cis*) and then between molecules on neighboring cells (*trans*) to mediate homophilic cell-cell adhesion (Fig. 6) (Van Roy and Berx 2008). A tripeptide consisting of histidine, alanine and valine (HAV, amino acids 79-81) within EC1 is an essential recognition sequence for cell adhesion in Type I cadherins (Ivanov *et al.* 2001). The classical cadherins display homophilic specificity; for example, E-cadherin binds E-cadherin more strongly than it binds other classical cadherins (Patel *et al.* 2003).

The cytoplasmic domains of classical cadherins interact with proteins involved in endocytosis and intracellular signaling, as well as with the actin cytoskeleton.

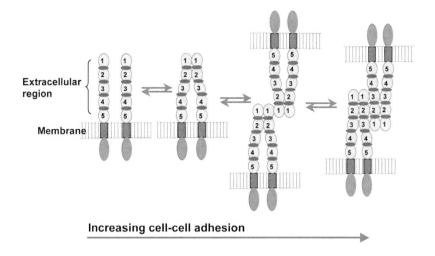

Extracellular region

Membrane

Increasing cell-cell adhesion

Fig. 6 Cadherin-mediated cell-cell adhesion. A proposed mechanism for cadherin-mediated cell-cell adhesion where cadherin molecules form *cis* dimers, which remain associated to form *trans* adhesive complexes with dimers on an opposing cell surface. Increasing interdigitation stabilizes adhesion. Adapted from Leckband and Prakasam (2006).

Interactions with actin are crucial for cadherins to cluster and form stable cell-cell adhesions. The cytoplasmic tail of classical cadherins has a juxtamembrane domain (JMD) and a C-terminal catenin-binding domain (CBD). The JMD binds to p120-catenin and stabilizes the molecule at the cell surface while the CBD binds to β-catenin, which links classical cadherins to the actin cytoskeleton via intermediates such as α-catenin and vinculin (Ivanov *et al.* 2001).

Type II Cadherins

Type II cadherins have a smaller pre-domain than classical cadherins, lack the HAV adhesion recognition sequence and have two conserved tryptophans in EC1 at positions two and four of the mature protein (W2 and W4), with a correspondingly larger homophilic pocket (Patel *et al.* 2006, Morishita and Yagi 2007). The EC1 domains of neighboring Type II cadherins dimerize similarly to Type I cadherins, with W2 and W4 inserting into an acceptor site on a neighboring molecule and a salt bridge forming between the N-terminus and a conserved glutamic acid (Patel *et al.* 2006). However, type II cadherins have a greater tendency towards heterophilic binding with other subfamily members than Type I cadherins (Patel *et al.* 2006). The intracellular domain of type II cadherins is similar to type I cadherins, except that the CBD binds γ-catenin rather than β-catenin (Patel *et al.* 2003).

Desmosomal Cadherins

In humans, there are seven desmosomal cadherins: three desmocollins (DSC 1-3) and four desmogleins (DSG 1-4). Both subfamilies have five EC domains, although the fifth domain is less well conserved (Angst *et al.* 2001). The EC1 domains of DSC and DSG proteins are structurally similar to classical cadherins (e.g., a conserved tryptophan [W2] and glutamic acid [E89]) and it is likely these proteins also dimerize via strand swapping (Posy *et al.* 2008). Desmosomal cadherins are believed to form *cis* dimers within a desmosome, then interact with dimers on a neighboring cell, with heterophilic interactions between DSC and DSG proteins favored (Angst *et al.* 2001).

The DSC and DSG proteins differ from classical cadherins and from each other in their cytoplasmic domains. Both groups contain a membrane proximal intracellular anchor domain, which corresponds to the JMD in type I cadherins, and an intracellular cadherin segment (ICS) similar to the CBD in classical cadherins. The ICS in DSC and DSG proteins, however, binds predominantly to γ-catenin and links to the intermediate filament cytoskeleton (Green and Simpson 2007).

Protocadherins

The protocadherins are the largest subgroup of the cadherin superfamily, with at least 70 members (Frank and Kemler 2002). Protocadherins have six or seven EC domains, a single transmembrane region and divergent cytoplasmic domains. Members of the protocadherin subfamily do not contain a conserved W2 residue or a hydrophobic acceptor site in their EC1 domain. Instead, protocadherins have two cysteine residues and a disulfide loop in the EC1 domain with an RGD motif (Morishita and Yagi 2007). Interestingly, protocadherins mediate homophilic and heterophilic adhesion in a *cis* conformation and may be involved in *trans* homophilic adhesion after forming complexes with other molecules (Frank and Kemler 2002, Morishita and Yagi 2007). Few intracellular binding partners have been identified for the cytoplasmic domains, but one is the tyrosine kinase Fyn, which binds to the constant domain of the α-protocadherin family (Frank and Kemler 2002).

SELECTINS

The selectins are C-type lectins expressed on the surface of leukocytes, platelets and activated endothelial cells. Selectins promote the tethering and rolling of leukocytes and platelets on the vascular endothelium and are important for lymphocyte homing to secondary lymphoid organs and for the recruitment of leukocytes to sites of inflammation (McEver 2002, Cummings and McEver 2009).

There are three selectins: L-selectin, expressed on all leukocytes except activated T-cells; E-selectin, expressed by cytokine-activated endothelial cells; and P-selectin, expressed by platelets and activated endothelial cells (Alberts *et al.* 2002). The selectins are involved in heterophilic binding and interact with glycoproteins and glycolipids that carry α2,3-linked sialic acid and α1,3-linked fucose structures on their terminal branches (McEver 2002, Smith 2008).

Selectin Structure

C-type lectins are carbohydrate-binding proteins defined by the presence of a highly conserved Ca^{2+}-dependent carbohydrate recognition domain (CRD) of about 120 amino acids. Selectins are rigid, extended molecules and their extracellular regions comprise an N-terminal CRD domain that mediates binding to carbohydrate moieties, an EGF-like domain and differing numbers of short consensus repeats (SCR). The size difference among selectins reflects the number of SCRs each contains: L-selectin has two SCRs, E-selectin has six and P-selectin has alternatively spliced forms containing eight or nine SCRs (Fig. 7) (Smith 2008, Cummings and McEver 2009). The transmembrane and cytoplasmic domains of the selectins are quite divergent, indicating they bind different intracellular proteins (Ivetic and Ridley 2004).

P-Selectin

P-selectin is constitutively expressed by megakaryocytes, but is stored in pre-formed cytosolic granules, which fuse with the cell membrane following activation of platelets and endothelial cells (Ivetic and Ridley 2004). P-selectin may be internalized via clathrin-coated pits and directed to secretory granules for recycling, or to lysosomes for degradation (Kaur and Cutler 2002). The main ligand for P-selectin is P-selectin glycoprotein ligand 1 (PSGL-1), a sialomucin expressed on microvilli-like projections of leukocytes. Interactions between P-selectin and PSGL-1 are critical for leukocytes to tether and roll on endothelial cells, or immobilized platelets, expressing P-selectin (Smith 2008). P-selectin also binds weakly to some forms of heparin and heparan sulfate and to some glycoproteins carrying the sialylated Lewis X (SLe^X) structure (Fig. 7B) (Cummings and McEver 2009).

E-Selectin

E-selectin is expressed *de novo* in cytokine-stimulated endothelial cells and recognizes a number of glycoproteins on leukocytes that display the SLe^x moiety, including PSGL-1 (Ivetic and Ridley 2004, Cummings and McEver 2009). The position of the SLe^x binding site is conserved between E- and P-selectin (Fig. 7B). The cytoplasmic tail of E-selectin interacts with cytoskeletal elements and signaling molecules, and E-selectin is internalized via clathrin-coated pits (Setiadi and McEver 2008).

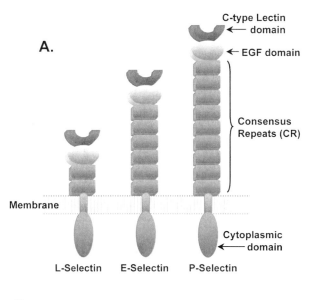

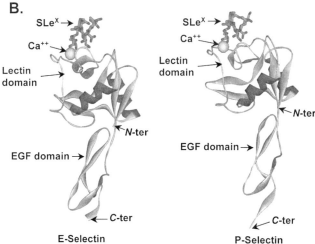

Fig. 7 The selectin family. (A) Schematics of L-selectin, E-selectin and P-selectin. (B) Structure of the C-type lectin and the EGF domains of E- and P-selectins (PDB codes: 1G1T and 1G1R) (Somers *et al.* 2000) depicted as ribbon diagrams represented according to secondary structure. Structural similarities are evident and, if optimally overlaid, bound Ca^{2+} ions are superimposed (Somers *et al.* 2000).

Color image of this figure appears in the color plate section at the end of the book.

L-Selectin

L-selectin is constitutively expressed on the microvilli of leukocytes. Cell surface levels of L-selectin are regulated by metalloproteinase-dependent cleavage of the extracellular domain following leukocyte activation (Smith 2008). The intracellular tail of L-selectin binds to calmodulin, and to the actin cytoskeletan via α-actinin and the ezrin/radixin/moesin protein family. The balance between these competing intracellular interactions may influence the proteolytic cleavage of the extracellular domain (Ivetic and Ridley 2004). The extracellular domain of L-selectin binds both PSGL-1 and glycoproteins found on specialized high endothelial venules of peripheral lymphoid tissue (Smith 2008).

Selectin Functions

The selectins act together to promote tethering and rolling of leukocytes along the vascular endothelium. This occurs under dynamic conditions, as blood flowing in the capillaries exerts shear forces on the leukocytes and platelets as they interact with each other and endothelial cells (McEver 2002, Smith 2008). At low shear rates, leukocytes are unable to roll, but as shear rates increase past a threshold, selectin-ligand bonds are strengthened, a phenomenon known as 'catch bonds' (Smith 2008). When shear rates increase beyond an optimal level, bonds weaken and shorten the lifetime of selectin-ligand binding. These interactions are called 'slip bonds' (Smith 2008). A combination of catch and slip bonds allows leukocytes to roll along endothelia between 100 and 1000 times slower than the mean rate of blood flow, with rolling sustained by the formation of new and the dissociation of old bonds (Smith 2008).

SUMMARY

- Cell adhesion receptors mediate cell communication and cell-cell and cell-matrix interactions.
- Without cell adhesion molecules, there would be no multicellular organisms.
- The main groups of cell adhesion receptors are the Ig superfamily, integrins, cadherins and selectins
- Cell-cell adhesion is mediated primarily by cadherins and is modulated by members of the Ig superfamily.
- Integrins are critical for cell-matrix interactions.
- Selectins mediate leukocyte tethering and rolling in the blood vasculature.

Acknowledgements

We acknowledge those colleagues whose work we were unable to cite due to space limitations.

Abbreviations

αI /αA	insertion domain of integrin α chains
C domain	constant immunoglobulin domain
CAM	cell adhesion molecule
CBD	catenin binding domain
CRD	carbohydrate recognition domain
C-terminal	carboxyl-terminal
DSC	desmocollin
DSG	desmoglein
EC domain	extracellular cadherin domain
ECM	extracellular matrix
EGF	epidermal growth factor
FNIII domain	fibronectin type III domain
I domain	intermediate immunoglobulin domain
ICAM-1	intercellular adhesion molecule 1
ICS	intracellular cadherin segment
Ig	immunoglobulin
IgSF	immunoglobulin superfamily
JMD	juxtamembrane domain
NCAM	neural cell adhesion molecule
N-terminal	amino-terminal
PSI domain	plexin-semaphorin-integrin domain
SCR	short consensus repeat
V domain	variable immunoglobulin domain

Definition of Terms

Cadherins: Cell surface proteins that mediate calcium dependent cell-cell adhesion, each containing an extracellular cadherin (EC) repeat.

Heterophilic binding: The binding by an adhesion molecule to a different molecule.

Homophilic binding: The binding of an adhesion to molecule itself on another cell.

Ig-like domain: The basic structural unit of IgSF members, consisting of two sheets of anti-parallel β strands which form a sandwich structure, stabilized by a disulfide bridge.

Immunoglobulin superfamily (IgSF): A family of proteins that contain at least one Ig-like domain.

Integrins: Cell adhesion molecules mediating cell-cell and cell-matrix adhesion and consisting of two non-covalently bound subunits (α and β).

Selectins: Carbohydrate-binding adhesion molecules, E-, L- and P-selectin, that mediate early stages of leukocyte migration from the vasculature.

References

Alberts, B. and A. Johnson, J. Lewis, M. Raff, K. Roberts, and P. Walter. 2002. Molecular Biology of the Cell. Garland Science, New York and London.

Angst, B.D. and C. Marcozzi, and A.I. Magee. 2001. The cadherin superfamily: diversity in form and function. J. Cell Sci. 114: 629-641.

Aricescu, A.R. and E.Y. Jones. 2007. Immunoglobulin superfamily cell adhesion molecules: zippers and signals. Curr. Opin. Cell Biol. 19: 543-550.

Arnaout, M.A. 2002. Integrin structure: new twists and turns in dynamic cell adhesion. Immunol. Rev. 186: 125-140.

Arnaout, M.A. and S.L. Goodman, and J.P. Xiong. 2007. Structure and mechanics of integrin-based cell adhesion. Curr. Opin. Cell Biol. 19: 495-507.

Askari, J.A. and P.A. Buckley, A.P. Mould and M.J. Humphries. 2009. Linking integrin conformation to function. J. Cell Sci. 122: 165-170.

Barclay, A.N. 2003. Membrane proteins with immunoglobulin-like domains—a master superfamily of interaction molecules. Semin. Immunol. 15: 215-223.

Boggon, T.J. and J. Murray, S. Chappuis-Flament, E. Wong, B.M. Gumbiner and L. Shapiro. 2002. C-cadherin ectodomain structure and implications for cell adhesion mechanisms. Science 296: 1308-1313.

Chen, X. and B.M. Gumbiner. 2006. Crosstalk between different adhesion molecules. Curr. Opin. Cell Biol. 18: 572-578.

Cummings, R.D. and R.P. McEver. 2009. C-type Lectins. Essentials of Glycobiology. A. Varki (Ed.). CSH Press, New York.

Frank, M. and R. Kemler. 2002. Protocadherins. Curr. Opin. Cell Biol. 14: 557-562.

Green, K.J. and C.L. Simpson. 2007. Desmosomes: new perspectives on a classic. J. Invest. Dermatol. 127: 2499-2515.

Harpaz, Y. and C. Chothia. 1994. Many of the immunoglobulin superfamily domains in cell adhesion molecules and surface receptors belong to a new structural set which is close to that containing variable domains. J. Mol. Biol. 238: 528-539.

Humphries, J.D. and A. Byron, and M.J. Humphries. 2006. Integrin ligands at a glance. J. Cell Sci. 119: 3901-3903.

Humphries, M.J. and E.J.H. Symonds, and A.P. Mould. 2003. Mapping functional residues onto integrin crystal structures. Curr. Opin. Struct. Biol. 13: 236-243.

Ivanov, D.B. and M.P. Philippova, and V.A. Tkachuk. 2001. Structure and functions of classical cadherins. Biochemistry 66: 1174-1186.

Ivetic, A. and A.J. Ridley. 2004. The telling tail of L-selectin. Biochem. Soc. Trans. 32: 1118-1121.

Kaur, J. and D.F. Cutler. 2002. P-selectin targeting to secretory lysosomes of Rbl-2H3 Cells. J. Biol. Chem. 277: 10498-10505.

Leckband, D. and A. Prakasam. 2006. Mechanism and dynamics of cadherin adhesion. Annu. Rev. Biomed. Eng. 8: 259-287.

Luo, B.H. and C.V. Carman, and T.A. Springer. 2007. Structural basis of integrin regulation and signaling. Annu. Rev. Immunol. 25: 619-647.

Morishita, H. and T. Yagi. 2007. Protocadherin family: diversity, structure, and function. Curr. Opin. Cell. Biol. 19: 584-592.

Parisini, E. and J.M.G. Higgins, J.H. Liu, M.B. Brenner and J.H. Wang. 2007. The Crystal Structure of Human E-cadherin Domains 1 and 2, and Comparison with other Cadherins in the Context of Adhesion Mechanism. J. Mol. Biol. 373: 401-411.

Patel, S.D. and C.P. Chen, F. Bahna, B. Honig, and L. Shapiro. 2003. Cadherin-mediated cell-cell adhesion: sticking together as a family. Curr. Opin. Struct. Biol. 13: 690-698.

Patel, S.D. and C. Ciatto, C.P. Chen, F. Bahna, M. Rajebhosale, N. Arkus, I. Schieren, T.M. Jessell, B. Honig, S.R. Price, and L. Shapiro. 2006. Type II cadherin ectodomain structures: implications for classical cadherin specificity. Cell 124: 1255-1268.

Posy, S. and L. Shapiro, and B. Honig. 2008. Sequence and structural determinants of strand swapping in cadherin domains: do all cadherins bind through the same adhesive interface? J. Mol. Biol. 378: 954-968.

Setiadi, H. and R.P. McEver. 2008. Clustering endothelial E-selectin in clathrin-coated pits and lipid rafts enhances leukocyte adhesion under flow. Blood 111: 1989-1998.

Shimaoka, M. and J. Takagi, and T.A. Springer. 2002. Conformational regulation of integrin structure and function. Annu. Rev. Biophys. Biomol. Struct. 31: 485-516.

Shimaoka, M. and T. Xiao, J.H. Liu, Y. Yang, Y. Dong, C.D. Jun, A. McCormack, R. Zhang, A. Joachimiak, J. Takagi, J.H. Wang, and T.A. Springer. 2003. Structures of the aL I domain and its complex with ICAM-1 reveal a shape-shifting pathway for integrin regulation. Cell 112: 99-111.

Smith, C.W. 2008. Adhesion molecules and receptors. J. Allergy Clin. Immunol. 121: S375-379.

Somers, W.S. and J. Tang, G.D. Shaw, and R.T. Camphausen. 2000. Insights into the molecular basis of leukocyte tethering and rolling revealed by structures of P- and E-selectin bound to SLe(X) and PSGL-1. Cell 103: 467-479.

Takada, Y. and X. Ye, and S. Simon. 2007. The integrins. Genome Biol. 8: 215.

Van Roy, F. and G. Berx. 2008. The cell-cell adhesion molecule E-cadherin. Cell Mol. Life Sci. 65: 3756-3788.

Yang, Y. and C.D. Jun, J.H. Liu, R. Zhang, A. Joachimiak, T.A. Springer, and J.H. Wang. 2004. Structural basis for dimerization of ICAM-1 on the cell surface. Mol. Cell 14: 269-276.

Yang, Y. and C.D. Jun, J.H. Liu, R. Zhang, A. Joachimiak, and T.A. Springer. 2004. Structural basis for dimerization of ICAM-1 on the cell surface. Mol. Cell. 14: 269-276.

Xiong, J.P. and T. Stehle, B. Diefenbach, R. Zhang, R. Dunker, D.L. Scott, A. Joachimiak, S.L. Goodman, and M.A. Arnaout. 2001. Crystal structure of the extracellular segment of integrin alpha V beta 3. Science 294: 339-345.

CHAPTER 2

Immunohistochemical Localization of Adhesion Molecules

H. Wayne Sampson[1],* and Alan R. Parrish[2]

[1]Ph.D., Department of Systems Biology and Translational Medicine, Texas A&M Health Science Center College of Medicine, 702 SW HK Dodgen Loop, Temple, Texas 76504 and Department of Orthopedic Surgery, Scott & White Hospital, E-mail: sampson@medicine.tamhsc.edu

[2]Ph.D., Department of Systems Biology and Translational Medicine, Texas A&M Health Science Center College of Medicine, 336 Joe H. Reynolds Medical Building, College Station, Texas 77843-1114, E-mail: parrish@medicine.tamhsc.edu

Departmental contact: Tina Mandoza, Department of Systems Biology and Translational Medicine, Texas A&M Health Science Center College of Medicine, 702 SW HK Dodgen Loop, Temple, Texas 76504, E-mail: trm@tamu.edu

ABSTRACT

Immunohistochemistry is the science of reacting specific antibodies attached to visible dyes or fluorochromes with specific proteins of interest in tissue, in order to identify their exact location. Although the methods of immunohistochemistry are briefly overviewed in this chapter, its primary purpose is to identify advances in our knowledge of adhesion molecules using immunohistochemical techniques. There are many types of adhesion molecules. Some, such as claudins and occludins, function almost exclusively in forming tight junctions for holding cells together. Others, such as integrins, cadherins and catenins, function not only in cell adhesion, but also in cell signaling, carrying chemical messages from outside the cell to the interior. These signaling messages can result in the dissolution of the cell-cell or cell-extracellular matrix attachments and even result in cell migration. This phenomenon is very important for cell migration during embryological

---

*Corresponding author
Key terms are defined at the end of the chapter.

development and occurs during the formation of cancers and their metastases. The movement and rearrangement of cells during the inflammatory process is possible because of the properties of adhesion molecules. Many pathologies can be explained by changes in adhesion molecule properties, such as acute renal failure, which has been reported to be associated with disruption of intercellular adhesions in the proximal tubules of the kidneys. Much of our knowledge concerning embryological migration, cell migration, and pathological events has come from immunohistochemical studies.

INTRODUCTION

Adhesion molecules not only form gap junctions holding cells together and to the extracellular matrix, but they also have an important role in development, regulation of cell behavior, cell signaling, pathologies and cancer formation (Sampson *et al.* 2007), all of which have been demonstrated or discovered with immunohistochemical techniques. The adhesion molecules include cadherins, integrins, claudins and occludins, as well as species that are less commonly studied. Claudins and occludins are special adhesion molecules involved in tight junctional sealing, as can be clearly seen in Fig. 1. Cadherins are trans-membrane molecules that connect the cell interior, especially the actin cytoskeleton, with the extracellular environment. Cadherins are named for their original source of discovery, for example, E-cadherin from epithelium, N-from neuronal tissue, P-from placental, R-from retina, VE-from vascular tissue, OB- from osteoblasts, M-from myotubule and T- or H-from heart, but tissues have been found to possess many of the various cadherins at various stages of their development, as recently demonstrated in the bony growth plate, which appears to contain essentially all of the cadherins in some spatial level (Sampson *et al.* 2007).

Integrins are trans-membrane cell adhesion molecules that link the extracellular matrix proteins, collagen, laminin and fibronectin, with the cytoskeleton. They are composed of two trans-membrane subunits called α and β. There are specialized attachments between integrins and actin filaments called focal contacts that allow cells to pull on the substratum to which they are attached. Figure 2 is a micrograph of focal attachments created by Total Internal Reflection Fluorescence (TIRF) imaging of a smooth muscle cell in culture using the antibody p-Tyr-FITC, Sigma #F3145 (clone #PT66). TIRF microscopy represents a method of exciting and visualizing fluorophores present in the near-membrane region of live cells grown on glass coverslips. TIRF microscopy is based on the total internal reflection phenomenon that occurs when light passes from a highly refractive medium (e.g., glass) into one with lower refractive index (e.g., water, cell). Owing to the difference between the refractive indices at the interface, only a short-range electromagnetic disturbance called evanescent field will pass into the medium of lower refractive index. This type of excitation can be used to obtain high contrast fluorescence images, with very low background and virtually no out-of-focus light

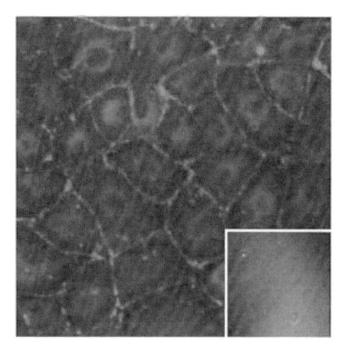

Fig. 1 Occludins tightly attaching adjacent cells. Occludins are special adhesion molecules in tight junctions between cells. Reprinted from Kaneda *et al.* (2006), with permission.
Color image of this figure appears in the color plate section at the end of the book.

(Trache and Meininger 2008). There has also been much research on the involvement of integrins as a sampler of the environment for mechanotransduction of endothelial cells.

OVERVIEW OF METHODS

Immunohistochemistry consists of attaching a visible or fluorescent dye to an antibody against a particular target antigen that one wants to identify within a cell or tissue. To conduct immunohistochemistry, the tissue must be prepared for histology, with the preparation of thin paraffin-embedded sections. Paraffin must be removed from the sections, usually by being placed in an oven overnight or a series of xylenes. They must then be rehydrated by being passed through a graded series of alcohols from absolute to 50%, followed by tap water. The availability of the antigen for interaction with a specific antibody can be maximized by enzymatic digestion, microwave irradiation or pressure cooking. Heat-induced epitope retrieval is the most commonly used method today. Following epitope retrieval in the appropriate pH solution, the slides are rinsed and ready for staining. Before a staining reaction is created, background proteins that might interfere with the

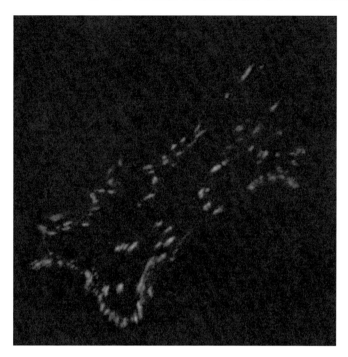

Fig. 2   Focal contacts demonstrated by TIRF. Focal contacts are specialized attachments between integrins and actin filaments that allow cells to pull on the substratum to which they are attached. This figure was taken using Total Internal Reflection Fluorescence (TIRF) imaging of a smooth muscle cell in culture using the antibody p-Tyr-FITC, Sigma #F3145 (clone #PT66). Courtesy of Soon-Mi Lim and Andreea Trache, Texas A&M Health Science Center College of Medicine.

Color image of this figure appears in the color plate section at the end of the book.

reaction must be blocked. The first of these reactions is a block against enzymatic endogenous peroxidase activity with a solution of hydrogen peroxide and sodium azide. This is very important because the step (see below) that attaches the visible dye reacts with peroxidase to produce a brown precipitate and it is critical that it react with only the intended molecule. Next is a casein and protein mixture, called a background Sniper, which is required to reduce nonspecific protein background staining. This is followed by an avidin-biotin blocker to suppress endogenous biotin. Avidin is a glycoprotein that has a very strong affinity for the protein biotin. This is important because the secondary antibody that is going to be applied later has a biotin molecule attached to it and must be the only biotin molecule 'visible' to the dye that will be used to color the reactive tissue. Once the tissue is prepared with all extraneous proteins blocked so that only the protein of choice remains available, the tissue is rinsed with buffer and the appropriate primary antibody is applied. This antibody will react with only the specific antigen to which it is antagonistic. It is important to omit the primary antibody step in some of the tissue as a negative control. Next is applied a secondary antibody against the primary antibody. The secondary antibody has a biotin molecule attached to it or is said to be biotinylated. The slides are then incubated in a conjugated streptavidin horseradish peroxidase solution. As discussed above, avidin is a glycoprotein that has a very strong affinity for the protein biotin, so there is a strong binding reaction between the streptavidin and the biotinylated secondary antibody. The final reactive step is to add a visible marker or dye. The most frequently used dye is 3,3′-diaminobenzidine, a visible dye, that binds to streptavidin and in the presence of peroxidase produces a brown precipitate (Fig. 3). There are other chromogens that yield red, purple, or blue final reaction colors, but brown is the most commonly used. The ability to use more than one color is very helpful for co-localizing different proteins in the same tissue or cell as in Fig. 4 (Schmetz *et al.* 2001). It is also possible to attach fluorescent dyes to the secondary antibody to produce brilliant colors under a fluorescent microscope (Fig. 5) or a confocal microscope (Fig. 4).

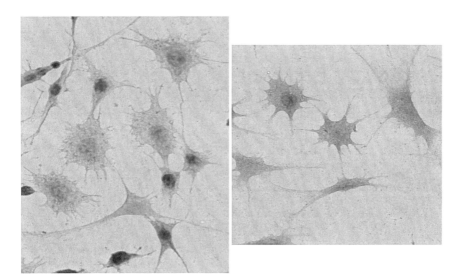

Fig. 3   Immunohistochemistry of p120 in MLO-Y4 osteocytes using a 3,3′-diaminobenzidine marker dye. Immunohistochemistry using monoclonal anti-mouse antibodies against p120 in MLO-Y4 cells. Left, a control with no primary antibody. Right, with a primary antibody and 3, 3′-diaminobenzidine marker dye. Original = 400 × (MLO-Y4 cells were kindly provided by Dr. Lynda Bonewald, Dept. Oral Biology, U. Missouri at KC School of Dentistry).

Color image of this figure appears in the color plate section at the end of the book.

Early Embryology

Development and differentiation of multicellular organisms is critically dependent on cell-cell adhesions (Trejo *et al.* 2008). Early 1–4-cell embryos mark the change

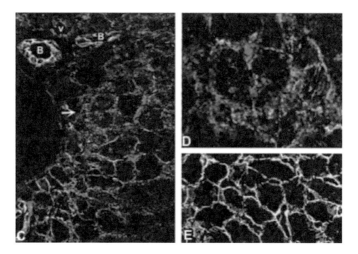

Fig. 4   Co-localizing of different proteins in the same tissue or cell. Double immunofluorescence staining for E-cadherin (red) and β-catenin (green) as seen with a laser scanning confocal microscope. Reprinted from Schmetz *et al.* (2001), with permission.

Color image of this figure appears in the color plate section at the end of the book.

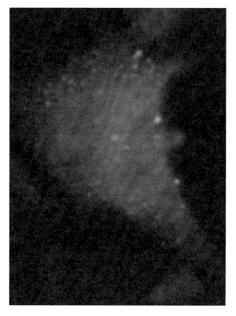

Fig. 5   Fluorescent immunohistochemistry of p120 in MLO-Y4 osteocytes using fluorescence microscopy. Fluorescent immunohistochemistry for labeled antibodies against mouse-p120 in MLO-Y4 cells. The primary antibody was mouse anti-p120 (BD Bioscience) with an Alexa Fluor 488 goat anti-mouse secondary antibody (Invitrogen). Yellow dots of labeled p120 are found scattered evenly throughout the cytoplasm of the MLO-Y4 cell.

Color image of this figure appears in the color plate section at the end of the book.

from maternal genome control to zygomatic genome with cell-cell adhesion and intercellular communication and adhesion molecules playing an important role in regulating embryonic cell development. This is a concept that has been clearly demonstrated by immunohistochemistry. Figure 6 uses immunohistochemistry to demonstrate a 4-cell embryo being held together by adhesion molecules. As the embryo grows, the adhesion molecules disappear, the cells rearrange and the adhesion molecules return to hold the cells together. The polarization and compaction of morula stage blastomeres leading to differentiation towards trophectoderm and inner mass seem to be closely linked with the formation and maintenance of calcium-dependent adhesion molecules on their surfaces (Reima 1990).

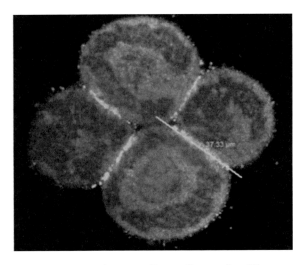

Fig. 6   Cell adhesion in the early embryo. E-cadherin adhesion of 4-cell hamster embryos. As the embryo grows, the adhesion molecules disappear, the cells rearrange and the adhesion molecules return to hold the cells together. Reprinted from Trejo *et al.* (2008), with permission.

Later Embryological Development

During tissue-organ system formation, cells undergo an epithelial-to-mesenchymal transition where the cells lose cell-to-cell contact mediated by adhesion molecules, reorganization of the cytoskeleton and the acquisition of a motile character (Taneyhill 2008). Following migration to their ultimate location, the cells coalesce once again and regain their adhesion molecules. As the neural tube forms and separates from the overlying ectoderm, it loses E-cadherin and acquires N-cadherin, but the ectodermal cells that were originally above it continue to express E-cadherin. As the cells migrate to their adult site, they lose the expression of N-cadherin during the migration but regain it in the final adult tissue (Hatta and Takeichi 1986, Nakagawa and Takeichi 1998). Various adhesion molecules might be required in a certain stage of development, and they disappear only to reappear in a latter stage. This requirement for adhesion molecules in the various stages of development can be easily demonstrated in the growth plate of long bones, which contains all of the stages of development in a small, confined region (Fig. 7).

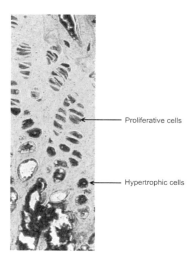

Proliferative cells

Hypertrophic cells

Fig. 7 Tibial growth plate demonstrating E-cadherin in various stages of development. E-cadherin seen in different stages of development in the tibial growth plate of a rat. Notice the nuclear localization in the zone of proliferation and cytoplasmic localization in the zone of hypertrophy.

Color image of this figure appears in the color plate section at the end of the book.

Heterogeneity of Adhesion Molecule Expression along the Nephron

The kidney nephron elegantly demonstrates the heterogenous expression of adhesion molecules in dictating organ function. In the kidney, the proximal tubule has a 'leaky' tight junction and both the transepithelial electrical resistance and complexity of the tight junction increase from the proximal tubule to the collecting duct. Therefore, efforts have focused on determining the spatial expression pattern of tight junction proteins along the nephron. In the rabbit kidney, immunofluorescence was used to detect occludin, ZO-1 and ZO-2 expression (Gonzalez-Mariscal *et al.* 2000). Occludin staining was quite weak in the proximal tubules, with much more intense staining in the distal tubules. ZO-1 was found in all tubules but was expressed at higher levels in distal segments than in proximal.

ZO-2 was diffusely expressed in the proximal tubules with weak cell border staining but was highly expressed at cell borders from Henle's loop to collecting ducts. In mouse kidney, claudins-1 and -2 are expressed in the Bowman's capsule, while claudins-2, -10 and -11 are found in proximal tubules (Kiuchi-Saishin *et al.* 2002). In the thin descending limb of Henle, claudin-2 is expressed and claudins-3, -4 and -8 are localized to the thin ascending limb of Henle. Claudins-3, -10, -11 and -16 are expressed in the thick ascending limb of Henle, while claudins-3 and -8 are present in the distal tubule and claudins-3, -4 and -8 in the collecting duct. It is postulated that claudin-2 might form a 'leakier' tight junction than claudin-3, based on localization along the nephron. The finding that enforced expression of claudin-2 is associated with the conversion from 'tight' to 'leaky' junctions in MDCK cells supports this hypothesis (Furuse *et al.* 2001). However, the higher expression of occludin, ZO-1, and ZO-2 in distal tubules than in proximal tubules likely also contributes to differences in the 'strength' of tight junctions along the nephron.

In adult mouse kidney, E-cadherin is detected everywhere but the initial segment where the proximal tubule joins Bowman's capsule (Piepenhagen *et al.* 1995). This pattern of expression contrasts with that of adult human kidney, where E-cadherin is not detected in the proximal tubule (Nouwen *et al.* 1993). N-cadherin is expressed in the proximal, but not distal, tubules (Nouwen *et al.* 1993). Given that E- and N-cadherin are highly homologous, the functional relevance of the localization of N-cadherin in the proximal tubules and E-cadherin in the distal tubules is unclear. α- and β-catenin are expressed in all nephron segments (adult mouse), while γ-catenin is detected only in the distal part of the nephron (Piepenhagen and Nelson 1995).

Cancer

Immunohistochemistry plays a critical role in the detection and progression of some cancers. In malignant tumors, intracellular adhesion decreases because of the loss of E-cadherin and the cells change from a stationary epithelial type cell with cell-to-cell adhesion to a mesenchymal cell with increased tumor cell motility and invasive properties necessary for distant metastases (Guarino *et al.* 2007). This mesenchymal cell type is reminiscent of early migratory embryological mesenchymal tissue. Stationary epithelial cells thus modulate their phenotype by progressively weakening their cell-cell connections and acquire migratory mesenchymal-like properties with the ability to migrate as individual cells (Guarino *et al.* 2007). This epithelial-mesenchymal transition or 'cadherin switch' is a phenomenon in which N-cadherin is overexpressed with the loss of E-cadherin and has been associated with advancing stages of breast, prostate and liver cancer (Chetty *et al.* 2008) (Fig. 8). The loss of cell-cell contact is not the only consequence of the 'cadherin switch', but E-cadherin is known to be intimately associated with

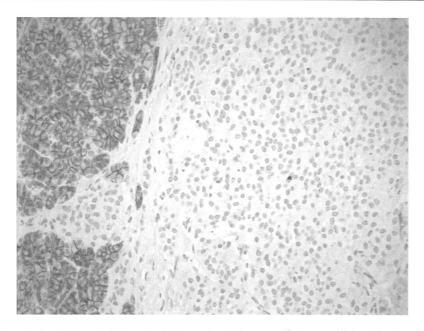

Fig. 8 'Cadherin switch'. E-cadherin is negative in the tumor (right two-thirds of the figure), whereas adjacent normal pancreatic parenchyma is heavily stained (left third of the figure). Reprinted from Chetty *et al.* (2008), with permission. © American Society for Clinical Pathology.

Color image of this figure appears in the color plate section at the end of the book.

β-catenin and p120. Loss of E-cadherin in invasive carcinoma causes re-localization of these molecules from the cell membrane to the cytoplasm where it can regulate members of the Rho GTPase and cell motility. P120, which stabilizes E-cadherin to the cell membrane, has been shown by immunohistochemical means to migrate from the cell membrane to the cytoplasm and nucleus, destabilizing the plasma membrane (Chetty *et al.* 2008) (Fig. 9). E-cadherin immunohistochemistry is considered diagnostic for some types of carcinoma (Kim *et al.* 2008).

β-Catenin also functions as a component of the Wnt signaling pathway and plays an important role in the regulation of gene expression, including tumor development and progression (Cadigan *et al.* 1997). Activation of Wnt signaling results in inhibition of GSK-3β activity via phosphorylation at ser9 (Daugherty and Gottardi 2007) causing an accumulation of cytoplasmic β-catenin, which translocates to the nucleus to regulate transcription via the Tcf/Lef family (Cadigan *et al.* 1997). Several genes regulated by β-catenin have been identified, including cyclin D1, c-myc, c-jun, and E-cadherin (Willert 1998). The nuclear localization of β-catenin may be a prognostic indicator in colorectal cancer (Wong *et al.* 2004).

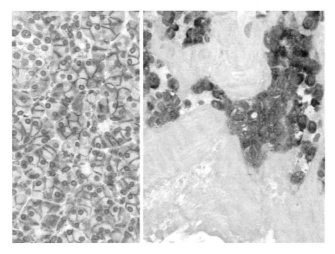

Fig. 9 Translocation of p120 catenin from the juxtamembrane position to cytoplasmic. Left, normal pancreatic parenchyma showing linear, crisp membrane staining for p120 in contrast to, right, the intense cytoplasmic positivity in a solid pseudopapillary tumor. Reprinted from Chetty *et al.* (2008), with permission. © 2008 American Society for Clinical Pathology.

Color image of this figure appears in the color plate section at the end of the book.

Inflammation

It is theorized that as squamous epithelial cells, such as cervical epithelium, mature, they undergo a downregulation of the genes responsible for the synthesis of adhesion molecules and are exfoliated at the surface (Politi *et al.* 2008). Immunohistochemistry methods have demonstrated that some infections inhibit the normal downregulation of the surface epithelial cells, resulting in clusters of cells still tightly bound to each other.

Pathology: Acute Renal Failure

Acute renal failure is characterized by a loss of cell polarity as assessed by translocation of $Na^+ K^+$-ATPase to the apical domain and loss of the barrier function of the tubular epithelium (Nissenson 1998); importantly, both of these changes involve disruption of intercellular adhesion, specifically the cadherin/catenin complex. Using an *in vitro* model, we have demonstrated that ischemia is associated with disruption of intercellular adhesions in the proximal tubules of the kidneys, as clearly demonstrated by immunofluorescence (Fig. 10). Reperfusion in a simulated ischemia model in normal rat kidney causes translocation of β-catenin to a perinuclear location, as seen in Fig. 11, but was not associated with activation of β-catenin signaling (Chen *et al.* 2007). This result is consistent with several studies suggesting that loss of cadherin function is not always associated with activation of β-catenin signaling (Vasioukhin *et al.* 2001). Disruption of cadherin/catenin complexes during acute renal failure is demonstrated *in vivo*, as evidenced

Control   Ischemia

E-Cadherin

N-Cadherin

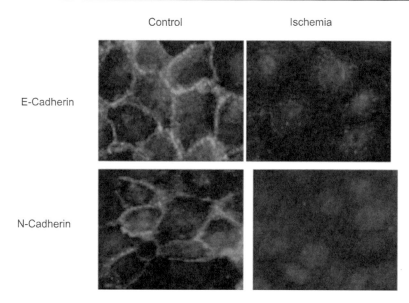

Fig. 10 Impact of ischemia on cadherin/catenin localization in pSM2 cells. pSM2 cells made ischemic by a mineral oil overlay model, demonstrating loss of E- and N-cadherin. Reprinted from Covington *et al.* (2006), with permission.

Color image of this figure appears in the color plate section at the end of the book.

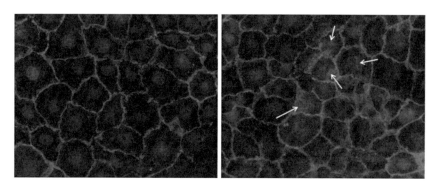

Fig. 11 Peri-nuclear translocation of β-catenin. Reperfusion (right) in a simulated ischemia model (left) in normal rat kidney causes translocation of β-catenin (red) to a perinuclear location (nucleus in blue).

Color image of this figure appears in the color plate section at the end of the book.

by the finding that HgCl$_2$-induced nephrotoxicity resulted in disruption of the N-cadherin/β-catenin/α-catenin complex in the proximal tubular epithelium (Jiang *et al.* 2004), while both bismuth and cadmium (Prozialeck and Lamar 2005) are also associated with alterations in the cellular localization of cadherins and catenins in the kidney.

CONCLUSIONS

Immunohistochemistry has contributed to our understanding of cell attachment, embryological development, cancer and metastases and some pathological conditions. A great deal of evidence demonstrates that immunohistochemistry is a valuable tool to investigate adhesion molecule expression and spatial distribution in both normal physiology and pathophysiology. Importantly, the critical role of adhesion molecules in development has been elucidated, in part, by immunohistochemical techniques. Cancer development, inflammation and acute renal failure are useful examples of this methodology in dissecting the contribution of adhesion molecules to pathophysiology. While several important questions have been addressed, the techniques are still limited by antibody specificity. However, progress in epitope unmasking has significantly advanced the field, as well as the development of more sensitive fluorescence-based detection. In addition, the ability to perform co-localization of molecules has proven to be a major advance that has facilitated the understanding of the indispensable involvement of adhesion molecules in normal and disease states.

SUMMARY

- Immunohistochemistry is a valuable technique that has been used to understand the function of adhesion molecules.
- Immunohistochemistry has advanced our understanding of cell attachment.
- Immunohistochemistry techniques led to the understanding of cell attachment and migration during embryological development.
- Immunohistochemistry demonstrated the change from solid tumors to migrating cells in cancer.
- Immunohistochemistry demonstrated the movement of some species of adhesion molecules from the cell membrane to the cytoplasm to the nucleus.
- Immunohistochemistry techniques have led to a better understanding of the pathophysiology of inflammation and kidney disease.

Key Facts about Immunohistochemistry and Localization of Proteins of Interest

1. Immunohistochemistry is used to visibly identify the location of specific proteins within a tissue.
2. Colored or fluorescent dyes can be attached to antibodies that will in turn attach to the protein of interest.

3. Protein localization by immunohistochemistry can be visualized by conventional microscopy, fluorescence microscopy, confocal microscopy or any other method that can visualize the dye used.
4. Different proteins can be visualized simultaneously by using different colored dyes for the secondary antibodies.

Definition of Terms

Antibody: A protein that reacts with or binds to another protein called an antigen.

Antigen: A protein that attracts and binds with an antibody.

Avidin: A glycoprotein that has a very strong affinity for the protein biotin, so there is a strong binding reaction between the streptavidin and the secondary antibody.

Avidin-biotin blocker: A chemical that blocks any background biotin in the tissue so it will not react with the streptavidin step of the process.

Background Sniper: A casein and protein mixture that is required to reduce nonspecific protein background staining.

Biotin: A B-vitamin that has a strong binding affinity for avidin, thus allowing for a strong attachment between streptavidin and a secondary protein.

Conjugated streptavidin: A compound containing streptavidin and a horseradish peroxidase tightly bound together

Endogenous: Normally found within the tissue.

Epitope retrieval: Exposure of the protein or portion of protein of interest so as to make it available to bind with the antibody.

Fluorescent dye: A dye that is visible under fluorescent light.

Horseradish peroxidase: An enzyme attached to streptavidin so that it will more readily bind to the dye and antibody.

Paraffin-embedded sections: Sections in which all water has been replaced with wax or paraffin.

Peroxidase: An enzyme commonly found in mammalian tissues that could bind to the streptavidin as background, making specific detection of a particular protein impossible.

Primary antibody: The antibody that attaches to the specific protein that is being sought.

Secondary antibody: The antibody that attaches to a streptavidin-dye complex and to the primary antibody.

References

Cadigan, K.M. and R. Nusse. 1997. Wnt signaling: a common theme in animal development. Genes Dev. 11: 3286-3305.

Chen, G. and A.D. Akintola, J.M. Catania, M.D. Covington, D.D. Dean, J.P. Trzeciakowski, R.C. Burghardt, and A.R. Parrish. 2007. Ischemia-induced cleavage of cadherins in NRK cells is not sufficient for β-catenin transcriptional activity. Cell Comm. Adhesion 14: 111-123.

Chetty, R. and D. Jain, and S. Serra. 2008. p120 Catenin reduction and cytoplasmic relocalization leads to dysregulation of E-cadherin in solid pseudopapillary tumors of the pancreas. Am. J. Clin. Pathol. 130: 71-76.

Covington, M.D. and R.C. Burghardt, and A.R. Parrish. 2005. Ischemia-induced vleavage of cadherins in NRK cells requires MT1-MMP (MMP-14). Am. J. Physiol. Renal Physiol. 290: F43-F51.

Daugherty, R.L. and C.J. Gottardi. 2007. Phospho-regulation of beta-catenin adhesion and signaling functions. Physiology 22: 303-309.

Furuse, M. and K. Furuse, H. Sasaki, and S. Tsukita. 2001. Conversion of zonulae occludentes from tight to leaky strand type by introducing claudin-2 into Madin-Darby canine kidney I cells. J. Cell Biol. 153: 263-272.

Gonzalez-Mariscal, L. and M.C. Namorado, D. Martin, J. Luna, L. Alarcon, S. Islas, L. Valencia, P. Muriel, L. Ponce, and J.L. Reyes. 2000. Tight junction proteins ZO-1, ZO-2, and occludin along isolated renal tubules. Kidney Int. 57: 2386-2402.

Guarino, M. and B. Rubino, and G. Ballabio. 2007. The role of epithelial-mesenchymal transition in cancer pathology. Pathology 39: 305-318.

Hatta, K. and M. Takeichi. 1986. Expression of N-cadherin adhesion molecules associated with early morphogenetic events in chick development. Nature 320: 447-449.

Jiang, J. and D. Dean, R.C. Burghardt, and A.R. Parrish. 2004. Disruption of cadherin/catenin expression, localization, and interactions during $HgCl_2$-induced nephrotoxicity. Toxicol. Sci. 80: 170-182.

Kaneda, K. and K. Miyamoto, S. Nomura, and T. Horiuchi. 2006. Intercellular localization of occludins and ZO-1 as a solute transport barrier of the mesothelial monolayer. J. Arti. Organs 9: 241-250.

Kim, M.J. and S.J. Jang, and E. Yu. 2008. Loss of E-cadherin and cytoplasmic-nuclear expression of β-catenin are the most useful immunoprofiles in the diagnosis of solid-pseudopapillary neoplasm of the pancreas. Human Pathol. 39: 251-258.

Kiuchi-Saishin, Y. and S. Gotoh, M. Furuse, A. Takasuga, Y. Tano, and S. Tsukita. 2002. Differential expression patterns of claudins, tight junction membrane proteins, in mouse nephron segments. J. Am. Soc. Nephrol. 13: 875-886.

Nakagawa, S. and M. Takeichi. 1998. Neural crest emigration from the neural tube depends on regulated cadherin expression. Development 125: 2963-2971.

Nissenson A.R. 1998. Acute renal failure: Definition and pathogenesis. Kidney Int. 53(S6): S7-S10.

Nouwen E.J. and S. Dauwe, I. van der Biest, and M.E. de Broe. 1993. Stage- and segment-specific expression of cell-adhesion molecules N-CAM, A-CAM, and L-CAM in the kidney. Kidney Int. 44: 147-158.

Piepenhagen P.A. and W.J. Nelson. 1995. Differential expression of cell-cell and cell-substratum adhesion proteins along the kidney nephron. Am. J. Physiol. 269: C1433-C1449.

Piepenhagen, P.A. and L.L. Peters, S.E. Lux, and W.J. Nelson. 1995. Differential expression of Na^+-K^+-ATPase, ankyrin, fodrin, and E-cadherin along the kidney nephron. Am. J. Physiol. 269: C1417-C1432.

Politi, E.N. and A.C. Lazaris, M. Kehriotis, T.G. Papathomas, E. Nikolakopoulou, and H. Koutselini. 2008. Altered expression of adhesion molecules in inflammatory cervical smears. Cytopathology 19: 172-178.

Prozialeck, W.C. and P.C. Lamar. 2005. Cadmium nephrotoxicity is associated with a loss of N-cadherin-mediated adhesion and alterations in epithelial polarity in the proximal tubule. Toxicol. Sci. 84: S327.

Reima, I. 1990. Maintenance of compaction and adherent-type junctions in mouse morula-stage embryos. Cell Diff. Dev. 29: 143-153.

Sampson, H.W. and A.C. Dearman, A.D. Akintola, W.E. Zimmer, and A.R. Parrish. 2007. Immunohistochemical localization of cadherin and catenin adhesion molecules in the murine growth plate. J. Histochem. Cytochem. 55: 845-852.

Schmetz, M. and V.J. Schmid, and A.R. Parrish. 2001. Selective disruption of cadherin/catenin complexes by oxidative stress in precision-cut mouse liver slices. Toxicol. Sci. 61: 389-394.

Taneyhill, L.A. 2008. To adhere, or not to adhere. Cell Adh. Migr. 2: 1-8.

Trache, A. and G. Meininger. 2008. Total Internal Reflection Fluorescence Microscopy. Unit 2A.2.1-2A.2.22. *In*: Coico, R., T. Kowalik, J.M. Quarles, B. Stevenson, and R.K. Taylor. [eds.] Current Protocols in Microbiology. Wiley and Sons, Inc. New Jersey, USA.

Trejo, A. and D. Ambriz, M.C. Navarro-Maldonado, E. Mercado, and A. Rosado. 2008. Presence and distribution of E-cadherin in the 4-cell golden hamster embryo. Effect of maternal age and parity. Zygote 16: 271-277.

Vasioukhin V. and C. Bauer, L. Degenstein, B. Wise, and E. Fuchs. 2001. Hyperproliferation and defects in epithelial polarity upon conditional ablation of alpha-catenin in skin. Cell 104: 605-617.

Willert, K. and R. Nusse. 1998. Beta-catenin: A key mediator of Wnt signaling. Curr. Opin. Genet. Dev. 8: 95-102.

Wong S.C. and E.S. Lo, K.C.Lee, J.K. Chan, and W.L. Hsiao. 2004. Prognostic and diagnostic significance of beta-catenin nuclear immunostaining in colorectal cancer. Clin Cancer Res. 10: 1401-1408.

Wong S.C. and E.S. Lo, A.K. Chan, K.C. Lee, and W.L. Hsiao. 2003. Nuclear beta catenin as a potential prognostic and diagnostic marker in patients with colorectal cancer from Hong Kong. Mol. Pathol. 56: 347-352.

CHAPTER 3

Junctional Adhesion Molecules (JAMs)

Klaus Ebnet[1,*], Volker Gerke[2] and Michel Aurrand-Lions[3]

[1]Institute of Medical Biochemistry, Center of Molecular Biology of Inflammation (ZMBE), University of Muenster, Von-Esmarch-Str. 56, D-48149 Münster, GERMANY, E-mail: ebnetk@uni-muenster.de

[2]Institute of Medical Biochemistry, Center of Molecular Biology of Inflammation (ZMBE), University of Muenster, Von-Esmarch-Str. 56, D-48149 Münster, GERMANY, E-mail: gerke@uni-muenster.de

[3]UMR891, Inserm, Centre de Recherche en Cancérologie de Marseille, Institut Paoli-Calmettes, 27, bd Leï Roure, F-13009 Marseille, FRANCE E-mail: Michel.Aurrand-Lions@inserm.fr

Departmental Contact: Institute of Medical Biochemistry, Center of Molecular Biology of Inflammation (ZMBE), University of Muenster, Von-Esmarch-Str. 56, D-48149 Münster, GERMANY, E-mail: hentrey@uni-muenster.de

ABSTRACT

Junctional adhesion molecules (JAMs) comprise a small family of immunoglobulin-like adhesion molecules. JAMs were originally identified in leukocytes, endothelial and epithelial cells, and their main function has been considered to be regulation of cell-cell interactions during inflammation. Meanwhile, the diversity of cell types that express JAMs has increased, and JAMs have been found on cells of the reproductive system, on cells of the nervous system, on fibroblasts and on stem cells. In addition, a number of extracellular ligands as well as intracellular binding partners were identified. Two major functions can be attributed to JAMs: (1) the regulation of vascular inflammation by mediating transient interactions of immune cells with endothelial cells and (2) the regulation of cellular polarization by mediating the targeting of cytoplasmic proteins to specific sites of cell-cell adhesion. Through the latter activity, JAMs are involved in processes such as the

---

* Corresponding author
Key terms are defined at the end of the chapter.

formation of tight junctions in epithelial and endothelial cells, the development of spermatids or the formation of functional nerves. Recent evidence indicates that JAMs play a role in cell migration and proliferation, suggesting that their functions are more diverse than anticipated. Here, we will review the role of JAMs in different biological systems.

INTRODUCTION

Junctional adhesion molecules (JAMs) are members of the immunoglobulin superfamily (IgSF) and belong to the CTX subfamily, which is characterized by a membrane-distal V-type Ig domain and a membrane-proximal C2-type Ig-domain (Fig. 1). The JAM family comprises 'sensu stricto' JAMs including JAM-A, JAM-B and JAM-C and 'sensu lato' JAMs including CAR, ESAM, JAM4 and JAM-L (Ebnet *et al.* 2004, Bradfield *et al.* 2007a). All JAMs undergo homophilic interactions that require cis-dimerization followed by trans-homophilic interactions (Fig. 2). In addition to their homophilic interactions, several JAMs undergo heterophilic interaction with either other JAM family members or other cell adhesion molecules (see below). Most of these interactions serve to regulate the recruitment of immune cells to sites of inflammation by mediating the adhesion of leukocytes to endothelial or epithelial cells (Weber *et al.* 2007). The homophilic binding between JAM-A, -B and -C is weak; the primary role of these homophilic and perhaps certain heterophilic interactions as well is probably to recruit cytoplasmic proteins to specific subcellular locations. In support of this notion, JAMs are expressed in a large variety of cell types where they localize to specific membrane domains. Cell types that express JAM proteins include leukocytes, platelets, epithelial and endothelial cells, but also germ cells or Schwann cells. In this chapter, we discuss the pleiotropic functions of JAMs as regulators of inflammation, as organizers of specific membrane domains and as signal transducing molecules. Owing to space limitations, we focus on the three classical JAM molecules JAM-A, -B and -C.

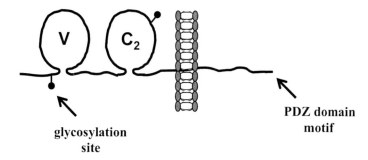

Fig. 1   Principal organization of JAMs. All JAM family members contain two Ig-like domains, a membrane distal V-type Ig-like domain and a membrane-proximal C2-type Ig-like domain. The extracellular domain contains a potential glycosylation site, the C-terminus bears a type II PDZ domain-binding motif.

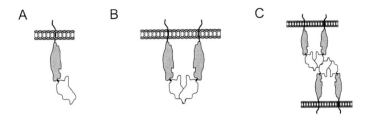

Fig. 2   Cis-dimerization and trans-homophilic interaction. A: Diagram of JAM-A monomer. The membrane-distal and membrane-proximal Ig-domains are depicted in light grey and dark grey, respectively. B: Membrane-bound JAM-A exists as dimer. Dimerization is mediated through a dimerization motif in the first Ig-like domain which is conserved in JAM-A, -B and -C. C: JAM dimers undergo trans-homophilic interactions.

JAMs REGULATE LEUKOCYTE-ENDOTHELIAL CELL INTERACTIONS DURING INFLAMMATION

The role of our immune system is to detect and eliminate pathogens or self-modified components from the body. Cells of the innate immune system recognize pathogens or self-modified components, which results in the eradication of the foreign agent at the site of injury or in the initiation of the adaptive immune response mediated by T and B cells in the draining secondary lymphoid organs. Indeed, tissue macrophages or dendritic cells take up antigens in the periphery and migrate to lymph nodes in which they present antigens to T cells. This evolutionary strategy has succeeded because the immune system has co-developed the structures necessary to communicate between blood, tissues and lymph nodes, namely vascular and lymphatic systems. An exquisite coordination exists between the initiation of the inflammatory reaction through activation of the innate immune system and the endothelial inflammatory response. Inflammation results in local modifications in the adhesive properties of endothelial cells, which are mediated by the upregulation of vascular adhesion molecules as well as changes in subcellular localization of pre-existing ones. JAMs most probably belong to this latter class of molecules, since no upregulation of JAM expression has been reported on inflamed endothelial cells *in vivo* except for JAM-A or JAM-C, which are upregulated by hyperlipidemic conditions or oxidized lipoproteins (Keiper *et al.* 2005). In contrast, several studies have shown that the junctional or apical localization of the JAM family members is affected by inflammatory stimulation of endothelial cells (Fig. 3). Therefore, when considering the function of the JAMs in leukocyte trafficking, one has to consider three different levels of regulation: (1) localization of the JAMs in the inter-endothelial junctions or apical surface, (2) coupling of the JAMs with intracellular adaptors and (3) engagement of the JAMs in trans depending on their cis-dimerization state.

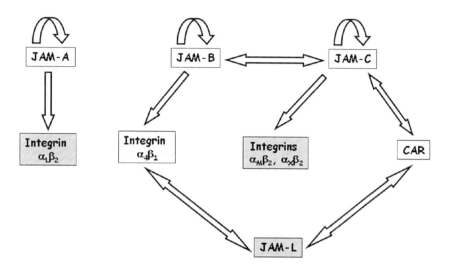

Fig. 3   Putative molecular interaction network of JAMs and integrins involved in leukocyte trans-endothelial migration. Grey boxes: Proteins exclusively expressed on circulating cells. Spotted grey boxes: Proteins expressed on both circulating and endothelial cells. White boxes: Proteins exclusively expressed on endothelial cells.

Evidence for a relocalization of JAM-A during inflammation came from studies with cultured endothelial cells in which combined treatment with IFN-γ and TNF-α results in JAM-A localization to the apical membrane domain, where it can serve as counter-receptor for the leukocyte integrin $\alpha_L\beta_2$ (Weber *et al.* 2007). These results confirmed the pioneering work by E. Dejana and coworkers demonstrating that JAM-A was essential for monocyte trans-endothelial migration *in vitro* and that treatment of mice with anti-JAM-A mAbs inhibits leukocyte recruitment in air pouch model and in cytokine-induced meningitis (Martin-Padura *et al.* 1998). Although the studies on JAM-A paved the way for understanding the function of JAM-B and JAM-C in leukocyte migration, the latter members of the family differ from JAM-A in terms of tissue distribution and inflammatory regulation. Indeed, JAM-B has been initially identified as an adhesion molecule highly expressed on High Endothelial Venules (HEVs) and later as a vascular ligand for JAM-C, which is also expressed on human leukocytes and platelets. JAM-C has been found to be transported from intracellular stores to the surface of human dermal microvascular endothelial cells upon VEGF or histamine stimulation (Orlova *et al.* 2006). Based on the pleiotropic homo- and heterophilic interactions of JAMs, many research groups focused on the adhesive function of the JAM proteins and their role during leukocyte trans-endothelial or trans-epithelial migration (Weber *et al.* 2007).

For example, using genetically engineered mice and antibodies directed against JAM-C, it has been shown that the protein contributes to the recruitment of leukocytes to inflammatory sites, is involved in the retention of monocytes within tissues and regulates the pool of myeloid progenitors present in the bone marrow (Aurrand-Lions *et al.* 2005, Bradfield *et al.* 2007b, Praetor *et al.* 2008). Such results have been interpreted in the light of the known homophilic and heterophilic trans-interactions of the protein. However, additional interactions in cis might further add to the complexity of the molecular network that regulates JAM-mediated cell-cell interactions (Fig. 4). For example, the interaction of JAM-B with the integrin $\alpha_4\beta_1$ requires the coexpression of JAM-C with the integrin, and the interaction of JAM-L with CAR occurs only if leukocytes lack $\alpha_4\beta_1$ integrin expression or express activated $\alpha_4\beta_1$ integrin (Cunningham *et al.* 2002, Luissint *et al.* 2008). This implicates a crosstalk between JAMs and other adhesion molecules like integrins, which influences the final outcome of the adhesive activities of JAMs.

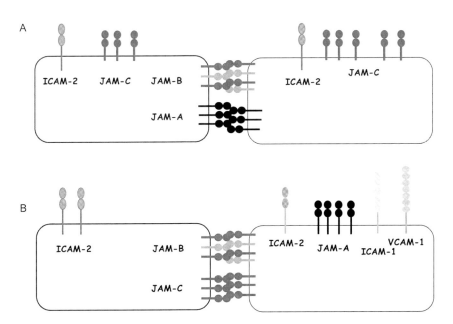

Fig. 4 Schematic representation of sub-cellular localizations of JAMs on resting and activated endothelial cells. A. Resting endothelial cells: JAM-A and JAM-B are concentrated at cell-cell adhesion sites, whereas JAM-C is also expressed on the apical surface. B. Activated endothelial cells: JAM-A is re-localized to the apical surface, whereas JAM-C is concentrated at cell-cell junctions. These re-localizations are cell type- and stimulus-dependent but may occur simultaneously with de novo expression of the inflammatory adhesion molecules ICAM-1 and VCAM-1.

JAMs REGULATE THE ORGANIZATION OF CELL-CELL CONTACTS IN VARIOUS CELL TYPES AND TISSUES

Through their cytoplasmic tails, JAMs directly interact with various cytoplasmic proteins. All these proteins contain one or several PDZ domains, a domain consisting of 80 to 90 AA that is used to mediate protein-protein interactions. PDZ domain-containing proteins, in particular those containing multiple PDZ domains, are frequently used to assemble multiprotein complexes at the membrane. JAM-A, -B and -C all contain a conserved motif that mediates the interaction with PDZ domains, and the interactions with the hitherto known direct binding partners, i.e., AF-6/afadin, ZO-1, Par-3 and MUPP1, are mediated through this motif and a PDZ domain of the binding partner (Fig. 5) (Ebnet *et al.* 2004).

The first clear evidence for a role of JAMs in the organization of cell-cell contacts came with the identification of the cell polarity protein Par-3 as direct binding partner for JAM-A (Ebnet *et al.* 2001). Par-3 forms a complex with Par-6 and atypical PKC (aPKC) that localizes to the tight junction (TJ). The TJs represent a specific structure at the most apical region of the lateral cell-cell contact that is implicated in the regulation of paracellular permeability but also in the separation of the apical from the basolateral membrane domain (Tsukita *et al.* 2001). A critical role for the Par-aPKC complex for the formation of functional TJs is suggested by the observations that overexpression of dominant-negative mutants or siRNA-mediated downregulation of individual components of the complex

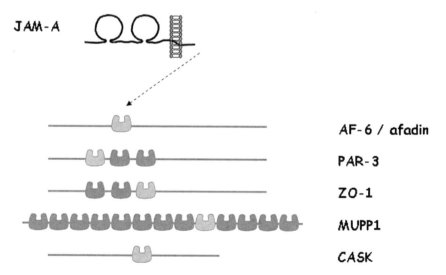

Fig. 5 Cytoplasmic proteins directly associating with JAM-A. JAM-A interacts with various cytoplasmic proteins through specific PDZ domain interactions. The PDZ domain involved in JAM-A interaction is shown in light gray. The interaction with ZO-1 and PAR-3 is conserved in JAM-B and JAM-C, interactions with AF-6/afadin and CASK have not been tested yet.

disturb or retard TJ formation in cultured epithelial cells. The function of JAM-A in TJ formation thus most likely resides in targeting the Par-aPKC complex to sites of cell-cell contacts where it is required to establish specific membrane domains.

This role of JAM-A might be of particular importance during cell-cell contact formation. In the absence of cell-cell adhesion, for example, during migration, cells form thin protrusions to scan the extracellular environment. The protrusions are dynamic in the absence of cell-cell adhesion, and they are stabilized upon cell-cell contact formation (Adams *et al.* 1998). Signals mediated by the adhesion molecules regulate the further maturation of the cell-cell contacts. Interestingly, JAM-A is among the proteins that are present at the earliest sites of cell-cell adhesion during contact formation, the so-called primordial, spot-like junctions or puncta. The Par-aPKC complex is recruited shortly after the puncta are formed. Thus, a primary role of JAM-A might be to recruit the Par-aPKC complex and thereby regulate the correct localization of this complex, which regulates the further cell contact formation that eventually leads to fully matured lateral cell-cell contacts with TJs localized at the apex and separated from AJs. In accordance with this view, ectopic expression of a JAM-A mutant that cannot bind Par-3 leads to defects in the formation of functional TJs and in the development of apico-basal polarity (Rehder *et al.* 2006).

Interestingly, a similar function albeit in a different cell type seems to be true for JAM-C as well as revealed in JAM-C knockout mice. The inactivation of the JAM-C gene in mice leads to male sterility (Gliki *et al.* 2004), and the reason for the sterility turned out to be a blockade in the development of spermatids. During normal spermatogenesis, spermatids undergo intimate contacts with Sertoli cells, and this interaction in part is mediated by JAM-C and JAM-B expressed by spermatids and Sertoli cells, respectively. Cell polarity proteins such as Par-6, aPKC and Cdc42 are localized in close proximity of the site of spermatid-Sertoli cell interaction. This polarized localization is lost in JAM-C knockout spermatids, resulting in a blockade of spermatid polarization and ultimately in sterility (Gliki *et al.* 2004). Thus, JAM-C expressed by spermatids provides another example of the functional role of JAMs: through homophilic and heterophilic interactions, JAMs are localized to specific subcellular sites, and through their cytoplasmic domains, they recruit their binding partners to the specific localizations where these are required to regulate developmental processes.

A recent report describes the localization of JAM-C at autotypic junctions of Schwann cells in myelinated nerves (Scheiermann *et al.* 2007). Schwann cells wrap around the axons to allow for saltatory conduction. At specific sites such as the paranodal loops and Schmidt-Lanterman incisures, the Schwann cells form tight interactions between membrane patches of the same cell, and these tight interactions are necessary for efficient electrical insulation. In the absence of JAM-C, the integrity of the myelin sheath is altered and nerve conduction is defective. It is not clear yet whether this defects results from a lack of adhesive interactions, from a lack of a specific membrane domain required to establish these tight interactions or from both. Nevertheless, this example adds another facet to the versatility of JAMs.

JAMs TRANSMIT SIGNALS TO THE INTERIOR OF THE CELL

There is accumulating evidence to support that homophilic JAM interactions also transmit signals that regulate cell-matrix contact formation and cell proliferation (Fig. 6). In endothelial cells, overexpression of JAM-A results in increased migration of endothelial cells on vitronectin as well as in increased proliferation and endothelial tube formation *in vitro* accompanied by increased MAP kinase activity (Naik *et al.* 2003). JAM-A exists in a complex with $\alpha_V\beta_3$ integrin that has been implicated in endothelial cell proliferation and angiogenesis, and bFGF, a physiological stimulus of angiogenesis that signals through $\alpha_V\beta_3$ integrin and dissociates JAM-A from $\alpha_V\beta_3$ integrin. Although the molecular mechanisms are not clear in detail, these findings indicate that JAM-A signaling contributes to endothelial migration and proliferation, which might have implications on pathophysiological angiogenesis as it occurs during tumor growth and metastasis formation.

A similar effect of JAM-A on cell migration has been observed in epithelial cells. Here, both JAM-A knockdown and overexpression of a dimerization-deficient JAM-A mutant lacking the membrane-distal Ig domain result in decreased adhesion to extracellular matrix proteins, changes in cell morphology and reduced levels of $\beta1$ integrins (Mandell *et al.* 2005). Knockdown of JAM-A also results in decreased levels of active Rap1 small GTPase, and siRNA-mediated knockdown of Rap1 results in similarly reduced levels of $\beta1$ integrins like knockdown of JAM-A, suggesting that JAM-A activates Rap1 to regulate $\beta1$ integrin-mediated adhesion and migration. As in endothelial cells, JAM-A seems to regulate epithelial cell proliferation. Deletion of the JAM-A gene in mice leads to increased intestinal inflammation in a model of inflammatory bowel disease accompanied by increased proliferation of colonic epithelial cells (Laukoetter *et al.* 2007). Thus, JAM-A regulates proliferation but, as opposed to endothelial cells, it has a negative regulatory effect on proliferation in epithelial cells.

A third line of evidence for the signaling role of JAM-A is suggested by observations that JAM-A serves as cell surface receptor for reovirus. Reovirus infection of target cells leads to activation of NF-κB and apoptosis in these cells. Interestingly, despite the ability of reovirus to infect cells in the absence of JAM-A (by binding to sialic acid residues), its ability to activate NF-κB and to induce apoptosis depends on JAM-A (Barton *et al.* 2001), implicating JAM-A in regulating the activity of transcription factors under certain circumstances.

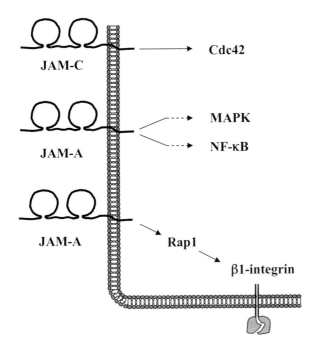

Fig. 6   JAMs transmit signals. JAM-A activates Rap1 to regulate β1 integrin-mediated adhesion and migration of epithelial cells. JAM-A also activates MAP kinase signaling in endothelial cells as well as NF-κB activation in response to reovirus binding. JAM-C can mediate Cdc42 activation.

JAMs AND CANCER

The predicted role of JAM family members in various biological processes involved in tumor growth, such as immune regulation, angiogenesis, cell-cell adhesion and polarity, rendered JAM proteins interesting candidates for targets to blunt tumor-induced angiogenesis (Dejana *et al.* 2001). A study in which JAM-C targeting with function-blocking antibodies inhibited neovacsularization and tumor growth in mice (Lamagna *et al.* 2005) provided a proof of concept for this hypothesis. This concept has been further reinforced by observations that JAM-A plays a role in βFGF-induced angiogenesis and that JAM-B is highly expressed in and around tumor foci and is differentially expressed on lymphatic endothelial cells isolated from tumors as compared with normal skin-derived lymphatic endothelial cells. Although these observations are suggestive, the function of JAM-A and JAM-B expressed on micro-environmental cells within tumor still awaits a formal demonstration in pre-clinical settings.

In contrast to these indirect effects of JAMs on tumor development, several reports suggest a direct effect of tumor cell-expressed JAMs. For example, expression of JAM-C on non-small lung cancer cells has been shown to promote tumor cell adhesion to JAM-C-expressing endothelial, suggesting that homophilic JAM-C interactions may be involved in tumor metastasis. This has been remarkably confirmed in two independent studies using the HT1080 human fibrosarcoma cell line. In one study, Fuse and collaborators showed a strong positive correlation between JAM-C expression and the metastatic potential of cancer cell lines (Fuse *et al.* 2007). In another study, JAM-C was identified by a proteomic approach as one of 47 molecules differentially expressed between two HT1080 variant cell lines differing in their ability to intravasate and disseminate (Conn *et al.* 2008). These findings strongly suggest a role for JAM-C in tumor mestastasis. In contrast to JAM-C, the level of JAM-A expression has been inversely correlated to the ability of breast cancer cells to intravasate (Naik *et al.* 2008). Finally, one single study on leukemic patients has shown that JAM-C is highly expressed in hairy cell leukemias and in marginal zone B-cell lymphomas but not on chronic lymphocytic leukemias, suggesting that JAM-C could be used as diagnosis marker in leukemias. Thus, differential expression of the JAMs is expected in molecular signatures of metastatic versus non-metastatic cancers and might be used as a prognosis marker in certain tumors. However, these preliminary observations await confirmation through a meta-analysis of gene and protein expression patterns in larger panels of the different tumors.

CONCLUSIONS

The ability of JAMs to undergo numerous homo- and heterophilic interactions highlights their role in mediating interactions between immune cells and endothelial or epithelial cells and suggests a critical role for JAMs during inflammations. Small molecule inhibitors or antibodies blocking homophilic JAM interactions and as a consequence heterotypic cell interactions might thus turn out to be useful in inhibiting unwanted immune reactions such as chronic inflammations. On the other hand, the identification of evolutionarily conserved cytoplasmic binding partners of JAMs, some of which are involved in the fundamental process of cellular polarization, indicates that JAMs participate in the regulation of cell polarity. Their wide expression in various cell types suggests a similar role in different cell types and organ systems. The involvement of JAM-C in spermatid development as well as in nerve conduction by mediating heterotypic and autotypic interactions, respectively, strongly supports this notion. Furthermore, there is good evidence for a role for JAM-A in endothelial and epithelial cell migration as well as in proliferation. The correlation of JAM-C expression with the tumorigenic potential of various cell lines suggests that misregulation of JAM-C expression might contribute to tumor and metastasis formation. As during inflammatory reactions, inhibitors blocking JAM interactions could turn out to be useful in inhibiting tumor growth. In future, it will be important to understand the molecular mechanisms underlying the regulation of cell polarization, migration and proliferation by JAMs.

SUMMARY

- JAMs undergo homophilic as well as various heterophilic interactions. Some of these interactions are probably subject to regulation by cis-interacting integrins.
- In the immune system, JAM-mediated cell-cell interactions serve to regulate inflammatory processes by mediating leukocyte binding to the surface of endothelial cells and probably also by regulating their trans-endothelial migration.
- Through their cytoplasmic tails, JAMs interact with various PDZ domain-containing scaffolding proteins known to regulate the assembly of protein complexes that regulate cell-cell contact formation and cell polarity.
- In other cellular systems, such as epithelial cells, spermatids, Sertoli cells or Schwann cells, homophilic and heterophilic JAM interactions serve to regulate the specific subcellular localization of these associated proteins.
- Accumulating evidence suggests that JAMs can activate signaling pathways leading to the formation of active small GTPases, to the activation of the MAP kinase pathway and to the activation of transcription factors.
- Recent studies implicate JAMs in tumor development and metastasis formation.

Acknowledgment

We apologize to authors of research in the field that could not be cited because of space limitations. This work was supported by grants from the German Research Foundation (DFG) and from the Medical Faculty of the University Münster (to KE) and by grants from the Swiss National Foundation (No 310000-112551/I) and the Thorn Foundation (to MAL).

Abbreviations

AJ	adherens junction
aPKC	atypical protein kinase-C
bFGF	basic fibroblast growth factor
CAR	coxsackie- and adenovirus receptor
Cdc42	cell division cycle-42

CTX	cortical thymocyte marker for *Xenopus*
GTPase	guanine nucleotide triphosphatase
IgSF	immunoglobulin-superfamily
JAM	junctional adhesion molecule
MAP kinase	mitogen activated protein kinase
MUPP1	multiple PDZ domain protein 1
Par	partitioning-defective
PDZ	Postsynaptic density-95—Discs large—Zonula Occludens-1
TJ	tight junction

Key Facts about Junctional Adhesion Molecules

	Molecule	Cell type	Function/ mechanism
Inflammation	JAM-A	endothelial cells epithelial cells leukocytes	homo- and heterophilic binding
	JAM-C	endothelial cells epithelial cells leukocytes	homo- and heterophilic binding
Tight junction formation	JAM-A	epithelial cells	organization of membrane polarity
	JAM-C	epithelial cells	organization of membrane polarity
Cell motility and migration	JAM-A	epithelial cells endothelial cells DC	integrin regulation integrin regulation unknown
Cell proliferation	JAM-A	epithelial cells endothelial cells	unknown MAPK activation
Germ cell development	JAM-C	spermatids	organization of membrane polarity
Virus infection	JAM-A	epithelial cells	reovirus receptor
Nerve conduction	JAM-C	Schwann cells	autotypic cell-cell interaction
Stem cell marker	JAM-A	HSC	unknown
	JAM-B	ESC, HSC	unknown
	JAM-C	HSC	unknown

ESC, embryonic stem cell; DC, dendritic cells; HSC, hematopoietic stem cell.

Definition of Terms

Atypical protein kinase C: One of the three subfamilies of the protein kinase C. This subfamily is characterized by the requirements for activation. As opposed to the other subfamilies (conventional PKCs, novel PKCs), the atypical PKCs do not require diacylglycerol (DAG) or Ca^{2+} ions for activation, and they are not activated by phorbol esters.

Cell adhesion molecules: Integral membrane proteins that mediate the interaction between cells. They can bind homophilically to the same molecule or heterophilically to a different molecule on the adjacent cells. The most prominent families of adhesion molecules are cadherins, immunoglobulin-superfamily adhesion molecules and integrins.

Cell polarity: The asymmetric distribution of organelles or proteins in the cells.

Inflammation: A process characterized by massive infiltration of leukocytes to sites of antigen deposition. Under normal circumstances, inflammation serves antigen clearance and is tightly regulated. When misregulated, it can lead to tissue damage.

Membrane polarity: Refers to different membrane identities regarding their composition of lipids and proteins. In epithelial and endothelial cells, the apical membrane (facing the lumen of an organ) and the basolateral domain (in contact with neighboring cells and the extracellular matrix) differ in their composition. Membrane polarity is important for the specific uptake and the vectorial transport of molecules.

Par proteins: A family of proteins originally discovered in *C. elegans* as regulators of the asymmetric distribution of cytoplasmic granules in the fertilized egg. These proteins turned out to be master regulators of cell polarity in many different organisms and cell types.

Sertoli cell: A cell type within the male reproductive tract. Undergoes homotypic cell-cell interactions to form the blood-testis barrier to prevent the interaction of blood-borne molecules or cell types with developing germ cells. Undergoes heterotypic interactions with developing sperm cells to provide nutrients and development signals.

Small GTPase: Protein that binds guanine nucleotides and shuttles between an inactive state (GDP bound) and an active state (GTP bound). Contains intrinsic activity to hydrolyze bound GTP to GDP. GTP hydrolysis deactivates the GTPase.

Stem cell: A cell type with the ability to renew itself and to give rise to different daughter cells by a process called asymmetric cell division. Hematopoietic stem cells give rise to all cells of the immune system.

Tight junctions: A structure at the cell-cell contacts between epithelial or endothelial cells that appears as discrete spots of membrane fusions by electron microscopy. The tight junctions regulate paracellular diffusion of molecules as well as lumen formation and prevent the intermixing of distinct membrane domains.

References

Adams, C.L. and Y.T. Chen, S.J. Smith, and W.J. Nelson. 1998. Mechanisms of epithelial cell-cell adhesion and cell compaction revealed by high-resolution tracking of E-cadherin-green fluorescent protein. J. Cell. Biol. 142: 1105-1119.

Aurrand-Lions, M. and C. Lamagna, J.P. Dangerfield, S. Wang, P. Herrera, S. Nourshargh, and B.A. Imhof. 2005. Junctional adhesion molecule-C regulates the early influx of leukocytes into tissues during inflammation. J. Immunol. 174: 6406-6415.

Barton, E.S. and J.C. Forrest, J.L. Connolly, J.D. Chappell, Y. Liu, F.J. Schnell, A. Nusrat, C.A. Parkos, and T.S. Dermody. 2001. Junction adhesion molecule is a receptor for reovirus. Cell 104: 441-451.

Bradfield, P.F. and S. Nourshargh, M. Aurrand-Lions, and B.A. Imhof. 2007a. JAM family and related proteins in leukocyte migration (Vestweber series). Arterioscler. Thromb. Vasc. Biol. 27: 2104-2112.

Bradfield, P.F. and C. Scheiermann, S. Nourshargh, C. Ody, F.W. Luscinskas, G.E. Rainger, G.B. Nash, M. Miljkovic-Licina, M. Aurrand-Lions, and B.A. Imhof. 2007b. JAM-C regulates unidirectional monocyte transendothelial migration in inflammation. Blood 110: 2545-2555.

Conn, E.M. and M.A. Madsen, B.F. Cravatt, W. Ruf, E.I. Deryugina, and J.P. Quigley. 2008. Cell surface proteomics identifies molecules functionally linked to tumor cell intravasation. J. Biol. Chem. 283: 26518-26527.

Cunningham, S.A. and J.M. Rodriguez, M.P. Arrate, T.M. Tran, and T.A. Brock. 2002. JAM2 interacts with alpha4beta1. Facilitation by JAM3. J. Biol. Chem. 277: 27589-27592.

Dejana, E. and R. Spagnuolo, and G. Bazzoni. 2001. Interendothelial junctions and their role in the control of angiogenesis, vascular permeability and leukocyte transmigration. Thromb. Haemost. 86: 308-315.

Ebnet, K. and A. Suzuki, Y. Horikoshi, T. Hirose, M.K. Meyer Zu Brickwedde, S. Ohno, and D. Vestweber. 2001. The cell polarity protein ASIP/PAR-3 directly associates with junctional adhesion molecule (JAM). Embo. J. 20: 3738-3748.

Ebnet, K. and A. Suzuki, S. Ohno, and D. Vestweber. 2004. Junctional adhesion molecules (JAMs): More molecules with dual functions? J. Cell. Sci. 117: 19-29.

Fuse, C. and Y. Ishida, T. Hikita, T. Asai, and N. Oku. 2007. Junctional adhesion molecule-C promotes metastatic potential of HT1080 human fibrosarcoma. J. Biol. Chem. 282: 8276-8283.

Gliki, G. and K. Ebnet, M. Aurrand-Lions, B.A. Imhof, and R.H. Adams. 2004. Spermatid differentiation requires the assembly of a cell polarity complex downstream of junctional adhesion molecule-C. Nature 431: 320-324.

Keiper, T. and N. Al-Fakhri, E. Chavakis, A.N. Athanasopoulos, B. Isermann, S. Herzog, R. Saffrich, K. Hersemeyer, R.M. Bohle, J. Haendeler, K.T. Preissner, S. Santoso, and T. Chavakis. 2005. The role of junctional adhesion molecule-C (JAM-C) in oxidized LDL-mediated leukocyte recruitment. Faseb J. 19: 2078-2080.

Lamagna, C. and K.M. Hodivala-Dilke, B.A. Imhof, and M. Aurrand-Lions. 2005. Antibody against junctional adhesion molecule-C inhibits angiogenesis and tumor growth. Cancer Res. 65: 5703-5710.

Laukoetter, M.G. and P. Nava, W.Y. Lee, E.A. Severson, C.T. Capaldo, B.A. Babbin, I.R. Williams, M. Koval, E. Peatman, J.A. Campbell, T.S. Dermody, A. Nusrat, and C.A. Parkos. 2007. JAM-A regulates permeability and inflammation in the intestine in vivo. J. Exp. Med. 204: 3067-3076.

Luissint, A.C. and P.G. Lutz, D.A. Calderwood, P.O. Couraud, and S. Bourdoulous. 2008. JAM-L-mediated leukocyte adhesion to endothelial cells is regulated in cis by alpha4beta1 integrin activation. J. Cell. Biol. 183: 1159-1173.

Mandell, K.J. and B.A. Babbin, A. Nusrat, and C.A. Parkos. 2005. Junctional adhesion molecule 1 regulates epithelial cell morphology through effects on beta1 integrins and Rap1 activity. J. Biol. Chem. 280: 11665-11674.

Martin-Padura, I. and S. Lostaglio, M. Schneemann, L. Williams, M. Romano, P. Fruscella, C. Panzeri, A. Stoppacciaro, L. Ruco, A. Villa, D. Simmons, and E. Dejana. 1998. Junctional adhesion molecule, a novel member of the immunoglobulin superfamily that distributes at intercellular junctions and modulates monocyte transmigration. J. Cell. Biol. 142: 117-127.

Naik, M.U. and S.A. Mousa, C.A. Parkos, and U.P. Naik. 2003. Signaling through JAM-1 and {alpha}v{beta}3 is required for the angiogenic action of bFGF: dissociation of the JAM-1 and {alpha}v{beta}3 complex. Blood 102: 2108-2114.

Naik, M.U. and T.U. Naik, A.T. Suckow, M.K. Duncan, and U.P. Naik. 2008. Attenuation of junctional adhesion molecule-A is a contributing factor for breast cancer cell invasion. Cancer Res. 68: 2194-2203.

Orlova, V.V. and M. Economopoulou, F. Lupu, S. Santoso, and T. Chavakis. 2006. Junctional adhesion molecule-C regulates vascular endothelial permeability by modulating VE-cadherin-mediated cell-cell contacts. J. Exp. Med. 203: 2703-2714.

Praetor, A. and J.M. McBride, H. Chiu, L. Rangell, L. Cabote, W.P. Lee, J. Cupp, D.M. Danilenko, and S. Fong. 2008. Genetic deletion of JAM-C reveals a role in myeloid progenitor generation. Blood. Epub ahead of print (DOI 10.1182/blood-2008-06-159574).

Rehder, D. and S. Iden, I. Nasdala, J. Wegener, M.K. Brickwedde, D. Vestweber, and K. Ebnet. 2006. Junctional adhesion molecule-A participates in the formation of apico-basal polarity through different domains. Exp. Cell. Res. 312: 3389-3403.

Scheiermann, C. and P. Meda, M. Aurrand-Lions, R. Madani, Y. Yiangou, P. Coffey, T.E. Salt, D. Ducrest-Gay, D. Caille, O. Howell, R. Reynolds, A. Lobrinus, R.H. Adams, A.S. Yu, P. Anand, B.A. Imhof, and S. Nourshargh. 2007. Expression and function of junctional adhesion molecule-C in myelinated peripheral nerves. Science 318: 1472-1475.

Tsukita, S. and M. Furuse, and M. Itoh. 2001. Multifunctional strands in tight junctions. Nat. Rev. Mol. Cell. Biol. 2: 285-293.

Weber, C. and L. Fraemohs, and E. Dejana. 2007. The role of junctional adhesion molecules in vascular inflammation. Nat. Rev. Immunol. 7: 467-477.

4

Molecular Imaging of Endothelial Adhesion Molecules

Alistair C. Lindsay[1], Martina A. McAteer[2] and Robin P. Choudhury[3], *

Department of Cardiovascular Medicine, Level 6, West Wing, John Radcliffe Hospital, University of Oxford, UK, OX3 9DU
[1]E-mail: Alistair.Lindsay@cardiov.ox.ac.uk
[2]E-mail: Martina.McAteer@cardiov.ox.ac.uk
[3]E-mail: Robin.Choudhury@cardiov.ox.ac.uk
Departmental contact: Eunice.Berry@ndm.ox.ac.uk.

ABSTRACT

Imaging of endothelial adhesion molecules spearheads the emerging field of molecular imaging. Fundamental to the emergence of this new field has been the development of specific molecular imaging probes that have permitted adhesion molecule imaging using ultrasound, magnetic resonance, nuclear and optical imaging techniques. A wide variety of imaging probes have developed from basic science work and been optimized for adhesion molecule detection using various (and often multiple) imaging modalities; they are described in detail in this chapter according to the methods used for their detection. The main animal experiments demonstrating the utility of these probes are highlighted, and the potential applications of adhesion molecule imaging to human disease are discussed. In addition, the chapter describes how the development of methods of molecular imaging not only offers the possibility of a new diagnostic era, where disease can be detected in its earliest stages, but will also add further insights into disease processes, and eventually may even lead to novel methods of delivering targeted treatments that cannot be delivered systemically. The main challenges facing the future of endothelial adhesion molecule imaging are highlighted throughout the chapter.

*Corresponding author
Key terms are defined at the end of the chapter.

INTRODUCTION

In clinical practice, medical imaging techniques such as X-ray angiography, computed tomography (CT), *magnetic resonance imaging (MRI),* and ultrasound have been used to diagnose diseases of the vascular system by identifying structural abnormalities. However, in recent years, opportunities for molecular imaging— defined as 'the visualisation, characterization, and measurement of biological processes at the molecular and cellular levels' (Mankoff 2007)—have been developed. *Imaging* of endothelial adhesion molecules has been at the forefront of this emerging field and has attracted great interest for two main reasons: (1) they play an important pathological role in a variety of disease conditions, including atherosclerosis, cancer, and autoimmune disease, and (2) these molecules are accessible to blood-borne contrast agents. The potential benefits for molecular imaging of vascular endothelial adhesion molecules include the ability to diagnose vascular inflammation at an early stage, reveal novel pathological mechanisms and monitor the efficacy of therapeutic interventions in drug development.

To date, a variety of endothelial adhesion molecules have been successfully imaged, including *vascular cell adhesion molecule 1* (VCAM-1, CD106), *intracellular adhesion molecule 1* (ICAM-1, CD54), *E-selectin* (CD62E, endothelial leukocyte adhesion molecule 1) and $\alpha_v\beta_3$ *integrin.* VCAM-1, which is known to be upregulated in response to several inflammatory stimuli, plays an important role in leukocyte tethering and therefore in the initiation of atherosclerosis and other disease processes (Hillis and Flapan 1998). Therefore, VCAM-1 is an attractive molecular imaging target of early vascular inflammation. E-selectin is another important mediator of early rolling recruitment of leukocytes to the endothelium and can also serve as a molecular imaging marker of endothelial activation. ICAM-1 plays a key role in leukocyte trafficking and has also been implicated in microvascular slow flow (Benson *et al.* 2007). Lastly, $\alpha_v\beta_3$ integrin is expressed on small blood vessels in cancer and atherosclerosis and has been shown to mediate leukocyte-endothelial interaction, for example, following ischemia-reperfusion injury (Ichioka *et al.* 2007). Angiogenesis plays a pivotal role in a number of disease states, including ischemia, inflammation, malignancy, infection, and immune disorders.

Central to the paradigm of *molecular imaging* is the use of molecular 'probes' in order to provide specific contrast. Such molecular 'probes' have many synonyms, including tracers, nano- and micro-particles, and contrast agents (Massoud and Gambhir 2003), but the majority share two fundamental components: (1) the actual contrast agent (tautologous to a 'signaling element') and (2) the targeting ligand.

A variety of techniques have been used for molecular imaging to date and are listed in Table 1. In this chapter, each imaging modality is explored in turn, highlighting the opportunities and limitations of each for the development of molecular imaging techniques that can visualize endothelial adhesion molecules.

Table I Summary of imaging modalities available for molecular imaging (with permission from Massoud and Gambhir (2003)).

Imaging technique	Portion of EM radiation spectrum used	Spatial resolution	Depth	Temporal resolution	Sensitivity	Amount of probe used
Positron Emission Tomography (PET)	High energy gamma rays	1-2 mm	No limit	10 sec to minutes	10-11 to 10-12 mole/L	Nanograms
Single photon emission computed tomography (SPECT)	Lower energy gamma rays	1-2 mm	No limit	Minutes	10-10 to 10-11 mole/L	Nanograms
Optical bioluminescence imaging	Visible light	3-5 mm	1-2 cm	Seconds to minutes	Likely 10-15 to 1-17 mole/L	Micrograms to milligrams
Optical fluorescence imaging	Visible light or near infra-red	2-3 mm	<1 cm	Seconds to minutes	Likely 10-9 to 10-12 mole/L	Micrograms to milligrams
Magnetic resonance imaging (MRI)	Radiowaves	25-100 μm	No limit	Minutes to hours	10-3 to 10-5 mole/L	Micrograms to milligrams
Computed tomography (CT)	X-rays	50-200 μm	No limit	Minutes	Not established	Not applicable
Ultrasound	High-frequency ultrasound	50-500 μm	mm to cm	Seconds to minutes	Not established	Micrograms to milligrams

ULTRASOUND IMAGING OF ENDOTHELIAL ADHESION MOLECULES

Much of the initial work in molecular imaging of endothelial adhesion molecules was performed using *ultrasound* detection, and real-time ultrasound imaging of endothelial adhesion molecules has been achieved in animal models of a number of clinical conditions, such as transplant rejection (Weller *et al.* 2003), ischemia-reperfusion injury (Villanueva *et al.* 2007), and atherosclerosis (Kaufmann *et al.* 2007), using acoustically active micro- or nano-particle contrast agents. The most intensively studied are gas-filled microspheres (or 'microbubbles') (2-8 μm in diameter); the size of these spheres ensures that they stay in the intra-vascular compartment and can adhere to their endothelial target (Fig. 1). Although gas-filled microbubbles are approved for myocardial contrast imaging in humans, no targeted ultrasound contrast agents have yet been approved for use in humans.

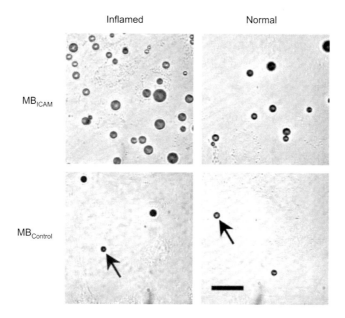

Fig. 1   Targeted microbubble adhesion. Micrographs of interleukin-activated (left) and normal (right) rat endothelial cells after exposure to ICAM-1-targeted microbubbles (top) and control microbubbles (bottom). ICAM-targeted microbubbles can be seen to adhere in good numbers to the activated endothelium (top left). Microbubbles such as this can be detected by ultrasound to give an indication of the distribution of adhesion molecule expression. With permission from Weller *et al.* (2003).

Various gas contents and sphere components have been used in the design of targeted microbubbles (Villanueva *et al.* 1998, 2001, Lindner *et al.* 2002). Examples of these include shells made from albumin, phospholipid or biodegradable polymers, mostly filled with either nitrogen or perfluorocarbon gas (Villanueva 2008). Typically, a targeted ligand is linked to the microbubble to allow adherence to a specific endothelial marker. Initial work used *monoclonal antibodies* directed against endothelial targets. However, the use of antibodies can be expensive to scale and is associated with immunological complications, so other moieties have more recently been evaluated, such as carbohydrates (Villanueva *et al.* 2007) and peptides (Weller *et al.* 2005). Endothelial attachment of a targeted microbubble is detailed by an increase in intensity on ultrasound, which persists even after the remaining microbubbles have passed from the circulation. An advantage of microbubble ultrasound imaging is the possibility of detecting molecular expression both in the microvascular circulation and in larger vessels. Limitations of the latter application, however, include a lower signal-noise ratio, and the need for high frequency ultrasound detection systems to achieve adequate spatial resolution (Villanueva 2008). Ongoing research aims to optimize microbubble composition to construct agents that can resonate at higher frequencies to allow higher spatial resolution detection (de Jong *et al.* 2002).

In an early proof-of-principle example of the feasibility of microbubble molecular imaging, Villanueva *et al.* (1998) constructed perfluorobutane gas-filled lipid-derived microspheres coupled with monoclonal antibody to ICAM-1. When the microbubbles were exposed to activated coronary artery endothelial cells *in vitro*, videomicroscopy confirmed specific binding (Villanueva *et al.* 1998). Subsequently, the potential of contrast-enhanced ultrasound using microbubbles has been investigated in a number of *in vivo* small animal experiments. For example, P-selectin upregulation in the context of ischemia-reperfusion injury has been demonstrated in both the kidney (Lindner *et al.* 2001) and the myocardium (Villanueva *et al.* 2007). ICAM-1 upregulation in rat myocardium during the onset of cardiac transplant rejection has also been demonstrated using lipid-based ICAM-1-targeted microbubbles (Weller *et al.* 2003).

In addition to their ability to detect the consequences of disease, ultrasound contrast agents have also been used to detect 'high-risk' disease, in particular, atherosclerotic plaques demonstrating high levels of *inflammation*. Hamilton *et al.* (2004) used echogenic immunoliposomes (ELIPs), phospholipid bilayer membrane vesicles with attached antibody targeted against a variety of endothelial adhesion molecules, to demonstrate endothelial injury using intra-vascular ultrasound in a swine model. VCAM-1-targeted microbubbles have also been used to detect inflammation in early atherosclerosis lesions of varying severity in apolipoprotein-E-deficient mice (Kaufmann *et al.* 2007). Importantly, this study also allowed quantification of VCAM-1 expression according to the intensity of signal received, and a more recent study has also described the use of ultrasound molecular imaging to assess the temporal progression of the inflammatory response in mice (Behm *et al.* 2008). Behm and colleagues targeted decafluorobutane lipid microbubbles to complement receptors on activated leukocytes, α_v-integrins, and VCAM-1, and injected each form of microbubble into the ischemic hindlimbs of mice in a randomized fashion. By repeating these injections at various stages of the ischemic process, the authors were able to determine the temporal expression of the proteins targeted and, in a novel application, to investigate the relationship between protein expression and blood flow by simultaneously performing perfusion imaging using untargeted lipid microbubbles.

Ultrasound molecular imaging has also shown promise outside of the potential applications to cardiovascular disease. By using microbubbles targeted to ICAM-1 and VCAM-1, Reinhardt *et al.* (2005) were able to both depict and quantify inflammation in the brain and spinal cords of rats in experimental autoimmune encephalitis (EAE) using an enhanced technique, Sensitive Particle Acoustic Quantification (SPAQ). In a separate model, Palmowski *et al.* (2008) used microbubbles linked to $\alpha_v\beta_3$ integrin binding ligand to examine angiogenesis in a mouse model of squamous cell carcinoma.

The main challenge to the future clinical translation of ultrasound molecular imaging is the need for further improvements in probe and transducer design; future research will be directed towards this end.

MOLECULAR MRI OF ENDOTHELIAL ADHESION MOLECULES

Molecular MRI (mMRI) is an attractive method for imaging of endothelial adhesion molecules for a number of reasons: it is non-invasive, offers high spatial resolution and signal-to-noise ratio, and provides excellent soft tissue contrast. The main challenge of mMRI, compared to nuclear or optical imaging techniques, is its inherently low sensitivity. However, the use of paramagnetic (Gd-based) or super-paramagnetic (iron oxide-based) nanoparticle MR contrast agents can overcome this limitation to generate positive or negative or contrast enhancement. Initial attempts to create a targeted mMRI contrast agent involved conjugating an antibody against the epitope of interest to a traditional MR blood-pool contrast agent, such as *gadolinium* diethylene-triamine-penta-acetic acid (Gd-DTPA). However, the major limitation of Gd-based agents is that only a small quantity of Gd can be delivered to an endothelial monolayer, resulting in modest contrast effects, and to address this more recent agents have used multiple Gd complexes per antibody-conjugated nanoparticle (Doiron *et al.* 2008). To further improve contrast signal, novel imaging agents and epitope-binding ligands with higher affinity have subsequently been developed to overcome the limitations of a Gd-based approach. These are described in detail below.

The conjugation of *microparticles* and *nanoparticles* with monoclonal antibodies shows promise as a method of detecting endothelial adhesion molecular expression using mMRI. Sipkins *et al.* (2000)reported the detection of ICAM-1 upregulation on the cerebral microvasculature of mice with EAE by *ex vivo* MRI (9.4 T) using antibody-conjugated paramagnetic liposomes (ACPLs). In 2002, Kang *et al.* successfully demonstrated the expression of E-selectin *in vitro* in human endothelial cells using the F_{ab} fragment of anti-human E-selectin monoclonal antibody conjugated to cross-linked iron oxide nanoparticles. More recently, McAteer *et al.*

(2007) have developed a novel, targeted *microparticle of iron oxide (MPIO)* probe that can identify VCAM-1 expression *in vivo* in mouse acute brain inflammation. MPIO (1 μm diameter), with reactive tosyl groups, were used for direct covalent conjugation of monoclonal VCAM-1 antibodies (VCAM-MPIO). The capacity of VCAM-MPIO for specific binding was first demonstrated *in vitro* using activated mouse endothelial cells (sEND-1) stimulated with graded doses of tumor necrosis factor alpha (TNF-α). The ability of VCAM-MPIO to demonstrate endothelial inflammation using *in vivo* mMRI was then tested in a mouse model of acute brain inflammation. Acute inflammation was induced by stereotactic injection of the proinflammatory interleukin 1β (IL-1β) into the left corpus striatum, while the right hemisphere received no injection and served as an internal control. After 3 hr, VCAM-MPIO or negative isotype control IgG-MPIO (4×10^8 MPIO) were intravenously injected via a tail vein and allowed to circulate for 1.5-2 hr prior to MRI. VCAM-MPIO generated highly specific, potent hypo-intense contrast effects in the IL-1β-activated hemisphere, which delineated the architecture of activated cerebral blood vessels, with minimal background contrast (Fig. 2). The specificity and potency of the contrast effects derived from a combination of targeted delivery of a large amount of iron oxide to sites of early inflammation and rapid clearance of MPIO from the blood, which minimized background signal.

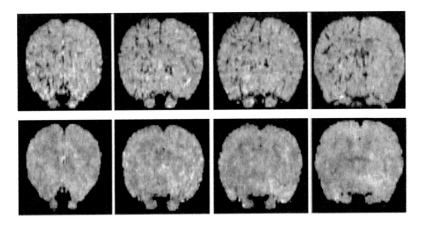

Fig. 2 Targeted microparticle of iron oxide detection using MRI. *In vivo* T2*-weighted coronal images from 3D gradient echo data sets of a mouse brain with ~90 μm isotropic resolution. Four slices are shown per brain. Top row: Mouse injected intrastriatally with 1 ng of IL-1β in 1 μl of saline 3 hr prior to IV injection of VCAM-MPIO. Areas of low signal (black) in the left hemisphere reflect MPIO retention on acutely inflamed endothelium. The opposite hemisphere shows almost no signal dropout. Bottom row: Identical experiment but using IgG-MPIO as a control. No signal dropout is seen in the inflamed hemisphere. With permission from McAteer *et al.* (2007).

Dual-ligand MPIO (4.5 μm diameter) was subsequently designed to mimic more closely leukocyte binding pathways *in vivo*; since *in vivo* leukocyte binding involves multiple receptor-ligand interactions, dual-ligand MPIOs were made for the detection of endothelial P-selectin and VCAM-1 expression in a mouse model of atherosclerosis (McAteer *et al.* 2008). Dual-ligand MPIO binding to arterial endothelium following *in vivo* intravenous injection was demonstrated by high resolution *ex vivo* MRI (9.4T). Biodistribution studies showed that MPIO were sequestered by the liver and spleen after 24 hr, with no evidence of tissue infarction, inflammation or hemorrhage.

Examples of other targeting ligands used with Gd-based nanoparticles include Sialyl Lewis^X (sLe^X), a tetrasaccharide carbohydrate associated with CD15 on the surface of leukocytes that binds to E-selectin expressed on activated endothelial cells. Sibson *et al.* (2004) successfully targeted E-selectin expression in rat brain *in vivo* using a sLe^X mimetic moiety conjugated to Gd-DTPA (Gd-DTPA-B(sLe^X) A). Gd-DTPA-B(sLeX)A, administered systemically 3-4 hr following injection of either IL-1β or TNF-α into the left striatum (to induce focal endothelial activation) produced hyper-intense contrast effects on the activated brain endothelium on spin-echo T_1-weighted images, at a stage when no pathological changes were apparent with conventional MRI. Gd-DTPA-sLe^{X}A has also been used to detect early endothelial activation following transient focal ischemia in a mouse model of middle cerebral artery occlusion using *in vivo* MRI (9.4T) (Barber *et al.* 2004).

To generate ligand diversity, *phage display* has been used to identify novel peptides that bind specifically to activated endothelium and are internalized by cells expressing VCAM-1 (Kelly *et al.* 2005, Nahrendorf *et al.* 2006). These include a VCAM-1 specific, cyclic peptide sequence (CVHSPNKKC; termed VHS peptide) (Kelly *et al.* 2005) and a linear peptide, VHPKQHR, termed VCAM-1 internalizing peptide (VINP) (Nahrendorf *et al.* 2006). The use of peptides for conjugation to nanoparticles enables a greater number of peptide ligands to be attached per nanoparticle, thereby increasing target affinity. For example, Tsourkas *et al.* (2005) reported that the number of VCAM-1 antibodies that can be conjugated to each nanoparticle is limited by steric constraints to 1-2 antibodies per nanoparticle. By contrast, the number of peptide ligands that could be attached to each nanoparticle was increased to 4 using cyclic VHS peptides (Kelly *et al.* 2005) and 20 using linear VINP peptides, respectively (Nahrendorf *et al.* 2006). In addition, Nahrendorf *et al.* (2006) demonstrated a 20-fold superior targeting affinity of the VINP peptide to cultured murine heart endothelial cells (MCEC), expressing high VCAM-1 levels, compared to VHS peptide. Both VHS- and VINP-targeted nanoparticles detected VCAM-1 expression in mouse atherosclerosis by *in vivo* MRI, and VINP-targeted nanoparticles also detected a reduction in VCAM-1 expression in response to statin treatment (Nahrendorf *et al.* 2006). However, one drawback of this method is that it requires the development of sophisticated cell-internalized probes that cannot easily be applied to detect other ligands.

Finally, mMRI molecular imaging probes have been developed that both visualize and treat disease. Winter *et al.* (2006) used paramagnetic perfluorocarbon nanoparticles targeted to $\alpha_v\beta_3$ to assess the development of angiogenesis, but also to deliver therapy with an anti-angiogenic drug, and to quantify the response to treatment. The authors used 1.5T MRI to first image the presence of angiogenesis in the vasa vasorum of the aorta of rabbits fed a high-cholesterol diet. Rabbits were then injected with a further round of nanoparticles of the same formulation but containing fumagillin, a known anti-angiogenic agent, before further nanoparticle imaging showed decreased vasa vasorum signal enhancement. Such local targeting approaches allow the use of agents, such as fumagillin, that may not be used systemically because of adverse side-effects.

A broad body of work therefore suggests that mMRI probes may be of future use in vascular endothelial adhesion molecule imaging. For such probes to become clinically useful, an important step will be ensuring the biocompatibility of these agents in humans. This in turn will require the ongoing development of imaging probes that are capable of providing high imaging contrast while remaining biodegradable and non-toxic. The development of micron-sized biodegradable contrast particles for clinical use is already underway.

RADIONUCLIDE BASED IMAGING: SPECT AND PET

In many ways, molecular imaging can be considered an extension of *nuclear imaging* techniques that use radiolabeled tracers to provide image contrast. More recently, novel radiolabeled tracers targeted to specific targets have been developed to study vascular disease (Langer *et al.* 2008). Compared to other imaging modalities, nuclear imaging techniques such as positron emission tomography (PET) and single photon emission computed tomography (SPECT) have the benefit of high sensitivity and are thus able to detect low levels (picomolar range) of a specific tracer (Saji 2004). A further advantage is the ability of nuclear imaging to provide functional quantitative information, such as the density of a specific protein marker (Langer *et al.* 2008). However, spatial resolution of the technique, and hence anatomical localization, are limited. In an early human application of molecular imaging, Jamar *et al.* (2002) coupled the Fab fragment of an antibody against E-selectin with ^{99m}Tc, and used this probe to detect inflammation in rheumatoid arthritis. ^{99m}Tc-Fab uptake was not seen in normal joints, but was taken up by joints with synovitis with a diagnostic accuracy of 88%. More recently, radionuclide imaging has been used to image many of the biological processes responsible for atherosclerosis, such as lipoprotein accumulation, apoptosis, and proteolysis (Langer *et al.* 2008), in animal models. Hua *et al.* (2005) constructed a ^{99m}Tc-labeled peptide targeted at $\alpha_v\beta_3$ integrin to image angiogenesis in an ischemic hindlimb mouse model. The peptide selectively localized to endothelial cells in regions of increased angiogenesis and could be used for serial tracking of the extent of angiogenesis. Outside of the cardiovascular system, radionuclide nanoparticles have also been used to image pulmonary endothelium and vascular brain tumors. For the former, fluorescein isothiocyanate (FITC) labeled polystyrene latex nanoparticles (100 nm in diameter) were conjugated to a mixture of ICAM-1 antibody and 64Cu-DOTA-IgG. The lungs of mice administered the anti-ICAM nanoparticles were clearly seen using small-animal PET up to 24 hr after administration (Rossin *et al.* 2008). For the lattter, abegrin, a monoclonal antibody to human $\alpha_v\beta_3$ integrin, was used to target a ^{90}Y radionuclide to glioblastoma multiforme in a mouse model.

Such widespread applications suggest that nuclear molecular imaging, in combination with anatomical imaging from MRI or CT, is likely to play a significant role in developing viable clinical methods of molecular imaging of vascular disease (Fig. 3).

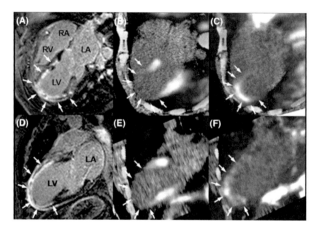

Fig. 3 Human Molecular Imaging using PET. Nuclear: Use of a novel $\alpha_v\beta_3$-targeting PET agent (F-Galakto-RGD) to assess integrin expression in a 35-year-old male several weeks following a myocardial infarction (C, F). Focal tracer retention is seen in the infarcted area previously defined using late-gadolinium-enhanced MRI (A, D) and perfusion PET (B, E) imaging. This signal may reflect angiogenesis within the healing area (arrows). With permission from Makowski (2008).

OPTICAL IMAGING

While ultrasound, MRI, or nuclear imaging may prove to be the first detection modalities to reach the clinical goal of non-invasive detection of adhesion molecule expression, optical imaging using fluorochromes will play an important role in the development of molecular imaging techniques. *Optical imaging* can be performed rapidly and at relatively low cost, and as such can be used to screen candidate probes before launching into more expensive and complicated *in vivo* imaging experiments, for example to allow quantification of probe specificity and sensitivity (Fig. 4).

Optical imaging requires the introduction of a fluorescent probe that is targeted to a specific adhesion molecule epitope. Several different reporter technologies can

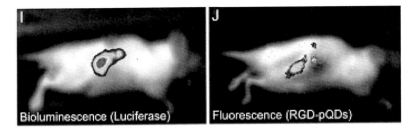

Fig. 4 Bioluminescence and fluorescence imaging of a mouse with a luciferase-expressing renal carcinoma tumor after injection of luciferin (left). The signal colocalizes with the fluorescence signal after IV injection of $\alpha_v\beta_3$-targeted quantum dot, a semiconductor nanocrystalloid particle with bright fluorescence (right). With permission from Mulder (2008).

then be used to detect signal (Ntziachristos 2006). *In vivo* immunofluorescence microscopy and intravital microscopy can allow probe specificity to be determined *in vivo*, non-invasive fluorescence transillumination, whereby light at an excitation wavelength is shone through the tissue and the fluorescence emitted is recorded on the other side with a highly sensitive charge-coupled device camera, has only more recently been described, and optical tomography uses light measurements collected at the tissue boundary to create a 3D reconstruction of fluorophore distribution.

Currently *fluorescence* imaging is limited to small animal studies, since the limited penetration depth of the optimal signal currently limits the use of this technique in everyday practice. Microscopic resolution can be obtained using invasive techniques, and sub-millimeter resolution can be achieved using fluorescence molecular tomography (Graves *et al.* 2003), which uses the deep tissue penetration ability of near infra-red fluorescent light (Nahrendorf *et al.* 2008).

Tsourkas *et al.* (2005) constructed a novel monocrystalline nanoparticle consisting of an iron oxide core and an animated cross-linked dextran coating. Cy5.5 fluorochrome was then attached to the particle to enable near-infra red fluorescent imaging. Monoclonal antibodies to VCAM-1 were then added to allow specific targeting of the protein's expression on murine endothelium using intravital fluorescence microscopy and MRI. A similar particle has been used by Funovics *et al.* (2005) to detect E-selectin expression in mouse lung carcinoma.

In the future, optical imaging promises to shed further insights into the mechanisms of disease. For example, Walls *et al.* (2008) used optical projection tomography of the ubiquitous vascular marker PECAM-1 to analyze vascular development in the mouse embryo. Analysis of the vascular tree at separate stages of development provided new information regarding the normal development of several different areas of the vascular tree. In a further example using fluorochromes that are excited at different wavelengths, and multi-channel detectors capable of receiving several signals at once, several biological processes were imaged simultaneously. To investigate the biology of aortic valve degeneration, the expression of VCAM-1, calcification, macrophage recruitment and protease activity were simultaneously imaged using a panel of near-infrared fluorescence imaging agents (Aikawa *et al.* 2007) (Fig. 5).

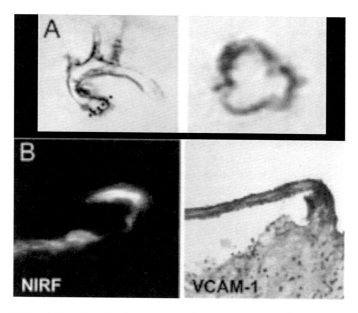

Fig. 5 Multimodal imaging of early aortic valve disease. Multimodal: (A) *Ex vivo* MRI of the mouse aortic arch and root. Negative signal enhancement (darkening) is seen at the areas of uptake of VCAM-1-targeted nanoparticles. Dotted line indicates the slice position of the short axis view (top right), where VCAM-1 expression can be located around the aortic ring. (B) The near-infrared signal (NIRF) from the aortic valve commissure, which is confirmed by immunohistochemical staining (bottom right, dark area). With permission from Aikawa *et al.* (2007).

SUMMARY

- Molecular imaging of endothelial adhesion molecules may lead to substantial changes in the way that many disease processes are diagnosed, treated and monitored.
- Ultrasound imaging is arguably the most extensively used modality to date and has been used to image adhesion molecule expression in a variety of animal disease models (e.g. transplant rejection, ischemia-reperfusion injury).
- Several probes have been developed for adhesion molecule imaging using MRI and have been used to successfully image vascular and cerebral inflammation, ischemia, and angiogenesis in animal models.

- Nuclear imaging offers the benefit of high sensitivity, permitting detection of even very low levels of adhesion molecule expression, but lacks detailed anatomical information. Nonetheless, its clinical application is already described.
- Various forms of optical imaging can currently be used to image endothelial adhesion molecules in experimental models, allowing verification of probe specificity before progression to larger imaging studies.
- Multimodal imaging, using a combination of the above techniques, can be used to verify the findings from one imaging modality, and when applied sequentially can also be used to examine patterns of adhesion molecule expression.
- The ongoing development of this field will require multi-disciplinary collaboration between biologists, chemists, engineers and physicians to optimize the design of future biodegradable molecular imaging probes and imaging sequences at clinical field strengths.
- Ultimately, attempts to perfect molecular imaging should provide new insights into the pathophysiological roles played by endothelial adhesion molecules in disease and allow novel methods for disease detection, and even treatment, to become apparent.

Abbreviations

ACPLS	Antibody-Conjugated Paramagnetic Liposomes
CT	Computed Tomography
EAE	Experimental Autoimmune Encephalitis
Gd	Gadolinium
ICAM	Intercellular Adhesion Molecule
IVUS	Intra-Vascular Ultrasound
mMRI	Molecular Magnetic Resonance Imaging
SPAQ	Sensitive Particle Acoustic Quantification
VCAM	Vascular Cell Adhesion Molecule

Key Facts about Molecular Imaging

1. Traditionally, medical imaging techniques have been able to detect only macroscopic changes suggestive of disease.
2. Molecular imaging is a new field that may eventually allow the detection and monitoring of the biological processes (e.g., endothelial adhesion molecule expression) that cause disease.

3. Molecular imaging techniques use uniquely designed imaging probes that are targeted to the biological process of interest.

4. A wide variety of biological processes can be imaged, including protein expression, gene upregulation, and apoptosis.

5. Molecular imaging probes have so far been detected with the use of ultrasound, computed tomography (CT), magnetic resonance, optical, and nuclear imaging techniques.

6. Much research is currently directed towards the design and application of molecular imaging probes, and the customization of detection modalities.

7. Although the majority of work done to date has been performed on animals, some initial human studies have shown promise.

8. In the future, molecular imaging techniques may provide novel insights into disease processes, in addition to providing a new means of diagnosis and even treatment.

Definition of Terms

Microbubble molecular imaging: The use of small gas-filled spheres (microbubbles), most commonly detected by ultrasound, to image specific molecules in living organisms.

Molecular MRI (mMRI): The use of magnetic resonance imaging to detect molecular probes or contrast agents targeted at specific biological processes.

Positron emission tomography (PET): A radionuclide-based technique that images the concentration of a positron-emitting tracer in the body using a specialized detection system.

Radionuclide-based imaging: A collective term for the medical imaging techniques that function by detecting radioactive decay from tracers.

Single photon emission computed tomography (SPECT): Similar to PET, but uses a different method of detection and the radioisotopes used are easier to make and have relatively longer half-lives.

References

Aikawa, E. and M. Nahrendorf, D. Sosnovik, V.M. Lok, F.A. Jaffer, M. Aikawa, and R. Weissleder. 2007. Multimodality molecular imaging identifies proteolytic and osteogenic activities in early aortic valve disease. Circulation 115(3): 377-386.

Barber, P.A. and T. Foniok, D. Kirk, A.M. Buchan, S. Laurent, S. Boutry, R.N. Muller, L. Hoyte, B. Tomanek, and U.I. Tuor. 2004. MR molecular imaging of early endothelial activation in focal ischemia. Ann. Neurol. 56(1): 116-120.

Behm, C.Z. and B.A. Kaufmann, C. Carr, M. Lankford, J.M. Sanders, C.E. Rose, S. Kaul, and J.R. Lindner. 2008. Molecular Imaging of Endothelial Vascular Cell Adhesion Molecule-1 Expression and Inflammatory Cell Recruitment during Vasculogenesis and Ischemia-mediated Arteriogenesis. Circulation 117: 2902-2911.

Benson, V. and A.C. McMahon, and H.C. Lowe. 2007. ICAM-1 in acute myocardial infarction: a potential therapeutic target. Curr. Mol. Med. 7(2): 219-227.

de Jong, N. and A. Bouakaz, and P. Frinking. 2002. Basic acoustic properties of microbubbles. Echocardiography 19(3): 229-240.

Doiron, A.L. and K. Chu, A. Ali, and L. Brannon-Peppas. 2008. Preparation and initial characterization of biodegradable particles containing gadolinium-DTPA contrast agent for enhanced MRI. Proc. Natl. Acad. Sci. 105(45): 17232-17237.

Funovics, M. and X. Montet, F. Reynolds, R. Weissleder, and L. Josephson. 2005. Nanoparticles for the optical imaging of tumor E-selectin. Neoplasia 7(10): 904-911.

Graves, E.E. and J. Ripoll, R. Weissleder, and V. Ntziachristos. 2003. A submillimeter resolution fluorescence molecular imaging system for small animal imaging. Med. Phys. 30(5): 901-911.

Hamilton, A.J. and S.L. Huang, D. Warnick, M. Rabbat, B. Kane, A. Nagaraj, M. Klegerman, and D.D. McPherson. 2004. Intravascular ultrasound molecular imaging of atheroma components in vivo. J. Am. Coll. Cardiol. 43(3): 453-460.

Hillis, G.S. and A.D. Flapan. 1998. Cell adhesion molecules in cardiovascular disease: a clinical perspective. Heart (British Cardiac Society) 79(5): 429-431.

Hua, J. and L.W. Dobrucki, M.M. Sadeghi, J. Zhang, B.N. Bourke, P. Cavaliere, J. Song, C. Chow, N. Jahanshad, N. van Royen, I. Buschmann, J.A. Madri, M. Mendizabal, and A.J. Sinusas. 2005. Noninvasive imaging of angiogenesis with a 99mTc-labeled peptide targeted at alphavbeta3 integrin after murine hindlimb ischemia. Circulation 111(24): 3255-3260.

Ichioka, S. and N. Sekiya, M. Shibata, and T. Nakatsuka. 2007. AlphaV beta3 integrin inhibition reduces leukocyte-endothelium interaction in a pressure-induced reperfusion model. Wound Repair Regen. 15(4): 572-576.

Jamar, F. and F.A. Houssiau, J.P. Devogelaer, P.T. Chapman, D.O. Haskard, V. Beaujean, C. Beckers, D.H. Manicourt, and A.M. Peters. 2002. Scintigraphy using a technetium 99m-labelled anti-E-selectin Fab fragment in rheumatoid arthritis. Rheumatology (Oxford, England) 41(1): 53-61.

Kang, H.W. and L. Josephson, A. Petrovsky, R. Weissleder, and A. Bogdanov. 2002. Magnetic resonance imaging of inducible E-selectin expression in human endothelial cell culture. Bioconjug. Chem. 13(1): 122-127.

Kaufmann, B.A. and J.M. Sanders, C. Davis, A. Xie, P. Aldred, I.J. Sarembock, and J.R. Lindner. 2007. Molecular imaging of inflammation in atherosclerosis with targeted ultrasound detection of vascular cell adhesion molecule-1. Circulation 116(3): 276-284.

Kelly, K.A. and J.R. Allport, A. Tsourkas, V.R. Shinde-Patil, L. Josephson, and R. Weissleder. 2005. Detection of vascular adhesion molecule-1 expression using a novel multimodal nanoparticle. Circ. Res. 96(3): 327-336.

Langer, H.F. and R. Haubner, B.J. Pichler, and M. Gawaz. 2008. Radionuclide imaging: a molecular key to the atherosclerotic plaque. J. Am. Coll. Cardiol. 52(1): 1-12.

Lindner, J.R. and J. Song, J. Christiansen, A.L. Klibanov, F. Xu, and K. Ley. 2001. Ultrasound assessment of inflammation and renal tissue injury with microbubbles targeted to P-selectin. Circulation 104(17): 2107-2112.

Lindner, J.R. and J. Song, A.R. Jayaweera, J. Sklenar, and S. Kaul. 2002. Microvascular rheology of Definity microbubbles after intra-arterial and intravenous administration. J. Am. Soc. Echocardiogr. 15(5): 396-403.

Makowski, M.R., U. Ebersberer, S. Nekolla *et al.* 2008. In vivo molecular imaging of angiogenesis, targeting $\alpha_v\beta_3$ integrin expression, in a patient after acute myocardial infarction. European Heart Journal 29(18): 2201.

Mankoff, D.A. 2007. A Definition of Molecular Imaging. J. Nucl. Med. 48(6), p. 18N.

Massoud, T.F. and S.S. Gambhir. 2003. Molecular imaging in living subjects: seeing fundamental biological processes in a new light. Genes Dev. 17(5): 545-580.

McAteer, M.A. and N.R. Sibson, C. von Zur Muhlen, J.E. Schneider, A.S. Lowe, N. Warrick, K.M. Channon, D.C. Anthony, and R.P. Choudhury. 2007. In vivo magnetic resonance imaging of acute brain inflammation using microparticles of iron oxide. Nat. Med. 13(10): 1253-1258.

McAteer, M.A. and J.E. Schneider, Z.A. Ali, N. Warrick, C.A. Bursill, C. von zur Muhlen, D.R. Greaves, S. Neubauer, K.M. Channon, and R.P. Choudhury. 2008. Magnetic resonance imaging of endothelial adhesion molecules in mouse atherosclerosis using dual-targeted microparticles of iron oxide. Arterioscler. Thromb. Vasc. Biol. 28(1): 77-83.

Mulder, W.J.M., K. Castermans, J.R. van Beijnum *et al.* 2009. Molecular imaging of tumor angiogenesis using $\alpha_v\beta_3$-integrin targeted multimodal quantum dots. Angiogenesis 12(1): 17-24.

Nahrendorf, M. and F.A. Jaffer, K.A. Kelly, D.E. Sosnovik, E. Aikawa, P. Libby, and R. Weissleder. 2006. Noninvasive vascular cell adhesion molecule-1 imaging identifies inflammatory activation of cells in atherosclerosis. Circulation 114(14): 1504-1511.

Nahrendorf, M. and D.E. Sosnovik, and R. Weissleder. 2008. MR-optical imaging of cardiovascular molecular targets. Basic Res. Cardiol. 103(2): 87-94.

Ntziachristos, V. 2006. Fluorescence molecular imaging. Ann. Rev. Biomed. Eng. 8: 1-33.

Palmowski, M. and J. Huppert, G. Ladewig, P. Hauff, M. Reinhardt, M.M. Mueller, E.C. Woenne, J.W. Jenne, M. Maurer, G.W. Kauffmann, W. Semmler, and F. Kiessling. 2008. Molecular profiling of angiogenesis with targeted ultrasound imaging: early assessment of antiangiogenic therapy effects. Mol. Cancer Ther. 7(1): 101-109.

Reinhardt, M. and P. Hauff, R.A. Linker, A. Briel, R. Gold, P. Rieckmann, G. Becker, K.V. Toyka, M. Maurer, and M. Schirner. 2005. Ultrasound derived imaging and quantification of cell adhesion molecules in experimental autoimmune encephalomyelitis (EAE) by Sensitive Particle Acoustic Quantification (SPAQ). NeuroImage 27(2): 267-278.

Rossin, R. and S. Muro, M.J. Welch, V.R. Muzykantov, and D.P. Schuster. 2008. In vivo imaging of 64Cu-labeled polymer nanoparticles targeted to the lung endothelium. J. Nucl. Med. 49(1): 103-211.

Saji, H. 2004, Development of radiopharmaceuticals for molecular imaging. Intl. Cong. Ser. 1264: 139-147.

Sibson, N.R. and A.M. Blamire, M. Bernades-Silva, S. Laurent, S. Boutry, R.N. Muller, P. Styles, and D.C. Anthony. 2004. MRI detection of early endothelial activation in brain inflammation. Magn. Reson. Med. 51(2): 248-252.

Sipkins, D.A. and K. Gijbels, F.D. Tropper, M. Bednarski, K.C. Li, and L. Steinman. 2000. ICAM-1 expression in autoimmune encephalitis visualized using magnetic resonance imaging. J. Neuroimmunol. 104(1): 1-9.

Tsourkas, A. and V.R. Shinde-Patil, K.A. Kelly, P. Patel, A. Wolley, J.R. Allport, and R. Weissleder. 2005. In vivo imaging of activated endothelium using an anti-VCAM-1 magnetooptical probe. Bioconjug. Chem. 16(3): 576-581.

Villanueva, F.S. and R.J. Jankowski, S. Klibanov, M.L. Pina, S.M. Alber, S.C. Watkins, G.H. Brandenburger, and W.R. Wagner. 1998. Microbubbles targeted to intercellular adhesion molecule-1 bind to activated coronary artery endothelial cells. Circulation 98(1): 1-5.

Villanueva, F.S. and E.W. Gertz, M. Csikari, G. Pulido, D. Fisher, and J. Sklenar. 2001. Detection of coronary artery stenosis with power Doppler imaging. Circulation 103(21): 2624-2630.

Villanueva, F.S. and E. Lu, S. Bowry, S. Kilic, E. Tom, J. Wang, J. Gretton, J.J. Pacella, and W.R. Wagner. 2007. Myocardial ischemic memory imaging with molecular echocardiography. Circulation 115(3): 345-352.

Villanueva, F.S. 2008 Molecular imaging of cardiovascular disease using ultrasound. J. Nucl. Cardiol. 15(4): 576-586.

Walls, J.R. and L. Coultas, J. Rossant, and R.M. Henkelman. 2008. Three-dimensional analysis of vascular development in the mouse embryo. PLoS ONE 3(8): e2853.

Weller, G.E. and E. Lu, M.M. Csikari, A.L. Klibanov, D. Fischer, W.R. Wagner, and F.S. Villanueva. 2003. Ultrasound imaging of acute cardiac transplant rejection with microbubbles targeted to intercellular adhesion molecule-1. Circulation 108(2): 218-324.

Weller, G.E. and Wong, M.K., Modzelewski, R.A., Lu, E., Klibanov, A.L., Wagner, W.R., and F.S. Villanueva. 2005. Ultrasonic imaging of tumor angiogenesis using contrast microbubbles targeted via the tumor-binding peptide arginine-arginine-leucine. Cancer Res. 65(2): 533-539.

Winter, P.M. and A.M. Neubauer, S.D. Caruthers, T.D. Harris, J.D. Robertson, T.A. Williams, A.H. Schmieder, G. Hu, J.S. Allen, E.K. Lacy, H. Zhang, S.A. Wickline, and G.M. Lanza. 2006. Endothelial alpha(v)beta3 integrin-targeted fumagillin nanoparticles inhibit angiogenesis in atherosclerosis. Arterioscler. Thromb. Vasc. Biol. 26(9): 2103-2109.

CHAPTER 5

Deleterious Roles of Inflammatory Adhesion Molecules during Aging

Yani Zou[1], Byung Pal Yu[2] and Hae Young Chung[3],[*]

[1]Department of Neurology and Neurological Sciences, School of Medicine,
Stanford University, Palo Alto, CA 94304, USA; E-mail: yanizou@stanford.edu
[2]Department of Physiology, University of Texas Health Science Center at San
Antonio, TX 78229-3900, USA; E-mail: yu6936@sbcglobal.net
[3]College of Pharmacy, Aging Tissue Bank, Molecular Inflammation Research
Center for Aging Intervention, Pusan National University,
Pusan 609-735, South Korea
Departmental contact: Department of Pharmacy, College of Pharmacy,
Molecular Inflammation Research Center for Aging Intervention,
Pusan National University, San 30, Jangjun-dong, Gumjung-gu,
Pusan 609-735, South Korea; E-mail: hyjung@pusan.ac.kr

ABSTRACT

Adhesion molecules (AMs) are trans-membrane proteins involved in diverse
biological activities related to cell-cell and cell-matrix interactions, cell migration,
differentiation, tumor metastasis, signal transduction, inflammation and wound
healing. Some AMs coordinate and regulate inflammatory response by triggering
leukocyte recruitment and migration through the extracellular cellular matrix.
Because of the essential roles that AMs play in multiple physiological processes,
dysregulated AMs are implicated in various pathological conditions including
diabetes, atherosclerosis, cancer, cardiovascular disease and neurodegenerative
disease, as seen during aging. The aging process itself is a major risk factor

---

Corresponding author: Molecular Inflammation Research Center for Aging Intervention, College of
Pharmacy, Pusan National University, 30 Jangjun-dong, Gumjung-gu, Pusan 609-735, South Korea;
E-mail: hyjung@pusan.ac.kr
Key terms are defined at the end of the chapter.

underlying various chronic inflammatory diseases. The deleterious roles of inflammatory AMs underline age-related pathophysiological conditions.

In this chapter, we review the deleterious roles of AMs by focusing on (1) altered AMs by age-related redox mediators, (2) alterations of AMs during aging, (3) mechanisms underlining inflammatory AM changes during aging, (4) the role of AMs in age-related diseases, and (5) possible interventions in age-related AM changes.

INTRODUCTION

Adhesion molecules (AMs) are a group of membrane proteins exposed on the cell surface that adhere to other cells or to the extracellular matrix in a process known as cell adhesion. AM proteins basically function as receptors and are composed of three domains: an intracellular domain that interacts with the cytoskeleton, a transmembrane domain, and an extracellular domain that interacts with either other AMs or with the extracellular matrix through homophilic or heterophilic binding. When undergoing proteolytic cleavage, AMs release the soluble ectodomain from the cell surface, resulting in circulating or a soluble form of AMs (sAMs). Most AMs are classified into four protein families: immunoglobulin (Ig) superfamily, integrins, cadherins and selectins.

One of the deleterious roles of AMs is their participation in the inflammation process. For instance, in response to pro-inflammatory cytokines and chemokines such as tumor necrosis factor-α (TNF-α) and interleukin 1-β (IL-1β), AMs direct leukocyte-endothelial cell (EC) interactions through a sequential process of tethering, rolling, adhesion and transmigration. When acting as macrophages, leukocytes pass through the tight junction between ECs and infiltrate into the vessel wall, where they release various inflammatory mediators, thereby triggering further inflammation. Consequently, prolonged, unchecked inflammatory AMs lead to chronic inflammatory conditions. To distinguish among the AMs, the AMs that participate in the inflammatory process are called inflammatory AMs in this chapter.

Aging is a complex biological phenomenon that is characterized as a progressive, physiological deterioration with time, accompanied by increased vulnerability to pathogenesis. Although the underlying cause and mechanisms of aging are unknown, the popular oxidative stress theory of aging provides molecular insights into possible causative factors for the aging process (Yu and Chung 2006). According to the tenets of this theory, the combined effects of accumulated oxidative damage and weakened antioxidative defense systems cause a disturbance in the organism's redox balance.

However, one of the most intriguing and important questions is, how does a disrupted redox balance increase the vulnerability to disease during aging? The answer could be found in the age-associated activation of redox-sensitive transcription factors that cause the inflammation process (Chung *et al.* 2006, 2009). It is likely that an oxidative stress-induced redox imbalance and a dysregulated immune system with age increase the systemic inflammatory status to activate a wide variety of inflammatory mediators, including AMs. The salient point is that unresolved, chronic inflammation during aging may act as the pathophysiological link that drives normal functional changes to become many of the age-related degenerative diseases (Chung *et al.* 2009).

ALTERED AMs BY AGE-RELATED REDOX MEDIATORS

AMs are well known to mediate interactions between blood cells (leukocyte, platelet) and ECs. The coordinated recruitment of leukocytes to sites of inflammation is largely governed by the time-course and magnitude of endothelial AM expression. The expression of AMs during inflammation is generally transient and quick, but a prolonged and dysregulated AM-mediated process of leukocyte recruitment often results in EC dysfunction.

It is well established that most of the regulations of inflammatory AMs in cells are at the transcriptional level. AMs are stimulated when cells are exposed to mediators such as the pro-inflammatory cytokines, TNF-α and IL-1β. Two transcription factors, NF-kappaB and AP-1, are identified as major regulators responsible for the upregulation of E-selectin, P-selectin, VCAM-1 and ICAM-1 on cytokine-treated ECs.

In addition to typical cytokine stimuli, various other AM regulators have been identified (Fig. 1) and most of these regulators are influenced by age-related oxidative stress. For instance, many studies have indicated that oxidized low-density lipoprotein (oxLDL) is an effective stimulus for the expression of AMs on ECs. Our recent study reported that the major components of oxLDL, lysophosphatidylcholine (LPC) directly induce the endothelial expression of VCAM-1 and P-selectin through a G-protein coupled receptor 4 pathway (Zou *et al.* 2007). It was shown that advanced glycation end products (AGEs) enhanced levels of mRNA and antigen for VCAM-1, ICAM-1, and E-selectin in primary cultures of human saphenous vein ECs through engagement of AGE receptor, thereby increasing adhesion of polymorphonuclear leukocytes to stimulated ECs.

Age-related oxidative stress is reported to play an important role in the upregulation of inflammatory AMs. Oxidative stress induced by either depletion of glutathione or generation of reactive species (RS) induced the upregulation of P-selectin and VCAM-1 in ECs. NADPH oxidase-derived RS induce expression of VCAM-1 and ICAM-1 in ECs (Dworakowski *et al.* 2008). Expressions of both P-selectin and VCAM-1 were induced in a dose dependent manner by XOD/xanthine, which generates $^\bullet O_2^-$ and H_2O_2, indicating the role of reactive oxygen species in triggering AM expression (Zou *et al.* 2006). Among several stimuli we tested using the VCAM-1 reporter vector, LPC, TNF-α and IL-1β were highly effective factors inducing VCAM-1 expression (unpublished data). H_2O_2 promotes adhesion of eosinophils, neutrophils, and monoblastoid U-937 cells via increased expression of integrin CD11β/CD18 molecules (Nagata 2005). Therefore, the expression of inflammatory AMs in cells is also regulated by redox status that is significantly changed during the aging process.

ALTERATIONS OF AMs DURING AGING

Expressions of AMs are altered variably during aging, as shown in Table 1. A key fact of regulation of AMs under normal aging is that inflammatory AMs are upregulated with age. Increased VCAM-1 has been detected in aorta from aged Fisher 344 (F344) rats, Fischer 344/Brown Norway F1 hybrid (F344xBN) rats and

Table I   Age-related changes in AMs

AMs	Major function	Changes with age
VCAM-1	Inflammation	↑
ICAM-1	Inflammation	↑
NCAM-1	Development, neural plasticity	↓
PECAM-1	Adhesion	–
MAdCAM-1	Immune control	–/↓
MAG	Myelination	↓
VLA-4	Inflammation	↑
LFA-1	Inflammation	↑
E-selectin	Inflammation	↑
P-selectin	Inflammation	↑
L-selectin	Inflammation	↑
E-cadherin	Cell junction	↓
N-cadherin	Synapse function	↓
OB-cadherin	Bone function	↓

↑: increased with age; ↓: decreased with age; –: no change with age.

from old LDL receptor knockout mice. The level of VCAM-1 in kidney was also increased in old F344 rats and in old (28 mon) mice.

ICAM-1 was shown to have increased in aged F344 rat kidney, vascular intima and retina, in F344xBN rat aorta (Miller *et al.* 2007) and brain, in the liver of old B6 mice (Ito *et al.* 2007) and also in the orbitofrontal cortex from elderly human subjects (Miguel-Hidalgo *et al.* 2007). Senescent human aortic ECs also exhibit increased ICAM-1 expression. In a wound repairing model, E-selectin was found strongly expressed in a perivascular distribution in elderly human subjects. P-selectin was increased in the aorta (Zou *et al.* 2006) and kidney from aged F344 rats. In addition, levels of ligands for inflammatory AMs also increase with age. VCAM-1 ligand very late antigen 4 (VLA-4, α4β1 integrin, CD49d) as well as expression of L-selectin (CD62L) are all found to increase in elderly subjects. ICAM-1 ligand lymphocyte-function-associated antigen 1 (LFA-1, CD11a) also increases with age in human T lymphocytes.

With respect to sAMs, plasma levels of soluble VCAM-1 (sVCAM-1) were found significantly higher in elderly persons (Richter *et al.* 2003), old mice and rats. With respect to circulating ICAM-1, an age-dependent elevation was found in aged people (Richter *et al.* 2003) and rats. The circulating E-selectin and P-selectin levels were also correlated with age in rats (Zou *et al.* 2004).

In contrast, non-inflammation-related AMs are reduced or show no changes during the aging process. Using germ line stem cell from *Drosophila* testis, somatic niche cells in testes from older males display reduced expression of the E-cadherin (Boyle *et al.* 2007). Jung *et al.* (2004) have reported an age-dependent decrease in renal N-cadherin expression, whose mRNA and protein expression decreased in parallel. Implicating a transcriptional mechanism in the age-dependent loss of expression. OB-cadherin (cadherin-11) is expressed in osteoblast and has a specific function in bone development and maintenance. It has been known that aged rabbits show a greater than 9-fold reduction in cadherin-11 specific gene expression as compared with mature rabbits.

The polysialylated NCAM (PSA-NCAM) spatial distribution in normal condition differs between young and old rats and the expression decreases markedly during development in the brain. The numbers of PSA-NCAM-expressing granule cells show a parallel age-dependent decrease during aging. In muscle, PSA-NCAM expression is upregulated following denervation, but this response is weakened in aging rats. Level of another neural cell AM, myelin-associated glycoprotein (MAG), was also reported reduced in the old rhesus monkey.

It is worth mentioning that levels of platelet endothelial cell adhesion molecule 1 and mucosal addressin cell adhesion molecule 1 do not change with age, although reduced expression with age was also reported.

MECHANISMS UNDERLINING INFLAMMATORY AM CHANGES DURING AGING

The important question to address is, why does the aging process increase the expression of inflammatory AMs? The answers may lie in enhanced oxidative stress and pro-inflammatory status, the two key features of the aging process.

Enhanced Oxidative Stress of Aging

Oxidative stress is described generally as a condition under which increased production of RS occurs, including singlet oxygen and reactive lipid peroxidative products, such as reactive aldehydes and peroxides (Yu and Chung 2006). Accumulated evidence shows increased oxidative stress during aging. Increased

RS have been found accumulated in aged animals, such as $^{\bullet}O_2^-$, hydrroxyl radical, and H_2O_2, reactive nitric oxide, and peroxynitrite. Higher lipid peroxidation and products were also determined in the old animals, including 4-hydroxyhexenal, malondialdehyde and LPC. In addition, protein oxidation was also detected; for instance, level of AGEs was reported increased with age, including argpyrimidine, fluorolink, pyrraline, and imidazolone derivatives of arginine. In contrast, the anti-oxidative defense systems are weakened by age, such as the reduced anti-oxidative capacity of serum. Therefore, the age-related redox imbalance occurring under oxidatively stressed conditions is likely caused by the net effect of weakened anti-oxidative defense systems and the incessantly increasing production of RS. As discussed above, most of these redox mediators, like RS, LPC and AGE, have been identified as potent stimuli for inflammatory AMs (Fig. 1).

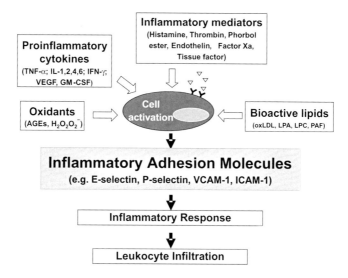

Fig. 1   Age-related stimuli that cause upregulation of inflammatory AMs. Diverse aging-related stimulators can upregulate expression of AMs on EC. Uncontrolled expression of AMs causes dysregulated leukocyte-EC interactions, leading to improper recruitment of leukocytes. (Unpublished.)

Pro-inflammatory Status of Aging

One inescapable cellular consequence of oxidative stress is the formation of a primary driving force for increased activation of redox-regulated transcription factors, such as NF-κB, which regulates the expression pro-inflammatory molecules (Fig. 2) (Chung *et al.* 2006). Most of the data indicated that the molecular events involved in age-related NF-κB activation requires phosphorylation by inhibitor of kappaB (IκB) kinase (IKK) / NF-κB-inducing kinase (NIK) and mitogen-activated

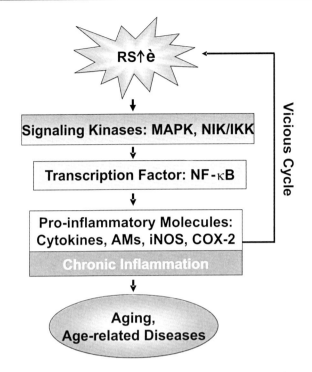

Fig. 2 Major molecular pro-inflammatory pathways involved in aging. (Unpublished.)

protein kinases (MAPK). For instance, 4-hydoxyhexenal triggered NF-κB activation by IκB phosphorylation via the IKK/NIK pathway, through increased p38 kinase and extracelluar signal-regulated kinase, but not c-jun kinase signaling.

Redox-sensitive activation of NF-κB-dependent genes is a major culprit responsible for the systemic inflammatory process during aging. In aged organisms, NF-κB-regulated inflammatory reactions lead to a chronic pro-inflammatory state as evidenced from the activation of pro-inflammatory gene expression. The activation of NF-κB is responsible for transcription of not only pro-inflammatory proteins such as TNF-α, interleukins such as IL-1, IL-2, and IL-6 as well as the chemokines iNOS and COX-2, but also AMs (Chung *et al.* 2006). The difference in the NF-κB activation observed during aging could be explained by a much faster and more extensive IκB degradation in old than young, leading to more abundant expression of ICAM-1 in the old vascular smooth muscle cells, for example. A greater activation of the NIK/IKK/IκB pathway in aorta from old rats was found; therefore, this pathway may be responsible for increased levels of aortic P-selectin and VCAM-1 in the old rats.

As proposed in the molecular inflammation hypothesis of aging, a state of chronic, low-grade inflammation mediated by redox-sensitive transcription factors is a possible converging process linking normal aging and the pathogenesis of age-related diseases (Chung *et al.* 2006).

AMs IN AGE-RELATED DISEASES

AMs have received much attention, not only for the essential roles they have in normal physiological processes, but also for their deleterious roles as modulators of uncontrolled cell-cell interactions, which contribute to the vascular dysfunction and tissue injury associated with diseases. Because of the increased incidence of these diseases with age, it is important to recognize the age-associated upregulation of AMs that plays a crucial role in the initiation and development of age-related diseases such as cardiovascular diseases, immunosenescence, cancer, and neurodegenerative diseases, as illustrated in a few of the following examples.

Atherosclerosis

Evidence clearly shows increased arterial susceptibility to atherogenetic stimuli with age. In a rabbit model of atherosclerosis, soon after initiating an atherogenic diet, microscopic examination revealed leukocyte attachment to the intima-lining ECs. VCAM-1 is proved to be involved in this adhesion at sites of atheroma initiation. An increased VCAM-1 level plus high plasma cholesterol was suspected as an underlying cause of atherosclerosis during aging. Experiments with hypomorphic variants of VCAM-1 introduced into mice that are rendered susceptible to atherogenesis (by inactivation of the apolipoprotein E gene) show reduced lesion formation. In addition to VCAM-1, P- and E-selectin also contribute to leukocyte recruitment in atherosclerosis-susceptible mice.

The mechanism of VCAM-1 induction after the initiation of an atherogenic diet likely involves the inflammatory process instigated by modified lipoprotein particles accumulating in the arterial intima in response to hyperlipidemia. Constituents of modified lipoprotein particles, among them certain oxidized phospholipids and short-chain aldehydes arising from lipoprotein oxidation, can induce transcriptional activation of the VCAM-1 gene mediated in part by NF-κB.

Cancer

The incidence of cancer increases with age in humans and in laboratory animals alike, and AMs are involved in the various aspects of cancer. VCAM-1, for example, has been suggested as a regulator of ovarian cancer peritoneal metastasis (Slack-Davis *et al.* 2009). VCAM-1 expression was observed by Slack-Davis and colleagues on the mesothelium of women with ovarian cancer. Blocking antibodies to, or small interfering RNA knockdown of, VCAM-1 or its ligand integrin α4β1 significantly decreases, but did not completely inhibit, transmigration of SKOV-3 cells through mesothelial monolayers. It is worth mentioning that higher serum ICAM-1 can be useful for diagnosis of non-small cell lung cancer, while E-selectin levels have prognostic significance and could be a potential prognostic factor in these cancer patients (Guney *et al.* 2008).

The detection of sICAM-1 was found to be higher in a malignant group of bladder cancer patients compared to the control group. In the cancer group, sICAM-1 was significantly correlated with the patient's age. Also, it was increased in advanced and poorly differentiated tumors (Aboughalia 2006). Serum ICAM-1 levels were also found to be related to tumor presence, clinical stages, and grade of colorectal cancer; therefore, sICAM-1 may be useful for monitoring malignant disease stages and for evaluating the effectiveness of various therapeutic approaches for colorectal carcinomas.

Alzheimer's Disease (AD)

AD is a disease of the aged populations in which the inflammatory process is involved (Wyss-Coray 2006). One study showed that Amyloid beta (Aβ) interaction with a monolayer of normal human brain microvascular ECs results in increased adherence and transmigration of monocytes and that this reaction was inhibited by antibodies to AGE receptor, and to AM PECAM-1 (Giri *et al.* 2002). Plasma sICAM-1 and sPECAM-1 levels were higher and the cerebrospinal fluid sVCAM-1 was lower in AD patients and in patients with dementia with Lewy bodies than in controls. Lewy bodies patients had higher cerebrospinal fluid sICAM-1, but lower cerebrospinal fluid sVCAM-1 (Nielsen *et al.* 2007).

The synaptic function of NCAM may be compromised by the increased Aβ levels in AD brain. Lack of NCAM signaling leading to inhibition of GSK3b would allow the progression of tangle pathology in AD. A significant increase of Ig superfamily adhesion molecules L1 and a strong tendency for increase of the soluble fragments of NCAM in the cerebrospinal fluid of Alzheimer patients were reported, compared to the normal control group (Strekalova *et al.* 2006). The proteolytic fragments of L1, but not NCAM, were also elevated in patients with vascular dementia and dementia of mixed types. Higher L1 concentrations were observed irrespective of age and gender. NCAM concentrations also were independent of gender, but positively correlated with age and, surprisingly, also with the incidence of multiple sclerosis (Strekalova *et al.* 2006).

POSSIBLE INTERVENTIONS IN AGE-RELATED AM CHANGES

Considering the effects of dysregulated AMs during aging on various age-associated diseases, to seek possible interventions in age-related AM changes by various means is important, and should be of great interest to researchers. In this section, we present a few known experimental paradigms by which the activation of AMs was effectively modulated.

Calorie restriction (CR) is the most powerful nutritional intervention of the aging process, and researchers have now accepted CR as the only established anti-aging experimental paradigm. Although the molecular mechanism of effects of CR is still not clearly known, its anti-aging effects are thought to be due mainly to its powerful resistance against oxidative stress and ability to maintain a proper cellular redox status. It has been clearly demonstrated that CR suppresses age-related diseases, modulates redox-sensitive transcription factors and inflammation, and restores various membrane-associated functions in experimental animals (Chung *et al.* 2006, Yu and Chung 2006).

Accumulating evidence indicates that anti-oxidative CR significantly attenuates NF-κB, TNF-α, interleukins (IL-1β, IL-2, and IL-6), chemokines (IL-8 and RANTES) and AMs (Chung *et al.* 2009). The increased levels of aortic AMs, VCAM-1 and P-selectin, as well as sAMs, E-selectin, P-selectin, VCAM-1 and ICAM-1 during aging in non-restricted old rats were effectively blunted by CR, and leukocyte infiltration in the old rats was also reduced by CR (Son *et al.* 2005). CR even initiated in late adulthood confers beneficial effects, such as the attenuation of oxidative stress, enhanced expression of HSP-70, neural plasticity markers NCAM, and PSA-NCAM, and reduced levels of GFAP (Kaur *et al.* 2008).

Antioxidants and anti-inflammatory treatments are also effective in controlling the level of AMs in the elderly. Polyphenols or vitamin E may assist in preventing cardiovascular disease, in part by decreasing EC expression of pro-inflammatory cytokines, AMs, and monocyte adhesions. Ferulate is a well-described natural antioxidant found in plants, and ferulate has exhibited its anti-oxidative action by reducing the NF-κB-induced, pro-inflammatory VCAM-1 and ICAM-1 in kidney from old Sprague-Dawley rats (Jung *et al.* 2009). Betaine suppresses pro-inflammatory signaling during aging including the VCAM-1 and ICAM-1 expression through its anti-oxidative effects as betaine is involved in glutathione metabolism.

Aberrant AMs with age could also be regulated by anti-inflammatory treatments. Increased ICAM-1 was decreased by etanercept treatment, which binds and inactivates TNF-α. Anti-TNF-α treatment exerts anti-aging, vasculoprotective effects (Csiszar *et al.* 2007). Aspirin is one of the most commonly used non-steroidal anti-inflammatory drugs. In a recent study, we investigated the effect of short-term, low dose aspirin intake on the modulation of pro-inflammatory NF-κB activation in old rats (Jung *et al.* 2006). The data showed that NF-κB activation in the old rats and its associated gene expressions, including VCAM-1 and ICAM-1, were all suppressed by the low dose aspirin supplementation through the inhibition of phosphorylation and degradation of IκBα via the NIK/IKK pathway. Peroxisome proliferator-activated receptors (PPARs) are transcription factors belonging to the nuclear hormone receptor superfamily. Several studies have demonstrated that PPARα and PPARγ inhibit the expression of inflammatory genes, such as cytokines, metalloproteases, and acute phase proteins. Recent data showed that age-related inflammation and oxidative stress, including the upregulation of VCAM-1 and P-selectin, was ameliorated by PPARγ activator 2,4-thiazolidinedione. Other anti-inflammatory reagents such as 3-methyl-1,2-cyclopentanedione also showed similar effects in regulating age-associated VCAM-1 upregulation.

Regarding those AMs that are downregulated by the aging process, new therapeutic strategies are undergoing development. For instance, new evidence showed that the synthetic molecule C3d, which is a peptide mimetic of NCAM, promotes choline acetyltransferase activity in cultured rat embryonic septal neurons. These findings on the possible involvement of AMs may be significant when considering new strategies aimed at stimulating cholinergic function and improving cognition in disorders such as Alzheimer's disease (Burgess *et al.* 2009).

CONCLUSION

In conclusion, AMs are unavoidably involved in many physiological and pathological processes, especially in triggering and coordinating inflammation responses. The aging process is accompanied by accumulated oxidative stress and micro-inflammatory changes that are two major factors triggering the upregulation of inflammatory AMs in the elderly. Dysregulated AMs contribute to the initiation and development of various age-associated diseases. Consideration of effective nutritional and therapeutic approaches to modulate age-related, deleterious AM changes is becoming increasingly important. Antioxidant supplementation and anti-inflammatory therapies may be beneficial for reducing upregulated AMs during the aging process, thereby helping to attenuate age-related degenerative diseases.

SUMMARY

- A brief description on oxidative stress hypothesis is presented as a possible mechanism underlying the aging process.
- Age-related oxidative stress and redox imbalance are described as major factors responsible for the pro-inflammatory status of the aged, as proposed by the molecular inflammatory hypothesis.

- Experimental data showing several key inflammatory AMs altered during aging are detailed.
- Increased NF-κB activation is highlighted as a major contributing factor for modulating the age-related upregulation of inflammatory AMs, including VCAM-1 and P-selectin.
- Molecular mechanisms on increased NF-κB activation are explored in relation to upregulation of inflammatory AMs.
- Evidence is presented implicating dysregulated expressions of AMs in age-related chronic diseases, such as atherosclerosis, cancer and Alzheimer's disease.
- Considerations on possible anti-oxidative and anti-inflammatory therapies based on the attenuation of upregulated inflammatory AMs during aging are discussed.

Definition of Terms

Aging: The gradual physiological and functional decline that is accompanied with the increased incidence of disease with time.

Age-related redox imbalance: An age-related change in biological condition in which the ratio on reductants/oxidants is disrupted in favor of oxidation.

Calorie restriction: A dietary regime that has proven to be the most effective anti-aging measure, and used as the gold standard for the life extesion study in the aging field.

Inflammatory AMs: AMs molecules that have been shown to cause or involved closely in inflammation.

Oxidative stress: Oxidative stress refers to a condition under which increased production of reactive species, coupled with weakened anti-oxidative defense systems, resulting in redox imbalance.

Reactive species: Reactive species include free radicals derived from oxygen, nitrogen species, and those reactive lipid spices, which are not free radicals.

Acknowledgements

This work was supported by National Research Foundation of Korea (NRF) grant funded by Korea government (MEST) (No. 20090093226). We are grateful to the 'Aging Tissue Bank' for supplying research resource.

Abbreviations

AD	alzheimer's disease
AGEs	advanced glycation end products
AM	adhesion molecule
Aβ	amyloid beta

AP-1	activator protein-1
COX	cyclooxygenase
CR	calorie restriction
EC	endothelial cell
ICAM-1	intercellular adhesion molecule 1
Ig	immunoglobulin
IκB	inhibitor of kappa B
IKK	IκB kinase
IL-1	interleukin 1
LFA-1	lymphocyte function-associated antigen 1
LPC	lysophosphatidylcholine
LPS	lipopolysaccharide
MAG	myelin-associated glycoprotein
MAPK	mitogen-activated protein kinases
NCAM	neural cell adhesion molecule
NF-κB	nuclear factor-κB
NIK	NF-κB inducing kinase
oxLDL	oxidized low-density lipoprotein
PPARs	peroxisome proliferator-activated receptors
RS	reactive species
sAM	soluble adhesion molecule
TNF-α	tumor necrosis factor-α
VCAM-1	vascular cell adhesion molecule 1
VLA-4	very late antigen 4

References

Aboughalia, A.H. 2006. Elevation of hyaluronidase-1 and soluble intercellular adhesion molecule-1 helps select bladder cancer patients at risk of invasion. Arch. Med. Res. 37: 109-116.

Boyle, M. and C. Wong, M. Rocha, and D.L. Jones. 2007. Decline in self-renewal factors contributes to aging of the stem cell niche in the Drosophila testis. Cell Stem Cell 1: 470-478.

Burgess, A. and S. Saini, Y.Q. Weng, and I. Aubert. 2009. Stimulation of choline acetyltransferase by C3d, a neural cell adhesion molecule ligand. J. Neurosci. Res. 87: 609-616.

Chung, H.Y. and B. Sung, K.J. Jung, Y. Zou, and B.P. Yu. 2006. The molecular inflammatory process in aging. Antioxid. Redox Signal. 8: 572-581.

Chung, H.Y. and M. Cesari, S. Anton, E. Marzetti, S. Giovannini, A.Y. Seo, C. Carter, B.P. Yu, and C. Leeuwenburgh. 2009. Molecular inflammation: underpinnings of aging and age-related diseases. Aging Res. Rev. 8: 18-30.

Csiszar, A. and N. Labinskyy, K. Smith, A. Rivera, Z. Orosz, and Z. Ungvari. 2007. Vasculoprotective effects of anti-tumor necrosis factor-alpha treatment in aging. Am. J. Pathol. 170: 388-398.

Dworakowski, R. and S.P. Alom-Ruiz, and A.M. Shah. 2008. NADPH oxidase-derived reactive oxygen species in the regulation of endothelial phenotype. Pharmacol. Rep. 60: 21-28.

Giri, R. and S. Selvaraj, C.A. Miller, F. Hofman, S.D. Yan, D. Stern, B.V. Zlokovic, and V.K. Kalra. 2002. Effect of endothelial cell polarity on beta-amyloid-induced migration of monocytes across normal and AD endothelium. Am. J. Physiol. Cell Physiol. 283: C895-904.

Guney, N. and H.O. Soydinc, D. Derin, F. Tas, H. Camlica, D. Duranyildiz, V. Yasasever, and E. Topuz. 2008. Serum levels of intercellular adhesion molecule ICAM-1 and E-selectin in advanced stage non-small cell lung cancer. Med. Oncol. 25: 194-200.

Ito, Y. and K.K. Sorensen, N.W. Bethea, D. Svistounov, M.K. McCuskey, B.H. Smedsrød, and R.S. McCuskey. 2007. Age-related changes in the hepatic microcirculation in mice. Exp. Gerontol. 42: 789-797.

Jung, K.J. and E.K. Go, J.Y. Kim, B.P. Yu, and H.Y. Chung. 2009. Suppression of age-related renal changes in NF-kappaB and its target gene expression by dietary ferulate. J. Nutr. Biochem. 20: 378-388.

Jung, K.J. and J.Y. Kim, Y. Zou, Y.J. Kim, B.P. Yu, and H.Y. Chung. 2006. Effect of short-term, low dose aspirin supplementation on the activation of pro-inflammatory NF-kappaB in aged rats. Mech. Ageing Dev. 127: 223-330.

Jung, K.Y. and D. Dean, J. Jiang, S. Gaylor, W.H. Griffith, R.C. Burghardt, and A.R. Parrish. 2004. Loss of N-cadherin and alpha-catenin in the proximal tubules of aging male Fischer 344 rats. Mech. Ageing Dev. 125: 445-453.

Kaur, M. and S. Sharma, and G. Kaur. 2008. Age-related impairments in neuronal plasticity markers and astrocytic GFAP and their reversal by late-onset short term dietary restriction. Biogerontology 9: 441-454.

Miguel-Hidalgo, J.J. and S. Nithuairisg, C. Stockmeier, and G. Rajkowska. 2007. Distribution of ICAM-1 immunoreactivity during aging in the human orbitofrontal cortex. Brain Behav. Immun. 21: 100-111.

Miller, S.J. and W.C. Watson, K.A. Kerr, C.A. Labarrere, N.X. Chen, M.A. Deeg, and J.L. Unthank. 2007. Development of progressive aortic vasculopathy in a rat model of aging. Am. J. Physiol. Heart Circ. Physiol. 293: H2634-2643.

Nagata, M. 2005. Inflammatory cells and oxygen radicals. Curr. Drug Targets Inflamm. Allergy 4: 503-504.

Nielsen, H.M. and E. Londos, L. Minthon, and S.M. Janciauskiene. 2007. Soluble adhesion molecules and angiotensin-converting enzyme in dementia. Neurobiol. Dis. 26: 27-35.

Richter, V. and F. Rassoul, K. Purschwitz, B. Hentschel, W. Reuter, and T. Kuntze. 2003. Circulating vascular cell adhesion molecules VCAM-1, ICAM-1, and E-selectin in dependence on aging. Gerontology 49: 293-300.

Slack-Davis, J.K. and K.A. Atkins, C. Harrer, E.D. Hershey, and M. Conaway. 2009. Vascular cell adhesion molecule-1 is a regulator of ovarian cancer peritoneal metastasis. Cancer Res. 69: 1469-1476.

Son, T.G. and Y. Zou, B.P. Yu, J. Lee, and H.Y. Chung. 2005. Aging effect on myeloperoxidase in rat kidney and its modulation by calorie restriction. Free Radic. Res. 39: 283-289.

Strekalova, H. and C. Buhmann, R. Kleene, C. Eggers, J. Saffell, J. Hemperly, C. Weiller, T. Müller-Thomsen, and M. Schachner. 2006. Elevated levels of neural recognition molecule L1 in the cerebrospinal fluid of patients with Alzheimer's disease and other dementia syndromes. Neurobiol. Aging 27: 1-9.

Wyss-Coray, T. 2006. Inflammation in Alzheimer's disease: driving force, bystander or beneficial response? Nat. Med. 12: 1005-1015.

Yu, B.P. and H.Y. Chung. 2006. The inflammatory process in aging. Rev. Clin. Gerontology 16: 179-187.

Zou, Y. and K.J. Jung, J.W. Kim, B.P. Yu, and H.Y. Chung. 2004. Alteration of soluble adhesion molecules during aging and their modulation by calorie restriction. FASEB J. 18: 320-322.

Zou, Y. and C.H. Kim, J.H. Chung, J.Y. Kim, S.W. Chung, M.K. Kim, D.S. Im, J. Lee, B.P. Yu, and H.Y. Chung. 2007. Upregulation of endothelial adhesion molecules by lysophosphatidylcholine. Involvement of G protein-coupled receptor GPR4. FEBS J. 274: 2573-2584.

Zou, Y. and S. Yoon, K.J. Jung, C.H. Kim, T.G. Son, M.S. Kim, Y.J. Kim, J. Lee, B.P. Yu, and H.Y. Chung. 2006. Upregulation of aortic adhesion molecules during aging. J. Gerontol. A Biol. Sci. Med. Sci. 61: 232-244.

CHAPTER 6

The PI3K/Akt-NF-kappa B Signaling Pathway and Hypoxia-induced Mitogenic Factor in Vascular Adhesion Molecule 1 Gene Expression

Dechun Li[‡]

MD, PhD, Associate Professor, Department of Internal Medicine, Division of Pulmonary, Critical Care and Sleep Medicine, Saint Louis University, Room R110, Doisy Hall, St. Louis, MO 63104, E-mail: dli2@slu.edu
7th Floor, Desloge Towers, St. Louis, MO 63110-2539

ABSTRACT

The phosphoinositide 3-kinase (PI3K) signaling transduction pathways play critical roles in regulating cellular activation, inflammatory responses, chemotaxis, and apoptosis; and NF-κB has been demonstrated to be a pivotal downstream transcription factor in adhesion molecule gene upregulation induced by cytokine or inflammatory agents. In particular, most of the adhesion molecule genes have NF-κB binding site in their promoter regions, indicating the significance of this signaling transduction pathway in the regulation of adhesion molecule gene expression in inflammatory cells and vascular endothelium. Activation of the PI3K/Akt-NF-κB signaling pathways in response to inflammatory stimuli will modify the expression levels of adhesion molecules in both leukocytes and vascular endothelium resulting in a coordinated process to facilitate the extravasation of specialized leukocytes into the inflammatory foci. We have discovered that the cytokine-like protein hypoxia-induced mitogenic factor (HIMF) stimulates the endogenous PI3K/Akt signaling pathways and promotes the activation of NF-κB. The latter directly binds to the VCAM-1 promoter and significantly enhances

Key terms are defined at the end of the chapter.

VCAM-1 gene expression, which then facilitates the monocyte adherence and emigration into the pulmonary parenchyma. This chapter reviews the current data on the role of the PI3K/Akt-NF-κB signaling pathways as a strong regulator for adhesion molecule gene expression and uses the HIMF-induced inflammatory responses as a typical model to explore the changes and importance of the PI3K/Akt-NF-κB in inflammation. Of greater importance, the data reviewed in this article strongly suggest that modulation of the PI3K/Akt-NF-κB signaling pathways can alter the expression levels of adhesion molecule. Thus, manipulation of the endogenous PI3K/Akt-NF-κB signaling pathway may represent novel preventive and therapeutic approaches to management of important diseases such as acute or chronic inflammation and cancer as well.

INTRODUCTION

Over the past decades, extensive research to explore the network and crosstalk among the different signaling transduction pathways has uncovered multiple levels of complexity dictating that no single pathway operates in isolation and that all pathways are small parts of fully integrated networks within the cell. This is certainly true of the PI3Ks-Akt pathway; and several studies have demonstrated that PI3K-Akt signaling can activate NF-κB transcription factor downstream of stimuli, such as TNF-α, and PDGF. PI3Ks are the prominent members of a unique and conserved family of intracellular lipid kinases that catalyze the phosphorylation of the 3′-hydroxyl group of phosphatidylinositol and phosphoinoditides in the inner leaflet of the plasma membrane (Engelman *et al.* 2006). This reaction leads to the activation of versatile intracellular signaling pathways that regulate diverse functions including cell proliferation, differentiation, survival, metabolism, and migration (Cantley 2002). The PI3K signaling transduction pathways play imperative roles in the maintenance of normal cell function as well as in disease conditions including cancer and inflammation. PI3Ks participate in the inflammatory processes by upregulating inflammatory related genes such as cytokines, chemokines, growth factors, and adhesion molecules via the activation of Akt and transcription factors (TFs), such as AP1, NF-κB. Vice versa, many of these inflammatory proteins also use the PI3K/Akt-NF-κB signaling pathways as their signaling carriers to execute their biological functions (Hacker and Karin 2006, Hawkins and Stephens 2007, Manning and Cantley 2007).

Inflammatory cell migration and sequestration into the airway walls and the parenchyma of the alveolar compartments are the characteristic features of acute and chronic inflammation in the pulmonary system. Recent studies have revealed that this process is a dynamically sequential cascade. Cooperative and efficient leukocyte-endothelium interaction is the prerequisite. Leukocyte extravasation is mediated by several cell adhesion molecules including selectins (P-, E- and L-), integrins, and members of the immunoglobulin gene (Ig) superfamily (ICAM-1 and

VCAM-1) expressed from both leukocytes and vascular endothelium (Carlos and Harlan 1994, Haverslag *et al.* 2008). During recent years, efforts have intensified to elucidate the links between activation of PI3K/Akt-NF-κB signaling transduction pathways and the regulation of gene expression of adhesion molecules. However, up to now, very few comprehensive studies have investigated the molecular signaling interaction between the PI3K and NF-κB pathways in inflammatory cytokine-elicited upregulation of VCAM-1 at the levels of gene promoter, PI3K activity, and NF-κB modulation, simultaneously. Further understanding of the crosstalk and interaction between vascular endothelial cells and inflammatory leukocytes will lead us to new avenues for the prevention and therapy of inflammatory and cardiovascular diseases as well as cancer. One of the crucial proteins is VCAM-1, a member of the Ig superfamily that mediates leukocyte binding to the endothelial cell through its interaction with the leukocyte's integrin counterreceptor very late activation antigen 4 (VLA-4). Because of the selective expression of VLA-4 on monocytes and lymphocytes, but not neutrophils, VCAM-1 plays an important role in mediating mononuclear leukocyte-selective adhesion. However, the molecular mechanisms of VCAM-1 upregulation in the inflammatory pulmonary system are still not completely understood. We investigated the effects of an inflammatory cytokine, HIMF, in the regulation of VCAM-1 expression and the results showed that VCAM-1 gene expression is closely controlled by HIMF-induced PI3K/Akt-NF-κB activation (Tong *et al.* 2006). Blocking PI3K/Akt activity or inhibiting NF-κB activation with, respectively, mutant dominant negative proteins or chemical inhibitors significantly prevented VCAM-1 gene upregulation. These findings certainly provided a typical example for inflammatory cytokine-induced adhesion molecule overproduction and further facilitate our understanding of the process of leukocyte-endothelium interaction and inflammatory cell migration.

BACKGROUND

The PI3K/Akt Signaling Transduction Pathway

PI3Ks are grouped into three classes (I-III) with different isoforms in class I as IA (including PI3Kα, β and δ) and IB (PI3Kγ), according to their substrate preference and sequence homology (Engelman *et al.* 2006). The coexistence of classes of PI3K and their isoforms in either the same or different type of cells has distinct functions in cellular signal transduction and is accountable for the diversified biological roles in those cells. Class IA PI3Ks, heterodimers that consist of a p85 regulatory subunit and a p110 catalytic subunit, are mainly involved in regulating cellular proliferation and survival in response to activation of protein tyrosine kinase-coupled growth factor receptors. Whereas class IB PI3K (PI3Kγ), a heterodimer that consists of a p101 regulatory subunit and a p110γ catalytic subunit, allows fast-acting, heterotrimeric G protein-coupled receptors to access PI3K signaling networks (Cantley 2002, Andrews *et al.* 2007, Hawkins and Stephens 2007). Members of class II PI3K consist of only a p101 regulatory subunit and a p110γ catalytic subunit. Class III PI3K consists of a single member, vacuolar protein-sorting defective 34 (Vps34), which functions for cellular membrane trafficking vesicles from the Golgi apparatus to the vacuole. Vps34 was also found to regulate mammalian target of rapamycin (mTOR) activity in response to amino acid availability and nutrient starvation (Cantley 2002). Recent studies have shown that PI3Kγ and δ are expressed predominantly, although not exclusively, in leukocytes and gene targeting experiments indicated that these two enzymes are involved in innate and adaptive immune responses. In particular, PI3Kγ functions as a chemokine sensor regulating leukocyte migration. Furthermore, this enzyme is found in vessels, in both smooth muscle and endothelial cells, and participates in the inflammatory process by enhancing neutrophil adhesion and subsequent transendothelial migration, then extravasation (Barberis and Hirsch 2008).

Most of the ligands to activate PI3Kγ are involved in the regulation of gene expression, secretion, adhesion, migration, and contraction in multiple cell types in the immune system and vasculature (Hawkins and Stephens 2007). More specifically, PI3Ks and the downstream serine/threonine kinase Akt regulate cellular activation, inflammatory responses, chemotaxis, cell metabolism, cytoskeletal rearrangements, and apoptosis (Hirsch *et al.* 2000, Williams *et al.* 2006, Andrews *et al.* 2007, Manning and Cantley 2007). Signaling proteins with pleckstrin homology (PH) domains accumulate at sites of PI3K activation by directly binding to PIP2 or PIP3. The interaction of PH domains on signaling proteins with PIP2 or PIP3 can modulate signaling and the intracellular localization of the signaling protein. Some examples of signaling proteins that interact with PIP2 and/or PIP3 include phosphoinositide-dependent kinase-1 (PDK1), PDK2, and Akt. PDK activates Akt by phosphorylation of Thr308 and Ser473. The activated Akt results in phosphorylation of a host of other proteins that modulates cell cycle entry, growth, and survival by regulating an array of specific gene expression, in particular to activate NF-κB to modulate adhesion molecule gene expression (Fry 1994, Engelman *et al.* 2006, Manning and Cantley 2007).

Nuclear Factor Kappa-B (NF-κB)

Members of the NF-κB family of transcription factors (TFs) regulate expression of a large number of genes involved in immune responses, inflammation, cell survival, and cancer. NF-κB is rapidly activated in response to various stimuli, including cytokines, infectious agents, and radiation-induced DNA damages (Barnes and Karin 1997, Hacker and Karin 2006). First identified as an expression regulator of the kappa light-chain gene in murine B lymphocytes but subsequently found in many different cells, NF-κB is a heterodimeric (p65, also called RelA, with p50, p52 or other subunits such as Rel, RelB, v-rel and p52, respectively), ubiquitously expressed TF that plays a critical role in regulating inducible gene expression in inflammatory responses (Baldwin 1996, Barnes and Karin 1997). Different forms of NF-κB activate various sets of target genes to exert its diversified functions. NF-κB increases gene expression of many cytokines, enzymes, and adhesion molecules in acute and chronic inflammatory conditions. In chronic inflammatory diseases, adhesion molecules recruit inflammatory cells from the circulation to the site of inflammation and NF-κB regulates the expression of these genes that encode adhesion molecules such as intercellular adhesion molecule 1 (ICAM-1), E-selectin, integrin, and VCAM-1. Interestingly, certain viruses use NF-κB activation to enhance viral replication, host cell survival, and evasion of immune responses. Simultaneously, it also evokes the adhesion molecule gene upregulation and results in specific leukocyte migration into the site of viral infection (Hiscott *et al.* 2001). Importantly, it has been confirmed that most but not all of these adhesion molecule genes possess at least one or more NF-κB binding sites in their promoter regions, indicating the importance of NF-κB activation in the modulation of expression of these genes (Voraberger *et al.* 1991, Schindler and Baichwal 1994, Wang *et al.* 1999, Melotti *et al.* 2001).

In resting cells, NF-κB is normally sequestered in the cytoplasm through its interaction with the IκB (inhibitor of NF-κB) family of inhibitory proteins. A major advance in the understanding of NF-κB regulation came with the identification of the multisubunit IKK kinase complex, which contains two catalytic subunits, IKK-α and IKK-β, and the regulatory subunit IKK-γ. The predominant form of the IKK complex is an IKK-α/IKK-β heterodimer associated with IKK-γ, and this association is mediated by the interaction of IKK-β and IKK-γ. IKK-γ is not a kinase per se but is absolutely essential for NF-κB activation by multiple activators. Since the identification of the IKK complex, attention was focused on the upstream kinases of different signal transduction pathways and how these pathways converge on the IKK complex. These kinases include NF-κB-inducing kinase (NIK), mitogen-activated protein kinase/extracellular signal-regulated kinase kinase 1 (MEKK1), TGF-β-activated kinase (TAK1), protein kinase R (PKR), PKC, PI3K-Akt (or PKB), and mixed-lineage kinase 3 (MLK3). Once activated, IKK-β becomes autophosphorylated at a carboxyl terminal serine cluster, which decreases IKK activity and prevents prolonged NF-κB activation by negative autoregulation (Hacker and Karin 2006). In response to external stimuli, IκB proteins undergo rapid phosphorylation by IKK complex on specific serine residues. Phosphorylation of IκBα on serines 32 and 36, and of IκBβ on serines 19 and 23 facilitates their ubiquitination on neighboring lysine residues, thereby targeting these proteins for rapid degradation by the proteosome (Baldwin 1996, Hacker and Karin 2006). Following the degradation of IκB, NF-κB is released and is free to translocate to the nucleus and to activate target inflammatory genes by binding to the specific sequences in the promoters.

Inflammatory Cytokines, Adhesion Molecule Activation, and Gene Expression via PI3K/Akt-NF-κB Signaling Pathways

Inflammatory cytokines dramatically and selectively modulate the transcription and translation of adhesion molecules and chemoattractant proteins in leukocytes and endothelial cells. Cytokines commonly found in inflammatory lesions, such as TNF-α and interleukin-1 (IL-1), induce the concurrent expression of VCAM-1, ICAM-1, and E-selectin in endothelial cells to increase its adhesiveness (Haraldsen *et al.* 1996, Sawa *et al.* 2007). IL-4 synergistically with other cytokines increases adhesion of lymphocytes and induces VCAM-1. It is likely that the precise mixture of chemoattractants and cytokine produced at inflammatory sites in vivo determines which types of leukocytes emigrate. The elevated and prolonged expression of VCAM-1, ICAM-1 and E-selectin has been observed in both experimental models and human inflammatory processes. Treatment of endothelial cells with bacterial lipopolysaccharide (LPS) in vitro as well as in vivo upregulates VCAM-1, ICAM-1, and E-selectin expression in endothelial cells (Van Kampen and Mallard 2001a, b, Sawa *et al.* 2007). However, the expression kinetics for these proteins varies with the stimulants. TNF-α and LPS induced similar VCAM-1 expression at 1 hr, followed by a significant increase at 3 hr that was maintained until 48 hr. Meanwhile, the expression of ICAM-1 mRNA peaked at 12-18 hr and then diminished but remained at above baseline level up to 72 hr. LPS-stimulated E-selectin mRNA expression peaked at 6 hr, followed by a decline to baseline by 24 hr. Conversely, TNF-α stimulated significant upregulation of E-selectin mRNA by 6 hr, followed by a gradual increase and eventually sharp increase between 18 and 72 hr. Furthermore, to investigate the role of the signaling transduction pathways participating in the regulation of adhesion molecule gene upregulation, the activation of the PI3K/Akt-NF-κB signaling pathways has been brought to the central arena. Studies have demonstrated that integrin upregulation is mediated by increased PI3K/Akt activities in cytohesin-1-induced β2-integrin production in Jurkat cells (Nagel *et al.* 1998). In platelets the activation of integrin is largely controlled by PI3K (Roberts *et al.* 2004, Schoenwaelder *et al.* 2007). As most of the adhesion molecule genes possess NF-κB binding site in their promoter region, suppression of NF-κB activation with its inhibitor MG-132, significantly attenuated TNF-α and IL-1β induced chemokine (including MGSA, RANTES, MCO-1, M-CSF) and ICAM-1 upregulation in human retinal pigment epithelial cells. Moreover, TNF-α and IL-1β also caused degradation of IκB, NF-κB nuclear translocation, and increased NF-κB DNA binding activity. Similar results of NF-κB activation were found in adenovirus-induced ICAM-1 induction in A549 cells (Voraberger *et al.* 1991). However, most of the studies mentioned above are limited either in the cellular or transcriptional and translational levels, which resulted in our current incomplete understanding of the exact molecular mechanisms of PI3K/Akt-NF-κB modulated leukocyte-endothelium interaction in the aspects of adhesion molecule gene expression during the inflammatory process.

HYPOXIA-INDUCED MITOGENIC FACTOR (HIMF)

HIMF and Inflammation in the Lung

Using a cDNA microarray from Incyte Genomics, we performed gene expression profiling in lungs from mice treated with hypoxia (10%, normal baric) for 4 d. Data analysis showed that there are 270 genes and expressed sequence tags (ESTs) significantly upregulated out of 9,415 gene elements. The highest upregulated ESTs with Genbank accession number AA712003 are increased more than 4-fold in the hypoxic lung compared with the normoxic controls. Further studies showed that AA712003 has identical sequence to FIZZ1 and RELMα. We named this homologue hypoxia-induced mitogenic factor (HIMF) because of its potent mitogenic action on PMVSMC and its upregulation by hypoxia. HIMF belongs to the resistin protein families (Holcomb *et al.* 2000), whose members encode proteins of 105-114 amino acids with three domains: (1) an N-terminal signal sequence, (2) a variable middle portion, and (3) a highly conserved C-terminal signature sequence that constitutes nearly half of the molecule. The signature region of HIMF contains a unique and invariant spacing of the cysteine residues: $C\text{-}X_{11}\text{-}C\text{-}X_8\text{-}C\text{-}X\text{-}C\text{-}X_3\text{-}C\text{-}X_{10}\text{-}C\text{-}X\text{-}C\text{-}X\text{-}C\text{-}X_9\text{-}CC\text{-}X_{3\text{-}6}\text{-}END$ (C represents cyteine and X is any amino acid residue). This is reminiscent of the so-called "EGF repeats" that are characteristic of a number of signaling molecules. Most important, studies from our laboratory have shown that HIMF enhances VCAM-1 gene production in MLE-12 and endothelial cells. In LPS-treated cells and lung, HIMF is highly upregulated accompanied by increased monocyte sequestration in the lung parenchyma and levels of VCAM-1 expression. Administration of recombinant HIMF into the lung induced robust upregulation of VCAM-1. These findings strongly indicate that HIMF may be an important cytokine-like protein participating in the inflammatory process via modulation of adhesion molecule expression in the lung.

HIMF Enhances VCAM-1 Expression in the Lung

To test the hypothesis that HIMF participates in regulation of VCAM-1 expression, recombinant HIMF protein was administered by intratracheal instillation, followed by western blot and immunohistochemical staining. As shown in Figs. 1 and 2, HIMF protein instillation significantly increased VCAM-1 production in the lungs and the increased VCAM-1 protein is localized to vascular endothelial, alveolar type II, and airway epithelial cells compared with control mouse lungs.

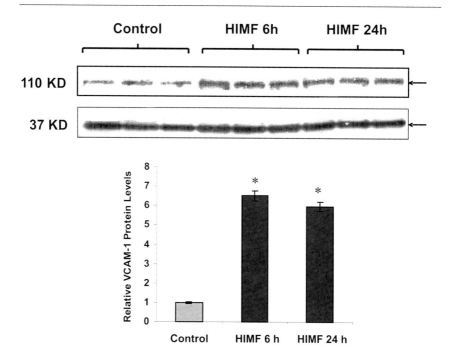

Fig. 1   Western blot for VCAM-1 protein after intratracheal instillation of HIMF in mouse lung. Intratracheal instillation of recombinant HIMF-induced upregulation of VCAM-1 protein in mouse lung. Results expressed as mean ± SEM. *Significant difference from the control group at both time points (D. Li, unpublished data).

Activation of NF-κB is Essential for HIMF-induced VCAM-1 Production

To search for the TFs involved in HIMF-induced VCAM-1 production, sequentially deleted 5′-flanking sequences of mouse VCAM-1 promoter were transfected into cultured endothelial cells, SVEC 4-10, and HIMF stable expression SVEC-HIMF (Fig. 4). The sequences TGGGTTTCCC at -73 bp and AGGGATTTCCC at -58 bp are identical to the consensus sequence (GGG R[C/A/T]TYYCC) of the NF-κB binding site. The results showed that treatment of HIMF induces VCAM-1 promoter-luciferase reporter activity significantly in SVEC-HIMF, which stably overexpresses HIMF. The highest VCAM-1 luciferase activity was found from a short construct pGLVCAM-1 (-0.3 luc) that contains two NF-κB binding sites within 329 bp of VCAM-1 promoter, whereas another deletion construct exhibited lower luciferase activity. These results suggest that a negative regulatory element might exist between the -0.7 and -0.3 region of the VCAM-1 promoter. It has been reported that NF-κB regulates VCAM-1 expression; it is unknown, however,

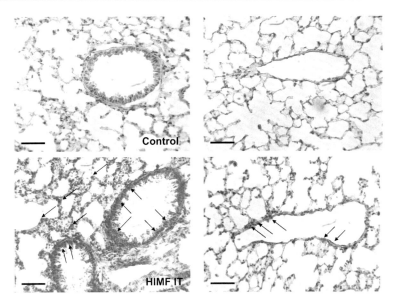

Fig. 2 Immunohistochemical staining for VCAM-1 in mouse lung tissue after intratracheal instillation of HIMF. There is marked increase of VCAM-1 protein expression in HIMF-treated mouse lungs. The upregulated VCAM-1 is mainly localized to airway epithelium and alveolar type II cells and vascular endothelium (arrows). Scale bars = 100 μm. Reprint from Tong *et al.* (2006), with permission.

whether HIMF enhances VCAM-1 expression through NF-κB pathway. To assess this possibility, transfection of several dominant-negative mutants in NF-κB pathway, IKKα (K44A), IKKβ (K44A) and IκBα (S32A/S36A), was performed in SVEC 4-10 and MLE-12 cells. In response to HIMF, IKK was phosphorylated, which in turn phosphorylates IκBα leading to NF-κB activation in both SVEC 4-10 and MLE-12 cells (Fig. 4).

PI-3K/Akt Pathway is Involved in HIMF-Induced NF-κB Activation and VCAM-1 Production

To test whether PI-3K/Akt participates in HIMF-mediated NF-κB activation, SVEC 4-10 and MLE-12 cells were treated with HIMF and western blot was performed. The results showed that HIMF strongly induced Akt phosphorylation at Ser473 and Thr308. Furthermore, only the PI-3K inhibitor LY294002 (10 μmol/L) inhibited HIMF-activated Akt phosphorylation (Fig. 5). However, incubation of cells with SB203580 (5 μmol/L), PD098059 (5 μmol/L) or U0126 (5 μmol/L), inhibitors against p38 and ERK1/2 MAPK pathways respectively, had no effect on HIMF-induced Akt phosphorylation. In addition, transfection of Δp85, a

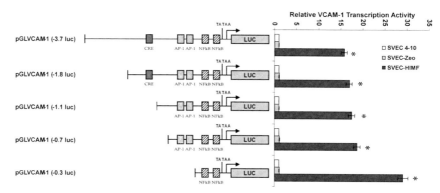

Fig. 3 Luciferase reporter activity of different deletion constructs for VCAM-1 promoter in SVEC 4-10 cells. Luciferase reporter activity of different deletion constructs for VCAM-1 promoter in SVEC 4-10 cells was measured after transfection. The results showed that NF-κB sites in the 5′-prime proximal regions play critical roles. However, other TFs may negatively regulate VCAM-1 expression in response to HIMF stimulation. Results are expressed as mean ± SEM. *Significant difference from the control group at both time points. Modified from Tong *et al.* (2006), with permission.

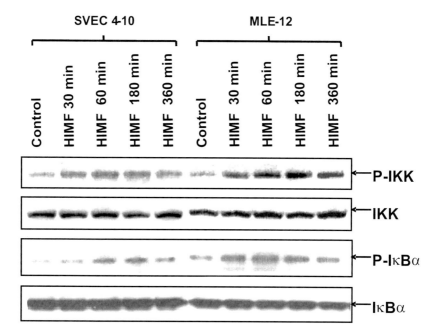

Fig. 4 HIMF induces IKK and IκBα phosphorylation in SVEC 4-10 and MLE-12 cells. To examine the effects of HIMF on IKK and IκBα phosphorylation, SVEC 4-10 and MLE-12 cells were treated with HIMF for up to 6 hr. The phosphorylation of these two proteins was examined. HIMF enhanced the phosphorylation IKK and IκBα (D. Li, unpublished data).

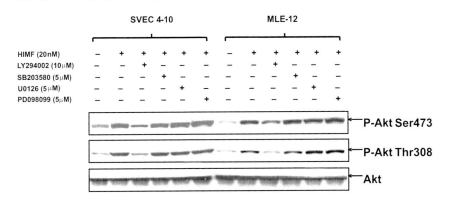

Fig. 5 HIMF stimulates Akt phosphorylation via PI3K activation. SVEC 4-10 and MLE-12 cells were treated with HIMF and the Akt phosphorylation was examined with specific antibodies for Ser473 and Thr308. The results showed that HIMF-stimulated Akt phosphorylation is inhibited by PI3K inhibitor LY294002, especially at Thr308. Reprint from Tong *et al.* (2006), with permission.

dominant-negative mutant of PI-3K, into SVEC 4-10 and MLE-12 cells, abolished HIMF-induced phosphorylation of IKK and IκBα and reduced the transcriptional activities for NF-κB and VCAM-1 (Figs. 6 and 7), and the subsequent NF-κB activation. These observations suggest that PI-3K/Akt pathway functions upstream of NF-κB in responding to HIMF signals. Together, these data suggest that the PI-3K/Akt pathway is involved in HIMF-mediated NF-κB activation and the subsequent VCAM-1 production.

Applications to Other Areas of Health and Disease

The signaling transduction pathways of PI3K and NF-κB play critical roles in adhesion molecule gene expression, especially in leukocyte extravasation induced by cytokine and inflammatory agents. The ubiquitous expression and coexistence of the PI3K and NF-κB signaling pathways in both inflammatory cells and in vascular endothelial cells determines the potential interaction and crosstalk between the two pathways in these cells and ensures there are coordinated responses to facilitate precise migration of specialized leukocytes into the inflammatory sites. Adhesion molecule upregulation is not only an important process in inflammatory response, but also a crucial step for cardiovascular diseases and cancer. To fully understand the molecular mechanisms of adhesion molecule expression under the controls of the PI3K/Akt-NF-κB signaling will certainly be advantageous for the prevention and treatment of these diseases in the future.

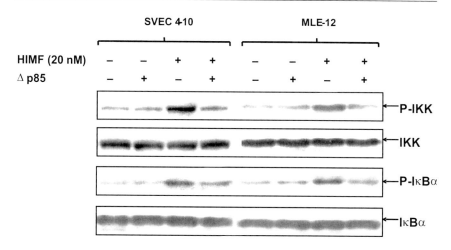

Fig. 6 HIMF-stimulated IKK and IκBα phosphorylation is inhibited by dominant negative PI3K Δp85. SVEC 4-10 and MLE-12 cells were transfected with PI3K dominant negative mutant Δp85 and then stimulated with HIMF. The phosphorylation of IKK and IκBα was examined with specific antibodies for phosphorylated IKK and IκBα. The results showed that HIMF-stimulated IKK and IκBα phosphorylation is inhibited by dominant negative PI3K Δp85. Reprint from Tong *et al.* (2006), with permission.

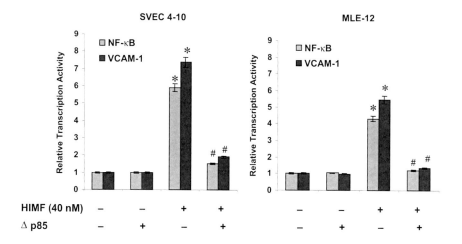

Fig. 7 HIMF stimulated NF-κB and VCAM-1 transcriptional upregulation is via PI3K activation. Luciferase reporter activity of NF-κB and VCAM-1 promoters in SVEC 4-10 and MLE-12 cells was measured after transfection. The results showed that HIMF induced NF-κB and VCAM-1 transcriptional increase is blocked by the dominant negative mutant of PI3K Δp85. Results are expressed as mean ± SEM. * and # represent significant difference from the control group at both time points. Reprint from Tong *et al.* (2006), with permission.

SUMMARY

- The PI3K/Akt signaling transduction pathway is widespread in leukocytes and vascular endothelial cells.
- There are crosstalk and interactions between the PI3K/Akt and the NF-κB signaling pathways to control the inflammatory protein gene expression.
- Most adhesion molecule genes possess one or more NF-κB biding sites in their promoter region, which is the basis for the upregulation of adhesion molecules in inflammatory response.
- Inflammatory proteins such as TNF-α and HIMF enhance adhesion molecule gene expression via the activation of the PI3K/Akt-NF-κB cascade.
- Manipulation of the activities of the PI3K/Akt-NF-κB signaling transduction pathways may directly affect the inflammatory response.

Abbreviations

ESTs	expressed sequence tags
HIMF	hypoxia-induced mitogenic factor
ICAM-1	intercellular adhesion molecule 1
Ig	superfamily immunoglobulin gene superfamily
LPS	lipopolysaccharide
MCP-1	monocyte chemoattractant protein 1
M-CSF	macrophage colony stimulating factor
MEKK1	mitogen-activated protein kinase/extracellular signal-regulated kinase kinase 1
MGSA/gro-α	melanoma growth stimulating activity gro-α
MLK3	mixed-lineage kinase 3
mTOR	mammalian target of rapamycin
NF-κB	nuclear factor-κB
NIK	NF-κB-inducing kinase
PDGF	platelet-derived growth factor
PI3K	phosphoinositide 3-kinase
PIP2	phosphatidyl-inositol (4,5)-biphosphate
PIP3	phosphatidyl-inositol (3,4,5)-triphosphate
PKR	protein kinase R
RANTES	regulated on activation normal T-cell expression and secreted
TAK1	TGF-β-activated kinase

TFs	transcription factors
TNF-α	tumor necrosis factor α
VACM-1	vascular adhesion molecule 1
VLA-4	very late activation antigen 4
Vps34	vacuolar protein-sorting defective 34

Key Facts about the PI3K/Akt-NF-κB Signaling Pathways

1. The PI3K/Akt-NF-κB signaling pathways are ubiquitously expressed, including inflammatory cells and vascular endothelium.
2. Cytokines, chemokines, and inflammatory agents activate PI3Ks, which generate PIP2 and PIP3 in the plasma membrane of target cells.
3. PIP2 and PIP3 bind to the pleckstrin homology (PH) domain of Akt and cause Akt activation.
4. Akt directly phosphorylates IKKα and the latter results in the phosphorylation and degradation of IκB in the NF-κB signaling pathway.
5. NF-κB is freed and translocated into the nucleus of the cell with increased NF-κB binding activity to the promoters of the adhesion molecule genes, such as VCAM-1, E-selectins, and integrins.
6. Together with other transcription factors, there is an upregulation of adhesion molecules, which will facilitate the interaction between the inflammatory cells and endothelial cells.
7. The enhanced adhesion molecule gene expression in these cells together with other chemoattractants produced by the local cells or tissues will facilitate the extravasation of the leukocyte into the inflammation foci.
8. Manipulation of PI3K and NF-κB activity in the cells with chemical agents or dominant negative proteins may regulate adhesion molecule gene expression and modulate the inflammatory process.

Definition of Terms

Heterodimer: A protein complex consisting of different units.

Homodimer: A protein complex formed by two identical units.

Inflammatory proteins: Cytokines, chemokins, adhesion molecules and other proteins that play critical roles in inflammatory response.

Leukocyte extravasation: Process by which leukocytes emigrate from the circulation into the inflammatory foci and tissues.

Signaling transduction pathways: Networks consisting of a series of different components in a certain order to transmit and execute a specific cell signal for its specified function or activities.

Signaling transduction pathway activation: Process by which the specific components in a certain pathway are modified with phosphorylation or dephosphorylation; then the protein activity is increased. Such a process is often seen in the activation of Akt or IKK for NF-κB activation.

Transcription factors: A group of proteins that participate in the gene expression by binding to the specific DNA sequences present in the promoter regions of specific genes.

Transcriptional activity: Description of gene expression, in particular to measure the TF binding and modification of mRNA expression levels in a reporter assay system. Most of them use luciferase activity as the endpoint.

References

Andrews, S. and L.R. Stephens, and P.T. Hawkins. 2007. PI3K class IB pathway. Sci. STKE 2007, cm2.

Baldwin, A.S. 1996. The NF-κB and IκB proteins: new discoveries and insights. Annu. Rev. Immunol. 14: 649-681.

Barberis, L. and E. Hirsch. 2008. Targeting phosphoinositide 3-kinase ? to fight inflammation and more. Thromb. Haemost. 99: 279-285.

Barnes, P.J. and M. Karin. 1997. Nuclear factor-kappaB: a pivotal transcription factor in chronic imflammatory diseases. N. Engl. J. Med. 336: 1066-1071.

Cantley, L.C. 2002. The phosphoinositide 3-kinase pathway. Science 296: 1655-1657.

Carlos, T.M. and J.M. Harlan. 1994. Leukocyte-endothelial adhesion molecules. Blood 84: 2068-2101.

Engelman, J.A. and J. Luo, and L.C. Cantley. 2006. The evolution of phosphatidylinositol 3-kinases as regulators of growth and metabolism. Nat. Rev. Genet. 7: 606-619.

Fry, M.J. 1994. Structure, regulation and function of phosphoinositide 3-kinases. Biochim. Biophys. Acta 1226: 237-268.

Hacker, H. and M. Karin. 2006. Regulation and function of IKK and IKK-related kinases. Sci. STKE 2006, re13.

Haraldsen, G. and D. Kvale, B. Lien, I.N. Farstad, and P. Brandtzaeg. 1996. Cytokine-regulated expression of E-selectin, intercellular adhesion molecule-1 (ICAM-1), and vascular cell adhesion molecule-1 (VCAM-1) in human microvascular endothelial cells. J. Immunol. 156: 2558-2565.

Haverslag, R. and G. Pasterkamp, and I.E. Hoefer. 2008. Targeting adhesion molecules in cardiovascular disorders. Cardiovasc. Hematol. Disord. Drug Targets 8: 252-260.

Hawkins, P.T. and L.R. Stephens. 2007. PI3K{gamma} is a key regulator of inflammatory responses and cardiovascular homeostasis. Science 318: 64-66.

Hirsch, E. and V.L. Katanaev, C. Garlanda, O. Azzolino, L. Pirola, L. Silengo, S. Sozzani, A. Mantovani, F. Altruda, and M.P. Wymann. 2000. Central role for G protein-coupled phosphoinositide 3-kinase gamma in inflammation. Science 287: 1049-1053.

Hiscott, J. and H. Kwon, and P. Genin. 2001. Hostile takeovers: viral appropriation of the NF-κB pathway. J. Clin. Invest. 107: 143-151.

Holcomb, I.N. and R.C. Kabakoff, B. Chan, T.W. Baker, A. Gurney, W. Henzel, C. Nelson, H.B. Lowman, B.D. Wright, N.J. Skelton, G.D. Frantz, D.B. Tumas, F.V. Peale Jr., D.L.

Shelton, and C.C. Hebert. 2000. FIZZ1, a novel cysteine-rich secreted protein associated with pulmonary inflammation, defines a new gene family. EMBO J. 19: 4046-4055.

Manning, B.D. and L.C. Cantley. 2007. AKT/PKB Signaling: navigating downstream. Cell 129: 1261-1274.

Melotti, P. and E. Nicolis, A. Tamanini, R. Rolfini, A. Pavirani, and G. Babrini. 2001. Activation of NF-κB mediates ICAM-1 induction in respiratory cells exposed to an adenovirus-derived vector. Gene Therapy 8: 1436-1442.

Nagel, W. and L. Zeitlmann, P. Schilcher, C. Geiger, J. Kolanus, and W. Kolanus. 1998. Phosphoinositide 3-OH kinase activates the beta 2 integrin adhesion pathway and induces membrane recruitment of cytohesin-1. J. Biol. Chem. 273: 14853-14861.

Roberts, M.S. and A.J. Woods, T.C. Dale, P. van der Sluijs, and J.C. Norman. 2004. Protein kinase B/Akt acts via glycogen synthase kinase 3 to regulate recycling of {alpha}v{beta}3 and {alpha}5{beta}1 integrins. Mol. Cell. Biol. 24: 1505-1515.

Sawa, Y. and Y. Sugimoto, T. Ueki, H. Ishikawa, A. Sato, T. Nagato, and S. Yoshida. 2007. Effects of TNF-{alpha} on leukocyte adhesion molecule expressions in cultured human lymphatic endothelium. J. Histochem. Cytochem. 55: 721-733.

Schindler, U. and V.R. Baichwal. 1994. Three NF-kappa B binding sites in the human E-selectin gene required for maximal tumor necrosis factor alpha-induced expression. Mol. Cell. Biol. 14: 5820-5831.

Schoenwaelder, S.M. and A. Ono, S. Sturgeon, S.M. Chan, P. Mangin, M.J. Maxwell, S. Turnbull, M. Mulchandani, K. Anderson, G. Kauffenstein, G.W. Rewcastle, J. Kendall, C. Gachet, H.H. Salem, and S.P. Jackson. 2007. Identification of a unique co-operative phosphoinositide 3-kinase signaling mechanism regulating integrin {alpha}IIbbeta3 adhesive gunction in platelets. J. Biol. Chem. 282: 28648-28658.

Tong, Q. and L. Zheng, L. Lin, B. Li, D. Wang, and D. Li. 2006. Hypoxia-induced mitogenic factor promotes vascular adhesion molecule-1 expression via the PI-3K/Akt-NF-{kappa}B signaling pathway. Am. J. Respir. Cell Mol. Biol. 35: 444-456.

Van Kampen, C. and B.A. Mallard. 2001a. Regulation of bovine E-selectin expression by recombinant tumor necrosis factor alpha and lipopolysaccharide. Vet. Immunol. Immunopathol. 79: 151-165.

Van Kampen, C. and B.A. Mallard. 2001b. Regulation of bovine intercellular adhesion molecule 1 (ICAM-1) and vascular cell adhesion molecule 1 (VCAM-1) on cultured aortic endothelial cells. Vet. Immunol. Immunopathol. 79: 129-138.

Voraberger, G. and R. Schafer, and C. Stratowa. 1991. Cloning of the human gene for intercellular adhesion molecule 1 and analysis of its 5′-regulatory region. Induction by cytokines and phorbol ester. J. Immunol. 147: 2777-2786.

Wang, X.C. and C. Jobin, J.B. Allen, W.L. Roberts, and G.J. Jaffe. 1999. Suppression of NF-kappaB-dependent proinflammatory gene expression in human RPE cells by a proteasome inhibitor. Invest. Ophthalmol. Vis. Sci. 40: 477-486.

Williams, D.L. and T. Ozment-Skelton, and C. Li. 2006. Modulation of the phosphoinositide 3-kinase signaling pathway alters host response to sepsis, inflammation, and ischemia/reperfusion injury. Shock 25: 432-439.

CHAPTER 7

Arginine and Adhesion Molecules

Sung-Ling Yeh

Professor, School of Nutrition and Health Sciences, Taipei Medical University,
250 Wu-Hsing Street, Taipei, Taiwan 110, Republic of China,
E-mail: sangling@tmu.edu.tw
Departmental contact: E-mail: aaa.ccc@msa.hinet.net

ABSTRACT

Arginine (Arg) is a non-essential amino acid for healthy adults. It is the precursor of nitric oxide (NO). Previous reports show that Arg and NO possess numerous physiological properties. Many studies have demonstrated the benefits of Arg supplementation on immune functions. Arg is considered to be an essential amino acid for patients with catabolic conditions. Currently, Arg is added to enteral formulas in an attempt to modulate immune function and improve clinical outcomes of critically ill patients. Adhesion molecules are important mediators of host defense and are localized in the earliest inflammatory lesions. Numerous studies have shown that Arg, possibly through the regulation of NO, modulates the expression of adhesion molecules and hence attenuates the inflammatory reaction. This chapter appraises several animal studies and current clinical evidence regarding the effects of Arg supplementation on adhesion molecule expression in various conditions. Some *in vitro* studies using an NO donor and NO synthase inhibitors to investigate the roles of NO in leukocyte adherence and immigration are also included. Most studies suggested that Arg and/or NO administration can reduce adhesion molecule expression and decrease leukocyte adherence and transmigration. The mechanisms through which NO decreases endothelial and leukocyte adhesion molecule expressions are postulated. However, conflicting results in some situations have been reported, and discrepancies between different experiments with Arg or NO administration are discussed.

Key terms are defined at the end of the chapter.

INTRODUCTION

Arginine (Arg) was first isolated and named in 1886. Its presence in animal protein was documented in 1895. Arg was classified as a semi-essential amino acid after a study by Rose (1937) observed that adult rats did not require dietary Arg for growth and maintenance of nitrogen balance, but that young growing rats demonstrated more rapid growth when receiving dietary Arg. This finding indicated that endogenous biosynthesis of Arg occurs but not at a sufficient rate for maximal growth in the young. This was also found to be true for humans. Arg has been shown to possess numerous physiological properties. It functions in the body as a free amino acid, a component of most proteins, and as a substrate for several non-protein, nitrogen-containing compounds. As a free amino acid, Arg is an integral constituent of the urea cycle. Arg contains a guanidine group, which is essential for the synthesis of urea in most mammals. In the cytosol of hepatocytes, arginase removes the guanidine group from Arg to produce urea and ornithine. Urea is then transported from the hepatocyte into the blood, and ornithin is used to regenerate Arg within hepatocytes. Another function of Arg is protein synthesis. Arg can be converted into proline, glutamate, and glutamine, all of which are common amino acids found within most proteins. Arg metabolism also generates several essential non-protein and nitrogen-containing compounds, including creatine, polyamines, and nitric oxide (NO). Synthesis of creatine is dependent on the guanidine group of Arg. Creatine functions as a carrier for phosphate and is needed for the rapid regeneration of adenosine triphosphate in muscles. Polyamines function in membrane transport and in cell growth, proliferation, and differentiation. Arg and products of Arg metabolism are necessary for both the regulation and synthesis of polyamine. Arg not only provides the substrate for polyamine synthesis, it also indirectly stimulates the release of growth hormone, which in turn promotes polyamine synthesis by increasing ornithine decarboxylase activity. Arg is a precursor of NO. NO, a short-lived small molecule produced by most cell types, has a variety of well-defined pathophysiological roles. NO is an antimicrobial agent that is effective against pathogens, parasites, and bacteria. NO is also a neurotransmitter and vasodilator. The enzyme that produces NO is nitric oxide synthase (NOS). There are three isoforms of NOS. NOS-1 (also known as nNOS) is constitutive and is predominantly located in neuronal tissue. NOS-2 (iNOS) is inducible and is located in a variety of tissues. NOS-3 (eNOS) is constitutive and is primarily localized in endothelial tissue.

ARGININE AND IMMUNITY

Previous reports showed that plasma Arg declines after severe injury, and a marked reduction in serum Arg is a predictor of mortality in septic patients (Freund *et al.* 1979). Arg supplementation significantly increases plasma Arg levels. Numerous studies have demonstrated the benefits of Arg supplementation on immune functions. Previous reports revealed that supplemental dietary Arg accelerated wound healing and improved nitrogen retention after trauma. Arg supplementation also increased peripheral lymphocyte mitogenesis in response to mitogens in healthy humans and in postoperative patients (Daly *et al.* 1988, Barbul *et al.* 1990). Previous work in our laboratory demonstrated that Arg supplementation attenuates oxidative stress, and a better in vitro macrophage response was observed in burned mice (Tsai *et al.* 2002). Also, enteral Arg supplementation before sepsis significantly enhances peritoneal macrophage phagocytic activity and intestinal immunoglobin A secretion in septic rats (Wang *et al.* 2003, Shang *et al.* 2004). Arg is considered to be an essential amino acid for patients with catabolic conditions. Currently, Arg is added to enteral formulas at pharmacological levels in an attempt to boost immune function and improve clinical outcomes of critically ill patients. However, the effects of Arg on various immune parameters are not consistent. Some studies showed no change (Torre *et al.* 1993) or a decrease (Gonce *et al.* 1990) in the thymus weight, splenocyte proliferation, or survival in animal studies. Nieves and Langkamp-Henken (2002) summarized previous clinical trials that examined the effect of Arg as a single variable on immune functions. They found that among six double-blind, randomized controlled studies, two of the trials showed an increase, while the other four trials showed no effect or a decrease in mitogen-induced lymphocyte proliferation. No studies have demonstrated improved outcomes with Arg supplementation. A meta-analysis included 22 randomized trials with a total of 2,419 surgical or critically ill patients to examine the relationship of enteral nutritional supplementation using immune-enhancing nutrients with infectious complication and mortality rates of patients. They found that compared with standard enteral nutrition, an Arg-containing formula had no effect on infectious complications and may increase mortality in critically ill patients (Heyland *et al.* 2001). The efficacy of Arg supplementation on critically ill patients remains controversial.

ARGININE AND ADHESION MOLECULES

Adhesion molecules play key roles in cell-cell interactions and cell-extracellular matrix interactions. These proteins are important mediators of host defense and are localized in the earliest inflammatory lesions. Overexpression of adhesion molecules by endothelial cells (ECs) and leukocytes may contribute to tissue injury and multiple organ dysfunction. Arg modulates the inflammatory response perhaps partly through regulating adhesion molecules.

Atherogenesis

Hypercholesterolemia is a risk factor for atherosclerosis, and monocyte adhesion to ECs is a key early event in atherogenesis. The earliest observable abnormality of vessel walls in hypercholesterolemic animals is enhanced monocyte adherence to the endothelium, which occurs within 1 wk of initiation of a high-cholesterol diet. This event is thought to be mediated by reduced activity of NO in accordance with increased expression of endothelial surface adhesion molecules and chemotactic proteins induced by hypercholesterolemia. A study by Tsao *et al.* (1994) investigated whether L-Arg attenuated endothelial adhesiveness in hypercholesterolemia. Rabbits were fed chow, a 1% high-cholesterol diet, a high-cholesterol diet supplemented with 2.25% L-Arg HCl in drinking water, or a normal diet with an NOS inhibitor. After 2 wk, thoracic aortas were harvested and placed in a culture dish with the endothelial surface exposed to medium containing monocytoid cells. They found that monocytoid cell binding to aortic endothelium was significantly increased in the high-cholesterol and NOS-inhibitor groups. Binding was markedly reduced in Arg-fed hypercholesterolemic rabbits. These results suggest that hypercholesterolemia enhances the adhesiveness of monocytes to the aortic endothelium, and this effect is attenuated by dietary L-Arg. Inhibition of NO synthesis enhances monocyte binding indicating that endothelium-derived NO plays an important role in regulating the adhesiveness of monocytes to the endothelium. In an in vitro study, Adams *et al.* (1997) investigated the effects of Arg on monocyte adhesion to ECs and endothelial expression of cell adhesion molecules. Human umbilical vein endothelial cells (HUVECs) and monocytes were obtained from healthy human volunteers who had no clinical evidence of cardiovascular disease. Confluent EC monolayers or monocytes were treated with various concentrations of Arg (0, 100, and 1000 µM) or 100 µM Arg with NG-monomethyl-L-arginine (L-NMMA, a non-selective NOS inhibitor) for 24 hr before the adhesion assays. Subsequently, monocytes were incubated with HUVECs for 1 hr at 37°C, and adhesion was measured by light microscopy. They found that, compared to the control, monocyte adhesion was reduced by Arg and increased by L-NMMA. Surface expression of intracellular adhesion molecule (ICAM)-1 by HUVECs was reduced at both concentrations of Arg compared to the control in both the basal and interleukin-1 beta-stimulated states, which correlated with decreased levels of mRNA. Expression of vascular cell adhesion molecule (VCAM)-1 was reduced in the stimulated state and only in the presence of 1000 µM Arg. In a prospective, double-blind, randomized crossover trial, 10 men with coronary atherosclerosis took Arg (7 g three times a day) or a placebo for 3 d each, with a washout period of 10 d. Serum from six of the ten subjects after Arg and placebo was then added to confluent monolayers of HUVECs for 24 hr, and monocytes were added and cell adhesion assessed. They found that adhesion was reduced following L-Arg administration compared to the placebo (Adams *et al.* 1997).

Ischemia/Reperfusion

Reperfusion of ischemic myocardium has been reported to cause rapid degeneration of endothelial function, characterized by a decreased release of NO in response to endothelium-dependent vasodilators. A previous study showed that NO donors given during reperfusion preserve coronary artery ring vasorelaxation and reduce myocardial injury associated with ischemia and reperfusion. A study by Engelman *et al.* (1995) evaluated whether the NO precursor, L-Arg, could reduce ischemia/reperfusion injury by preventing leukocyte-endothelial interactions. They induced regional ischemia in an open-chest pig heart for 30 min followed by 90 min of reperfusion. A preischemic 10-min intravenous infusion of 4 mg/kg/min of Arg was compared to control pigs. The results showed that L-Arg administration reduced plasma levels of ICAM-1, E-selectin, and VCAM-1. Myocardial stunning and arrhythmias were also reduced. A study by Hayashida *et al.* (2000) investigated the effect of Arg on myocardial reperfusion injury. Isolated hearts were perfused with blood at 37°C from a support rat. After 20 min of aerobic perfusion, the hearts were arrested for 60 min with warm blood cardioplegia given at 20 min intervals. This was followed by 60 min of reperfusion. The hearts were divided into three groups according to the supplemental drugs added to the cardioplegic solution. The control group received warm-blood cardioplegia. The Arg group received warm blood supplemented with Arg, and the third group received warm blood supplemented with Arg and NG-nitro-L-arginine methyl ester (L-NAME, an iNOS inhibitor). The results showed that compared to the other two groups, the Arg group showed early recovery of lactate metabolism and greater coronary blood flow during reperfusion. Levels of myocardial release of circulating ICAM-1 and E-selectin were lower in the Arg group.

Diabetes

Cardiovascular complications represent 80% of the causes of death in patients with type 2 diabetes. Reactive oxygen and NO have recently been considered to be involved in the cardiovascular complications of patients with type 2 diabetes. A recent double-blind study tested the effects of L-Arg and N-acetylcysteine administration in type 2 diabetic patients. Arg was used to enhance NO production, and N-acetylcysteine was administered to ameliorate the antioxidant defense and increase intracellular nitrosothiol concentrations, thus increasing NO availability. Twenty-four male patients with type 2 diabetes and hypertension were randomly divided into two groups of 12 patients and received either oral supplementation of a placebo or N-acetylcysteine plus Arg for 6 mon. They found that N-acetylcysteine plus Arg treatment caused reductions in blood pressure, high sensitivity C-reactive protein, ICAM, VCAM, fibrinogen, and plasminogen activator inhibitor 1. These results suggest that N-acetylcysteine plus Arg administration improves endothelial function and reduces inflammatory markers in hypertensive patients with type 2 diabetes (Martina *et al.* 2008).

Smoking

Cigarette smoking is a major risk factor for the development of atherosclerosis. Cigarette smoke is associated with abnormal endothelial function in both active and passive young adult smokers. A study performed by Adams *et al.* (1997) assessed the effects of cigarette smoking on adhesion of human monocytes to HUVECs and measured the effects of Arg supplementation on this interaction. They collected serum from eight smokers with no other coronary risk factors and eight age- and gender-matched non-smokers. Serum was added to HUVECs and incubated for 24 hr. Human monocytes were then added to HUVECs for 1 hr, and adhesion was analyzed by microscopy. To assess the reversibility, monocyte/endothelial cell adhesion was measured for 2 hr after 7 g of Arg was orally administered. The results showed that monocyte/endothelial adhesion and EC expression of ICAM-1 increased in smokers compared to control subjects. After oral Arg administration, monocyte/endothelial adhesion was reduced in smokers, as was EC expression of ICAM-1. These results indicated that cigarette smoking is associated with increased monocyte-EC adhesion, and this abnormality is acutely reversible by oral Arg supplementation.

Surgery and Sepsis

Surgery and infection result in inflammatory reactions and stimulate the production of a variety of endogenous mediators. These mediators initiate immune responses and metabolic alterations that are integral to a response by a host to injury. The activation of leukocytes and ECs and the expressions of adhesion molecules are important events in the pathogenesis of the inflammatory response. A previous study performed by our laboratory investigated the effects of different Arg levels on adhesion molecule expression by ECs and leukocytes, and the transendothelial migration of polymorphonuclear neutrophils (PMNs) through ECs stimulated by biological fluid from surgical patients was examined. We treated ECs and PMNs with different Arg concentrations, including low (50 μM), approximately physiological (100 μM), and high (1000 μM) Arg levels in an *in vitro* study stimulated with patient's peritoneal drain fluid. The results showed that ECs and PMNs were activated after peritoneal drain fluid stimulation. A low Arg concentration comparable to a catabolic condition resulted in higher ICAM-1 and VCAM-1 expressions by ECs. Also, the IL-8 receptor and CD11a/CD18 expressed by neutrophils were enhanced. Arg administration at levels similar to or higher than physiological concentrations reduced IL-8 and CAM expressions, and decreased PMN transmigration after stimulation by peritoneal drain fluid from surgical patients (Fig. 1). Inactivation of NO by L-NMMA results in high CAM expression, indicating that NO is an important inhibitor of CAM expression and EC-PMN interactions (Yeh *et al.* 2007). In an established septic animal study also performed by our laboratory, mice were assigned to a control group or an Arg group. The control group was fed a common semipurified diet, whereas 2% of total calories in the diet was provided by Arg in the Arg group. After 3 wk, sepsis was induced in both groups of mice by cecal ligation and puncture. The results showed that compared to the control group, pretreatment with an Arg-supplemented diet resulted in higher plasma ICAM-1 levels and leukocyte CD11a/CD18 and CD11b/CD18 expression during sepsis (Table 1). The Arg group had higher myeloperoxidase activities in various organs at 24 hr after sepsis. Myeloperoxidase is a neutrophil-specific enzyme and is considered to be an index of neutrophil infiltration. These results indicated that Arg administration may aggravate the inflammatory reaction and increase neutrophil infiltration into tissues (Yeh *et al.* 2006). The adverse effects of Arg on the inflammatory reaction observed in that study were distinct from previous Arg supplementation studies with beneficial results. Many factors may contribute to the inconsistent effects of enteral Arg supplementation. Previous reports suggested that NO can have differential effects on different stages of leukocyte recruitment and CAM expression, depending on the levels produced. The amount and timing of Arg supplementation and the characteristics and the severity of various diseases may result in different levels of NO and NO-related cytotoxic substances. This may partly explain the discrepancies between experiments.

Table I Effects of Arg on leukocyte CD11a/CD18 and CD11b/CD18 expression during sepsis

	CD 11a/CD 18%	CD 11b/CD 18(%)
NC	5.23 ± 1.65	9.56 ± 1.98
0 h		
Control	4.87 ± 1.61	10.78 ± 0.26
Arg	4.08 ± 0.24	11.38 ± 1.74
6 h		
Control	3.6 ± 1.51	$19.58 \pm 1.25^{\dagger,\ddagger}$
Arg	$14.65 \pm 1.96^{*,\dagger}$	$17.2 \pm 1.94^{\dagger}$
12 h		
Control	3.06 ± 1.79	$13.68 \pm 0.54^{\dagger,\ddagger}$
Arg	$25.38 \pm 4.5^{*,\dagger}$	$25.22 \pm 2.16^{*,\dagger,\ddagger}$
Control	6.13 ± 2.45	9.83 ± 0.59
Arg	$43.03 \pm 3.28^{*,\dagger,\ddagger}$	$26.2 \pm 1.39^{*,\dagger,\ddagger}$

Leukocyte integrins CD11a/CD18 and CD11b/CD18 expressions were significantly higher in the Arg group 12, and 24 hr after sepsis than those of the corresponding control and the normal control (NC) group. Results are presented as mean ± SD. *Different from the control group at the same time point. Reprinted from Yeh *et al.* (2006), with permission.

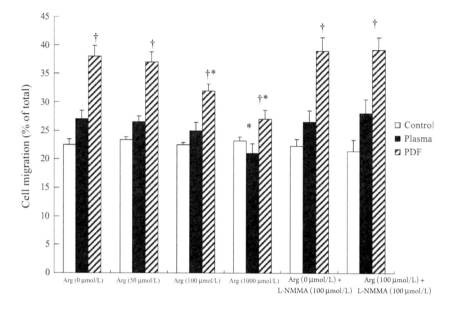

Fig. 1 Plasma and peritoneal drain fluid (PDF)-stimulated migration of PMNs across HUVECs cultured with the addition of different Arg and L-NMMA concentrations. Among the groups stimulated with patients' plasma and PDF, PMN migration was the lowest with 1000 μM Arg compared with the other Arg levels. In addition, PMN migration was lower with 100 μM than with 0 μM and 50 μM Arg. The reduced PMN migration was abrogated when L-NMMA was administered. Reprinted from Yeh *et al.* (2007), with permission.

NO and Adhesion Molecules

Evidence from previous reports pointed out that the influences of Arg on inflammatory mediators including adhesion molecules may mainly depend on regulation of NO. Numerous studies used an NO donor and NOS inhibitors to investigate the roles of NO in leukocyte adherence to the vascular endothelium and emigration. A study by Kubes *et al.* (1991) designed an experiment to assess the direct effects of L-NAME and L-NMMA on PMN adherence. Cat mesenteric venules were used to observe leukocyte adherence to the venular endothelium by intravital video microscopy. They found that both inhibitors of NO production increased leukocyte adherence, and leukocyte emigration was also enhanced. Incubation of isolated cat neutrophils with L-NMMA resulted in direct upregulation of the leukocyte adhesion glycoprotein, CD11/CD18. L-NAME-induced adhesion was inhibited by L-Arg but not D-Arg. These findings suggest that endothelium-derived NO is an important endogenous modulator of leukocyte adherence, and impairment of NO production results in leukocyte adhesion and emigration. A study performed by Armstead *et al.* (1997) added L-NAME and an NO donor to human cultured iliac vein ECs, and P-selectin mRNA expression was quantified. They found that L-NAME caused increased expression of P-selectin mRNA. The stimulatory effect of L-NAME was reversed by the addition of L-Arg. The effects of the NO donor and L-Arg were also paralleled by decreases in P-selectin protein synthesis and in decreased adherence of human neutrophils to human iliac venous ECs. Another study investigated plasma NO levels and the expression of P-selectin on platelets in preeclampsia. Studies were carried out in the third trimester of normal pregnant women and women with preeclampsia. Also, the effects of the inhibition of NO synthesis on the expression of P-selectin on platelets *in vitro* were examined. The results showed that nitrate levels and the expression of P-selectin in preeclampsia were higher than those in a normal pregnancy. NO synthesis inhibition *in vitro* significantly increased the expression of P-selectin in normal pregnancy and preeclampsia. The findings of that study were consistent with a modulatory role of NO in dampening excessive platelet activation in normal and preeclamptic pregnant women (Yoneyama *et al.* 2002). However, an *in vitro* study performed by Beauvais *et al.* (1995) found that when human neutrophils were placed in a gradient of an NO donor, directed locomotion was induced, as evidenced by experiments of chemotaxis under agarose indicating that exogenous NO elicits chemotaxis of neutrophils. An animal study investigating the effects of NO on neutrophil sequestration in lung tissue found that endogenous NO might have no association with the expression levels of L-selectin, P-selectin, or CD18 by neutrophils (Roman and McGahren 2006). The results of those studies mentioned above suggest that NO has both pro- and anti-inflammatory activities. Depending on the levels produced, NO may be released to influence nearby cells or may act as an intracellular messenger regulating the activity of cells responsible for its generation. The discrepancies between experiments dealing with the inhibition of NOS versus exogenous application of NO-generating compounds may be partly explained by localized versus widespread activities of NO.

POSSIBLE MECHANISMS OF NO IN REGULATING CAM EXPRESSIONS

According to a study by Spiecker *et al.* (1997), the mechanisms by which NO decreases endothelial and leukocyte adhesion molecule expressions possibly occurs through inhibition of the pleiotropic transcription factor, nuclear factor-κB (NF-κB). Previous studies suggested that NF-κB is required for the transcriptional induction of EC adhesion molecules. The activation of NF-κB involves the degradation of its cytoplasmic inhibitor, IκB. After administration of external stimulants, including lipopolysaccharide and cytokines, IκB is phosphorylated. Phosphorylation of IκB targets IκB for ubiquitination and rapid degradation. The degradation of IκB allows the unbound NF-κB to be translocated to the nucleus, where it can transactivate the enhancer elements of many pro-inflammatory genes.

The phosphorylation of IκB is a key regulatory step in the activation of NF-κB. Spiecker *et al.* (1997) determined how NO inhibits NF-κB and examined the fate of IκBα following tumor necrosis factor (TNF)-α stimulation in the presence of NO donors. They found that activation of NF-κB by TNF-α occurred within a short period of time and coincided with the rapid degradation of IκBα. Co-treatment with NO donors did not prevent IκBα degradation. However, NO donors inhibited NF-κB activation and augmented IκBα resynthesis and nuclear translocation after 2 hr of TNF-α stimulation. This correlated with a reduction in TNF-α-induced VCAM-1 expression. They also used murine macrophage-like RAW 264.7cells, co-cultured with ECs to examine the effects of endogenously derived NO on cell adhesion molecule expression. The results showed that induction of RAW 264.7-derived NO inhibited lipopolysaccharide-induced endothelial VCAM-1 expression, which was reversed by the addition of the NOS inhibitor, L-NMMA. Those findings indicate that NO inhibits NF-κB activation and VCAM-1 expression by increasing the expression and nuclear translocation of IκBα. A study by Nolan *et al.* (2008) used a simplified in vitro system to identify the mechanism for regulating neutrophil migration by NO. They found that inhibition of constitutive NO production may lead to increased neutrophil migration through microparticle formation and thus propagation of the inflammatory response. Microparticles are formed from the membrane of activated immune cells and express adhesion molecules that enable them to influence cell-cell interactions and cell recruitment. Neutrophil-derived microparticles have been shown to contain inflammatory mediators such as platelet-activating factor, CD11a, CD11b, L-selectin, and P-selectin glycoprotein ligand 1 (PSGL-1) and to activate ECs causing the release of pro-inflammatory cytokines. In that study, they found that regulation of adhesion by NO may involve direct effects on leukocytes. Pretreatment of microparticles with antibodies to L-selectin and PSGL-1 significantly inhibited neutrophil migration. The ability of L-NAME-induced microparticles to enhance migration was found to be dependent on the number of microparticles produced. Those data showed that NO can modulate neutrophil migration by regulating microparticle formation.

SUMMARY

- Arg plays important roles in maintaining health and immunity.
- The influences of Arg on modulating immune response may mainly depend on NO.
- Most studies have suggested that Arg or NO administration reduces adhesion molecule expression and decreases leukocyte adherence and transmigrations.
- Some conflicting results showed that the use of Arg or NO enhances adhesion molecule expression and thus aggravates inflammatory reaction.

- NO can be treated as a pro- or anti-inflammatory mediator under different physiological conditions.
- The application of Arg in critically ill patients should be carefully evaluated.

Abbreviations

Arg	arginine
EC	endothelial cell
E-selectin	endothelial leukocyte adhesion molecule
HUVECs	human umbilical vein endothelial cells
ICAM	intracellular adhesion molecule
L-NAME	NG-nitro-L-arginine methyl ester
L-NMMA	NG-monomethyl-L-arginine
NF-κB	nuclear factor-κB
NO	nitric oxide
NOS	nitric oxide synthase
PMN	polymorphonuclear neutrophil
VCAM	vascular cell adhesion molecule

Key Facts about Arginine

1. Arg is a semi-essential amino acid for humans, but it is now considered an essential amino acid for patients with catabolic conditions.
2. Plasma Arg concentration for healthy adults is approximately 100 μM.
3. Many studies have demonstrated the benefits of Arg supplementation on immune functions.
4. Arg is the precursor of NO.
5. The influences of Arg on inflammatory mediators including adhesion molecules may mainly depend on regulation of NO.
6. Adverse effects of Arg have been reported in some inflammatory situations.
7. The efficacy of Arg supplementation on critically ill patients remains controversial.

Definition of Terms

Leukocyte-endothelial interactions: Leukocyte adherence to vascular endothelium. An initial step of inflammation results from endothelial dysfunction.

Leukocyte transmigration: Transendothelial movement of leukocyte to the site of injury. This procedure plays an important role in the pathogenesis of inflammatory reaction.

NF-κB: A factor present in the cytoplasm in an inactive state and activated by lipopolysaccharide and cytokines. Activation of NF-κB triggers the transcription of genes involved in the inflammatory response.

NO: A small molecule produced by most cell types. NO is an antimicrobial agent, a neurotransmitter and a vasodilator.

NOS: The enzyme that produces NO. The substrate for this reaction is Arg and the products are NO and citrulline.

NOS inhibitor: Substance inhibits the NOS activities and inhibition of NO production. L-NMMA is a non-selective NOS inhibitor. L-NAME is an inducible NOS inhibitor.

References

Adams, M.R. and W. Jessup, and D.S. Celermajer. 1997. Cigarette smoking is associated with increased human monocyte adhesion to endothelial cells: reversibility with oral L-arginine but not vitamin C. J. Am. Coll. Cardiol. 29: 491-497.

Adams, M.R. and W. Jessup, D. Hailstones, and D.S. Celermajer. 1997. L-arginine reduces human monocyte adhesion to vascular endothelium and endothelial expression of cell adhesion molecules. Circulation 95: 662-668.

Adams, M.R. and R. McCredie, W. Jessup, J. Robinson, D. Sullivan, and D.S. Celermajer. 1997. Oral L-arginine improves endothelium-dependent dilatation and reduces monocyte adhesion to endothelial cells in young men with coronary artery disease. Atherosclerosis 129: 261-269.

Armstead, V.E. and A.G. Minchenko, R.A. Schuhl, R. Hayward, T.O. Nossuli, and A.M. Lefer. 1997. Regulation of P-selectin expression in human endothelial cells by nitric oxide. Am. J. Physiol. 273: H740-746.

Beauvais, F. and L. Michel, and L. Dubertret. 1995. Exogenous nitric oxide elicits chemotaxis of neutrophils in vitro. J. Cell. Physiol. 165: 610-614.

Barbul, A. and S.A. Lazarou, D.T. Efron, H.L. Wasserkrug, and G. Efron. 1990. Arginine enhances wound healing and lymphocyte immune responses in humans. Surgery 108: 331-336.

Daly, J.M. and J. Reynolds, A. Thom, L. Kinsley, M. Dietrick-Gallagher, J. Shou, and B. Ruggieri. 1988. Immune and metabolic effects of arginine in the surgical patient. Ann. Surg. 208: 512-523.

Engelman, D.T. and M. Watanabe, N. Maulik, G.A. Cordis, R.M. Engelman, J.A. Rousou, J.E. Flack 3rd, D.W. Deaton, and D.K. Das. 1995. L-Arginine reduces endothelial inflammation and myocardial stunning during ischemia/reperfusion. Ann. Thorac. Surg. 60: 1275-1281.

Freund, H. and S. Atamian, J. Holroyde, and J.E. Fischer. 1979. Plasma amino acids as predictors of the severity and outcome of sepsis. Ann. Surg. 190: 571-576.

Gonce, S.J. and M.D. Peck, J.W. Alexander, and P.W. Miskell. 1990. Arginine supplementation and its effect on established peritonitis in guinea pigs. J. Parenter. Enter. Nutr. 14: 237-244.

Hayashida, N. and H. Tomoeda, T. Oda, E. Tayama, S. Chihara, K. Akasu, T. Kosuga, E. Kai, and S. Aoyagi. 2000. Effects of supplemental L-arginine during warm blood cardioplegia. Ann. Thorac. Cardiovasc. Surg. 6: 27-33.

Heyland, D.K. and F. Novak, J.W. Drover, M. Jain, X. Su, and U. Suchner. 2001. Should immunonutrition become routine in critically ill patients? A systematic review of the evidence. JAMA 286: 944-953.

Kubes, P. and M. Suzuki, and D.N. Granger. 1991. Nitric oxide: an endogenous modulator of leukocyte adhesion. Proc. Natl. Acad. Sci. USA 88: 4651-4655.

Martina, V. and A. Masha, V.R. Gigliardi, L. Brocato, E. Manzato, A. Berchio, P. Massarenti, F. Settanni, L. Della Casa, *et al.* 2008. Long-term N-acetylcysteine and L-arginine administration reduces endothelial activation and systolic blood pressure in hypertensive patients with type 2 diabetes. Diabet. Care 31: 940-944.

Nieves, C., Jr. and B. Langkamp-Henken. 2002. Arginine and immunity: a unique perspective. Biomed. Pharmacother. 56: 471-482.

Nolan, S. and R. Dixon, K. Norman, P. Hellewell, and V. Ridger. 2008. Nitric oxide regulates neutrophil migration through microparticle formation. Am. J. Pathol. 172: 265-273.

Roman, A. and E.D. McGahren. 2006. L-NAME-induced neutrophil accumulation in rat lung is not entirely because of interactions with L- and P-selectins or CD18. J. Pediatr. Surg. 41: 1743-1749.

Rose W.C. 1937. The nutritive significance of the amino acids and certain related compounds. Science 86: 298-300.

Shang, H.F. and Y.Y. Wang, Y.N. Lai, W.C. Chiu, and S.L. Yeh. 2004. Effects of arginine supplementation on mucosal immunity in rats with septic peritonitis. Clin. Nutr. 23: 561-569.

Spiecker, M. and H.B. Peng, and J.K. Liao. 1997. Inhibition of endothelial vascular cell adhesion molecule-1 expression by nitric oxide involves the induction and nuclear translocation of IκBα. J. Biol. Chem. 272: 30969-30974.

Torre, P.M. and A.G. Ronnenberg, W.J. Hartman, and R.L. Prior. 1993. Oral arginine supplementation does not affect lymphocyte proliferation during endotoxin-induced inflammation in rats. J. Nutr. 123: 481-488.

Tsai, H.J. and H.F. Shang, C.L. Yeh, and S.L. Yeh. 2002. Effects of arginine supplementation on antioxidant enzyme activity and macrophage response in burned mice. Burns 28: 258-263.

Tsao, P.S. and G. Theilmeier, A.H. Singer, L.L. Leung, and J.P. Cooke. 1994. L-arginine attenuates platelet reactivity in hypercholesterolemic rabbits. Arterioscler. Thromb. 14: 1529-1533.

Wang, Y.Y. and H.F. Shang, Y.N. Lai, and S.L. Yeh. 2003. Arginine supplementation enhances peritoneal macrophage phagocytic activity in rats with gut-derived sepsis. J. Parenter. Enter. Nutr. 27: 235-240.

Yeh, C.L. and C.S. Hsu, S.C. Chen, Y.C. Hou, W.C. Chiu, and S.L. Yeh. 2007. Effect of arginine on cellular adhesion molecule expression and leukocyte transmigration in endothelial cells stimulated by biological fluid from surgical patients. Shock 28: 39-44.

Yeh, C.L. and C.S. Hsu, W.C. Chiu, Y.C. Hou, and S.L. Yeh. 2006. Dietary arginine enhances adhesion molecule and T helper 2 cytokine expression in mice with gut-derived sepsis. Shock 25: 155-160.

Yoneyama, Y. and S. Suzuki, R. Sawa, A. Miura, D. Doi, Y. Otsubo, and T. Araki. 2002. Plasma nitric oxide levels and the expression of P-selectin on platelets in preeclampsia. Am. J. Obstet. Gynecol. 187: 676-680.

CHAPTER 8

Adhesion Molecules in Stem Cells

Hiroko Hisha[1,2,3], Xiaoli Wang[1,4] and Susumu Ikehara[1,2,3,*]

[1]First Department of Pathology
[2]Department of Transplantation for Regeneration Therapy (sponsored by Otsuka Pharmaceutical Co. Ltd.)
[3]Regeneration Research Center for Intractable Diseases, Kansai Medical University, Moriguchi City, Osaka, Japan
[4]Division of Hematology/Oncology, Tisch Cancer Institute, Departments of Medicine, Mount Sinai School of Medicine, 1 Gustave L. Levy Place, Box 1079, New York, NY, USA
Hiroko Hisha, Ph.D., First Department of Pathology, Kansai Medical University, 10-15 Fumizono-cho, Moriguchi City, Osaka 570-8506, Japan
E-mail: hishah@takii.kmu.ac.jp
Xiaoli Wang, M.D., Ph.D., Division of Hematology/Oncology, Tisch Cancer Institute, Department of Medicine, Mount Sinai School of Medicine, 1 Gustave L. Levy Place, Box 1079, New York, NY 10029
E-mail: xiaoli2004osaka@gmail.com
Susumu Ikehara, M.D., Ph.D., First Department of Pathology, Kansai Medical University, 10-15 Fumizono-cho, Moriguchi City, Osaka 570-8506, Japan
E-mail: ikehara@takii.kmu.ac.jp

ABSTRACT

Tissues and organs are composed of tissue-specific dormant stem cells, rapidly proliferating progenitor cells and differentiated cells. The stem cells are localized in special regions (niches) where their stemness is maintained and controlled. Direct physical interaction between the stem cells and the niche cells, mediated by adhesion molecules, is essential for the stem cells to maintain stemness, proliferate and differentiate. The interaction between the adhesion molecules and extracellular matrix also plays an important role.

---

*Corresponding author
Key terms are defined at the end of the chapter.

Recently, there has been great progress in the isolation of tissue-specific stem cells. Hematopoietic stem cells (HSCs) have been most extensively studied, and highly purified HSCs can now be obtained. The niche of the HSCs is also well elucidated. The HSCs are known to express adhesion molecules of cadherin family (N-cadherin), integrin family (VLA-4 and VLA-5, etc.), immunoglobulin superfamily (ICAM-1 and VCAM-1, etc.), CD44 family and sialomucin family (CD34), by which the HSCs interact with the niche cells. Mesenchymal stem cells (MSCs) have the capacity to differentiate into mesodermal, endodermal and ectodermal lineage cells. The cells can proliferate extensively *in vitro* and have the important ability to suppress the proliferation of lymphocytes. The approach by which MSCs are used for the treatment of graft-versus-host diseases is now gaining support. It is known that MSCs express the immunoglobulin superfamily (ICAM-1, ICAM-3, VCAM-1 and ALCAM), CD44 and P-selectin. The purification of neural stem cells (NSCs), hepatic stem cells (HpSCs) and hair follicle stem cells (HFSCs) remains inadequate despite having been extensively studied. Cadherin and integrin molecules are expressed by NSCs, HpSCs and HFSCs and contribute to the maintenance of stemness, the proliferation and the migration of these stem cells.

The final goal of regenerative medicine is to isolate stem cells from their niche, expand them *in vitro* and implant them into patients. For this purpose, it is necessary to understand the mechanism by which the stem cells localize in their niche and how their self-renewal and differentiation are regulated. Thus, the characterization of adhesion molecules on stem cells will provide insights for a wide range of clinical applications.

INTRODUCTION

Pluripotent stem cells are defined as cells that have the ability to self-renew (generating new pluripotent stem cells) and to differentiate into multilineage cells. Pluripotent stem cells can be found in special areas in tissues and organs, where their stemness is maintained and controlled. Such areas are called niches. The stem cells need a niche as a scaffold for their self-renewal and differentiation. In the niche, stem cells receive signals through their interaction with niche cells: adhesion molecules, cytokines produced by the niche cells and cell matrix molecules (such as collagen, laminin and teneicin). Several molecules (e.g., Notch-Jagged system and Tie2/angiopoietin system) play an important role in stem cell dormancy.

In recent decades, many tissue-specific stem cells have been purified and identified in adult humans and animals, including hematopoietic stem cells (HSCs), mesenchymal stem cells (MSCs), neural stem cells (NSCs), hepatic stem cells (HpSCs), and hair follicle stem cells (HFSCs). Many other tissue-specific stem cells have also been identified and characterized. However, the problem of insufficient analysis of adhesion molecules remains in the case of some stem cells. In this chapter, we provide an overview and a description of our current understanding regarding the interaction between adhesion molecules expressed by several tissue-specific adult stem cells and the stem cell niche.

HEMATOPOIETIC STEM CELLS

Hematopoietic stem cells (HSCs) were first identified in the mouse bone marrow by Till and McCulloch in the 1960s and are probably the most extensively studied of the stem cells. Hematopoiesis is maintained by a few pluripotent hematopoietic stem cells (P-HSCs) (the frequency is less than 0.1% of bone marrow mononuclear cells) having extensive developmental and self-renewal potential. The P-HSCs are in close contact with the niche cells in the bone marrow and are controlled by interactions with cytokines produced from the niche cells and adhesion molecules on the niche cells. Thus, the fate of the P-HSCs is determined by extrinsic and intrinsic mechanisms. The P-HSCs generate two kinds of cell populations after cell division: dormant P-HSCs and cytokine-reactive HSCs that have lost their self-renewal ability. The cytokine-reactive HSCs differentiate into lineage-specific hematopoietic progenitor cells (HPCs) under the influence of various cytokines produced by bone marrow niche cells, macrophages, fibroblasts and endothelial cells, and finally differentiate into mature blood cells (Fig. 1A).

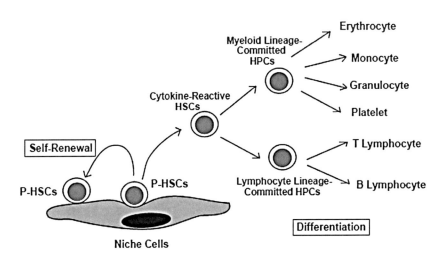

Fig. 1A Differentiation pathway of HSCs. P-HSCs are in close contact with niche cells and generate two kinds of cell populations: dormant P-HSCs and cytokine-reactive HSCs. The cytokine-reactive HSCs differentiate into mature blood cells under the influence of various cytokines.

The concept of a niche for HSCs was proposed by Schofield in the 1970s, but the niche itself was poorly characterized until recently, when, as a result of developments in molecular biological techniques, huge advances have been made. It has been shown that osteoblasts contacting the endosteal surface of the bones constitute the niche (Calvi *et al.* 2003, Zhang *et al.* 2003). These cells, which were named spindle-shaped N-cadherin-positive cells (SNO cells) by Zhang *et al.*, are closely associated with HSCs expressing N-cadherin. More recently, sinusoidal endothelium has been proposed as another niche (Sugiyama *et al.* 2006). This concept is supported by the observation that osteoblasts do not exist in spleen and liver but that extramedullary hematopoiesis can be induced in the case of hypoplasia in the bone marrow. At present, it is speculated that the endosteum/ osteoblast niche contributes mainly to the maintenance of P-HSCs, whereas the sinusoidal endothelium niche contributes to the proliferation/differentiation of HSCs and their mobilization into the periphery as well as their maintenance (Fig. 1B).

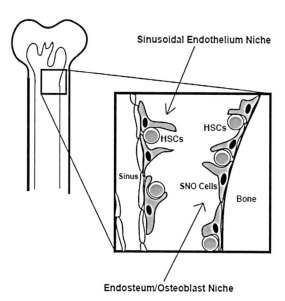

Fig. 1B   Regulation of self-renewal and differentiation of HSCs in their niche: Endosteum/ osteoblast niche and sinusoidal endothelium niche. Osteoblasts in contact with the endosteal surface of the bones constitute endosteum/osteoblast niche. Sinusoidal endothelium is also proposed as another niche.

It has been demonstrated that many adhesion molecules are expressed by HSCs and regulate the adhesion, motility, survival and differentiations of HSCs (Fig. 1C). Based on their structures and manner of adhesion, the adhesion molecules expressed by HSCs are divided into at least five families: cadherin family, integrin family, immunoglobulin superfamily, CD44 family and sialomucin family. In this section, we show representative adhesion molecules on HSCs.

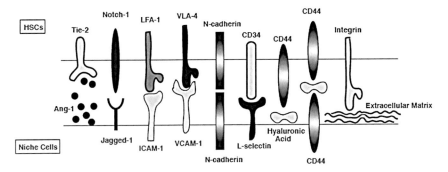

Fig. 1C Regulation of self-renewal and differentiation of HSCs in their niche: Interaction between HSCs and niche cells through cell surface molecules. The maintenance of stemness, self-renewal and differentiation of HSCs are controlled by the interaction with niche cells.

N-cadherin is expressed by both HSCs and the niche cells. Homophilic binding of the molecules induces the adhesion, stretching and migration of HSCs, because the intracellular domain of N-cadherin is associated with β catenin connecting with the actin cytoskeleton. Although many functions of N-cadherin have been proposed, the main function of this molecule might be to physically associate HSCs with niche cells at the proper position.

Integrins (VLA-4, VLA-5 and LFA-1) can interact with cell adhesion molecules belonging to the immunoglobulin superfamily (VCAM-1 and ICAM-1) expressed by vascular endothelial cells and niche cells. The trafficking of HSCs from the blood to the bone marrow cavity and their subsequent adhesion to niche cells (this process is designated as homing) can thus be induced. Integrins can also interact with extracellular matrix components (e.g., collagen, laminin and fibronectin). Such interactions between integrins and their ligands are important for the trafficking and homing of HSCs during embryogenesis and adult hematopoiesis, since HSCs obtained from integrin-deficient fetal or adult mice have a normal differentiation capacity, but cannot seed fetal or adult hematopoietic tissues (Potocnik *et al.* 2000).

CD34 (a member of the sialomucin family, mucosialin, HCPA-1) is a well-known HSC/HPC-marker and has been used for HSC/HPC-purification in mice and humans, although the biological significance of the molecule remains to be clearly defined; there have been some reports indicating that CD34-knockout mice did not show evident decrease in absolute number of mature cells (Cheng *et al.* 1996), and that HSCs with a long-term marrow reconstitution capacity were found almost exclusively in a CD34-negative population in mice (Wang *et al.* 2003). Fackler *et al.* (1995) reported that CD34 played a role in maintaining a murine myeloid cell line (M1 cells) in an immature state, since enforced surface expression of CD34 on the M1 cells inhibited their differentiation into macrophages. This observation is consistent with the well-known phenomenon that immature HSCs and HPCs are highly positive for CD34 but the expression of CD34 molecules is downregulated according to their differentiation. CD34 has also been shown to be associated with regulating the adhesive and migratory properties of HSCs (Gangenahalli *et al.* 2006) through interaction with L-selectin (CD62L) expressed by endothelial cells and niche cells. However, more extensive studies are necessary to investigate the exact role of the CD34 molecule in hematopoiesis.

MESENCHYMAL STEM CELLS

Mesenchymal stem cells (MSCs) were first identified in mouse bone marrow by Friedenstein *et al.* in 1976. They were originally isolated from bone marrow by subcultures of a heterogeneous adherent population obtained from the primary culture of bone marrow cells. Nowadays, there are many reports showing that MSCs can be isolated from various adult and fetal tissues (e.g., adipose tissue, scalp tissue, dermal tissue, peripheral blood, umbilical cord blood, amnion, placenta). MSCs were originally described as stem cells capable of differentiating into adipocytes, osteoblasts and chondrocytes. Recently, the cells were further shown to be able to differentiate into mesodermal, endodermal and ectodermal lineages: e.g., myocytes, cardiac cells, hepatocytes and neurons. In addition to this pluripotency, MSCs have an outstanding expansion capacity; the cells can continue to proliferate after 20 or more passages. Recently, MSCs have been attracting considerable attention because they facilitate the treatment of osteogenesis imperfecta, graft-versus-host disease and myocardial infarction. Indeed, it has been shown that MSCs secrete some immunosuppressive cytokines and hormones, such as TGFβ, HGF and prostaglandin-E. Thus, the pluripotency, expansion capacity and immunosuppressive effect of MSCs might provide important clues for the treatment of intractable diseases, based on the assumption that the mobilization of MSCs into damaged tissues and their subsequent differentiation into functional mature tissue cells could be induced.

Table 1 shows the cell surface markers expressed by MSCs. Hematolymphoid markers (CD4, CD8 and CD45, etc.) are not expressed, but the expression of many adhesion molecules has been reported. Recent analyses indicate the expression of NCAM in mouse and monkey MSCs (Wang *et al.* 2006, Kato *et al.* 2008).

Table I   Cell surface markers of MSCs

	Positive	*Negative or very low expression*
Adhesion molecules	CD44, CD73, CD105, ICAM-1, ICAM-3, VCAM-1, NCAM, ALCAM, VLA-4, VLA-5, P-selectin, E-selectin	PECAM-1, L-selectin
Cytokine receptors	FGFR, PDGFR, IL-1R, IFNγR	IL-2R
Hematolymphoid markers	CD34*	CD4, CD8, CD11b, CD14, CD45
Others	MHC class I	MHC class II

*Positive in mice but negative in humans. FGFR, Fibroblast growth factor-receptor. PDGFR, Platelet-derived growth factor-receptor. IL-1R, Interleukin 1-receptor. IL-2 R, Interleukin 2-receptor. IFNγR, Interferon γ-receptor (unpublished).

However, no specific markers for MSCs have yet been discovered, thereby making it difficult to obtain a purified population of MSCs. Very recently, the prospective isolation of MSCs from whole bone marrow mononuclear cells was attempted using antibody against SSEA-4 molecule (Gang *et al.* 2007). MSCs share many adhesion molecules and cytokines with the niche cells supporting hematopoiesis (Fig. 1C). For example, cell adhesion molecules such as ICAM-1, VCAM-1 and CD44 have been reported to be expressed on mouse and human MSCs. It is known that cytokines such as IL-6, SCF and M-CSF are produced by MSCs. Indeed, it has been demonstrated that MSCs have a hematopoiesis-supporting ability. Therefore, the question arises as to whether the niche cells have differentiated from MSCs or whether MSCs are the cells constituting the niche, and extensive studies to answer this question are awaited.

The rolling of MSCs on vascular endothelial cells is an important process before homing into tissues. Ruster *et al.* (2006) have demonstrated that MSCs express P-selectin but a very low level of L-selectin, indicating that MSCs perform their rolling on the vascular endothelial cells using P-selectin. Zhu *et al.* (2006) reported that PDGF stimulation upregulated CD44 expression on rat MSCs, thereby markedly promoting the migration of the MSCs through hyaluronic acid-coated membrane. They considered such a migration mechanism was critical for the recruitment of MSCs into wound sites for tissue generation. A more recent trial indicated that integrin β1 (CD29) was responsible for the migration of MSCs within the ischemic myocardium in mice, but not integrin α4 (CD49d) or CXC chemokine receptor 4 (CXCR4) (Ip *et al.* 2007).

NEURAL STEM CELLS

In mammals, it had been believed that neurogenesis did not continue in the adult brain. However, recent studies have overturned that theory and have shown that a supply of new neurons is continued even into adulthood in two distinct regions of the brain: the subventricular zone of lateral ventricles (SVZ) and the dentate gyrus of the hippocampus. Neural stem cells are defined as cells having the capacity to self-renew and to differentiate into neurons, astrocytes and oligodendrocytes, and are considered to exist in the SVZ, although the prospective isolation of NSCs has not been achieved because of the lack of NSC-specific markers. It has been reported that the differentiation capacity of the NSCs in the SVZ is skewed only towards neurons (but not towards astrocytes and oligodendrocytes) in the adult brain, whereas there is a report showing that oligodendrocytes are also generated from these NSCs (Menn *et al.* 2006). The NSC-niche in the SVZ is composed of several types of cells (Fig. 2): slow-proliferating/glial fibrillary acidic protein (GFAP)-positive astrocytes (considered as NSCs), rapidly proliferating/transient-amplifying precursors (derived from the NSCs), neuronal progenitors migrating towards the olfactory bulb (immature neurons) and ependymal cells lining the lateral ventricles. The ependymal cells are epithelial cells and come into close contact with the NSCs expressing Notch and integrin β1 (CD29). The ependymal cells are therefore regarded as niche cells. Campos *et al.* (2006) reported that integrin β1 affected the Notch signaling pathway through its interaction with the Notch intracellular domain.

In the dentate gyrus of the hippocampus, GFAP-positive neuron precursors can be found and these cells are considered to have a limited self-renewal and differentiation capacity.

Recent studies have shown that neuronal progenitors migrating towards the olfactory bulb (immature neurons) express integrin β1 and polysiaric acid-binding NCAM (PSA-NCAM), and that these adhesion molecules contribute to their smooth migration in the restral migratory stream (Cazal *et al.* 2000, Belvindrah *et al.* 2007). Thus, adhesion molecules contribute greatly to both the generation of neurons and the migration of the immature neurons towards the olfactory bulb.

HEPATIC STEM CELLS

The liver is a unique organ, having an extensive regeneration capacity. Oval cells, small hepatocytes and hepatoblasts have been proposed as candidates of hepatic stem cells (HpSCs). The oval cells, having an oval-shaped nucleus and scant cytoplasm, reside in the canals of Heling (the meeting point of hepatocyte canaliculi and the bile duct) when the liver is in a quiescent state, but the cells begin to proliferate, move to the parenchyma and give rise to hepatocytes and bile epithelial cells when the liver is in a regenerative state (Fig. 3). The oval cells share

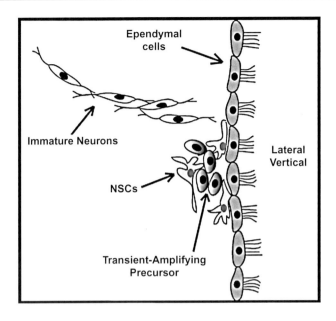

Fig. 2 Neurogenesis in subventricular zone of lateral ventricles (SVZ). NSCs differentiate into transient-amplifying precursors and immature neurons. Ependymal cells, lining the lateral ventricles, are considered to be niche cells.

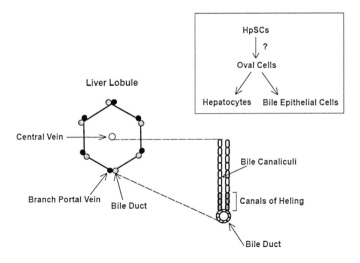

Fig. 3 Structure of liver lobule and differentiation pathway of HpSCs. Oval cells, one of the candidates of HpSCs, reside in the canals of Heling. The oval cells move to the parenchyma and differentiate to hepatocytes and bile epithelial cells when the liver is in a regenerative state.

some molecules with fetal hepatoblasts [e.g., α-fetoprotein, delta-like-1 homolog (Dlk-1)], biliary epithelial cells (cytokeratin-7, -8, -18, -19) and neuroepithelial cells (chromogranin A and NCAM). The expression of the neuroepithelial cell markers suggests that the oval cells are under the control of the central nervous system and form a neuroendocrine compartment in the liver. Alternatively, it has been proposed that the oval cells might be derived from bone marrow and contribute to liver generation, since they also have HSC/HPC-markers such as Thy-1 (CD90), c-kit (CD117) and CD34. These observations might indicate that the oval cells are composed of a heterogeneous population.

Recently, Yovchev *et al.* (2007) identified novel oval cell surface markers in adult rat liver: CD24, CD44, CD133 and epithelial cell adhesion molecule (EpCAM, CD326). They could not, however, detect the expression of some previously identified oval cell markers. In a very recent work, Zhang *et al.* (2008) isolated two populations from human adult liver: the HpSCs and their progenitors (hepatoblasts, α-fetoprotein-positive). The HpSCs proved to have a self-renewal capacity and to give rise to hepatoblasts by culture and transplantation studies of immunoselected HpSCs. The hepatoblasts are transient-amplifying cells and can differentiate into bile epithelial cells and mature hepatocytes. The HpSCs uniquely express EpCAM, NCAM and cytokeratin19, but they are only weakly positive for albumin and are negative for α-fetoprotein.

EpCAM is a transmembrane cell adhesion molecule associated with benign and malignant cell proliferations. It has been reported that the molecule controls oval cell migration by modulating the cell-to-cell interactions mediated by cadherin (Winter *et al.* 2003). CD24 is a ligand for P-selectin, and CD44 is suggested to have a role in oval cell proliferation and migration. Thus, great progress has been made in the identification and characterization of the HpSCs themselves, whereas the niche cells supporting the HpSCs remain to be analyzed.

HAIR FOLLICLE STEM CELLS

Hair follicles repeat the cycle of growth, decline and diapauses throughout life, and therefore the presence of hair follicle stem cells (HFSCs) has been speculated. Previously, the HFSCs were thought to reside in the bulbar region of the hair follicle. Cotsarelis *et al.* (1990), however, proposed that the HFSCs existed in the bulge area, a contiguous part of the outer root sheath of rodent hair follicles, since slow-cycling cells localized to the region (Fig. 4). Subsequent data have confirmed that the bulge is the niche of the HFSCs contributing to both hair follicle cycling and the repopulation of interfollicular epidermis/sebaceous epithelium. The HFSCs generate transient-amplifying cells that migrate to the sebaceous glands, epidermis and hair bulb and contribute to the formation of the hair follicle.

Recent works have shown that the HFSCs express integrin β1 (CD29), integrin α6 (CD49f), keratin 19, CD71, and CD34 to a greater extent than their progenitor

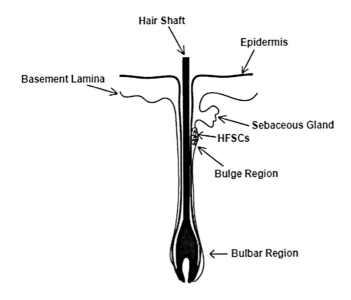

Fig. 4 Structure of hair follicle. The bulge region of the hair follicle is the niche of HFSCs. The HFSCs generate transient-amplifying cells that contribute to the formation of the hair follicle.

cells. The HFSCs, however, lack unique markers to distinguish them from their transient-amplifying progeny, similar to other stem cells. Accordingly, it is still difficult to obtain a pure population of HFSCs. The high expression of integrin on HFSCs suggests that strong adherence to the basement membrane is required for them to localize in the niche and maintain their stemness. Tumbar *et al.* (2004) reported that a purified HFSC population showed positive staining of integrin β1, integrin β4 and integrin α6, although their expression was not constant and fluctuated by hair cycle. Very recently, Kloepper *et al.* (2008) demonstrated that immunohistochemical staining did not show significant difference in the expression of integrin β1 between the bulge region and other parts of the outer root sheath. More work is required to clarify the contribution of integrin molecules to the hair follicle niche.

It has been demonstrated that CD34 expression co-localized with slow-cycling and keratin 15-positive cells in the bulge region on the mouse hair follicle (Trempus *et al.* 2003). These researchers sorted CD34 and integrin α6-double positive cells, and these cells were predominantly in the G_0/G_1 phase and showed a high proliferative potential *in vitro*. In contrast, CD34-negative cells were in the G_2/M and S phase, and the expression of integrin α6 was low. These results indicate that CD34 can be used as a specific marker for the HFSCs. In contrast, there are several reports showing that CD34-negative cells had stem cell properties. Such conflicting results must be clarified by extensive studies.

HFSCs have been used in preparing skin equivalent and can form epithelium when the cells are implanted in deep burn wounds. This result suggests that HFSCs can be used in preparing composite skin substitutes containing the epidermis and dermis.

SUMMARY

- Tissue-specific stem cells are defined as cells that have the ability to self-renew and to differentiate into multilineage cells.
- The stem cells reside in special areas in tissues and organs where their stemness is maintained and controlled (such areas being called niches).
- The physical adhesion between the stem cells and the niche cells, mediated by adhesion molecules, is essential for the maintenance of stemness, self-renewal and differentiation of stem cells.
- Hematopoietic stem cells (HSCs) have been most extensively studied and are known to express adhesion molecules of the cadherin family, integrin family, immunoglobulin superfamily, CD44 family and sialomucin family.
- Mesenchymal stem cells (MSCs) have the capacity to differentiate into mesodermal, endodermal and ectodermal lineages and can also proliferate extensively *in vitro* and suppress the immuno-response. The cells express immunoglobulin superfamily (ICAM-1, ICAM-3, VCAM-1, ALCAM), CD44 and P-selectin, etc.
- The purification of neural stem cells (NSCs), hepatic stem cells (HpSCs) and hair follicle stem cells (HFSCs) remains inadequate despite extensive studies.
- Cadherin and integrin molecules are expressed by NSCs, HpSCs and HFSCs and contribute to the maintenance of stemness, the proliferation and the migration of these stem cells.

Acknowledgments

We thank Mr. H. Eastwick-Field and Ms. K. Ando for manuscript preparation.

Abbreviations

ALCAM (CD166)	activated leukocyte cell adhesion molecule
EpCAM (CD326)	epithelial cell adhesion molecule
GFAP	glial fibrillary acidic protein
HFSCs	hair follicle stem cells
HGF	hepatocyte growth factor
HPCs	hematopoietic progenitor cells

HpSCs	hepatic stem cells
HSCs	hematopoietic stem cells
ICAM-1 (CD54)	intercellular adhesion molecule 1
ICAM-3 (CDw50)	intercellular adhesion molecule 3
IL-6	interleukin-6
LFA-1 (CD11a/CD18)	lymphocyte function-associated antigen 1
M-CSF	macrophage-colony stimulating factor
MSCs	mesenchymal stem cells
NCAM (CD56)	neural cell adhesion molecule
NSCs	neural stem cells
PDGF	platelet-derived growth factor
P-HSCs	pluripotent hematopoietic stem cells
PSA-NCAM (CD56)	polysiaric acid-binding neural cell adhesion molecule
SCF	stem cell factor
SSEA-4	stage-specific embryonic antigen 4
SVZ	subventricular zone of lateral ventricles
TGFβ	transformation growth factor β
Tie2	tyrosine kinases that contain immunoglobulin-like loops and epidermal growth factor-similar domain 2
VCAM-1 (CD106)	vascular cell adhesion molecule 1
VLA-4 (CD49d/CD29)	very late antigen 4
VLA-5 (CD49e/CD29)	very late antigen 5

Definition of Terms

Extracellular matrix: A meshwork of proteins occupying the space between cells and contributing to cell structure, cell adhesion, and other important functions.

Graft-versus-host disease: A complication after allogeneic hemopoietic cell transplantation in which donor functional immune cells recognize the recipient as "foreign" and attack the recipient tissues.

Homing: A process by which stem cells exit blood circulation and migrate to (home-in on) their niche in the extravascular space of tissues.

Pluripotent stem cells: Cells having the ability to self-renew (generating new pluripotent stem cells) and to differentiate into multilineage cells.

Rolling: The first step of homing process. Cells in the bloodstream interact with endothelial cells in the blood vessel wall and then roll along the endothelial cell surface, followed by their firm adhesion to the endothelial cells.

Stem cell niche: A special area in tissue and organs, where stem cells reside and their stemness is maintained and controlled.

References

Belvindrah, R. and S. Hankel, J. Walker, B.L. Patton, and U. Müller. 2007. β1 integrins control the formation of cell chain in the adult rostal migratory stream. J. Neurosci. 27: 2704-2717.

Calvi, L.M. and G.B. Adams, K.W. Weibrecht, J.M. Weber, D.P. Olson, M.C. Knight, R.P. Martin, E. Schipani, P. Divieti, F.R. Bringhurst, L.A. Milner, H.M. Kronenberg, and D.T. Scadden. 2003. Osteoblastic cells regulate the hematopoietic stem cell niche. Nature 425: 841-846.

Campos, L.S. and L. Decker, V. Taylor, and W. Skarnes. 2006. Notch, epidermal growth factor receptor, and B1-integrin pathway are coordinated in neural stem cells. J. Biol. Chem. 281: 5300-5309.

Chazal, G. and P. Durbec, A. Jankovski, G. Rougon, and H. Cremer. 2000. Consequence of neural cell adhesion molecule deficiency on cell migration in the rostral migratory stream of the mouse. J. Neurosci. 20: 1446-1457.

Cheng, J. and S. Baumhueter, G. Cacalano, K. Carver-Moore, H. Thibodeaux, R. Thomas, H.E. Broxmeyer, S. Cooper, N. Hague, M. Moore, and L.A. Lasky. 1996. Hematopoietic defects in mice lacking the sialomucin CD34. Blood 87: 479-490.

Cotsarelis, G. and T.T. Sun, and R.M. Lavker. 1990. Label-retaining cells reside in the bulge area of pilosebaceous unit: implications for follicular stem cells, hair cycle, and skin carcinogenesis. Cell 61: 1329-1337.

Fackler, M.J. and D.S. Krause, O.M. Smith, C.I. Civin, and W.S. May. 1995. Full length but not truncated CD34 inhibits hematopoietic cell differentiation of M1 cells. Blood 85: 3040–3047.

Gang, E.J. and D. Bosnakovski, C.A. Figueiredo, J.W. Visser, and R.C.R. Perlingerio. 2007. SSEA-4 identifies mesenchymal stem cells from bone marrow. Blood 109: 1743-1751.

Gangenahalli, G.U. and V.K. Singh, Y.K. Verma, P. Gupta, R.K. Sharma, R. Chandra, and P.M. Luthra. 2006. Hematopoietic stem cell antigen CD34: role in adhesion or homing. Stem Cells Dev. 15: 305-313.

Ip, J.E. and Y. Wu, J. Huang, L. Zhang, R.E. Pratt, and V.J. Dzau. 2007. Mesenchymal stem cells use integrin 1 not CXC chemokine receptor4 for myocardial migration and engraftment. Mol. Biol. Cell. 18: 2873-2882.

Kato, J. and H. Hisha, X. Wang, T. Mizokami, S. Okazaki, Q. Li, C.Y. Song, M. Maki, N. Hosaka, Y. Adachi, M. Inaba, and S. Ikehara. 2008. Contribution of neural cell adhesion molecule (NCAM) to hemopoietic system in monkeys. Ann. Hematol. 87: 797-807.

Kloepper, J.E. and S. Tiede, J. Brinckmann, D.P. Reinhardt, W. Meyer, R. Faessler, and R. Paus. 2008. Immunophenotyping of the human bulge region: the quest to define useful in situ markers for human epithelial hair follicle stem cells and their niche. Exp. Dermatol. 17: 592-609.

Menn, B. and J.M. Garcia-Verdugo, C. Yaschine, O. Gonzalez-Perez, D. Rowitch, and A. Alvarez-Buylla. 2006. Origin of oligodendrocytes in the subventricular zone of the adult brain. J. Neurosci. 26: 7907-7918.

Potocnik, A.J. and C. Brakebusch, and R. Fassler. 2000. Fetal and adult hemopoietic stem cells require β1 integrin function for colonizing fetal liver, spleen, and bone marrow. Immunity 12: 635-663.

Ruster, B. and S. Gottig, R.J. Ludwig, R. Bistrian, S. Muller, E. Seifried, J. Gille, and R. Henschler. 2006. Mesenchymal stem cells display coordinated rolling and adhesion behavior on endothelial cells. Blood 108: 3938-3944.

Sugiyama, T. and H. Kohara, M. Noda, and T. Nagasawa. 2006. Maintenance of the hemopoietic stem cell pool by CXCL12-CXCR4 chemokine signaling in bone marrow stromal cell niches. Immunity 25: 977-988.

Trempus, C.S. and R.J. Morris, C.D Bortner, G. Cotsarelis, R.S. Faircloth, J.M. Reece, and R.W. Tennant. 2003. Enrichment for living murine keratinocytes from the hair follicle bulge with the cell surface marker CD34. J. Invest. Dermatol. 120: 501-511.

Tumbar, T. and G. Guasch, V. Greco, C Blanpain, W.E. Lowry, M. Rendl, and E. Fuchs. 2004. Defining the epithelial stem cell niche in skin. Science 303: 359-363.

Wang, J. and T. Kimura, R. Asada, S. Harada, S. Yokota, Y. Kawamoto, Y. Fujimura, T. Tsuji, S. Ikehara, and Y. Sonoda. 2003. SCID-repopulating cell activity of human cord blood-derived CD34- cells assured by intra-bone marrow injection. Blood 101: 2924-2931.

Wang, X. and H. Hisha, S. Taketani, Y. Adachi, Q. Li, W. Cui, Y. Cui, J. Wang, C. Song, T. Mizokami, S. Okazaki, Q. Li, T. Fan, H. Fan, Z. Lian, M.E. Gershwin, and S. Ikehara. 2006. Characterization of mesenchymal stem cells isolated from mouse fetal bone marrow. Stem Cells 24: 482-493.

Winter, M.J. and B. Nagelkerken, A.E. Mertens, H.A. Rees-Bakker, I.H. Briaire-de Bruijn, and S.V. Litvinov. 2003. Expression of EP-CAM shifts the state of cadherin-mediated adhesion from strong to week. Exp. Cell Res. 285: 50-58.

Yovchev, M.I. and P.N. Grozdanov, B. Joseph, S. Gupta, and M.D. Dabeva. 2007. Novel hepatic progenitor cell surface markers in the adult liver. Hepatology 45: 139-149.

Zhang J. and C. Niu, L. Ye, H. Huang, X. He, W.G. Tong, J. Ross, J. Haug, T. Johnson, J.Q. Feng, S. Harris, L.M. Wiedemann, Y. Mishina, and L. Li. 2003. Identification of the hemopoietic stem cell niche and control of the niche size. Nature 425: 836-841.

Zhang. L. and N. Theise, M. Chua and L.M. Reid. 2008. The stem cell niche of human livers: Symmetry between development and regeneration. Hepatology 48: 1598-1607.

Zhu, H. and N. Mitsuhashi, A. Klein, L.W. Barsky, K. Weinberg, M.L. Barr, A. Demetriou, and G.D. Wu. 2006. The role of the hyaluronan receptor CD44 in mesenchymal stem cell migration in the extracellular matrix. Stem Cells 24: 928-935.

CHAPTER 9

Adhesion Molecules and Neutrophil Migration

Beth A. McCormick

Department of Molecular Genetics and Microbiology, University of Massachusetts
School of Medicine, 55 Lake Avenue North, Worcester, MA.
E-mail: Beth.McCormick@umassmed.edu
Alternative contact: E-mail: Karen.Mumy@umassmed.edu

ABSTRACT

Neutrophil infiltration of mucosal surfaces is a common event in many disease states and is a critical component of the host defense response. Activation and migration of neutrophils into tissues also contribute to inflammatory tissue injury and remodeling of tissue architecture. For example, in the kidney and bladder, the migration of neutrophils across tubular or transitional epithelia accompanies pyelonephritis and cystitis, respectively. In inflammatory pulmonary disorders, acute inflammation of the airway is characterized by the transmigration of neutrophils across the bronchial epithelium. In the alimentary tract, active inflammatory disease is characterized by the migration of neutrophils across the epithelial lining. While these disorders are diverse with distinct causes and pathologies, the underlying immune response to infection is very similar. In most instances, there is a large and often excessive influx of neutrophils.

This inflammatory process consists of several steps: initial emigration of neutrophils from the microcirculation, subsequent migration of neutrophils across the subepithelial matrix, and finally transepithelial migration. Thus, neutrophils must undergo specific and highly coordinated interactions with the endothelium, as well as epithelium. Knowledge of the mechanisms of neutrophil migration has

---

Key terms are defined at the end of the chapter.

greatly expanded in recent years and has significant clinical implications. This work highlights general mechanisms of the adhesive interactions involved in the migration of neutrophils to sites of injury.

INTRODUCTION

Polymorphonuclear leukocyte (neutrophil) migration across the vessel walls of the endothelium and into the surrounding tissue is not only a characteristic feature of the inflammatory response during host defense (site of injury or infection) but also occurs under pathological conditions critical to both acute and chronic inflammatory disease states. It is well recognized that neutrophil migration from the vasculature occurs as a multi-step process, governed by the sequential activation of adhesion proteins and their ligands on both neutrophils and the endothelium (Ley *et al.* 2007, Zarbock and Ley 2008). Initially, this process begins by the tethering to and rolling of neutrophils on the vascular epithelium in response to locally produced pro-inflammatory mediators (cytokines, chemokines, tumor necrosis factor (TNF)-α, interleukin (IL)-1, and lipopolysaccharide (LPS)). Additional stimulation by endothelial bound chemokines results in the rapid activation of β_1 and β_2 integrins, leading to the arrest of neutrophils on the endothelium and subsequent transmigration into tissues. However, central to the function of neutrophils in host defense is their ability to diapedes across the microvasculature and migrate through the tissue to the site of infection. A critical initial step in this process is the attachment of the neutrophils to the basolateral surface of the epithelium. Once the neutrophils have firmly adhered, they begin to migrate across the epithelial monolayer by using a paracellular route. It is also during this leg of the journey that unrestrained activation of neutrophils may result in the release of cytotoxic compounds that can cause harm to neighboring cells.

The molecular mechanisms underlying neutrophil transendothelial migration have been well characterized and are far better understood than the events that mediate neutrophil transepithelial migration. Nevertheless, what is evident is that distinct mechanisms govern the migration of neutrophils across endothelial surfaces as compared to epithelial surfaces. From an anatomical perspective, such differences are perhaps predicted given that transendothelial migration of neutrophils occurs in the apical-to-basolateral direction, whereas neutrophil migration across epithelial surfaces proceeds in the basolateral-to-apical direction. In addition, neutrophil migration across the epithelium (≥ 20 μm in length) is significantly longer than migration across the endothelium ($\leq$ a few micrometers), suggesting that neutrophils interact very intimately with the epithelium (Parkos 1997). Another important anatomical difference is that neutrophil adhesion to the endothelium, but not the epithelium, is dependent on an environment of shear force caused by blood flow. Given such anatomical differences, it is not surprising that different adhesion interactions facilitate neutrophil movement across endothelial surfaces as compared to epithelial surfaces.

The recruitment of neutrophils into tissues is a pivotal step that underlies the pathogenesis of inflammatory diseases at mucosal surfaces as well as the host defense response to injury and infection. The mechanisms underlying the movement of neutrophils from the bloodstream to sites of infection or injury are complex and involve numerous adhesion molecules, cytokines, and chemoattractants that function to direct neutrophils through endothelial barriers, basement membranes, and epithelial barriers. There is substantial interest in defining the mechanisms and functional consequences of neutrophil-endothelial and neutrophil-epithelial interactions, as understanding the molecular mechanisms by which neutrophils interact at these critical surfaces may lead to the development of novel therapeutics. Thus, this chapter highlights the current understanding of the major adhesive interactions that control the fundamental innate immune response of neutrophil migration to sites of injury or infection, or during chronic states of inflammation.

GENERAL MECHANISMS UNDERLYING NEUTROPHIL TRANSENDOTHELIAL MIGRATION

During inactive states, circulating neutrophils roll along the walls of the blood vessels. However, once an appropriate stimulus is present, the neutrophils adhere firmly to endothelial cells and are primed to migrate from the bloodstream to the site of injury. Because a detailed description of this process is beyond the scope of this chapter, a brief overview highlighting the salient features of neutrophil migration from the endothelium is discussed. (Figure 1).

The initial attachment of neutrophils and their movement along vessel walls is due to reversible binding to transmembrane glycoproteins (termed selectins) on neutrophils and endothelial cells (Pettersen and Adler 2002, Ley *et al.* 2007, Zarbock and Ley 2008). L-selectin is expressed solely on leukocytes, and although this selectin is expressed constitutively, neutrophils do not express L-selectin uniformly. For instance, neutrophils that are newly released from the bone marrow contain higher levels of L-selectin than older, circulating neutrophils. L-selectin binding is also rapid and short-lived. Additionally, P-selectin and E-selectin participate in neutrophil-endothelial cell adhesion, but these molecules require an appropriate inflammatory stimulus for expression. As an example, P-selectin is stored intracellularly in Weibel-Palade bodies of endothelial cells and in α-granules of platelets but is very quickly mobilized to the endothelial cell surface upon exposure to an appropriate inflammatory mediator (i.e., histamine, thrombin, oxidants). Once at the endothelial cell surface, P-selectin is available to interact with P-selectin glycoprotein ligand (PSGL)-1, a disulfide-bonded homodimer that is expressed on rolling neutrophils and is capable of binding two P-selectin ligands simultaneously. P-selectin binding causes slower rolling velocities, which results in the eventual tethering of neutrophils to vessel surfaces. Conversely, E-selectin is stimulated by cytokines (i.e., TNF-α, LPS, and endotoxins) and requires *de novo*

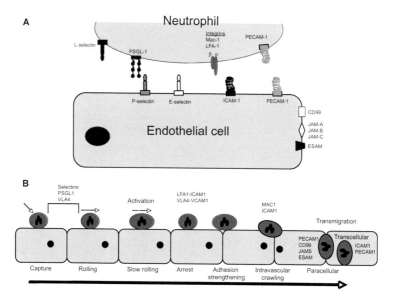

Fig. 1 The neutrophil adhesion cascade. A. Diagram of the major adhesion molecules and respective ligands involved in neutrophil–endothelial interactions during diapedesis. B. The key steps involved in neutrophil migration across the endothelium. This process begins by capture of the neutrophils to the wall of the blood vessel, followed by rolling (regulated by selectins), slow rolling, and then arrest, which is mediated by β2 integrins. Subsequent to arrest, the neutrophils undergo adhesion strengthening and intravascular crawling. Lastly, neutrophils diapedes across the endothelium by either a paracellular route (mediated by CD99, JAMs, ESAM, or PECAM-1) or a transcellular route (ICAM-1, PECAM-1). Adapted from Ley *et al.* (2007) and Wagner and Roth (2000). JAMs, junctional adhesion molecules; ESAM, endothelial cell-selective adhesion molecule; PECAM-1, platelet adhesion molecule 1; ICAM-1, intracellular adhesion molecule 1.

mRNA and protein synthesis. It is inferred that E-selectin is responsible for maintaining neutrophil rolling after downregulation of P-selectin. Interestingly, L-selectin and P-selectin require shear stress to support adhesion (Yago *et al.* 2007), which may explain why selectins are not involved in neutrophil-epithelial interactions.

Subsequent to rolling is 'slow rolling', a process mediated by selectin-triggered signaling (Ley *et al.* 2007, Zarbock and Ley 2008, Zemans *et al.* 2009). Slower rolling velocities prolong the time of neutrophil interaction with endothelial cells, which in turn conducts the proper activation of neutrophils and successful arrest. The arrest of neutrophils at the endothelial surface is an essential step and involves the interaction of β_1 and β_2 integrins and their cognate binding partners. Thus, when rolling neutrophils receive the appropriate trigger through selectin and/or chemokine receptor engagements, integrin activation is initiated. Integrins are a group of heterodimeric transmembrane glycoproteins that are expressed on neutrophils and other hematopoietic cells, and function to coordinate cell-to-cell and cell-to-extracellular matrix adhesions. There are 16 α subunits and 8 β subunits that give rise to a variety of integrins. In the case of β_2 integrins, there is only a common β chain (CD18) and a single, but variable, α chain (CD11a, b, c, or d). Only macrophage antigen (Mac)-1 (i.e., αMβ2 and CD11b/CD18) and lymphocyte-associated function antigen (LFA)-1 (αLβ2 and CD11a/CD18) mediate neutrophil binding to the endothelium. Neutrophils store Mac-1 in specific gelatin granules and in secretory vesicles, but Mac-1 can be rapidly mobilized to the cell surface in response to various inflammatory mediators, such as the bacterial peptide formyl-methionyl-leucyl-phenylalanine (fMLP). An important endothelial ligand for Mac-1 is intracellular adhesion molecule 1 (ICAM-1), which belongs to the Ig superfamily. ICAM-1 exhibits low constitutive expression on endothelial cell membranes but it is markedly induced by exposure to cytokines. LFA-1 can also bind to ICAM-1 but it has a higher affinity to the related protein ICAM-2. In addition, other members of the Ig superfamily are involved in leukocyte adhesion. For example, vascular cell adhesion molecule (VCAM)-1 binds to very late antigen (VLA)-4 on activated human and rat neutrophils.

The next stage in neutrophil emigration from the endothelium involves adhesion strengthening, which is mediated by integrins through 'outside-in' signaling pathways (Ley *et al.* 2007). It is well recognized that integrins generate intracellular signals that regulate various cellular functions, including cell motility, proliferation, and apoptosis (Shattil 2005). Ligand-induced integrin clustering and allosteric conformational changes likely contribute to the initiation of outside-in signaling and the formation of signalosomes, which are required for the efficient recruitment of protein tyrosine kinases and the initiation of the full repertoire of signaling pathways (Liu *et al.* 2002a). For instance, recent studies suggest that the induction of conformational changes to the cytosolic tail of LFA-1 heterodimer upon ICAM-1 binding may play a role in the rapid arrest of leukocytes under flow (Shamri *et al.* 2005). Two SRC-like protein tyrosine kinases (FGR and hemopoietic cell kinase) are important inducers of outside-in signaling by LFA-1 and Mac-1 (Giagulli *et al.* 2006). It is also inferred that the lack of outside-in signaling mediated by the β_2 integrin chain greatly accelerates the detachment of adherent neutrophil under flow (Giagulli *et al.* 2006). For further details regarding outside-in signaling in the adhesion-strengthening step involved in neutrophil emigration, see Ley *et al.* (2007) for review.

Intravascular crawling follows the adhesion-strengthening step, and this step results in the migration of neutrophils across the endothelium (Ley *et al.* 2007). This stage of neutrophil migration is known as transendothelial migration and involves several adhesion molecules that regulate paracellular trafficking, such as CD99, platelet adhesion molecule 1 (PECAM-1), as well as junctional adhesion molecules (JAM-A, JAM-B, JAM-C), and endothelial cell-selective adhesion molecule (ESAM) (Muller, 2003). Indeed, JAM-A is a ligand for CD11a/CD18 (Ostermann *et al.* 2002), whereas JAM-C is a ligand for CD11b/CD18 (Santoso *et al.* 2002). However, the underlying mechanisms by which these molecules mediate the process of transendothelial migration are not well understood. Quite remarkably, neutrophils can also take a transcellular route across the endothelium and this process involves ICAM-1, PECAM-1, and caveolins (Feng *et al.* 1998, Engelhardt and Wolburg 2004).

GENERAL MECHANISMS UNDERLYING NEUTROPHIL TRANSEPITHELIAL MIGRATION

Much of our understanding with regard to the adhesive interactions that govern neutrophil transepithelial migration are derived from studies using models of intestinal epithelia, and the adhesion molecules involved in such migration of neutrophils across epithelial barriers has been investigated using imposed gradients of N-formyl-methionyl-leucyl-phenylalanine (fMLP) as a chemoattractant. Unlike the interactions at the endothelium, selectins do not mediate neutrophil adherence to the epithelium. Rather, the first step in this process is the adherence of neutrophils to the basolateral surface involving β_2 integrins (Parkos *et al.* 1991). Although CD11b/CD18 has been described as a key neutrophil adhesion molecule involved in the initial adherence to the epithelial basolateral surface, CD18-independent mechanisms of neutrophil transepithelial migration have also been documented (Vedder and Harlan, 1988, Blake *et al.* 2004). Thus, it appears that CD18 dependency, which underlies the initial event of neutrophil adhesion, may be restricted to certain tissues (i.e., the intestine exhibits CD18 dependency, whereas the lung does not) and may also depend on the particular neutrophil chemoattractant encountered (see the next section) (Zemans *et al.* 2009). CD44, a glycosylated membrane receptor, is another neutrophil adhesion molecule that plays a role in neutrophil migration across epithelial monolayers whereupon activation this molecule negatively regulates the transepithelial migration of neutrophils (Si-Tahar *et al.* 2001). The molecular mechanism involved in such negative signaling following its activation may include modulation of outside-in cell signaling.

Although the epithelial counter-receptor for CD11b/CD18 on the basolateral epithelial surface has yet to be determined, other epithelial ligands for neutrophil β_2 integrins have been recognized to play an important role in events that govern neutrophil transepithelial migration. For instance, junctional adhesion molecules (JAMs) serve as epithelial ligands for β_2 integrins and JAM-C, in particular, plays an essential role in facilitating neutrophil movement across the epithelium, where it binds to neutrophil CD11b/CD18 and mediates neutrophil adhesion and subsequent migration. However, JAM-C is expressed at the level of the desmosomes, thus its binding to CD11b/CD18 occurs at a site distal to the initial adhesive interaction at the epithelial basolateral surface (Ostermann *et al.* 2002, Chavakis *et al.* 2004).

As CD11b/CD18 can also bind to carbohydrates it is not surprising that several carbohydrate interactions are recognized to be involved in neutrophil transepithelial migration. However, the role of carbohydrates in neutrophil transmigration occurs after firm adhesion has been established. Fucoidin, for example, was found to potently inhibit epithelial cell binding to CD11b/CD18, implying that fucosylated proteoglycans are expressed on the epithelial cell surface (Colgan *et al.* 1995). At present, however, the specific molecules that bind to CD11b/CD18 have yet to be identified.

As neutrophils migrate across epithelial monolayers, they exclusively travel paracellularly. An important mediator of this transmigration step is CD47, which is expressed on the basolateral surface of epithelial monolayers as well as on neutrophils (Parkos *et al.* 1996). While both the epithelial cell and neutrophil CD47 are involved in directing neutrophil migration across the epithelium, the role of epithelial CD47 in facilitating this migration event has yet to be determined. However, during transepithelial migration, neutrophil CD47 has been found to be redistributed to the neutrophil cell membrane where subsequent ligation of CD47 triggers downstream signaling pathways that enhance neutrophil migration by modifying the epithelial cell cytoskeleton. In addition, CD47 binds SIRPα *in cis* and such interactions, to a certain extent, regulate neutrophil transepithelial migration. For example, SIRPα interacts with SHP-1 and SHP-2 and it is inferred that these interactions induce signal transduction cascades that enhance the rate of neutrophil transepithelial migration. Moreover, SIRPα family members may also mediate neutrophil transepithelial migration in a CD47-independent fashion as SIRPα1, which is not a CD47 ligand, regulates neutrophil transmigration (Liu *et al.* 2002b). It has also been proposed that neutrophil SIRPα binds epithelial CD47 *in trans*, which may also initiate the responses in the epithelium to facilitate paracellular movement of the neutrophils. Adding yet another layer of complexity, neutrophil transepithelial migration is additionally mediated by JAM-like protein (JAML) (Moog-Lutz *et al.* 2003). Recent studies have found that JAML binds to the coxsackie and adenovirus receptor (CAR), an Ig superfamily receptor expressed at epithelial tight junctions (Zen *et al.* 2005).

Once the neutrophils have completely traversed the epithelial monolayer they are retained on the apical surface of the epithelium and then cleared. Retention of neutrophils at the epithelial cell surface serves as a defense barrier, which allows for the destruction of invading microorganisms. Thus, neutrophils must participate in adhesive interactions at the apical epithelial surface. As ICAM-1 is expressed at the apical surface of epithelial cells and this adhesion molecule is a known ligand for CD11b/CD18, it is tempting to speculate that ICAM-1 tethers neutrophils at mucosal surfaces. Lastly, neutrophils must be cleared from the apical surface of epithelial cells since prolonged exposure to activated neutrophils may be detrimental to the host. To this end, decay accelerating factor (DAF, CD55) functions to clear neutrophils at the mucosal surface. DAF is a complement regulatory protein that plays a crucial role in inhibiting complement-mediated cell lysis and is highly expressed on the epithelial apical surface of at least the lung and intestine. In addition, DAF most likely competes with ICAM-1 to dislodge neutrophils away from the epithelial surface (Louis *et al.* 2005). CD97 on neutrophils has also been shown to bind to DAF (Zemans *et al.* 2009).

ADHESION MOLECULES AND NEUTROPHIL CHEMOATTRACTANTS

Recruitment of neutrophils across epithelial surfaces, such as the intestinal epithelium, is dependent not only on specific adhesion molecules but also on neutrophil chemoattractants that diffuse from the intestinal lumen (Figure 2). An intriguing and emerging concept is that neutrophils utilize distinct adhesion molecules during transepithelial migration depending on the chemoattractant gradient imposed. Early studies determined that many neutrophil chemoattractants have the ability to modulate the expression of certain adhesion molecules. For example, fMLP has been shown to cause an increase in the expression of CD18/ CD11b on the surface of neutrophils (Vedder and Harlan, 1988). More recently, Blake *et al.* (2004) demonstrated that although neutrophil migration across T84 cell monolayers in response to an imposed gradient of fMLP is dependent upon CD18/ CD11b, neutrophils are capable of migrating across T84 monolayers in a CD18-independent manner in response to gradients of alternative chemoattractants such as IL-8, complement component C5a, and leukotriene (LT)B$_4$. Such observations underscore a versatility by which neutrophils may utilize distinct adhesion molecules depending on the chemoattractant gradient sensed during transepithelial migration.

In support of this emergent view, our group has established a discreet role for the neutrophil chemoattractant hepoxilin A$_3$ (HXA$_3$) in the process of neutrophil recruitment (Pazos *et al.* 2008). HXA$_3$ is an eicosanoid derived from the 12-lipoxygenase pathway, which is produced by epithelial cells and secreted from the apical surface, resulting in the guidance of neutrophils across the epithelial monolayer from the basolateral to the apical side. In studying the adhesion molecules required for neutrophil transepithelial migration to imposed gradients of HXA$_3$, we found that the adhesion interaction profile of neutrophil transepithelial migration in response to HXA$_3$ differs from that exhibited by the structurally related eicosanoid LTB$_4$, suggesting that these neutrophil chemoattractants play fundamentally different roles in the recruitment process (Hurley *et al.* 2008). Furthermore, unique to neutrophil transepithelial migration induced by gradients of HXA$_3$, but not LTB$_4$ or fMLP, was the critical dependency of four major surface adhesion molecules (CD18, CD47, CD44, and CD55). While neutrophil migration

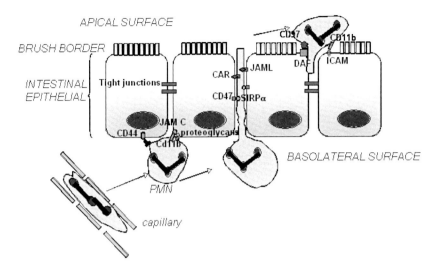

Fig. 2 Adhesive interactions involved in neutrophil migration across epithelial surfaces. Neutrophils adhere to the basolateral surface through the ligation of CD11b/CD18 to several molecules, including fucosylated glycoproteins and JAMs. However, the epithelial ligand to CD11b/CD18 has yet to be identified. Following adhesion, neutrophils transmigrate across the epithelium by taking a paracellular route that is primarily mediated by CD47. Neutrophil JAML also binds CAR. Once the neutrophils have completely migrated across the cell monolayer, they adhere to the epithelial surface. This process is mediated by the binding of CD11b/CD18 to ICAM-1 and DAF to CD97. JAM, junctional adhesion molecules; CAR, coxsackie and adenovirus receptor; ICAM-1, intracellular adhesion molecule 1; DAF, decay accelerating factor; CD, cellular differentiation number.

across intestinal epithelial cell monolayers has served as the model for the study of neutrophil–epithelial interactions, we also observed that both the type of epithelial cell monolayer and the particular chemoattractant gradient imposed play important roles in the adhesion molecules involved in transepithelial migration (Hurley *et al.* 2008). Specifically, our studies showed that neutrophil transepithelial migration across airway epithelial cells (as compared to intestinal epithelial cells) was more dependent upon CD44 and CD47 in response to imposed gradients of fMLP and LTB$_4$ (Hurley *et al.* 2008). Thus, our findings imply that the particular chemoattractant gradient imposed, together with the type of epithelial monolayer, somehow contribute to determining which adhesion molecules are involved in transepithelial migration of neutrophils.

NEUTROPHIL ADHESION IN HEALTH AND DISEASE

Inflammation is a localized protective response in vascular tissues induced by microbial infection or cell and tissue injury, and the neutrophil is a key effector cell in this process. Under normal conditions, the role of the neutrophil as a first line of defense is to kill and eliminate endogenous and exogenous stimuli such as airborne pollutants, allergens, and microbes by a variety of mechanisms (i.e., phagocytosis, respiratory burst, the release of cytotoxic mediators and proteases). However, many bacterial infections and a variety of human diseases in their active stages are characterized by the migration of neutrophils across epithelial surfaces. Thus, the transmigration of neutrophils across epithelial surfaces represents a shared phenomenon among a diverse array of inflammatory mucosal conditions. For instance, salmonellosis, shigellosis, and pneumonia, as well as autoimmune/idiopathic states such as Crohn's disease, ulcerative colitis, and bronchitis all culminate in the destructive breach of the protective outer epithelium by activated PMNs. Furthermore, the severity and clinical outcome of these inflammatory diseases correlate with the extent of PMN infiltration.

Expanding the knowledge base of the adhesion molecules involved in the process of neutrophil recruitment will ultimately lead to new targets for pharmacological intervention. The initial work in this regard has been encouraging, as several therapeutic manipulations of the adhesion and migratory processes have been successful in animal models. Furthermore, developmental endothelial locus (Del)-1 has recently been reported to be an anti-adhesive factor that interferes with LFA-1-dependent leukocyte-endothelial adhesion, and thus could provide a basis for targeting leukocyte-endothelial interactions in disease (Choi *et al.* 2008). However, further development is required before such studies can translate and be maximally effective in humans.

SUMMARY

- The primary objective of neutrophils as first responders is to phagocytose and destroy pathogenic microorganisms.
- Neutrophils represent an important effector of the acute (and chronic) inflammatory response.
- A common feature underlying active states of inflammation is the migration of neutrophils from the circulation and across a number of tissue barriers in response to pathogenic insult or injury.
- A complex series of neutrophil adhesive and de-adhesive events with both the endothelium and epithelium allows for successful migration.
- As the endothelium and epithelium offer potential direct targets for therapy, a better understanding of such adhesive interactions with neutrophils will lead to the discovery of novel anti-inflammatory drugs.

Abbreviations

CAR	coxsackie and adenovirus receptor
CD	cellular differentiation number
DAF	decay accelerating factor
ESAM	endothelial cell-selective adhesion molecule
fMLP	formyl-methionyl-leucyl-phenylalanine
HXA_3	hepoxilin A_3
ICAM-1	intracellular adhesion molecule 1
IL	interleukin
JAM	junctional adhesion molecules
LFA	lymphocyte-associated function antigen
LPS	lipopolysaccharide
LT	leukotriene
Mac	macrophage antigen
neutrophil	polymorphonuclear leukocyte
PECAM-1	platelet adhesion molecule 1
PSGL	P-selectin glycoprotein ligand
TNF	tumor necrosis factor
VCAM -1	vascular cell adhesion molecule 1
VLA	very late antigen

Key Features of Neutrophil Migration

1. Neutrophils are first captured to the wall of the blood vessel.
2. Once captured, neutrophils roll along the vessel wall (mediated by selectins) until they become arrested (mediated by $\beta2$ integrins).
3. Neutrophils then undergo adhesion strengthening and intracellular crawling.
4. Neutrophils diapedes across the endothelium, travel through the extracellular matrix, and arrive at the base of the tissue.
5. Neutrophils bind to the basolateral surface of epithelial cells and transmigrate across the epithelium.
6. Once the neutrophils have completely transmigrated across the epithelium they adhere to the epithelial surface.

Definition of Terms

Cytokines: A soluble, hormone-like protein produced by white blood cells that acts as a messenger between cells.

Epithelium: The outside layer of cells that covers all the free, open surfaces of the body including the skin, and mucous membranes that communicate with the outside of the body.

Endothelium: The thin layer of cells that line the interior surface of blood vessels, forming an interface between circulating blood in the lumen and the rest of the vessel wall.

Inflammation: The complex biological response of vascular tissues to harmful stimuli, such as pathogens, damaged cells, or irritants. It is a protective attempt by the organism to remove the injurious stimuli as well as initiate the healing process for the tissue. Inflammation can be classified as either acute or chronic.

Neutrophil: The most abundant type of white blood cell in humans, forming an essential part of the immune system as the first line of defense. Neutrophils react within an hour of tissue injury and are the hallmark of acute inflammation.

References

Blake, K.M. and S.O. Carrigan, A.C. Issekutz, and A.W. Stadnyk. 2004. Neutrophils migrate across intestinal epithelium using beta2 integrin (CD11b/CD18)-independent mechanisms. Clin. Exp. Immunol. 136: 262-268.

Chavakis, T. and T. Keiper, R. Matz-Westphal, K. Hersemeyer, U.J. Sachs, P.P. Nawroth, K.T. Preissner, and S. Santoso. 2004. The junctional adhesion molecule-C promotes neutrophil transendothelial migration in vitro and in vivo. J. Biol. Chem. 279: 55602-55608.

Choi, E.Y. and E. Chavakis, M.A. Czabanka, H.F. Langer, L. Fraemohs, M. Economopoulou, R.K. Kundu, A. Orlandi, Y.Y. Zheng, D.A. Prieto, C.M. Ballantyne, S.L. Constant, W.C. Aird, T. Papayannopoulou, C.G. Gahmberg, M.C. Udey, P. Vajkoczy, T. Quertermous, S. Dimmeler, C. Weber, and T. Chavakis. 2008. Del-1, an endogenous leukocyte-endothelial adhesion inhibitor, limits inflammatory cell recruitment. Science 322: 1101-1104.

Colgan, S.P. and C.A. Parkos, D. McGuirk, H.R. Brady, A.A. Papayianni, G. Frendl, and J.L. Madara. 1995. Receptors involved in carbohydrate binding modulate intestinal epithelial–neutrophil interactions. J. Biol. Chem. 270: 10531-10539.

Engelhardt, B. and H. Wolburg. 2004. Mini-review: Transendothelial migration of leukocytes: through the front door or around the side of the house? Eur. J. Immunol. 34: 2955-2963.

Feng, D. and J.A. Nagy, K. Pyne, H.F. Dvorak, and A.M. Dvorak. 1998. Neutrophils emigrate from venules by a transendothelial cell pathway in response to FMLP. J. Exp. Med. 187: 903-915.

Giagulli, C. and L. Ottoboni, E. Caveggion, B. Rossi, C. Lowell, G. Constantin, C. Laudanna, and G. Berton. 2006. The Src family kinases Hck and Fgr are dispensable for inside-out, chemoattractant-induced signaling regulating beta 2 integrin affinity and valency in neutrophils, but are required for beta 2 integrin-mediated outside-in signaling involved in sustained adhesion. J. Immunol. 177: 604-611.

Hurley, B.P. and A. Sin, and B.A. McCormick. 2008. Adhesion molecules involved in hepoxilin A3-mediated neutrophil transepithelial migration. Clin. Exp. Immunol. 151: 297-305.

Ley, K. and C. Laudanna, M.I. Cybulsky, and S. Nourshargh. 2007. Getting to the site of inflammation: the leukocyte adhesion cascade updated. Nat. Rev. Immunol. 7: 678-689.

Liu, S. and W.B. Kiosses, D.M. Rose, M. Slepak, R. Salgia, J.D. Griffin, C.E. Turner, M.A. Schwartz, and M.H. Ginsberg. 2002a. A fragment of paxillin binds the alpha 4 integrin cytoplasmic domain (tail) and selectively inhibits alpha 4-mediated cell migration. J. Biol. Chem. 277: 20887-20894.

Liu, Y. and H.J. Buhring, K. Zen, S.L. Burst, F.J. Schnell, I.R. Williams, and C.A. Parkos. 2002b. Signal regulatory protein (SIRPalpha), a cellular ligand for CD47, regulates neutrophil transmigration. J. Biol. Chem. 277: 10028-10036.

Louis, N.A. and K.E. Hamilton, T. Kong, and S.P. Colgan. 2005. HIF-dependent induction of apical CD55 coordinates epithelial clearance of neutrophils. Faseb J. 19: 950-959.

Moog-Lutz, C. and F. Cave-Riant, F.C. Guibal, M.A. Breau, Y. Di Gioia, P.O. Couraud, Y.E. Cayre, S. Bourdoulous, and P.G. Lutz. 2003. JAML, a novel protein with characteristics of a junctional adhesion molecule, is induced during differentiation of myeloid leukemia cells. Blood 102: 3371-3378.

Muller, W.A. 2003. Leukocyte-endothelial-cell interactions in leukocyte transmigration and the inflammatory response. Trends Immunol. 24: 327-334.

Ostermann, G. and K.S. Weber, A. Zernecke, A. Schroder, and C. Weber. 2002. JAM-1 is a ligand of the beta(2) integrin LFA-1 involved in transendothelial migration of leukocytes. Nat. Immunol. 3: 151-158.

Parkos, C.A. 1997. Cell adhesion and migration. I. Neutrophil adhesive interactions with intestinal epithelium. Am. J. Physiol. 273: G763-768.

Parkos, C.A. and S.P. Colgan, T.W. Liang, A. Nusrat, A.E. Bacarra, D.K. Carnes, and J.L. Madara. 1996. CD47 mediates post-adhesive events required for neutrophil migration across polarized intestinal epithelia. J. Cell. Biol. 132: 437-450.

Parkos, C.A. and C. Delp, M.A. Arnaout, and J.L. Madara. 1991. Neutrophil migration across a cultured intestinal epithelium. Dependence on a CD11b/CD18-mediated event and enhanced efficiency in physiological direction. J. Clin. Invest. 88: 1605-1612.

Pazos, M. and D. Siccardi, K.L. Mumy, J.D. Bien, S. Louie, H.N. Shi, K. Gronert, R.J. Mrsny, and B.A. McCormick. 2008. Multidrug resistance-associated transporter 2 regulates mucosal inflammation by facilitating the synthesis of hepoxilin A3. J. Immunol. 181: 8044-8052.

Pettersen, C.A. and and K.B. Adler. 2002. Airways inflammation and COPD: epithelial–neutrophil interactions. Chest 121: 142S-150S.

Santoso, S. and U.J. Sachs, H. Kroll, M. Linder, A. Ruf, K.T. Preissner, and T. Chavakis. 2002. The junctional adhesion molecule 3 (JAM-3) on human platelets is a counterreceptor for the leukocyte integrin Mac-1. J. Exp. Med. 196: 679-691.

Shamri, R. and V. Grabovsky, J.M. Gauguet, S. Feigelson, E. Manevich, W. Kolanus, M.K. Robinson, D.E. Staunton, U.H. von Andrian, and R. Alon. 2005. Lymphocyte arrest requires instantaneous induction of an extended LFA-1 conformation mediated by endothelium-bound chemokines. Nat. Immunol. 6: 497-506.

Shattil, S.J. 2005. Integrins and Src: dynamic duo of adhesion signaling. Trends Cell. Biol. 15: 399-403.

Si-Tahar, M. and S. Sitaraman, T. Shibahara, and J.L. Madara. 2001. Negative regulation of epithelium-neutrophil interactions via activation of CD44. Am. J. Physiol. Cell. Physiol. 280: C423-432.

Vedder, N.B. and and J.M. Harlan. 1988. Increased surface expression of CD11b/CD18 (Mac-1) is not required for stimulated neutrophil adherence to cultured endothelium. J. Clin. Invest. 81: 676-682.

Yago, T. and V.I. Zarnitsyna, A.G. Klopocki, R.P. McEver, and C. Zhu. 2007. Transport governs flow-enhanced cell tethering through L-selectin at threshold shear. Biophys. J. 92: 330-342.

Zarbock, A. and K. Ley. 2008. Mechanisms and consequences of neutrophil interaction with the endothelium. Am. J. Pathol. 172: 1-7.

Zemans, R.L. and S.P. Colgan, and G.P. Downey. 2009. Trans-Epithelial Migration of Neutrophils: Mechanisms and Implications for Acute Lung Injury. Am. J. Respir. Cell. Mol. Biol. 40: 519-535.

Zen, K. and Y. Liu, I.C. McCall, T. Wu, W. Lee, B.A. Babbin, A. Nusrat, and C.A. Parkos. 2005. Neutrophil migration across tight junctions is mediated by adhesive interactions between epithelial coxsackie and adenovirus receptor and a junctional adhesion molecule-like protein on neutrophils. Mol. Biol. Cell. 16: 2694-2703.

CHAPTER 10

Adhesion Molecules in Leukocytes and Their Reactivity

Maria Bokarewa and Piotr Mydel

Department of Rheumatology and Inflammation Research,
University of Göteborg, 41346 Sweden

ABSTRACT

Adhesion molecules belong to central regulators of leukocyte functions mediating cell–cell and cell–matrix interactions. In leukocytes, adhesion molecules mediate canonical functions participating in surface binding, transendothelial and interstitial migration. Additionally, each leukocyte type constitutively expresses or gains the expression of adhesion molecules characteristic for its functions. The effect on cell function depends on the type of adhesion molecule involved, the receptor partner it mobilizes, and the pathway of intracellular transduction it activates. Neutrophils and eosinophils have a limited panel of adhesion molecules which helps their rapid mobilization to the site of infectious or allergenic entrance, phagocytosis and degranulation. In monocytes, adhesion molecules facilitate phagocytosis and interstitial migration and support the broad differentiation capacity forming the subset of macrophages required in a particular tissue environment. Adhesion molecules on lymphocytes are key regulators of T and B cell maturation, tissue-specific trafficking and antigen presentation. The major groups of adhesion molecules identified in leukocytes include selectins together with their ligands mucins, and integrins with their specific ligands. Ligation of integrins by their counterparts or by the proteins of extracellular matrix containing RGD-motif induces conformation changes and clustering of integrins in cell membrane followed by intracellular signals (outside-in signaling). Stimulation of leukocytes through growth factor receptors, G-protein coupled receptors, and immunoreceptors potentiates the affinity of adhesion molecules and their reactivity (inside-out signaling). A net of intracellular signaling initiated by these receptors is mediated by a combination of adaptor molecules specific for each receptor followed by activation of PI3-kinase or protein tyrosine kinases. Coordination of inside and outside integrin signaling is proved to be crucial in the pathogenesis of acute and chronic inflammatory and metabolic conditions including sepsis, atherosclerosis, rheumatoid arthritis, multiple sclerosis and diabetes mellitus. Treatment modalities targeting adhesion molecules are on the way to becoming clinical reality.

CANONICAL FUNCTIONS OF ADHESION MOLECULES COMMON FOR ALL LEUKOCYTES

Leukocytes require morphological flexibility and rapid migratory behavior to fulfil their role as immunological cells. Adhesion molecules play a key role in controlling these activities. The first step is mobilization of leukocytes from bone marrow into the bloodstream. All circumstances and factors inducing leukocyte trafficking from bone marrow to peripheral blood are yet to be clarified. Expression of L-selectin has been proposed to serve as a retention factor for leukocyte in the bone marrow. Downregulation of L-selectin expression is indicated as the main factor regulating leukocyte egress out of the bone marrow (for more information about L-selectin see Table 1). During infection or allergic reaction, neutrophils and eosinophils mobilized from the bone marrow exhibit very low levels of L-selectin. Following bacterial challenge, L-selectin expression on the surface of neutrophils may be downregulated within minutes. Molecular basis of rapid downregulation of L-selectin is the proteolytic cleavage of its extracellular domain by disintegrins of ADAM family.

After leaving the bone marrow with the blood, monocytes, neutrophils and lymphocytes reach lymphoid and peripheral tissue and move toward their target. Velocity of migration differs between the cell types, being rapid in neutrophils and eosinophils, while lymphocyte migration is many times slower. Migration of leukocytes with the blood to peripheral tissues is directed by chemotaxis, a process of leukocyte polarization and migration to chemoattractant gradients (Friedl and Weigelin 2008). Various compounds including chemokines, cytokines, lipid mediators, bacterial fragments, and degradation products of extracellular matrix (ECM) induce chemotaxis (Bromley *et al.* 2008). Chemotactic signals are often transmitted through a family of heterodimeric G-protein coupled receptors (GPCRs, see below) including the fMLP (N-formyl-Met-Leu-Phe) receptor, complement receptors, multiple chemokine receptors including CCR7, CXCR4, CXCR5, and CCR3, and the leukotriene B4 receptor. Among the chemokines known to induce activation of β_1-, β_2- and β_7-integrin-dependent adhesion of leukocytes are CCL2, CCL3, CCL4, CXCL10, CXCL9, CCL5, CCL19, CXCL12, CCL20, CCL21 and CCL22. Ligation of GPCRs induces activation of integrins.

Table I   Adhesion molecules expressed by leukocytes and their ligands

L-selectin	*Selectin ligands*
Expressed by majority of leukocytes. Expression on the surface regulated by proteolytic cleavage near the cell surface by disintegrins of ADAM family. Cytoplasmic domain of L-selectin consists of 17 amino acid residues and interacts: α-actinin, calmodulin and members of ezrin/radixin/moesin (ERM) family. Calmodulin interacts with the cytoplasmatic tail of L-selectin in resting leukocytes negatively regulating its shedding. Activation of leukocytes leads to a release of calmodulin from the cytoplasmatic tail of L-selectin followed by the concomitant shedding of L-selectin.	Tetrasaccharide sialyl Lewis^X (CD15s) binds all selectins and it has been identified as a prototype selectin ligand. L-selectin recognizes heavily glycosylated mucin-like proteins: glycosylation-dependent cell adhesion molecule 1 (GlyCAM-1), high endothelial venules ligand (sgp200), CD34, and mucosal addressin (MadCAM-1) and P-selectin glycoprotein ligand-1 (PSGL-1). MadCAM-1 is a dominant physiological ligand for the $\alpha_4\beta_7$-integrin and for L-selectin and has high degree of homology with other immunoglobulin family members, ICAM-1 and VCAM-1.
Integrins	*Integrin ligands*
Membrane-bound heterodimeric glycoproteins consisting of noncovalently associated α and β subunits. In humans, 18 α subunits and 8 β subunits have been described giving rise to 24 integrin α/β dimers. Expression pattern of heterodimers varies among different cell types. Integrins control leukocyte homing to hematopoetic organs and also to sites of inflammation. The binding of integrins to their ligands is facilitating antigen recognition.	Intercellular adhesion molecules (ICAMs) are surface glycoproteins and structurally belong to the immunoglobulin (Ig) family and are ligands for the β_2-integrins. Expression of ICAMs is often constitutive and spread between multiple cell types. Some of ICAM subtypes (ICAM-2 and ICAM-3) appear only on lymphocytes and monocytes, which led to the proposal of their role in initiating immune responses.

Integrins are cell membrane heterodimeric glycoproteins consisting of noncovalently associated α and β subunits (Fig. 1). In response to stimulation, integrins signal through a conformation change of extracellular domains leading to polymerization, clustering and exposure of different binding epitopes. The extracellular domains of α- and β-integrins are formed by series of globular domains that incorporate genu, a folding point of extracellular domains. In the bend position, the knee is folded, and the integrins assume a compact structure in which the ligand binding site is close to the membrane. In the extend conformation state of the integrin legs, the ligand-binding site is projected away from the membrane and has limited availability for ligands (Fig. 2). This conformation probably depicts most closely the conformation of integrins in circulation (Hynes 2002).

Neutrophils and eosinophils can be rapidly recruited to the sites of inflammation. In contrast, monocytes and natural killer cells migrate at low levels. T lymphocytes are recruited to the inflammation site selectively from the nearest lymph nodes. Once arrived at their destination, leukocytes enter tissues through the vascular endothelium. Leukocyte interaction with vascular endothelial cells involve multiple

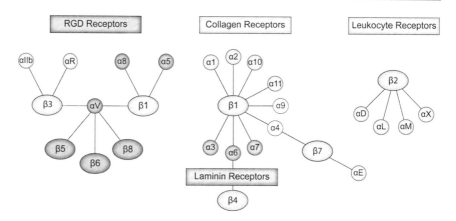

Fig. 1 Integrin family of adhesion molecules. The integrin family of molecules may be divided according to the extracellular matrix protein or sequence they bind (collagen, laminin, vitronectin, RGD). β_2-integrins may be found predominantly on leukocytes, and the most commonly studied of them are $\alpha L\beta_2$ (LFA-1) and $\alpha M\beta_2$ (Mac-1).

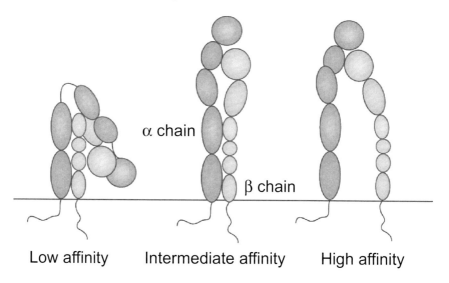

Fig. 2 Activation phases of integrin heterodimers. Depending on the activation state of leukocytes, integrins are present in three different conformational forms. (1) On resting leukocytes, the closed conformational form of integrins dominates. The α/β chains are bent down to the cell membrane forming a genu, knee (low affinity form). (2) On activated leukocytes, integrins are in the extended form, where the head of the chains is moved away from the cell membrane while the α/β chains are tightly attached to each other (intermediate affinity form). (3) Further activation of integrins results in dissociation of the α and β chains followed by external angle of the β-chain. This conformational change of β-chain exposes additional binding epitopes (high affinity form).

steps, including the capture of flowing leukocytes with their subsequent rolling, firm adhesion, and diapedesis or amoeboid migration (Fig. 3). The amoeboid mechanism of cell migration is fast, lacks strong adhesive interactions with the tissue and commonly preserves tissue integrity rather than degrading it.

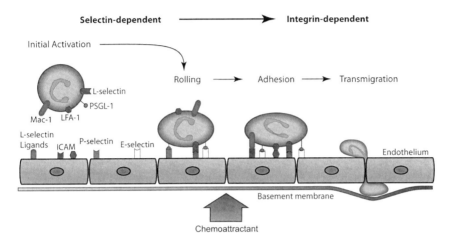

Fig. 3 The role of adhesion molecules in transendothelial migration of leukocytes. Selectin family of molecules and their ligands predominantly function during tethering (capture) and rolling of leukocytes along the endothelial surface forming unstable binding to their counterparts. During slow rolling, integrins are activated forming stable interactions with ICAM molecules, initiating adhesion and spreading of leukocytes. Activated leukocytes exhibit high degree of asymmetry and flexibility, changing their shape from spherical to amoeboid and forming a polarized edge penetrating the endothelial cell layer.

The first contact between leukocytes and endothelium is known as capture or tethering. It is mediated by selectins and their counter receptors. Selectin binding and presentation of chemokines by endothelial cells to chemokine receptors induce activation of signaling pathways in neutrophils (see below) that causes changes in integrin conformation. The process of conformational changes of extracellular integrin domains in response to intracellular rearrangements is called 'inside-out signaling'. Depending on the integrin conformation, binding to their counter receptors causes either slow rolling or arrest. The rolling capacity of leukocytes along endothelium is attributed to the β_2-integrins $\alpha L\beta_2$ and $\alpha M\beta_2$. Efficient conversion from rolling to firm adhesion is dependent on the time the leukocyte spends in close contact with the endothelium. Upon arrest, integrins stabilize the adhesion (post-adhesion strengthening), activate tension of α-actin, leading subsequently to crawling and transmigration through endothelium. Transmigration through endothelium is characterized by an amoeboid movement lacking strong adhesive interactions with the tissue, preserving its integrity, and supporting the ability to sense and integrate signals from the extracellular environment.

Transmigration through endothelium is followed by the processes of interstitial migration of leukocytes. This process is slow, generates stronger adhesion sites as compared to endothelial rolling or amoeboid transmigration, and causes proteolytic remodelling of ECM. The integrins facilitating interstitial migration belong to the family of β_1-integrins (very late antigens, VLA)—$\alpha_1\beta_1$ and $\alpha_2\beta_1$ (collagen adhesion), $\alpha_4\beta_1$ (fibronectin and VCAM-1), $\alpha_5\beta_1$ (fibronectin), and $\alpha_6\beta_1$ (laminin) (Fig. 1). Extracellular matrix proteins and short peptides containing Arg-Gly-Asp (RGD) sequence bind extracellular domains of integrins inducing rearrangement of cytoskeleton and regulating leukocyte migration, survival, and differentiation (Fig. 4). The integrin-induced rearrangements of cytoskeleton affect expression of a wide variety of inflammatory genes such as proteases, prostanoids, and cytokines. This process of intracellular signaling of integrins is called 'outside-in signaling'. The outside-in signaling from β_1- and β_3-integrins is mediated through integrin-linked kinases (ILK, see below). All circumstances and factors inducing leukocyte trafficking into the tissues are yet to be clarified. The function of leukocytes in various tissues and/or in the lymphoid organs is

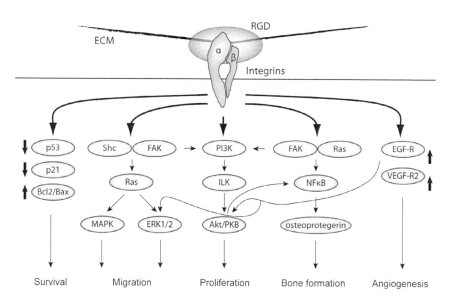

Fig. 4 Activation of intracellular pathways as a consequence of integrin ligation. Binding of integrins to the proteins of extracellular matrix is realized through the RGD (Arg-Gly-Asp) sequence of amino acids. This interaction activates intracellular network of protein tyrosine kinase (Scr, Erk, FAK), small GTPases (Rac, RhoA, Cdc42), and PI3-kinases. Activation of this broad intracellular network makes integrins essential regulators of migration, proliferation, survival, bone formation and angiogenesis. Ligation of integrins is followed by an increased expression of receptor tyrosine kinases (growth factor receptors), urokinase receptor, and chemokine receptors. Also it modulates transcription of proteins regulating cell cycle/apoptosis (Bcl2/Bax, p53, p21).

modulated by site- and environment-specific stimuli. Integrin activation through the receptors associated with these second stimuli and followed by a consequent change of integrin expression and clustering is known as 'inside-out signaling'. It becomes increasingly clear that the cooperation of inside-out and outside-in signaling events determines the effects of a particular leukocyte type and subset.

ADHESION MOLECULES ON NEUTROPHILS

Neutrophils are rapidly mobilized to the site of inflammation. An activated neutrophil manifests a number of functional responses as spreading, transmigration, phagocytosis, superoxide production, and degranulation. Neutrophils express PSGL-1, L-selectin and β_2-integrins, $\alpha L\beta_2$, $\alpha M\beta_2$, and $\alpha X\beta_2$, as well as low levels of $\alpha_4\beta_1$. Neutrophil interaction with endothelium involves multiple molecules including P-selectin glycoprotein ligand-1 (PSGL-1), L-selectin, G-protein coupled receptors, and integrins leading to activation of different intracellular pathways.

Expression of L-selectin and PSGL-1 regulates mobilization and capture of neutrophils by endothelium in conditions of share stress. Neutrophil is the only leukocyte type expressing PSGL-1. PSGL-1 is located in lipid rafts on microvilli of leukocytes serving as a ligand to L-selectin and P-selectin. Elimination of PSGL-1 gene reduces the number of neutrophils interacting with endothelium, alters rolling velocity and results in poor recruitment of neutrophils to the inflammation site. Ligation of PSGL-1 results in phosphorylation of its cytoplasmic tail and mobilization of the proteins of ezrin/radixin/moesin family, functioning as a link between plasma membrane and the actin cytoskeleton. Additionally, the N-terminal region of ezrin/radixin/moesin proteins interacts with spleen tyrosine kinase (Syk) regulating $\alpha L\beta_1$ activation (inside-out signaling).

Activation of neutrophils results in a rapid shedding of L-selectin from the cell surface by disintegrins of ADAM family (Ivetic and Ridley 2004). Blocking of L-selectin shedding *in vivo* leads to an increase in neutrophil arrest and reduces neutrophil rolling velocity. Cross-linking of L-selectin by antibodies results in increased intracellular Ca^{+2} levels, Scr-dependent tyrosine phosphorylation and activation of MAP-kinases followed by O_2-radicals production.

Slow rolling of neutrophils along the inflamed vessel wall is mediated by β_2-integrins $\alpha L\beta_2$ and $\alpha M\beta_2$. Ligation of PSGL-1 and exposure to chemokines induces activation of β_2-integrins. Binding of paxillin to the cytoplasmic tail of the α-integrin results in the dissociation of α and β subunits, conformational changes of the β_2-integrin I domain and the exposure of high affinity binding sites for its ligands. This leads to an efficient conversion from rolling to firm adhesion of neutrophils. During neutrophil migration, $\alpha L\beta_2$ forms ring-like clusters at the neutrophil–endothelial junction. $\alpha M\beta_2$ is stored in neutrophil granuli and readily released following activation. In addition to ICAM-1, $\alpha M\beta_2$ has several other ligands including bacterial and fungal glycoproteins, heparin, fibrinogen, coagulation factor X and complement C3.

In addition to direct contact activation through adhesion molecules, neutrophils are activated by soluble chemoattractants, most of which bind to GPCRs (Fu *et al.* 2007).

ADHESION MOLECULES ON EOSINOPHILS

Human blood eosinophils express two β_1-integrin dimers, $\alpha_4\beta_1$ and $\alpha_6\beta_1$, four β_2-integrins, $\alpha M\beta_2$, $\alpha L\beta_2$, $\alpha X\beta_2$, $\alpha D\beta_2$, and a β_7-integrin, $\alpha_4\beta_7$. The reactive pattern of eosinophils is similar to that of neutrophils including rapid mobilization into the asthmatic lung, interacting with a diversity of ligands on bronchial endothelium and degranulation within tissues.

$\alpha_4\beta_1$- and $\alpha M\beta_2$-integrins are the two most important integrins mediating eosinophil adhesion and movement. $\alpha_4\beta_1$-integrin is efficiently mobilized to the eosinophil membrane following antigen recognition. Recognition of VCAM-1 mediated by $\alpha_4\beta_1$ is an important selective mechanism by which eosinophils are preferred over neutrophils in asthma. $\alpha_4\beta_1$ mediates rolling, firm adhesion and migration of eosinophils to VCAM-1 potentiating granule release and generation of superoxide anions. The β_1 subunit localizes in podosomes and binds eosinophil to VCAM-1, ICAM-1, fibrinogen, vitronectin or albumin. $\alpha M\beta_2$ is present on peripheral blood eosinophils in conformational state that is constitutively less active than $\alpha_4\beta_1$. Eosinophils from peripheral blood do not easily adhere or migrate on the ligands presumably because the dominating conformation of integrins on the cell membrane is inactive and closed (Fig. 2). Conformation-sensitive antibodies showed that treatment of eosinophils with IL-5 and TNF-α changed the shape of integrins, converting them from bend and low affinity shape into the extended and opened, exposing the activation-induced epitope in the N-terminal region of β_1-integrin. Exposure of the activation-induced epitope is important in movement of primed blood eosinophils out in the circulation and into the airways in response to segmental antigen challenge. The activation of integrins on eosinophils can be achieved following incubation of blood with IL-5, GM-CSF, fMLP, or platelet activating factor in a GPCR-dependent mechanism. $\alpha_4\beta_7$ of blood eosinophils supports static adhesion on MAdCAM-1 and mediates rolling on MAdCAM-1 and VCAM-1 under flow. Cross-linking of β_7 on eosinophils by soluble VCAM-1 increases eosinophil survival in the airway lumen.

ADHESION MOLECULES ON T AND B LYMPHOCYTES

Unlike other leukocyte types, lymphocytes do not make stable focal adhesion and focal complexes. Role of adhesion molecules T and B cells have been mainly studied in differentiation, organ-specific trafficking, and antigen presentation.

The level of L-selectin expression is shown as a major recognition marker regulating retention of lymphocytes in bone marrow and migration into the secondary lymphoid organs (Lefrançois 2006). Shedding of L-selectin on $CD34^+$ stem cells is a prerequisite of their mobilization into the bloodstream. Naïve T cells lose expression of L-selectin when leaving thymus for further differentiation into effector cells in peripheral lymph nodes. L-selectin expressing T cells home more efficiently to lymph nodes. Re-expression of L-selectin is observed in central memory T cells and is shown to prolong their survival.

Besides L-selectin, lymphocytes express four leukocyte-specific β_2-integrins ($\alpha L\beta_2$, $\alpha M\beta_2$, $\alpha X\beta_2$ and $\alpha D\beta_2$), where the $\alpha L\beta_2$-integrin, leukocyte function-associated antigen 1 (LFA-1), is most abundant and widespread in expression. In common with many other cell types, lymphocytes express β_1-integrins mediating binding to ECM (α_1-$\alpha_6\beta_1$), and the two α_7-integrins ($\alpha_4\beta_7$ and $\alpha E\beta_7$). It has been shown that $\alpha L\beta_2$- and $\alpha_4\beta_1$-integrins promote T cell interaction with endothelial cells playing a crucial role for T cell recirculation and recruitment to inflammatory sites. $\alpha L\beta_2$ facilitates antigen-dependent interaction of T cells with antigen presenting cells and target cells. The interface between a T cell and an antigen presenting cell is characterized by the immunological synapse, where the outer ring defined by $\alpha L\beta_2$-integrin and the integrin-associated protein talin surrounds clustered TCRs. Localization of integrins to a lipid raft compartment is essential for integrin-mediated T cell adhesion. Active $\alpha L\beta_2$ and $\alpha_4\beta_1$ colocalize with GM1-enriched rafts and control binding and internalization of the proteins of Rho family GTPase (Rac, Rho and Cdc42) to rafts. More information about collaboration between integrins and TCR is given below.

Several integrin-specific interactions mediate lymphocyte trafficking and retention at specific anatomical locations. The migration of T cells to lymph nodes is regulated by a tight collaboration between integrins and a panel of chemokine receptors. Chemokine receptor CCR7 ligands CCL19 and CCL21, as well as the CXCR5 ligand CXCL13, have been identified as central conductors of these events. It begins with adhesive interaction of T cells with endothelial cells in the high endothelial venules. This adhesion is L-selectin dependent, which in combination with $\alpha L\beta_2$ and CCR7 is required for T cells to efficiently enter lymph nodes (Bromley *et al.* 2008). Expression of CCR7 and L-selectin regulates formation of B cell follicles and T cell areas in secondary peripheral organs.

Lymphocytes expressing $\alpha_4\beta_7$ home to skin and intestine through recognition of the mucosal addressin MAdCAM-1 expressed by high endothelial venules of Peyer's patches (Lefrançois 2006). Expression αE-integrin, which mediates lymphocyte binding to epithelial cell E-cadherin, divides regulatory T cells (Treg) into separate subpopulations. Expression of αE is associated with expression of inflammatory chemokine receptors and adhesion molecules that allow their migration into nonlymphoid tissues. αE-positive Treg cell subtype is able to suppress a delayed-type hypersensitivity response mediated by adoptive transferred antigen-specific Th1 cells.

INTEGRIN SIGNALING IN MONOCYTES/MACROPHAGES

Blood monocytes represent precursors for tissue-specific macrophages and dendritic cells. By expression of L-selectin, monocytes can be grouped into subsets with distinct functions *in vivo* (Geissmann *et al.* 2008). The first subset expressing L-selectin in combination with M-CSF receptor (CD115), Ly6c, CCR2 has inflammatory phenotype. L-selectin expressing cells are selectively recruited to inflamed tissues and lymph nodes and differentiate into inflammatory dendritic cells. These monocytes function as critical effector cells in primary and secondary immune responses. It has also been shown that L-selectin-positive subset could be expanded and polarized to inhibit T cell-mediated immunity. The second subset of monocytes is smaller and lacks expression of L-selectin, Ly6c, and CCR2. The L-selectin-negative subset is called resident or patrolling monocytes. These monocytes require firm binding to endothelium mediated by the β_2-integrins.

Binding of monocytes to ECM is accomplished by β_1-integrins represented by $\alpha_1\beta_1$, $\alpha_2\beta_1$, $\alpha_4\beta_1$, $\alpha_5\beta_1$, and $\alpha_6\beta_1$. Additionally, $\alpha M\beta_2$-integrin is present in all monocytes. The constitutive expression is observed for $\alpha_4\beta_1$, $\alpha_5\beta_1$, and $\alpha_6\beta_1$, while activation of monocytes results in the expression of $\alpha_1\beta_1$- as well as $\alpha v\beta_3$- and $\alpha_4\beta_7$-integrins. Adhesion of monocytes to ECM proteins changes the expression pattern of integrins, induces expression of pro-inflammatory molecules through downstream activation of NF-κB pathway and also promotes cell migration and phagocytosis (Fig. 4).

Macrophages express a vast number of receptors that mediate recognition of extracellular and intracellular pathogens followed by phagocytosis/endocytosis. The opsonic receptors include complement receptors (CR3, $\alpha M\beta_2/\alpha X\beta_2$) and Fc receptors (FcRs). Macrophages express also non-opsonic receptors mediating pathogen recognition, namely toll-like receptors (TLR) (Dale *et al.* 2008). By interacting with these receptors, integrins mediate innate and acquired immunity, participating in recognition of external danger signals and in activating FcRs by previously produced antibodies. Accumulation of this information in monocytes/macrophages makes further rearrangements in primary and secondary T cell responses (Geissmann *et al.* 2008). For interaction between integrins and FcRs and TLR, see below.

Several subpopulations of macrophages have been produced from the common mononuclear progenitor identified upon expression of $\alpha M\beta_2/\alpha X\beta_2$ by direct stimulation with colony stimulating factors, including M-CSF, GM-CSF, and Flt3-ligand. Functional role of $\alpha M\beta_2/\alpha X\beta_2$ for potentiating effects from growth factors has been suggested (Shi and Simon 2006). A direct implication of $\alpha X\beta_2$ in macrophage maturation and antigen presentation has been recently proved using $\alpha X\beta_2$-diphtheria toxin receptor transgenic mice. Monocytic precursors of inflammatory dendritic cells are defined as a population expressing $\alpha X\beta_2$. However, the role of $\alpha X\beta_2$ in differentiation and antigen presenting function of dendritic cells has been recently questioned.

Multinuclear osteoclasts are derived from αM-positive mononuclear cells in bone marrow and in circulation. Stimulation through αVβ_3-integrin is suggested to be essential for osteoclast differentiation. αVβ_3-integrin binds several ECM proteins including virtonectin, osteopontin, and bone sialopontin. RGD-containing peptides and blocking antibodies to αVβ_3 were shown to inhibit bone resorption *in vivo* and *in vitro* suggesting its important role in regulating osteoclast development and function.

INTRACELLULAR SIGNALING

Integrins differ from other receptors by being able to conduct signals not only from extracellular stimuli to induce intracellular changes (outside-in signaling) but also from intracellular stimuli to cause extracellular changes (inside-out signaling) (Fig. 5). Intracellular tails of integrins are short and have no tyrosine kinase activity of their own. Signals from integrins inside the cell are dependent on binding to accessory molecules that contribute further to activation of intracellular pathways. At present more than 150 signaling, structural and adaptor molecules for integrins are identified. The interaction of β_2-integrins with their ligands (ICAM-1) results in complex formation and phosphorylation transducing signals through Grb2, SOS and Shc tyrosine kinases followed by MAPK activation. The outside-in signaling through β_1- and β_3-integrins is mediated by integrin-linked kinases (ILK). ILK is a multifunctional protein binding a number of adaptor proteins that participate in cytoskeletal dynamics and cell signaling and assembling them into functional complexes. Binding of β-integrins to ILK stimulates kinase activity of ILK controlling phosphorylation of Akt at Ser473 in a PI3-kinase dependent manner. ILK activity may be also upregulated by hypoxia and several growth factors as insulin and PGDF and downregulated by a specific phosphatase (Dedhar 1999). Full functional ILKs have been shown to be essential for leukocyte adhesion, migration and proliferation. Talin is a component of the leading edge of activated lymphocytes and of the immunological synapse. Talin binding is believed to be the common final step in integrin activation. Talin binds to cytoplasmic tails of β-integrins, disrupting interface between the α and β subunits. Activated integrins, talin, vinculin and actin filaments form a link between the cytoskeleton and ECM. Less is known about signal transduction net of α-integrins. The α_4-integrin is directly associated with adaptor molecule paxillin. Paxillin binding to α-integrin tail has inhibitory effects blocking Rac. Dissociation of α_4-integrin/paxillin complex removes Rac suppression (Rose 2007).

For regulation of cell migration, survival, proliferation and tissue-specific gene expression, integrins cooperate with several groups of receptors (Huveneers 2007): chemotactic receptors (G-protein coupled receptors, GPCR), growth factor receptors (receptor tyrosine kinase), immunoreceptors (FcRs and T cell receptor), and, recently, pathogen-sensitive TLRs.

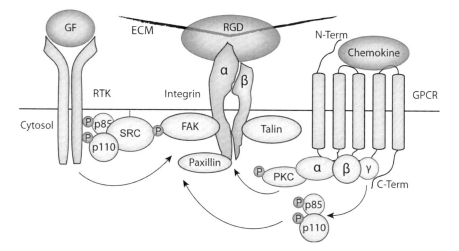

Fig. 5 Bidirectional signaling of integrins. Cooperation of integrins with growth factor receptors and chemokine receptors. Ligation of chemokine receptors (most of which are G-protein couples receptors, GPCR) results in activation of phospholipase C, Ca^{+2} influx followed by the GTPase proteins, Rac and Cdc42 PI3-kinase. Additionally, PI3-kinase p110γ catalytic subunit is activated and, consequently, Akt is activated. Both processes support translocation and binding of talin to β-chain while FAK activation leads to dissociation of paxillin from the α-chain. These processes together result in the formation of high affinity α/β-integrin heterodimer, increasing its affinity to proteins of extracellular matrix. Simultaneously, ligation of integrin gives rise to activation of FAK and Scr tyrosine kinases potentiating signaling from growth factor receptors (receptor tyrosine kinases, RTK), causing activation of Akt, MAPKs and cyclin-dependent kinases.

INTEGRIN SIGNALING IN COMPLEX WITH UROKINASE RECEPTOR (uPAR)

uPAR (CD87) is a glycosyl-phosphadylinositol-anchored protein expressed on most of leukocytes and serves as an extracellular adaptor for β-integrins. uPAR binds plasminogen activator urokinase and converts plasminogen to plasmin, promoting degradation of all components of ECM including fibrinogen, fibronectin and vitronectin. Depending on the cell context, uPAR association with β-integrins could yield positive and negative effects on cell behavior (Tang and Wei 2008). It has been shown that complex formation between uPAR and $\alpha_5\beta_1$ changes the integrin binding site of fibronectin from RGD to HepII, which results in a shift of prevalent tyrosine kinase activation from focal activation kinase (FAK)/Src to Rac 1 and consequences on MMP-9 production. Interaction between uPAR and β_1-integrins potentiates autotyrosine phosphorylation of EGFR and GPCR. Interaction between uPAR and $\alpha_5\beta_1$ mediates PI3-kinase-Akt-MAPK dependent production of uPA and MMPs, as well as anti-apoptotic factor Bcl-xl.

SIGNALING FROM GPCRs TO INTEGRINS

Integrin-mediated migration of leukocytes is directed by a heterogeneous group of components known as chemoattractants. Many chemoattractants transmit signals through heterodimeric G-protein coupled receptors (Fig. 5). GPCRs transmit their signals through the α-subunit of Gia, which may be inhibited by pertussis toxin. A key pathway mediated by GPCRs is signaling through class Ib phosphatylinositol-3-kinase (PI3-kinase), containing the p110γ catalytic subunit. PI3k-γ is recruited to the inner leaflet of the cell membrane by the G-protein βγ sbunit, where it becomes activated. The second pathway linked to PI3-kinase activation is induced by the fMLP receptor on neutrophils and leads to the activation of p38 MAPK and downstream activation of Rac. Rac induces actin polymerization through actin-binding proteins WAVE (Scar) and Arp2/3. The third, PI3-kinase-dependent pathway is common with that induced through TCR and FcRs and involves tyrosine kinases Lck and Zap-70 followed by signaling to class Ia PI3-kinases and its p110∂ subunit, and activation of Akt as well as GTPases Rac and Cdc42.

In addition to PI3-kinase activation, GPCRs mediate their signals through small GTPase (Ley *et al.* 2007). In monocytes, ligation of GPCRs results in rapid activation of PLC leading to intracellular influx of Ca^{2+}. Ca^{2+} and diacylglycerol activate the proteins of small GTPase family—Rac, activated by guanine nucleotide exchange factors, Ras-related protein 1 (Rap1) and Ras homologue gene-family member A (RhoA)—required for $\alpha L\beta_2$ affinity.

IMMUNORECEPTORS AND IMMUNORECEPTOR-LIKE MOLECULES

Receptors mediating adaptive immune responses are called immunoreceptors. This group includes lymphocyte antigen receptors (TCR, and B cell receptor, BCR) and the receptors binding Fc-portions of immunoglobulins (FcR on neutrophils and macrophages) and uses a common signal transduction mechanism (Bezman and Koretzky 2007) (Fig. 6).

For initiation of intracellular signaling, the cytoplasmic part of the receptor binds to a transmembrane adaptor molecule containing double tyrosine residues called immunoreceptor tyrosine-based activation motif (ITAM). Ligand binding and receptor clustering leads to phosphorylation of these tyrosines by Src-family tyrosine kinases, Lck and Fyn in T cells, and Lyn Fyn, or Src in other hematopoietic cells (Grande *et al.* 2007). The phosphorylated ITAM recruits the Syk family of tyrosine kinases (BCR and FcR) or the related ZAP-70 tyrosine kinase (TCR) through their dual phosphotyrosine-binding SH2 domains and forming the SLP76-Gads-LAT complex (Fig. 5). It has been shown that β_2- and β_3-integrins also signal by an ITAM-based, Syk-dependent mechanism. Syk, spleen tyrosine kinase is a central component of immunoreceptor signaling and was shown to be activated

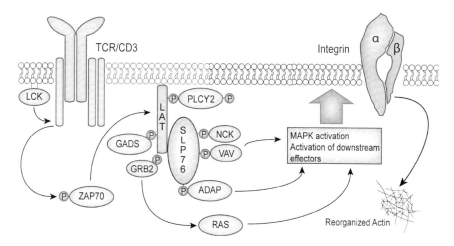

Fig. 6 Participation of integrins in antigen presentation. The binding of antigen to T cell receptor (TCR) activates tyrosine kinases of Src and Syk family, which phosphorylate the immunoreceptor tyrosine-based activation motif, and recruit the adaptor protein ZAP-70. ZAP-70 phosphorylates a transmembrane adaptor protein LAT and cytosolic adaptor protein SLP-76. The formation of LAT/SLP-76 complex is critical for integrin activation. $\alpha L\beta_2$ (LFA-1) is the major integrin being activated in response to TCR ligation. Activation of $\alpha L\beta_2$ associates with conformational changes and binding to CD3, expression of costimulatory molecules CD28 and CD69 potentiating antigen signaling.

through β_1-, β_2- and β_3-integrins. Syk activation appeared to be independent of cytoskeletal rearrangements and FAK. There is strong evidence of a critical role for Syk in β_2- and β_3-integrin-mediated differentiation and functions of leukocytes including neutrophils, macrophages and osteoclasts.

The ligation of TCR generates intracellular signals that increase $\alpha L\beta_2$-mediated cell adhesion. Signaling through TCR can cause lateral interactions of integrins with other transmembrane proteins leading to an increased integrin-mediated adhesion. Studies on transgenic and knockout mice identified small GTPase Rap1 as a key player in the pathway between TCR activation and $\alpha L\beta_2$. β_1- and β_2-intgerins are shown to translate information from TCR by binding SLP76 partners ADAP and SKAP-55. ADAP-deficient mice have poor integrin-dependent clustering and adhesion. SKAP-55 deficiency affects building of immunological synapse and antigen presentation in an $\alpha L\beta_2$-dependent manner. Present understanding of the integrin impact on function of B cells is mostly adopted from the studies on T cells (Batista *et al.* 2007). Ligation of BCR results in activation of $\alpha L\beta_2$- and $\alpha_4\beta_1$-integrins. In B cells and macrophages, the activation of β_1- and β_2-integrins is regulated by SKAP analogue SKAP-HOM. ADAP-SKAP complex activates Rap-1, a small GTPase molecule connecting the pathways of integrin activation of GPCR.

The adaptor protein cytohesin-1 binds selectively to β_2-subunit, while paxillin binds directly to the cytoplasmic tail of α_4 subunit and promotes activation of kinases FAK and Pyk2 (Batista *et al.* 2007).

In addition to the 'activation FcR', which carries ITAM sequence, B cells, neutrophils, monocytes and dendritic cells express the 'inhibiting FcR', which carry ITIM sequence, and are represented by FcγRIIB. The adhesion molecule PECAM has been described in modulating its activity.

SIGNALING THROUGH GROWTH FACTOR RECEPTORS

Cooperation of integrins with receptor protein tyrosine kinases (RTK) potentiates response to different growth factors (Fig. 5), such as insulin, PDGF, EGF, FGF, and VEGF, resulting in magnification of proliferative signals and in regulating survival and migration. RTK signaling mediates cross-talk with integrins through three different ways. (1) Direct phosphorylation of RTK intracellular domain by β-integrin occurs for several RTK, including EGFR, which interacts with β_1-, β_3- and β_4-integrins, HGF-R which binds to $\alpha_6\beta_4$, PDGFb-R, insulin-like growth factor 1 and VEGFR, which binds to $\alpha v\beta_3$-integrin, in the absence of any ligand. (2) Integrin-dependent RTK phosphorylation occurs by integrin assembly with Src and the adaptor protein p130Cas in a multimeric complex with integrins and RTK (Clemmons and Maile 2005). In the alternative mechanism, integrins cause recruitment of tyrodine phosphatase (SHP-2) in close proximity with PDGF-R, decreasing Ras-GAP binding and potentiating Ras signaling. Integrin-dependent EGFR activation leads to adhesion-dependent activation of ERK1/ERK2, MAPK, and Akt activation. Activation of Rac occurs through PI3-kinase activation involving GTP loading on Vav2. Joint integrin-RTK signaling causes expression of cyclin D-dependent kinases (Cdk) and degradation of Cdk inhibitor p27. This way is used by HGF to activate $\alpha M\beta_2$-integrins in neutrophils and $\alpha_4\beta_1$ in B cells (Cabodi *et al.* 2004). (3) Activation of RTK by their natural ligands alters the expression of α/β components of the integrin pair, regulating the effect on migration, proliferation and monocyte differentiation.

INTEGRIN SIGNALING IN COMPLEX WITH TLRs

Toll-like receptors are a family of cellular membrane structures responsible for the recognition of microbial patterns that are distinct from host molecules. Activation of TLRs initiates intracellular signaling that results in inflammation (Fisher and Ehlers 2008). Mechanisms facilitating the recognition of microbial patterns before their encounter with TLRs are poorly defined. Integrins have been reported to promote TLR signaling. Lipopolysaccharide (LPS) can directly interact with β_2-integrins, which together with αM-integrin forms the complement receptor type 3. Integrin αM mediates membrane recruitment of the Toll-interleukin 1 receptor domain containing adaptor protein (TIRAP) by controlling turnover of phosphadylinositol 4,5-bisphosphate (PIP2). Binding to PIP2 is required for TIRAP localization to the plasma membrane and transduction of TLR4 signaling through MyD88-NF-κB signaling pathway (Kagan and Medzhitov 2006). TLR2 is the main innate immune receptor responsible for detecting the anchored motif of bacterial lipoproteins consisting of a tri- or di-palmitoyl-S-glycerylcysteine and a short peptide. It has been recently reported that the preformed complex between TLR2 and β_3-integrins exists on non-stimulated human monocytes. Activation of TLR2 with lipoproteins results in a dissociation of β_3-integrin suggesting that $\alpha V\beta_3$ promotes the internalization of bacteria by presenting bacterial lipoproteins to TLR2 through vitronectin binding site (Gerold *et al.* 2008). Additionally, β_3-integrin coordinates TLR2 responses to bacterial lipoproteins and other agonists such as lipoteichoic acid and yeast zymosan.

SUMMARY

- Our present knowledge of adhesion molecules is far beyond simple surface binding facilitating migration. Adhesion molecules control the processes of leukocyte maturation, development and regulate their functions in pathogen recognition and antigen presentation.
- Adhesion molecules orchestrate leukocyte fate and function in coordinated fashion with receptors to growth factors, chemokines, as well as immunoreceptors.
- The diversity of signals transferred by integrins in bidirectional fashion from extracellular stimuli inside the cells and from intracellular stimuli to the extracellular domains transforms integrins into centres filtrating and redistributing information facilitating communication of leukocytes with the world outside the cell membrane. The mechanisms regulating selectivity of signals in each particular type of leukocyte are not completely recognized.
- Multiple clinical trails modulating leukocyte behavior by interacting with integrins have shown the benefits of this strategy in combating autoimmune inflammatory processes.

References

Batista, F.D. and E. Arana, P. Barral, Y.R. Carrasco, D. Depoil, J. Eckl-Dorna, S. Fleire, K. Howe, A. Vehlow, M. Weber, and B. Treanor. 2007. The role of integrins and coreceptors in refining thresholds for B-cell responses. Immunol. Rev. 218: 197-213.

Bezman, N. and G.A. Koretzky. 2007. Compartmentalization of ITAM and integrin signaling by adapter molecules. Immunol. Rev. 218: 9-28.

Bromley, S.K. and T.R. Mempel, and A.D. Luster. 2008. Orchestrating the orchestrators: chemokines in control of T cell traffic. Nat. Immunol. 9: 970-980.

Cabodi, S. and L. Moro, E. Bergatto, E. Boeri Erba, P. Di Stefano, E. Turco, G. Tarone, and P. Defilippi. 2004. Integrin regulation of epidermal growth factor (EGF) receptor and of EGF-dependent responses. Biochem. Soc. Trans. 32(Pt3): 438-442.

Clemmons, D.R. and L.A. Maile. 2005. Interaction between insulin-like growth factor-I receptor and alphaVbeta3 integrin linked signaling pathways: cellular responses to changes in multiple signaling inputs. Mol. Endocrinol. 19(1): 1-11.

Dale, D.C. and L. Boxer, and W.C. Liles. 2008. The phagocytes: neutrophils and monocytes. Blood 112(4): 935-945.

Dedhar, S. and B. Williams, and G. Hannigan. 1999. Integrin-linked kinase (ILK): a regulator of integrin and growth-factor signalling. Trends Cell. Biol. 9(8): 319-323.

Fischer, M. and M. Ehlers. 2008. Toll-like receptors in autoimmunity. Ann. NY. Acad. Sci. 1143: 21-34.

Friedl, P. and B. Weigelin. 2008. Interstitial leukocyte migration and immune function. Nat. Immunol. 9(9): 960-969.

Fu, H. and J. Karlsson, J. Bylund, C. Movitz, A. Karlsson, and C. Dahlgren. 2006. Ligand recognition and activation of formyl peptide receptors in neutrophils. J. Leukoc. Biol. 79(2): 247-256.

Geissmann, F. and C. Auffray, R. Palframan, C. Wirrig, A. Ciocca, L. Campisi, E. Narni-Mancinelli, and G. Lauvau. 2008. Blood monocytes: distinct subsets, how they relate to dendritic cells, and their possible roles in the regulation of T-cell responses. Immunol. Cell Biol. 86(5): 398-408.

Gerold, G. and K.A. Ajaj, M. Bienert, H.J. Laws, A. Zychlinsky, and J.L. de Diego. 2006. A Toll-like receptor 2-integrin beta3 complex senses bacterial lipopeptides via vitronectin. Nat. Immunol. 9(7): 761-768.

Grande, S.M. and G. Bannish, E.M. Fuentes-Panana, E. Katz, and J.G. Monroe. 2007. Tonic B-cell and viral ITAM signaling: context is everything. Immunol. Rev. 218: 214-234.

Huveneers, S. and H. Truong, and H.J. Danen. 2007. Integrins: signaling, disease, and therapy. Int. J. Radiat. Biol. 83(11-12): 743-751.

Hynes, R.O. 2002. Integrins: bidirectional, allosteric signaling machines. Cell 110(6): 673-687.

Ivetic, A. and A.J. Ridley. 2004. The telling tail of L-selectin. Biochem. Soc. Trans. 32(Pt 6): 1118-1121.

Kagan, J.C. and R. Medzhitov. 2006. Phosphoinositide-mediated adaptor recruitment controls Toll-like receptor signaling. Cell. 125(5): 834-836.

Lefrançois, L. 2006. Development, trafficking, and function of memory T-cell subsets. Immunol. Rev. 211: 93-103.

Ley, K. and C. Laudanna, M.I. Cybulsky, and S. Nourshargh. 2007. Getting to the site of inflammation: the leukocyte adhesion cascade updated. Nat. Rev. Immunol. 7(9): 678-689.

Rose, D.M. and R. Alon, and M.H. Ginsberg. 2007. Integrin modulation and signaling in leukocyte adhesion and migration. Immunol Rev. 218: 126-134.

Shi, C. and D.I. Simon. 2006 Integrin signals, transcription factors, and monocyte differentiation. Trends Cardiovasc. Med. 16(5): 146-152.

Streuli, C.H. and N. Akhtar. 2009. Signal co-operation between integrins and other receptor systems. Biochem. J. 418(3): 491-506.

Tang, C.H. and Y. Wei. 2008. The urokinase receptor and integrins in cancer progression. Cell. Mol. Life Sci. 65(12): 1916-1932.

CHAPTER 11

Adhesion Molecules in Decompression Sickness and Ischemia Reperfusion Injury

Nancy J. Bigley[1,*] and Barbara E. Hull[2]

[1]Professor of Microbiology and Immunology, Department of Neuroscience, Cell Biology, and Physiology; and the Department of Pathology, Boonshoft School of Medicine, 3640 Colonel Glenn Highway, Dayton, Ohio 45435, E-mail: nancy.bigley@wright.edu

[2]Professor of Biological Sciences, Department of Biological Sciences; and the Department of Medicine, Boonshoft School of Medicine, Wright State University, 3640 Colonel Glenn Highway, Dayton, Ohio 45435, E-mail: barbara.hull@wright.edu

Departmental contact: Kimberly Hagler, Department of Neuroscience, Cell Biology and Physiology, Boonshoft School of Medicine, Wright State University, 3640 Colonel Glenn Highway, Dayton, Ohio 45435, E-mail: kimberly.hagler@wright.edu

ABSTRACT

Cell adhesion molecules (CAMs) are the receptors and co-receptor ligands orchestrating the orderly migration of hematopoietic cells throughout the body and through endothelia in normal homeostasis and in pathological states. In this chapter, we examine the role of tissue injury as the trigger for decompression sickness (DCS) and ischemia-reperfusion (I/R) injury, pathologies in which tissue hypoxia and inflammation initiate changes in CAMs. Expression of CAMs is upregulated on injured/inflamed vascular endothelia, and the transmigration into the injured tissue sites of leukocytes bearing co-receptor ligands. The various inducers of increased CAM expression by vascular endothelia include noxious tissue injury products (reactive oxygen species) as well as molecules secreted by the inflammatory blood cells and inflamed endothelia cells (inflammatory cytokines).

*Corresponding author

A model is proposed that integrates the role of CAMs in inflammatory injury and mononuclear cell transmigration through endothelium into injured tissue sites in both I/R injury and DCS. In this model, tissue damage results from the rapid release of reactive oxygen species, nitric oxide, inflammatory cytokines from injured endothelium as well as from blood mononuclear cells. These molecules induce upregulation of CAMs on endothelium and subsequent transmigration of blood leukocytes into ischemic (hypoxic) site. The role of adenosine and a specific tissue receptor for adenosine ($A_{2A}R$) in mitigating the inflammatory damage in I/R injury and DCS is exciting and promises a strategy for intervention in these pathological states.

INTRODUCTION

In this brief review, we examine the role of cell adhesion molecules (CAMs) in two clinical entities characterized by tissue hypoxia: decompression sickness (DCS) and the better understood ischemia-reperfusion (I/R) injury. DCS is also called decompression syndrome, bends, or caisson disease. As defined by the Merriam-Webster Medical Dictionary/Medline Plus, DCS is a sometimes fatal disorder that is marked by neuralgic pains and paralysis, distress in breathing, and often collapse that is caused by the release of gas bubbles (as of nitrogen) in tissue upon too rapid decrease in air pressure after a stay in a compressed atmosphere. The restoration of circulation after a period of ischemia results in inflammation and oxidative damage through the induction of oxidative stress (Pacher *et al.* 2008, Jang *et al.* 2009). Nitric oxide (NO) produced during reperfusion reacts with superoxide to produce the potent reactive species peroxynitrite (Pacher *et al.* 2008). These molecules and reactive oxygen species or ROS (molecular oxygen, superoxide and hydroxyl radicals) attack cell membrane lipids, proteins, and glycosaminoglycans, causing further damage (Pacher *et al.* 2008, Jang *et al.* 2009).

The endothelium is intimately involved in a variety of pathologies including I/R injury, DCS, inflammation, oxidative stress, edema, thrombosis and hemorrhage. In exploring the role of CAMs in DCS and I/R injury, we examined their participation in inflammatory tissue injury. Leukocytes arriving at the injured areas release a plethora of inflammatory mediators (interleukins, free radicals, etc. (Pacher *et al.* 2008, Jang *et al.* 2009). The reintroduction of oxygen into the damaged sites initiates a cascade of reactions—damage to cellular proteins, DNA and plasma membranes—which may then cause a release of more free radials and cellular apoptosis. Leukocytes may collect in and damage small capillaries leading to thrombi formation and ischemia.

CAMs AND TRANSMIGRATION

Expression of intercellular adhesion molecule 1 (ICAM-1) and vascular adhesion molecule 1 (VCAM-1) on endothelial cells causes extravasation of T cells and monocytes/macrophages into tissue sites. The expression of ligands for these molecules (Tables 1, 2) (Rahman and Fazal 2009) helps to determine sites of adhesion followed by extravasation. The choice of a paracellular versus transcellular leukocyte migration pathway depends upon the tightness of the intercellular junctions of the endothelial cells, the type of stimuli, the type of leukocytes and the endothelial cell source. Rahman and Fazal note that transcellular leukocyte migration appears to occur more prominently in vascular areas where endothelial junctions are particularly tight, such as the blood-brain barrier, while paracellular migration of leukocytes occurs in vascular areas in which the endothelial junctions are less tight, such as postcapillary venules. Activation of endothelium may result from release of inflammatory cytokines (IL-1, TNF-α, IL-6) by endothelial cells and/or monocytes/macrophages following endothelial injury during reperfusion or by the gas bubbles causing DCS (Table 3). In the development of DCS, the surface of nitrogen bubbles that appear in the blood, extracellular space, and intracellular space during decompression may trigger an inflammation and activation of a cytokine cascade (Ersson *et al.* 1998, 2002) that increases ICAM-1/VCAM-1 expression.

Key Points Endothelium is activated by inflammatory molecules called cytokines. Blood leukocytes expressing certain CAMs interact with endothelial co-ligands, which permits transmigration of blood leukocytes into the injured tissue site.

Table 1 Cell adhesion molecule pairs on endothelia and blood mononuclear cells

CAM (immunoglubulin superfamily structure)	Ligand (integrins $\alpha_2\beta_1$)
ICAM-1; ICAM-2	LFA-1
VCAM-1	VLA-4

This information describes the CAMs found increased on endothelia and their corresponding co-ligands.

Table 2 Distribution of cell adhesion molecules involved in transmigration of blood leukocytes through injured/inflamed endothelia

ICAM-1	Activated endothelia, lymphocytes, dendritic cells
ICAM-2	Resting endothelia, lymphocytes, dendritic cells
VCAM-1	Activated endothelia
LFA-1	Monocytes, macrophages, T lymphocytes, neutrophils, dendritic cells
VLA-4	Monocytes, macrophages, lymphocytes

Table 3 Inducers of increased endothelial expression of ICAM-I and/or VCAM-I in leukocyte adhesion and activation (Ersson *et al.* 1998, 2002, Duan *et al.* 2008, Zang *et al.* 2009)

Interleukin-1 (IL-1); IL-1β
Interleukin-6 (IL-6)
Interleukin-18 (IL-18)
Tumor necrosis factor-α (TNF-α)
Reactive oxygen species (ROS including oxygen-free radical)
Interferon-γ (IFN-γ)

Cytokines and ROS in leukocyte activation and adhesion to endothelia.

Definitions

Inflammatory Molecules

a. Cytokines called interleukins (ILs)

- IL-1 causes endothelial activation in inflammation/coagulation, fever, acute phase protein synthesis by liver; it is produced mainly by macrophages, endothelial cells, and some epithelial cells.

- IL-6 causes the liver to produce acute phase proteins associated with inflammation and proliferation of antibody-producing B cells; it is produced mainly by macrophages, endothelial cells, and T cells.

- IL-17 causes macrophages and endothelial cells to produce increased amounts of chemokine (IL-8); it causes epithelial cells to produce granulocyte-monocyte-colony stimulating factor (GM-CSF) and granulocyte-colony-stimulating factor (G-CSF). IL-17 is produced by T cells (δ:γ T cells) and a subset of CD4 T cells.

- IL-8 is a chemoattractant molecule (chemokine) now designated as CXCL8, an effector of neutrophil recruitment; it is produced by blood monocytes.

- IL-18 is a macrophage product that synergizes with IL-12 to stimulate production of IFN-γ by NK and T cells.

- IL-12 activates NK cells and CD4 T cells to produce IFN-γ and promotes development of T_H1 cells (CD4 T cells that produce IFN-γ and TNF); produced by macrophages and dendritic cells.

- TNF (tumor necrosis factor) causes activation of endothelial cells in inflammation/coagulation, fever, and acute phase protein synthesis by liver, and it affects many cell types, causing apoptosis; it is mainly produced by macrophages and T cells. It is also stimulated by complement components (membrane attack complex) binding to endothelial cells (Table 5).

- IFN-γ activates monocytes/macrophages, increases expression of receptors for TNF on somatic cells, and increases expression of class I and class II MHC molecules; it is produced mainly by CD4 T cells, NKT cells and mast cells.
- IP-10 is a chemoattractant (chemokine) now designated as CXCL10, an effector of T cell recruitment; it is induced by IFN.
- Reactive oxygen species are ions or very small molecules that include oxygen ions, free radicals, and peroxides and nitric oxide .

b. Selectins
- E-selectin is a CAM expressed only on endothelial cells activated by cytokines (CD62E).
- L-selectin is a CAM found on white blood cells (leukocytes) (CD62L).
- P-selectin is a CAM found in granules in endothelial cells and in activated platelets (CD62P);

c. ICAM-1 – Intracellular Adhesion Molecule 1 (CD54).
- It facilitates adhesion between leukocytes and endothelial cells during immune and inflammatory responses.

2. Blood Cells

- α:β T cells are T cells in which the TCR is composed of an alpha (α) and a beta (β) chain. The antigen-binding pocket is compose of VJ amino acid segments in the α chain and of VDJ amino acid segments in the β chain amino (N)-terminal regions. α:β T cells comprise the majority of T cells in our bodies.
- γ:δ T cells are T cells in which the TCR is composed of a gamma (γ) and a delta (δ) chain. The antigen-binding pocket is compose of VDJ amino acid segments in the δ chain and of VJ amino acid segments in the γ chain amino (N)-terminal regions. T cells have more limited diversity for antigen than do α:β T cells. γ:δ T cells appear to recognize molecular patterns and those that produce IL-17 (an inducer of neutrophil migration from bone marrow) function in early inflammation.
- NK cells are natural killer lymphocytes, large granular cells lymphocytes that recognize altered-host cells and recognize loss of MHC class I molecules from host cell membranes.
- NKT cell are lymphocytes that recognize lipids presented by cells bearing CD1 molecules and can also act as NK cells.
- iNKT cells (invariant NK cells) are a subset of NKT cells in which the alpha (α) chain of the T cell receptor has a limited diversity characterized by a unique Vα-Jα rearrangement.

- CD4 T cells are helper T lymphocytes; they recognize peptide antigens resented by MHC II molecules on professional antigen-presenting cells (macrophages, dendritic cells, B cells).
- CD8T cells are cytotoxic T lymphocytes; they recognize peptide antigens presented by MHC I molecules.
- CD1–MHC I-b_type of glycoproteins are non-classical type I MHC-like molecules (found on monocytes/macrophages, some B cells, and thymocytes), which present lipid antigens to NKT cells.

CAMs AND I/R INJURY

Based on the paucity of information about changes in CAMs during DCS other than our own study (Bigley *et al.* 2008), we examined the features of another well-known model (I/R injury) of tissue hypoxia leading to changes in CAMs. In a murine hindlimb ischemia model, endothelial progenitor cells homed to and bound to endothelial cells through ICAM-1/β-2 integrin interaction; ICAM-1 was overexpressed on ischemic muscle (Yoon *et al.* 2006). As shown in Table 2, lymphocyte function-associated antigen (LFA-1), a ligand for ICAM-1, and very late antigen 4 (VLA-4), a ligand for VCAM-1, are involved in vasoendothelial adhesion and transendothelial migration of high proliferative potential endothelial progenitor cells to ICAM-1/2 and VCAM-1 expressing bone marrow endothelial cells (Duan *et al.* 2006, 2008). Endothelial cell expression of ICAM-1/2 and VCAM-1 increased after activation with cytokines IL-1β and TNF-α (8). Duan *et al.* observed that LFA-1 and VLA-4 are involved in the homing of these progenitor cells to ischemic tissues. Anti-ICAM-1 antibody effectively inhibited early inflammatory processes and reperfusion-induced injury in a rat middle cerebral artery occlusion model and in lung (Kanemoto *et al.* 2002, Chiang *et al.* 2006). Anti-TNF-α antibody has therapeutic and preventive effects on I/R lung injury (Chiang *et al.* 2006).

The significance of these CAMs and the complexity of the cells and molecules (Table 4) involved in inflammation in acute kidney injury are highlighted by Kinsey and Okusa (2008). Ischemia-reperfusion is the traumatic event inducing changes in endothelia, leukocytes and renal tubular epithelial cells and involves a variety of bone marrow-derived cells, endothelia, epithelia, and complement pathways contributing to hypoxia (Tables 4 and 5). Ischemia-reperfusion injury is a major problem in intestinal (Watson *et al.* 2008), heart (Haverslag *et al.* 2008, Lange *et al.* 2008) and lung transplantation (Ellman *et al.* 2008, Zang *et al.* 2009). It is an important factor in morbidity and mortality following lung transplantation (Zang *et al.* 2009). In renal transplantation, increased expression of ICAM-1 appears to be an early marker of acute rejection (Kinsey and Osuka 2008). In a mouse model of cardiac isografts, ICAM-1 mRNA peaked at 3 hr following

Table 4 Putative roles of bone marrow-derived and kidney cells in ischemia-reperfusion (acute kidney) injury (Rahman and Fazal 2008)

1. Bone marrow-derived cells cause increases in infiltration, activation, cytokine production.
   a. neutrophils produce reactive oxygen and nitrogen species
   b. monocytes/macrophages produce reactive oxygen and nitrogen species and inflammatory cytokines (IL-1, IL-6, TNF-α) and chemokines (IL-8)
   c. Resident dendritic cells are prominent secretors of TNF-α
   d. NK cells cause cytokine secretion
   e. Invariant NKT (iNKT) cells result in cytokine (IL-4, IL-10, IFN-γ) production
   f. Gamma delta (γδ) T cells
   g. CD4 T cells: T_H1 subsets damaging; T_H2 subsets protective
   h. B-1 B cells
2. Endothelial cells show increases in vascular permeability and adhesion molecule expression.
3. Epithelial cells display increases in complement deposition, toll-like receptor TLR2/4 expression. TLR signaling is involved in mediating renal damage.
4. Renal dendritic cells are involved in producing increases in cytokine production and in antigen presentation by draining lymph node.

Table 5 Role of innate immune complement pathways in hypoxic injury (Rahman and Fazal 2008)

1. A monoclonal antibody against Factor B (alternative complement pathway) protects mice from renal tubular injury and apoptosis after I/R.
2. Activation of lectin pathway triggered by pattern recognition receptors (mannose-binding lectin and ficolin), which bind endogenous ligands expressed on necrotic and apoptotic host cells, also bind cytokeratin exposed on hypoxic endothelia.
3. The membrane attack complex of all complement pathways (C5-C9) deposits on epithelial cells to stimulate production of TNF-α.

The complement pathways contribute to production of TNF-α following deposition of membrane attack complex on endothelial and epithelial cell membranes.

transplantation (2 hr following ischemia) and increases in ICAM-1 protein were apparent by immunohistochemistry after 6 hr of reperfusion (Wang *et al.* 1998).

A kappa-opiod receptor (κOR) agonist protected cardiomyocytes and neuron cells against I/R injury through activation of protein kinase C (Yu *et al.* 2009). Treatment of male Spraque Dawley rats with the KOR agonist significantly reduced ($p < 0.001$) the number of leukocytes adhering and transmigration in protecting the microcirculation of skeletal muscle from I/R injury. Expression of ICAM-1 in the cremaster muscle of the protected rats was reduced (Yu *et al.* 2009).

Key Point Blood leukocytes expressing certain CAMs interact with endothelial co-ligands, e.g. ICAM-1/β-2 integrin interaction, which permits transmigration of blood leukocytes into the injured tissue site.

Definition

ICAM-1 – Intracellular Adhesion Molecule 1 (CD54) facilitates adhesion between leukocytes to endothelial cells during immune and inflammatory responses.

CAMs IN DCS

We characterized early blood and tissue markers predictive of DCS during the 24 hr period immediately following compression-decompression of female Sprague-Dawley rats (Bigley *et al.* 2008). Control animals were maintained at 1 atmospheres absolute pressure ATA (101 kPascal). In animals exposed to 2 (202 kPascal), 3 (303 kPascal), or 4 (404 kPascal) ATA, followed by rapid decompression, increased levels of TNF-α and interferon gamma (IFN-γ) were found in the circulation at 6 hr after decompression, while increased levels of only IL-6 were observed at both 6 and 24 hr following decompression. Significant increases in expression of E-selectin and L-selectin as well as ICAM-1 were observed immunohistochemically in the lungs and brains of the rats 6 hr after decompression. These levels dropped by 24 hr. Greater increases in expression of E-selectin and L-selectin around vessels and connective tissue were seen at 24 hr after decompression in the quadriceps of rats exposed to either 3 or 4 ATA. In cardiovascular disorders and reperfusion damage, the multistep vascular extravasation process is mediated by P-selectin, E-selectin, L-selectin, ICAM-1 and VCAM-1 (Haverslag *et al.* 2008).

Macrophages were seen in high abundance in lung sections of experimental rats (Bigley *et al.* 2008), suggesting that the endothelial vasculature was activated by leukocytes and that both the macrophages and activated endothelium were sources of TNF-α and IL-6. IFN-γ-inducible proteins, IP-10 and ICAM-1, were elevated in the lungs and brains of all experimental rats at both 6 and 24 hr after rapid decompression, with higher expression seen at the earlier observation time. IP-10 was expressed only in the lungs of rats by 24 hr after decompression, with the 4 ATA rats showing the most marked expression. ICAM-1 and E-selectin, markers of endothelial activation, were observed in microvessels from the cerebral cortex within 24 hr after exposure of male Wistar rats to hypobaric hypoxia (Dore-Duffy *et al.* 1999). Increased expression of integrin αv, which activates latent transforming growth factor-β (TGF-β) on epithelium in lungs, was found only in the brains of rapidly decompressed rats (Bigley *et al.* 2008).

Notable expression of ICAM-1 was seen only in muscle from rats 24 hr following rapid decompression after exposure to 3 or 4 ATA. This observation implied that the muscle response to the inflammatory stimuli requires a period of time longer than 6 hr. Muscle cells present antigens, produce cytokines, and upregulate expression of ICAM-1 (Nagaraju 2001), suggesting that the inflammatory response could be modulated locally or by immune cells directed to this site. DCS can simultaneously affect endothelium at multiple areas, such as the brain, lung, liver, and skeletal muscle (Dore-Duffy *et al.* 1999). Lungs (the ultimate filter of venous blood) and the liver (the ultimate filter of portal blood flow) are especially targeted by leukocyte aggregation in severe neurological DCS in pigs (Nyquist *et al.* 2004). In our study, macrophages were found near vessels, suggesting the migration of immune cells to the site of inflammation. A significant increase in ICAM-1 levels by 24 hr after decompression may also suggest a local response by individual muscle cells. Sections of quadriceps were stained with hematoxylin and eosin (H&E). H&E-stained sections of the quadriceps of rats sacrificed at 24 hr post decompression showed macrophages located near blood vessels, which was not common in control slides (Fig. 1); the blood vessels appeared swollen in comparison with those seen in muscle tissue from control rats (rats held in the chamber at 1 ATA but not decompressed). Capillaries in quadriceps tissue of rats exposed to 4 ATA and rapidly decompressed appeared swollen (diameter 10 μm) at 24 hr following decompression. This swelling of the vessels was not observed in the vessels of control rats, which remained approximately 7 μm in diameter, or in vessels from skeletal muscle tissues from rats exposed to 2 or 3 ATA (Fig. 1). The Masson's Trichrome stain used to stain sections of the lungs contains four different dyes (fast green, hematoxylin, acid fuchsin and xylidine ponceau) enabling observation of specific cell types and organization of connective tissue. The trichrome stain of the lung tissues of the experimental and control rats revealed many cell types such as macrophages, neutrophils and eosinophils (Fig. 2). The lung tissues from control rats as well as experimental rats had many macrophages located in the alveoli. Notably the alveolar walls of rats exposed to 2 (approximately 3 μm), 3 or 4 ATA (approximately 10 μm), rapidly decompressed and sacrificed at 6 or 24 hr, were swollen when compared to alveoli from control rats (< 1 μm).

Key Points

- We (Bigley *et al.* 2008) identified changes in inflammatory mediators during the 24 hr period immediately following compression-decompression of female Sprague-Dawley rats.
- Increased levels of inflammatory cytokines, TNF-α, IL-6, and IFN-γ were detected in the circulation 6 hr after decompression, while increased levels of only IL-6 were observed at 24 hr.
- Significant increases in expression of E-selectin, L-selectin, and ICAM-1 were observed immunohistochemically in the lungs and brains of rapidly decompressed rats 6 hr after exposure to 2 , 3, or 4 ATA. These levels drop by 24 hr.
- Greater increases in expression of E-selectin and L-selectin around vessels and connective tissue were seen at 24 hr after decompression in the quadriceps of rats exposed to either 3 or 4 ATA.

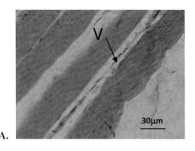

A.

Control Skeletal Muscle

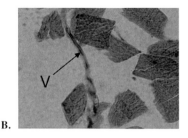

B.

Skeletal Muscle 24 hours after exposure
to 4 ATA

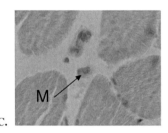

C.

Skeletal Muscle 24 hours after exposure
to 3 ATA

Fig. 1 H&E staining of skeletal muscle from control rat and rat exposed to 4 ATA and sacrificed at 24 hr after rapid decompression. Control rats held in chamber (1ATA) without pressure and no decompression (sham treatment). Arrows in Panels A and B point to endothelia; note the swelling in B. The arrow in C is pointing at a macrophage.

UNIFYING MODEL

In synthesizing the role of CAMs, especially ICAM-1, in inflammatory injury and mononuclear cell transmigration through endothelium into injured tissue sites in both I/R Injury and DCS, we propose the unified model of the early initiating events depicted in Fig. 3. Injury is initiated by gas bubbles in DCS or thrombi formation in I/R injury forming ischemic foci. Reperfusion of the injured site via small capillaries from blood causes tissue damage and rapid release of ROS, NO, inflammatory cytokines from injured endothelium as well as from monocytes-macrophages, and NKT cells. These molecules induce upregulation of ICAM-1 and VCAM-1 on endothelium and subsequent transmigration of blood leukocytes (neutrophils, monocytes-macrophages, NKT cells) into ischemic (hypoxic) site.

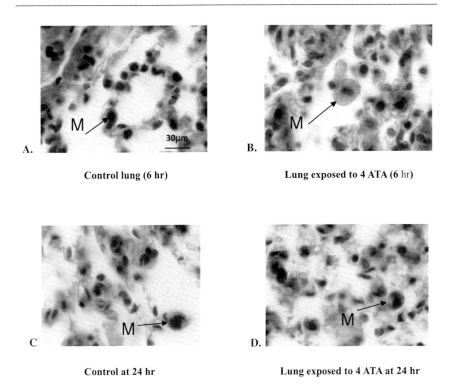

Fig. 2 Trichrome staining of sections of rat lung exposed to 4 ATA for 6 or 24 hr followed by rapid decompression. Lung tissue from sham-treated (rats held in chamber (1ATA) without pressure and no decompression) at 6 hr (A) and 24 hr (C) after removal from chamber. Lung tissue sections from rats exposed to 4 ATA sacrificed at 6 hr (B) and 24 hr (D) post-decompression. M = macrophage.

THE ROLE OF ADENOSINE AND ADENOSINE A$_{2A}$ RECEPTOR (A$_{2A}$R) IN RESTORING HOMEOSTASIS TO INJURY TISSUE SITES

Adenosine minimizes I/R injury in lungs, heart, liver, and kidney in addition to its anti-inflammatory effects (Cronstein *et al.* 1990, Chiang 2006, Ellman *et al.* 2008, Kinsey and Osukka 2008). Activation of the A$_{2A}$R modulates inflammation by downregulating production of inflammatory cytokines (Cronstein *et al.* 1992, Wessler 1996, Ohta and Sitkovsky 2001, Sullivan 2003, Reece *et al.* 2005, Thiel *et al.* 2005, Lynge and Hellsten 2009). In a rabbit lung transplant model, Gazoni *et al.* (2008) found that activation of A$_{2A}$R protected rabbits against lung I/R injury when administered before ischemia and during reperfusion. Similarly, A$_{2A}$R activation reduced infarct size in isolated perfused murine hearts by inhibiting

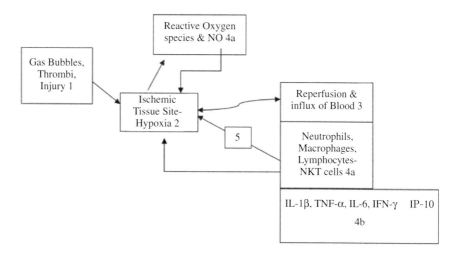

Fig. 3 Early events leading to transmigration of leukocytes from the blood into ischemic or inflamed tissue site. Initiation of injury via gas bubbles in DCS or injury and thrombin formation in I/R injury forming ischemic foci (1). Ischemia and hypoxia at tissue site (2) cause reperfusion of site via small capillaries and influx of blood (3). Tissue damage ensues releasing reactive oxygen species and nitric oxide (4a) and inflammatory cytokines from monocytes-macrophages and NKT cells (4b). Cytokines induce upregulation of ICAM-1 and VCAM-1 on vascular endothelium and transmigration of blood leukocytes (neutrophils, monocytes-macrophages, NKT cells) into ischemic (hypoxic) site (5).

resident cardiac mast cell degranulation. In our DCS study, significant increases in expression of the $A_{2A}R$ were detected only in the quadriceps but not in the lungs and brains of rats removed at 24 hr after decompression from 4 ATA. Type I DCS is described as joint and musculoskeletal pain, which may result from inflammation due to local ischemia. Our study demonstrated that rapid decompression induces the release of mediators of inflammation and resulting tissue inflammation cascades as well as a compensating modulation of a protective anti-inflammatory response (Bigley *et al.* 2008).

The $A_{2A}R$ is found in the cytosol and plasma membrane of skeletal muscle and in the endothelial cells within the connective tissue surrounding the muscle cells (Lynge and Hellsten 2009). In our study, the $A_{2A}R$ at 24 hr after rapid decompression was predominately membrane associated and occasionally clustered in a manner reminiscent of the motor endplate. Significant increases in $A_{2A}R$ expression were observed only in the decompressed rats exposed to 4 ATA, consistent with their increased symptoms of gait imbalance, compared with rats decompressed from lower pressure levels. This marker may be predictive of the development of DCS, as type I DCS symptoms may include musculoskeletal inflammation. Oxygenation disrupts the hypoxia-driven and $A_{2A}R$-mediated anti-inflammatory pathway in lungs (Thiel *et al.* 2005). Acute neutrophilic inflammation was inhibited by hypoxia in mice. The significant increase in the $A_{2A}R$ in rats exposed to 4 ATA suggests that $A_{2A}R$ may dampen inflammatory damage associated with decompression sickness. It would be interesting to determine whether inducers of adenosine minimize the effects of DCS as they have I/R injury.

Key Points

- Activation of $A_{2A}R$ at the onset of hypoxic tissue injury would lessen the damaging effects of inflammation.
- Significant increases in expression of $A_{2A}R$, which modulates inflammation by downregulating production of these cytokines, were detected only in the quadriceps removed at 24 hr after decompression from 4 ATA (Bigley *et al.* 2008).

Definitions

Adenosine is a nucleoside composed of a molecule of adenine attached to a ribose sugar molecule. Adenosine plays an important role in energy transfer, in signal transduction as cyclic adenosine monophosphate, as an inhibitory neurotransmitter. Adenosine is also a cryoprotector in preventing tissue damage from hypoxia and ischemia. Adenosine signals via four known receptors (A_1, A_{2A}, A_{2B}, and A_3), but adenosine activation of the A2A receptor is classified as anti-inflammatory, suppressing production of inflammatory cytokines (Sullivan 2003).

SUMMARY

- Tissue injury triggers tissue hypoxia and inflammation, which, in turn, initiate CAM changes in DCS and I/R injury.
- Expression of CAMs is upregulated on injured/inflamed vascular endothelia.
- Blood leukocytes transmigrate into the injured tissue sites through endothelia by attaching to endothelial co-receptor ligands for specific CAMs.
- Inducers of increased CAM expression by vascular endothelia include noxious tissue injury products (reactive oxygen species) as well as molecules secreted by the inflammatory blood cells and inflamed endothelia cells (inflammatory cytokines).
- A model is proposed that integrates the role of CAMs in inflammatory injury and mononuclear cell transmigration through endothelium into injured tissue sites.

- Adenosine and a specific tissue receptor for adenosine ($A_{2A}R$) dampen the inflammatory damage in I/R injury and DCS.

Acknowledgements

Our decompression study in rats was initiated by a seed grant from the School of Medicine, Wright State University. We want to express our appreciation to Dr. Jay B. Dean, University of South Florida College of Medicine, Tampa, Florida, for his encouragement and interest in decompression sickness when he was a faculty member at Wright State University.

Abbreviations

$A_{2A}R$	A_{2A} Receptor, one of four receptors for adenosine
ATA	atmospheres absolute pressure
CAM	cell adhesion molecule
CD	clusters of differentiation
CD1	MHC I-b type of glycoprotein
CD4 T cells	helper T lymphocytes
CD8T cells	cytotoxic T lymphocytes
DCS	decompression sickness
E-selectin	a CAM expressed only on endothelial cells activated by cytokines (CD62E)
GM-CSF	granulocyte-monocyte colony-stimulating factor
G-CSF	granulocyte colony stimulating factor
ICAM-1	intercellular adhesion molecule 1 (CD54)
ICAM-2	intercellular adhesion molecule 2 (CD102)
IFN-γ	interferon-gamma
IL	interleukin
I/R	ischemia reperfusion
kPascal	kiliPascal (1 kPascal = 101A)
L-selectin	a CAM found on white blood cells (leukocytes) (CD62L)
LFA-1	lymphocyte function-associated antigen-1 (CD11a)
MHC	major histocompatibility antigen complex
NK cells	natural killer lymphocyte
NKT cell	invariant natural killer T cells
NO	nitric oxide

P-selectin	a CAM found in granules in endothelial cells and activated platelets (CD62P)
TCR	T lymphocyte receptor
α:β T cells	T cells in which the TCR is composed of an alpha (α) and a beta (β) chain.
γ:δT cells	T cells in which the TCR is composed of a gamma (γ) and a delta (δ) chain.
TNF	tumor necrosis factor
ROS	reactive oxygen species
VCAM-1	vascular cell adhesion molecule 1 (CD106)
VLA-4	very late antigen 4 (CD49d)

References

Bigley, N.J. and H. Perymon, G.C. Bowman, B.E. Hull, H.F Stills, and R.A. Henderson. 2008. Inflammatory cytokines and cell adhesion molecules in a rat model of decompression sickness. J. Interferon. Cytokine Res. 28: 55-63.

Chiang, C.H. 2006. Effects of anti-tumor necrosis factor-alpha and anti-intercellular adhesion molecule-1 antibodies on ischemia/reperfusion lung injury. Chin. J. Physiol. 49: 266-274.

Cronstein, B.N. and L. Daguma, D. Nichols, A.J. Hutchison, and M. Williams. 1990. The adenosine/neutrophil paradox resolved: Human neutrophils possess both A1 and A2 receptors that promote chemotaxis and inhibit O2 generation, respectively. J. Clin. Invest. 85: 1150-1157.

Cronstein, B.N. and R.I. Levin, M. Philips, R. Hirschhorn, S.B. Abramson, and G. Weissmann. 1992. Neutrophil adherence to endothelium is enhanced via adenosine A1 receptors and inhibited via adenosine A2 receptors. J. Immunol. 148: 2201-2206.

Dore-Duffy, P. and R. Balabanov, T. Beaumont, M.A. Hritz, S.I. Harik, and J.C. LaManna. 1999. Endothelial activation following prolonged hypobaric hypoxia. Microvasc. Res. 57: 75-85.

Duan, H. and L. Cheng, X. Sun, Y. Wu, L. Hu, J. Wang, H. Zhao, and G. Lu. 2006. LFA-1 and VLA-4 involved in human high proliferative potential-endothelial progenitor cells homing to ischemic tissue. Thromb. Haemost. 96: 807-815.

Duan, H.X. and G.X. Lu, and L.M. Cheng. 2008. LFA-1 and VLA-4 involved in vasoendothelial adhesion and transendothelial migration of human high proliferative potential endothelial progenitor cells. Zhongguo Shi Yan Xue Ye Xue Za Zhi 16: 671-675.

Ersson, A. and M. Walles, K. Ohlsson, and A. Ekholm. 2002. Chronic hyperbaric exposure activates proinflammatory mediators in humans. J. Appl. Physiol. 92: 2375-2380.

Ersson, A. and C. Linder, K. Ohlsson, and A. Ekholm. 1998. Cytokine response after acute hyperbaric exposure in the rat. Undersea Hyperb. Med. 25: 217-221.

Gazoni, L.M. and V.E. Laubach, D.P. Mulloy, A. Bellizzi, E.B. Unger, J. Linden, P.I. Ellman, T.C. Lisle, and I.L. Kron. 2008. Additive protection against lung ischemia-reperfusion injury by adenosine A2A receptor activation before procurement and during reperfusion. J. Thorac. Cardiovasc. Surg. 135: 156-165.

Haverslag, R. and G. Pasterkamp, and I.E. Hoefer. 2008. Targeting adhesion molecules in cardiovascular disorders. Cardiovasc. Hematol. Disord. Drug Targets 8: 252-260.

Jang, H.R. and H. Rabb. 2009. The innate immune response in ischemic acute kidney injury. Clin. Immunol. 130: 41-50.

Kanemoto, Y. and H. Nakase, N. Akita, and T. Sakaki. 2002. Effects of anti-intercellular adhesion molecule-1 antibody on reperfusion injury induced by late reperfusion in the rat middle cerebral artery occlusion model. Neurosurgery 51: 1034, 1041; discussion 1041-1042.

Kinsey, G.R. and L. Li, and M.D. Okusa. 2008. Inflammation in acute kidney injury. Nephron Exp. Nephrol. 109: e102-107.

Lange, V. and A. Renner, M.R. Sagstetter, M. Lazariotou, H. Harms, J.F. Gummert, R.G. Leyh, and O. Elert. 2008. Heterotopic rat heart transplantation (lewis to F344): Early ICAM-1 expression after 8 hours of cold ischemia. J. Heart Lung Transplant. 27: 1031-1035.

Lynge, J. and Y. Hellsten. 2000. Distribution of adenosine A1, A2A and A2B receptors in human skeletal muscle. Acta Physiol. Scand. 169(4): 283-290.

Nagaraju, K. 2001. Immunological capabilities of skeletal muscle cells. Acta Physiol. Scand. 171: 215-223.

Nyquist, P.A. and E.J. Dick Jr., and T.B. Buttolph. 2004. Detection of leukocyte activation in pigs with neurologic decompression sickness. Aviat. Space Environ. Med. 75: 211-214.

Ohta, A. and M. Sitkovsky. 2001. Role of G-protein-coupled adenosine receptors in downregulation of inflammation and protection from tissue damage. Nature 414: 916-920.

Pacher, P. and C. Szabo. 2008. Role of the peroxynitrite-poly(ADP-ribose) polymerase pathway in human disease. Am. J. Pathol. 173: 2-13.

Rahman, A. and F. Fazal. 2008. Hug tightly and say goodbye: Role of endothelial ICAM-1 in leukocyte transmigration. Antioxid. Redox. Signal. Sept. 22. [Epub ahead of print]

Reece, T.B. and P.I. Ellman, T.S. Maxey, I.K. Crosby, P.S. Warren, T.W. Chong, R.D. LeGallo, J. Linden, J.A. Kern, C.G. Tribble, *et al.* 2005. Adenosine A2A receptor activation reduces inflammation and preserves pulmonary function in an in vivo model of lung transplantation. J. Thorac. Cardiovasc. Surg. 129:1137-1143.

Sullivan, G.W. 2003. Adenosine A2A receptor agonists as anti-inflammatory agents. Curr. Opin. Investig. Drugs 4: 1313-1319.

Thiel, M. and A. Chouker, A. Ohta, E. Jackson, C. Caldwell, P. Smith, D. Lukashev, I. Bittmann, and M.V. Sitkovsky. 2005. Oxygenation inhibits the physiological tissue-protecting mechanism and thereby exacerbates acute inflammatory lung injury. PLoS Biol. 3: e174.

Wang, C.Y. and Y. Naka, H. Liao, M.C. Oz, T.A. Springer, J.C. Gutierrez-Ramos, and D.J. Pinsky. 1998. Cardiac graft intercellular adhesion molecule-1 (ICAM-1) and interleukin-1 expression mediate primary isograft failure and induction of ICAM-1 in organs remote from the site of transplantation. Circ. Res. 82: 762-772.

Watson, M.J. and B. Ke, X.D. Shen, F. Gao, R.W. Busuttil, J.W. Kupiec-Weglinski, and D.G. Farmer. 2008. Intestinal ischemia/reperfusion injury triggers activation of innate toll-like receptor 4 and adaptive chemokine programs. Transplant. Proc. 40: 3339-3341.

Wessler, I. 1996. Acetylcholine release at motor endplates and autonomic neuroeffector junctions: A comparison. Pharmacol. Res. 33: 81-94.

Yoon, C.H. and J. Hur, I.Y. Oh, K.W. Park, T.Y. Kim, J.H. Shin, J.H. Kim, C.S. Lee, J.K. Chung, Y.B. Park, *et al.* 2006. Intercellular adhesion molecule-1 is upregulated in ischemic muscle, which mediates trafficking of endothelial progenitor cells. Arterioscler. Thromb. Vasc. Biol. 26: 1066-1072.

Yu, Y.M. and C.H. Lin, H.C. Chan, and H.D. Tsai. 2009. Carnosic acid reduces cytokine-induced adhesion molecules expression and monocyte adhesion to endothelial cells. Eur. J. Nutr. Jan 13. [Epub ahead of print]

Zhang, X. and D. Wu, and X. Jiang. 2009. ICAM-1 and acute pancreatitis complicated by acute lung injury. JOP 10: 8-14.

CHAPTER 12

Endothelial Adhesion Molecules in Diabetes

Yiqing Song[1, *], Cuilin Zhang[2] and Simin Liu[3]

[1]Division of Preventive Medicine, Brigham and Women's Hospital, Harvard Medical School, 900 Commonwealth Avenue East, Boston, Massachusetts, 02148, USA, E-mail: ysong3@rics.bwh.harvard.edu
[2]Epidemiology Branch, Division of Epidemiology, Statistics, and Prevention Research, Eunice Kennedy Shriver National Institute of Child Health and Human Development, National Institutes of Health, Bethesda, Maryland, USA
E-mail: zhangcu@mail.nih.gov
[3]Program on Genomics and Nutrition, School of Public Health, UCLA, Box 951772, 650 Charles E. Young Drive South, Los Angeles, CA 90095-1772, USA, E-mail: siminliu@ucla.edu

ABSTRACT

A large body of evidence supports the contention that endothelial dysfunction is an important pathological basis for the metabolic or insulin resistance syndrome and may thus play a pivotal role in the pathogenesis of diabetes mellitus (type 1 and type 2) and related complications. Measurable levels of some soluble cellular adhesion molecules including E-selectin, intercellular adhesion molecule 1 (ICAM-1), and vascular cell adhesion molecule 1 (VCAM-1) may reflect the degree of endothelial activation (i.e., released by activated endothelial cells to the general circulation, these molecules are considered useful indicators of endothelial dysfunction/activation). Elevated levels of soluble cellular adhesion molecules have been associated with insulin resistance and its associated metabolic abnormalities including diabetic complications. Recent evidence from several large prospective studies has shown that elevated levels of endothelial biomarkers, especially E-selectin, VCAM-1, and

---

*Corresponding author

Key terms are defined at the end of the chapter.

ICAM-1, were strong independent predictors of type 2 diabetes in populations with diverse ethnic backgrounds. This chapter aims to summarize both clinical and epidemiological studies relating plasma levels of endothelial adhesion molecules to diabetes mellitus and interpret these data in the context of emerging biological evidence. The available evidence indicates that elevated circulating levels of some, but not all, markers correlate not only with prediabetic insulin resistance and type 2 diabetes, but also with complications in both type 1 and type 2 diabetes. Future studies are warranted to evaluate the clinical utility of endothelial markers in comprehensively assessing and managing the development and prognosis of diabetes in the context of other diabetes biomarkers and traditional risk factors.

INTRODUCTION

Endothelial dysfunction has been closely related to insulin resistance, implicating its etiologic role in the development of type 2 diabetes as well as diabetic complications (Price and Loscalzo 1999, Ross 1999). Endothelial function can readily be measured by circulating levels of endothelial soluble adhesion molecules. In response to several inflammatory cytokines, endothelial cells secrete cellular adhesion molecules including E-selectin, intercellular adhesion molecule 1 (ICAM-1), and vascular cell adhesion molecule 1 (VCAM-1) on the cell surface (Price and Loscalzo 1999, Ross 1999). Soluble forms of these molecules are released from shedding or proteolytic cleavage from the endothelial cell surface and may reflect over-expression of their respective membrane-bound forms (Price and Loscalzo 1999). Measurable levels of these soluble cellular adhesion molecules have modest but also meaningful correlations with the direct assessment of endothelial function and are thus considered useful indicators of endothelial dysfunction/ activation.

Epidemiological evidence, primarily from cross-sectional studies, has shown that an elevation in circulating levels of E-selectin, ICAM-1, and VCAM-1 is associated with not only prediabetic insulin resistance, obesity, hypertension, and dyslipidemia among non-diabetic individuals but also type 2 diabetes patients with or without vascular complications (Schram and Stehouwer, 2005). However, these cross-sectional or retrospective studies cannot establish the time-to-event sequence for the associations observed. Few prospective studies have specifically examined the relationship between levels of endothelial adhesion molecules and the development of diabetes mellitus. Findings from several well-designed prospective studies suggested that elevated levels of soluble adhesion molecules, especially E-selectin, independently predicted risk of type 2 diabetes among initially non-diabetic individuals. Such findings have been confirmed in an ethnically diverse cohort of US post-menopausal women.

ADHESION MOLECULES AND RISK OF TYPE 2 DIABETES

Pinkney *et al.* (1997) first put forth the hypothesis that endothelial dysfunction is a common antecedent of the insulin resistance/metabolic syndrome and is intrinsically related to many of its key clinical features. Several biological mechanisms have subsequently been proposed to explain the complex and reciprocal relationships between endothelial dysfunction and insulin resistance (shown in Fig. 1). Briefly, endothelial dysfunction could directly promote the development and progression of insulin resistance. Alternatively, impaired insulin action may also directly exacerbate endothelial dysfunction. Several studies in non-diabetic individuals have suggested that mildly impaired fasting glucose levels (though within the normoglycemic range) accelerate the impairment of endothelial function via adverse effects on oxidative stress, formation of advanced glycation end products, and elevated levels of free fatty acids (Pinkney *et al.* 1997, Kim *et al.* 2006).

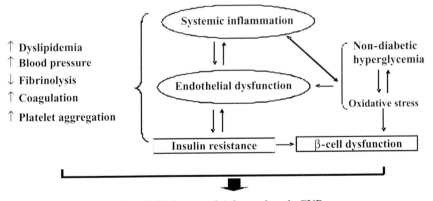

Fig. 1 Diagram of the 'common soil' hypothesis for reciprocal relationships between endothelial dysfunction and insulin resistance and other metabolic disorders for the pathogenesis of type 2 diabetes and atherosclerotic cardiovascular disease.

A number of cross-sectional studies in non-diabetic individuals have shown a strong positive relationship between levels of E-selectin, ICAM-1, and VCAM-1 and insulin resistance and/or glucose intolerance (Pinkney *et al.* 1997, Schram and Stehouwer 2005). However, the cross-sectional evidence does not prove causality. Endothelial dysfunction characterizes all phases of insulin resistance and its related metabolic abnormalities, lending support to the 'common soil' hypothesis that endothelial dysfunction, as reflected by elevated levels of soluble endothelial adhesion molecules, may be one of the common antecedents for the pathogenesis of both atherosclerotic cardiovascular disease (CVD) and type 2 diabetes (Fig. 1). Although there is some evidence to link several direct measures of endothelial dysfunction (such as flow-mediated dilation of the brachial artery or retinal arteriolar narrowing) to subsequent risk of type 2 diabetes, the role of elevated levels of endothelial adhesion molecules in predicting diabetes risk remains inconclusive because of limited prospective data.

Results from all prospective studies available by February 2009 are summarized in Table 1. A few prospective studies have directly evaluated the role of endothelial biomarkers in predicting future risk of developing type 2 diabetes but have yielded mixed results (Krakoff *et al.* 2003, Meigs *et al.* 2004, Herder *et al.* 2006, Thorand *et al.* 2006, Song *et al.* 2007). In the first nested case-control study of Pima Indians, none of the endothelial biomarkers, E-selectin, ICAM-1, and VCAM-1, was significantly associated with type 2 diabetes. The small sample size of this study may limit its statistical power to detect a significant association. By contrast, in a larger case-control study from the Nurses' Health Study, Meigs *et al.* (2004) reported that elevated levels of E-selectin and ICAM-1 were independent predictors of incident diabetes in initially non-diabetic Caucasian women and VCAM-1 was not associated with diabetes risk. Further adjustment for baseline levels of C-reactive protein (CRP), fasting insulin, and hemoglobin A1c did not change these associations. Consistent with these findings, E-selectin but not ICAM-1 was an independent predictor of type 2 diabetes risk in a population-based case-cohort study of middle-aged German men and women (Thorand *et al.* 2006). E-selectin remained significantly associated with diabetes risk in men and women after additionally controlling for age, body mass index, smoking, alcohol, physical activity, SBP, total cholesterol/HDL cholesterol ratio, parental history of diabetes, and CRP. In addition, two prospective cohort studies have presented data on ICAM-1 levels only (Herder *et al.* 2006, Sattar *et al.* 2009) and their results have also been inconsistent (Table 1).

In a case-control study nested within the Women's Health Initiative Observational Study (WHI-OS), an ethnically diverse cohort of US post-menopausal women including whites, blacks, Hispanics, and Asians and Pacific Islanders, we found ethnic differences in plasma levels of endothelial adhesion molecules and confirmed the predictive role of both E-selectin and ICAM-1 in all ethnic groups (Fig. 2) (Song *et al.* 2007). When these three endothelial markers were mutually adjusted, VCAM-1 was no longer predictive of diabetes risk in two studies, and E-selectin remains the strongest diabetes predictor among these three biomarkers (Fig. 3). Elevated levels of endothelial adhesion molecules may to some extent reflect a chronic inflammatory state. However, incremental changes in circulating levels of E-selectin remain independently associated with future diabetes risk irrespective of CRP levels at baseline.

Taken together, prospective data appear to support a robust association between E-selectin and incident diabetes. There are several plausible explanations. First,

Table I Prospective studies relating circulating adhesion molecules to incident type 2 diabetes

Study	Study cohort	Age, yr	Incident diabetes cases	Follow-up, yr	Main results
Krakoff *et al.* (2003)	Pima Indian, 2,530	Mean 32-33	71 cases (71 controls)	7	E-selectin, ICAM-1 and VCAM-1 were not associated with diabetes risk.
Meigs *et al.* (2005)	NHS, 32,826 women	30 to 55	737 cases (785 controls)	10	Adjusted RR in the top quintile vs. the bottom quintile: E-selectin: 5.43 (95% CI: 3.47-8.50); ICAM-1: 3.56 (95% CI: 2.28-5.58); VCAM-1: 1.12 (95% CI, 0.76-1.66).
Thorand *et al.* (2006)*	MONICA/KORA, 10,718 participants (5,382 men and 5,336 women)	35 to 74	532 cases (1,712 controls)	7-18	Adjusted HR in the high tertile vs. the low tertile: E-selectin: 2.79 (95% CI: 1.91-4.09) for men and 1.72 (95% CI: 1.07-2.75) for women; ICAM-1: 1.50 (95% CI: 1.02-2.19) for men and 1.16 (95% CI: 0.72-1.88) for women.
Herder *et al.* (2006)	DPS, 522 participants (172 men and 350 women)	40 to 65	86-107 cases	4	No significant association with ICAM-1 in the intervention group (n = 265; P for trend = 0.396) and in the control group (n = 257; P for trend = 0.316). (RR not shown).
Donahue *et al.* (2007)**	WNYS, 1,455 participants	39 to 79	Women: 52 cases (156 controls); Men: 39 cases (117 controls)	6	Age-, BMI- and HOMA-1R-adjusted mean levels for cases vs. controls: E-selectin: 51.2 vs. 45.5 ng/ml for women (P = 0.06); and 40.5 vs. 43.2 ng/ml for men (P = 0.37) ICAM-1: 273 vs. 255 ng/ml women (P = 0.04); and 255 vs. 264 ng/ml for men (P = 0.37).

contd...

Song et al. (2007)	WHI-OS, 82,069 post-menopausal women	50 to 79	1,584 cases: 749 whites, 366 blacks, 152 Hispanics, and 98 Asian/Pacific Islanders and 2,198 matched controls	6	Adjusted RR in the highest quartile vs. the lowest quartile: E-selectin, 3.46 (95% CI: 2.56-4.68); (95% CI: 1.75-3.13); VCAM-1, 1.48 (95% CI: 1.07-2.04). No significant ethnic differences.
Sattar et al. (2009)	PSPER, 4,945 men and women	70 to 82	292 cases	3	3 Adjusted HR per unit increase in log [ICAM-1]: 1.84 (95% CI: 1.26-2.69)

NHS, the Nurses' Health Study; MONICA/KORA, the Monitoring of Trends and Determinants in Cardiovascular Disease/Cooperative Research in the Region of Augsburg Study; DPS, The Finnish Diabetes Prevention Study; WNYS, The Western New York Study; WHI-OS, The Women's Health Initiative-Observational Study; PSPER, The Prospective Study of Pravastatin in the Elderly at Risk trial. CRP, C-reactive protein; HbA1c, hemoglobin A_{1c}; RR, relative risk; HR, hazard ratio.

*The exact number of incident diabetes was not reported in the paper.

**Incident cases were defined as those who progressed from normoglycemia to pre-diabetes.

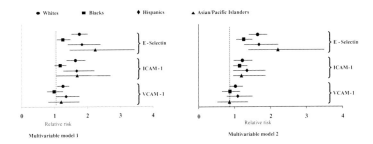

Fig. 2   Ethnicity-specific diabetes risk per 1 standard deviation increment in endothelial markers in four US ethnic groups. Based on values from Table 3 published in Song *et al.* (2006). Each standard deviation increase equals 20.4 ng/ml for E-selectin, 89.3 ng/ml for ICAM-1, and 245 ng/ml for VCAM-1. Model 1 was adjusted for matching factors (age, clinical center, time of blood draw, and follow-up duration), body mass index, alcohol intake, physical activity, cigarette smoking, post-menopausal hormone therapy, and family history of diabetes. Model 2 was simultaneously adjusted for E-selectin, ICAM1, and VCAM-1 in Model 1.

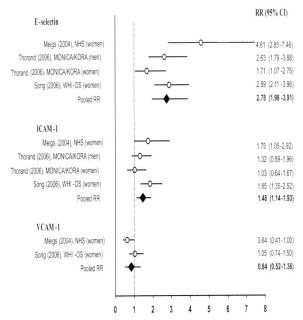

Fig. 3   Meta-analysis plot of relative risk (RR) and 95% confidence interval (CI) of type 2 diabetes for endothelial biomarkers in the highest quantities versus the lowest quantities. Unpublished meta-analysis results. Shaded diamonds indicate the pooled relative risks. The results were obtained from multivariable-adjusted models in each study when these biomarkers were adjusted simultaneously. The MONICA/KORA study included only E-selectin and ICAM-1. NHS, the Nurses' Health Study; MONICA/KORA, the Monitoring of Trends and Determinants in Cardiovascular Disease/Cooperative Research in the Region of Augsburg Study; WHI-OS, The Women's Health Initiative-Observational Study.

the specificity of soluble E-selection as a reflection of its membrane-bound form in the activated endothelium makes it a better surrogate than soluble ICAM-1 or VCAM-1 levels. Biologically, E-selectin is expressed exclusively by endothelial cells, while ICAM-1 is constitutively expressed on a number of cells including endothelium and leukocytes (Price and Loscalzo 1999, Ross 1999). VCAM-1 is expressed on activated endothelium and vascular smooth muscle cells (Price and Loscalzo 1999, Ross 1999). E-selectin may thus represent a more specific marker of the membrane-bound form of endothelium than other endothelial adhesion molecules. In addition, E-selectin levels appear to have a lower degree of intraindividual variability than other adhesion biomarkers. We cannot exclude the possibility that E-selectin simply has better assay measurement properties, allowing it to be measured accurately in the stored blood specimen.

ADHESION MOLECULES AND COMPLICATIONS OF TYPE 2 DIABETES

A plethora of *in vitro* and animal experimental studies and human studies support the central role of hyperglycemia in endothelial dysfunction (Price and

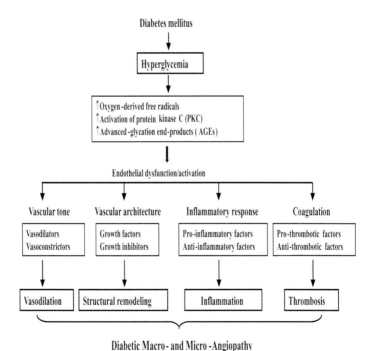

Fig. 4 Proposed pathophysiological pathways linking endothelial dysfunction to the development or progression of diabetic vascular complications.

Loscalzo 1999, Kim *et al.* 2006). Although the underlying mechanisms are not well understood, several pathways, including increased superoxide production and oxidative stress, activation of protein kinase C (PKC), and production of advanced-glycation end products (AGEs), have been proposed to explain the influence of hyperglycemia-mediated endothelial dysfunction on the development and progression of diabetic angiopathy (Fig. 4).

Numerous studies have shown increased levels of E-selectin, ICAM-1, and VCAM-1 in type 2 diabetes patients with microangiopathy or macroangiopathy compared to non-diabetic controls (Schram and Stehouwer 2005). First, elevated levels of adhesion molecules have been associated with hyperglycemia and/or insulin resistance in patients with type 2 diabetes. Several other cardiovascular risk factors such as cigarette smoking, hypertension, and dyslipidemia have also been associated with elevated levels of adhesion molecules. Since these factors tend to coexist to influence vascular function among diabetic patients, elevated levels of endothelial adhesion molecules may reflect the net involvement of several pathophysiological changes underlying insulin resistance and/or progressive β-cell dysfunction, including alterations of innate immune response, oxidative stress, thrombosis, and other metabolic homeostasis. It is tempting to further explore the temporal relation of endothelial function as reflected by adhesion molecules and diabetic complications; however, a few prospective studies have yielded inconsistent results. The Hoorn study reported that elevated VCAM-1 levels were significantly associated with the development of elevated urinary albumin excretion rate during a follow-up of 6.1 yr (Jager *et al.* 2002). This finding was confirmed by another prospective study of 328 type 2 diabetic patients, in which VCAM-1 and E-selectin were associated with increased urinary albumin excretion in patients with type 2 diabetes with 9-yr follow-up (Stehouwer *et al.* 2002). In these two studies, increased levels of VCAM-1 were significantly associated with increased risk of cardiovascular mortality (Jager *et al.* 2000) and total mortality in type 2 diabetes (Stehouwer *et al.* 2002). These associations were independent of inflammatory marker levels. Previous studies of endothelial adhesion molecules and retinopathy in type 2 diabetes have yielded conflicting results. In the Hoorn study of 625 individual with and without type 2 diabetes, baseline levels of ICAM-1, von Willebrand factor, and urinary albumin:creatinine ratio were combined to calculate a summary z score for endothelial dysfunction (van Hecke *et al.* 2005). This score was significantly associated with retinopathy among diabetic individuals but not in non-diabetic individuals (van Hecke *et al.* 2005). In the latter study, E-selectin was associated with the progression of retinopathy but not independently of HbA1c (Spijkerman *et al.* 2007). Given the complexity of the interactions involved in the pathophysiological cascade of diabetes with multiple abnormalities, differences in the population samples and basic characteristics may help account for inconsistencies in the literature. Whether and to what extent endothelial adhesion molecules have a direct causal impact on the multifactorial etiology of diabetic angiopathy remains elusive and requires further evaluation.

ADHESION MOLECULES AND OTHER TYPES OF DIABETES MELLITUS

Type 1 diabetes is an autoimmune disease characterized by absolute insulin deficiency. Hyperglycemia and its associated metabolic dysfunctions including endothelial damage could affect both macro- and microvasculature, leading to the development and progression of complications in patients with type 1 diabetes. Increased levels of soluble forms of various cell adhesion molecules, such as ICAM-1, VCAM-1, and ELAM-1 (endothelial leukocyte adhesion molecule 1), have been associated with type 1 diabetic patients with microangiopathy (Nowak *et al.* 2008). First, circulating levels of ICAM-1 appeared to be consistently associated with the presence or progression of retinopathy in patients with type 1 diabetes (Blann and Lip 1998). The association between ICAM-1 and type 1 diabetes has been supported by the findings from *in vitro* and *in vivo* data showing that ICAM-1 may regulate immune activation and inflammatory response (Blann and Lip 1998). Second, some but not all studies have reported elevated levels of E-selectin in patients with type 1 diabetes with microangiopathy. Due to sparse data, it remains controversial whether E-selectin is a reliable marker for type 1 diabetes with or without complications. Third, inferences from available studies regarding the link of soluble VCAM-1 to type 1 diabetes were hindered by their small sample sizes and inconsistent results. Finally, it is worth mentioning that endothelial damage may be involved in the pathogenesis of type 1 diabetes, but prospective data are lacking.

Gestational diabetes mellitus (GDM) was defined as glucose intolerance with onset or first recognition during pregnancy (Buchanan and Xiang 2005). Most women who developed GDM had underlying insulin resistance to which the insulin resistance of pregnancy was partly additive (Buchanan and Xiang 2005). There is no evidence suggesting that the formation of placental vascular bed causes alterations in endothelial function as reflected by the shedding of soluble adhesion molecules into maternal circulation during normal pregnancy (Chaiworapongsa *et al.* 2002). However, there is some evidence indicating a more significant rise in circulating levels of endothelial cell adhesion molecules in pregnant women with GDM as compared with healthy pregnant women. For instance, E-selectin and VCAM-1 levels have been observed to be significantly higher in pregnant women with GDM than in normal pregnant controls in some but not all studies (Kautzky-Willer *et al.* 1997, Lawrence *et al.* 2002, Telejko *et al.* 2009). Some studies have reported that levels of ICAM-1 and VCAM-1 among GDM women were similar to those in healthy pregnant women, although these endothelial markers were slightly lower in healthy pregnant women than in GDM women (Kautzky-Willer *et al.* 1997). Overall, findings from sparse and inclusive data indicated that endothelial dysfunction as reflected by persistent elevation of circulating adhesion molecules may be useful in predicting the risk of GDM and subsequent risk of type 2 diabetes in women with history of GDM. Nevertheless, further research is warranted to characterize the longitudinal changes of circulating endothelial cell adhesion molecules during pregnancy and to compare the levels of these molecules between GDM pregnancy and normoglycemic pregnancy.

IMPLICATIONS AND CONCLUSIONS

There are many important questions that must be answered before we formally incorporate measures of circulating adhesion molecules into diabetes risk prediction and treatment. First, endothelial dysfunction is a multifaceted process, and measurement of the three soluble adhesion molecules (i.e., E-selectin, ICAM-1, and VCAM-1) may not fully reflect the extent to which the vasculature has been altered by endothelial dysfunction. Soluble forms of these molecules are released from either passive shedding or active proteolytic cleavage from the endothelial cell surface (Price and Loscalzo 1999). The mechanisms regulating the release and clearance of soluble endothelial adhesion molecules are more likely to be variable for different adhesion molecules and remain incompletely defined. Second, additional research is needed to elucidate the physiological functions of soluble cellular adhesion molecules in circulation. Third, elevated plasma CRP levels may be a marker of an inflammatory component associated with a cluster of metabolic risk factors. Whether endothelial adhesion molecules have a direct causal impact on the multifactorial etiology of CVD and diabetes and the extent of such an impact remain elusive. The non-specific nature of endothelial dysfunction among insulin-resistant individuals with and without diabetes, and among patients with and without vascular complications, makes circulating cellular adhesion molecules less than ideal as a marker of diabetes or diabetic complications. Fourth, circulating levels of endothelial markers were assessed only once and had varying coefficients of variation. For example, the observed random measurement errors, especially for VCAM-1, may partially explain the null findings for VCAM-1 with diabetes in some studies. Fifth, owing to the lack of accurate assessment of preclinical atherosclerosis such as carotid artery intima-media thickness or coronary calcification, we cannot completely rule out the possibility that elevation in circulating adhesion molecule levels may be secondary to the presence of preclinical atherosclerosis, which is highly prevalent in both prediabetic individuals and diabetic patients. Finally, there is a notable lack of data from prospective studies regarding the practical clinical and public health significance of incorporating endothelial biomarker measurements into diabetes prevention, diagnosis and treatment, including type 1 and type 2 diabetes and GDM. Although careful consideration should be given to the incremental value of adding endothelial biomarker measures, more research in this area is needed.

In conclusion, epidemiological evidence for a relationship between circulating levels of endothelial adhesion molecules and type 1 and type 2 diabetes and their complications is substantial, although most previous studies have been cross-sectional. Nevertheless, findings from prospective data, though limited, have been consistent with those from cross-sectional studies, indicating that circulating levels of endothelial adhesion molecules may add significantly to clinical risk stratification and prognostic information for type 2 diabetes. Such findings, coupled with a large body of experimental work, advance our understanding of the pathophysiological role of endothelial dysfunction in the development and progression of diabetes mellitus. Although further research is needed, the assessment of circulating levels of endothelial adhesion molecules may have clinical value in identifying individuals who are at high risk and for whom pharmacological therapy is justified for diabetes risk reduction.

SUMMARY

- Early endothelial dysfunction is a common antecedent of the metabolic or insulin resistance syndrome.
- Elevated circulating levels of some endothelial adhesion molecules reflect structural microvascular changes related to the pathophysiological processes of impaired insulin action and secretion.
- Prospective data indicate that endothelial biomarkers, particularly E-selectin, may independently predict the risk of type 2 diabetes.
- Circulating levels of endothelial adhesion molecules appear to have significantly clinical prognostic value beyond inflammatory marker-CRP in predicting future risk of type 2 diabetes and diabetic complications.
- The clinical utility of assessing these adhesion molecules for diabetes risk stratification and management deserves further investigation.

Abbreviations

AGEs	advanced-glycation end products
CRP	C-reactive protein
CVD	cardiovascular disease
DPS	The Finnish Diabetes Prevention Study
GDM	gestational diabetes mellitus
HbA1c	hemoglobin A1c
ICAM-1	intercellular cell adhesion molecule 1
NHS	the Nurses' Health Study
PKC	protein kinase C
VCAM-1	vascular cell adhesion molecule 1
WHI-OS	The Women's Health Initiative-Observational Study

Key Facts about Diabetes Mellitus

1. Diabetes mellitus is a group of metabolic diseases characterized by high blood glucose levels that result from impairments in insulin action and/or secretion. In 2000, at least 171 million people worldwide were suffering from diabetes, or 2.8% of the population. It is estimated that by the year 2030, these numbers will double.

2. Type 1 diabetes usually occurs in children and young adults. In type 1, the pancreas makes little or no insulin. About 5-10% of all diabetes cases are type 1.

3. Type 2 diabetes is the most common form of diabetes. It is caused by impaired insulin action and/or insulin secretion.

4. Gestational diabetes is defined as glucose intolerance with onset or first recognition during pregnancy and places both the mother and baby at increased risk for certain health problems later in life, including type 2 diabetes.

5. Type 2 diabetes and CVD may share the pathophysiological mechanisms underlying the metabolic syndrome/insulin resistance syndrome.

6. Diabetic angiopathy is a major cause of morbidity and mortality in diabetes mellitus. Cardiovascular complications are the leading cause of death in diabetes.

Definition of Terms

Cross-sectional study: A study done at one time, not over the course of time.

Diabetes risk factors: Factors contributing to a person's risk of developing diabetes.

Endothelial dysfunction: The loss of proper endothelial function, which is very common in insulin resistance syndrome, diabetes mellitus, hypertension, or cardiovascular disease.

Metabolic syndrome/insulin resistance syndrome: A cluster of cardiovascular risk factors including abdominal obesity, high blood pressure, high blood glucose, high triglycerides and low levels of high-density lipoprotein (HDL) cholesterol. Insulin resistance: A condition in which the body is unable to use available insulin effectively.

Prospective study: A study in which the subjects are identified and then followed forward in time.

Systemic inflammation: The body's overall inflammatory response to infection, injury, or metabolic disorders.

References

Blann, A.D. and G.Y. Lip. 1998. Endothelial integrity, soluble adhesion molecules and platelet markers in type 1 diabetes mellitus. Diabet. Med. 15: 634-642.

Buchanan, T.A. and A.H. Xiang. 2005. Gestational diabetes mellitus. J. Clin. Invest. 115: 485-491.

Chaiworapongsa, T. and R. Romero, J. Yoshimatsu, J. Espinoza, Y.M. Kim, K. Park, K. Kalache, S. Edwin, E. Bujold, and R. Gomez. 2002. Soluble adhesion molecule profile in normal pregnancy and pre-eclampsia. J. Matern. Fetal Neonatal Med. 12: 19-27.

Herder, C. and M. Peltonen, W. Koenig, I. Kraft, S. Muller-Scholze, S. Martin, T. Lakka, P. Ilanne-Parikka, J.G. Eriksson, H. Hamalainen, S. Keinanen-Kiukaanniemi, T.T. Valle, M. Uusitupa, J. Lindstrom, H. Kolb, and J. Tuomilehto. 2006. Systemic immune mediators and lifestyle changes in the prevention of type 2 diabetes: results from the Finnish Diabetes Prevention Study. Diabetes 55: 2340-2346.

Jager, A. and V.W. van Hinsbergh, P.J. Kostense, J.J. Emeis, G. Nijpels, J.M. Dekker, R.J. Heine, L.M. Bouter, and C.D. Stehouwer. 2000. Increased levels of soluble vascular cell adhesion molecule 1 are associated with risk of cardiovascular mortality in type 2 diabetes: the Hoorn study. Diabetes 49: 485-491.

Jager, A. and V.W. van Hinsbergh, P.J. Kostense, J.J. Emeis, G. Nijpels, J.M. Dekker, R.J. Heine, L.M. Bouter, and C.D. Stehouwer. 2002. C-reactive protein and soluble vascular cell adhesion molecule-1 are associated with elevated urinary albumin excretion but do not explain its link with cardiovascular risk. Arterioscler. Thromb. Vasc. Biol. 22: 593-598.

Kautzky-Willer, A. and P. Fasching, B. Jilma, W. Waldhausl, and O.F. Wagner. 1997. Persistent elevation and metabolic dependence of circulating E-selectin after delivery in women with gestational diabetes mellitus. J. Clin. Endocrinol. Metab. 82: 4117-4121.

Kim, J.A. and M. Montagnani, K.K. Koh, and M.J. Quon. 2006. Reciprocal relationships between insulin resistance and endothelial dysfunction: molecular and pathophysiological mechanisms. Circulation 113: 1888-1904.

Krakoff, J. and T. Funahashi, C.D. Stehouwer, C.G. Schalkwijk, S. Tanaka, Y. Matsuzawa, S. Kobes, P.A. Tataranni, R.L. Hanson, W.C. Knowler, and R.S. Lindsay. 2003. Inflammatory markers, adiponectin, and risk of type 2 diabetes in the Pima Indian. Diabetes Care 26: 1745-1751.

Lawrence, N.J. and E. Kousta, A. Penny, B. Millauer, S. Robinson, D.G. Johnston, and M.I. McCarthy. 2002. Elevation of soluble E-selectin levels following gestational diabetes is restricted to women with persistent abnormalities of glucose regulation. Clin. Endocrinol. 56: 335-340.

Meigs, J.B. and F.B. Hu, N. Rifai, and J.E. Manson. 2004. Biomarkers of endothelial dysfunction and risk of type 2 diabetes mellitus. JAMA 291: 1978-1986.

Nowak, M. and T. Wielkoszynski, B. Marek, B. Kos-Kudla, E. Swietochowska, L. Sieminska, D. Kajdaniuk, J. Glogowska-Szelag, and K. Nowak. 2008. Blood serum levels of vascular cell adhesion molecule (sVCAM-1), intercellular adhesion molecule (sICAM-1) and endothelial leucocyte adhesion molecule-1 (ELAM-1) in diabetic retinopathy. Clin. Exp. Med. 8: 159-164.

Pinkney, J.H. and C.D. Stehouwer, S.W. Coppack, and J.S. Yudkin. 1997. Endothelial dysfunction: cause of the insulin resistance syndrome. Diabetes 46 (Suppl 2): S9-S13.

Price, D.T. and J. Loscalzo. 1999. Cellular adhesion molecules and atherogenesis. Am. J. Med. 107: 85-97.

Ross, R. 1999. Atherosclerosis—an inflammatory disease. N. Engl. J. Med. 340: 115-126.

Sattar, N. and H.M. Murray, P. Welsh, G.J. Blauw, B.M. Buckley, A.J. de Craen, I. Ford, N.G. Forouhi, D.J. Freeman, J.W. Jukema, P.W. Macfarlane, M.B. Murphy, C.J. Packard, D.J. Stott, R.G. Westendorp, and J. Shepherd. 2009. Are elevated circulating intercellular adhesion molecule 1 levels more strongly predictive of diabetes than vascular risk? Outcome of a prospective study in the elderly. Diabetologia 52: 235-239.

Schram, M.T. and C.D. Stehouwer. 2005. Endothelial dysfunction, cellular adhesion molecules and the metabolic syndrome. Horm. Metab. Res. 37 (Suppl 1): 49-55.

Song, Y. and J.E. Manson, L. Tinker, N. Rifai, N.R. Cook, F.B. Hu, G.S. Hotamisligil, P.M. Ridker, B.L. Rodriguez, K.L. Margolis, A. Oberman, and S. Liu. 2007. Circulating levels of endothelial adhesion molecules and risk of diabetes in an ethnically diverse cohort of women. Diabetes 56: 1898-1904.

Spijkerman, A.M. and M.A. Gall, L. Tarnow, J.W. Twisk, E. Lauritzen, H. Lund-Andersen, J. Emeis, H.H. Parving, and C.D. Stehouwer. 2007. Endothelial dysfunction and low-grade inflammation and the progression of retinopathy in Type 2 diabetes. Diabet. Med. 24: 969-976.

Stehouwer, C.D. and M.A. Gall, J.W. Twisk, E. Knudsen, J.J. Emeis, and H.H. Parving. 2002. Increased urinary albumin excretion, endothelial dysfunction, and chronic low-grade inflammation in type 2 diabetes: progressive, interrelated, and independently associated with risk of death. Diabetes 51: 1157-1165.

Telejko, B. and A. Zonenberg, M. Kuzmicki, A. Modzelewska, K. Niedziolko-Bagniuk, A. Ponurkiewicz, A. Nikolajuk, and M. Gorska. 2009. Circulating asymmetric dimethylarginine, endothelin-1 and cell adhesion molecules in women with gestational diabetes. Acta. Diabetol. (in press).

Thorand, B. and J. Baumert, L. Chambless, C. Meisinger, H. Kolb, A. Doring, H. Lowel, and W. Koenig. 2006. Elevated markers of endothelial dysfunction predict type 2 diabetes mellitus in middle-aged men and women from the general population. Arterioscler. Thromb. Vasc. Biol. 26: 398-405.

van Hecke, M.V. and J.M. Dekker, G. Nijpels, A.C. Moll, R.J. Heine, L.M. Bouter, B.C. Polak, and C.D. Stehouwer. 2005. Inflammation and endothelial dysfunction are associated with retinopathy: the Hoorn Study. Diabetologia 48: 1300-1306.

CHAPTER 13

Adhesion Molecules in Obesity and Metabolic Syndrome

Michelle A. Miller

Associate Professor (Reader) of Biochemical Medicine, Clinical Sciences
Research Institute (University of Warwick), Warwick Medical School,
UHCW Campus, Clifford Bridge Road, Coventry CV2 2DX (UK),
E-mail: michelle.miller@warwick.ac.uk
Departmental contact: Ms Patricia McCabe, Personal Assistant,
Clinical Sciences Research Institute, Warwick Medical School,
UHCW Campus, Clifford Bridge Road, Coventry CV2 2DX (UK),
E-mail: Patricia.McCabe@warwick.ac.uk

ABSTRACT

Adhesion molecules are important in atherosclerotic plaque development. Soluble
adhesion molecules that lack cytoplasm and membrane spanning domains can
be found in the circulation. Their levels are monitored as markers of endothelial
activation and studies have reported an association between some soluble
adhesion molecule levels and both cardiovascular and coronary heart disease, the
prevalence of which is more common in individuals with metabolic syndrome
(MetS). Relationships between soluble adhesion molecules, measures of obesity
and other risk factors for MetS (including serum lipids, blood pressure and
insulin) are adhesion molecule- and gender-specific. This may in part explain
inconsistencies reported in the literature. However, robust associations have been
observed between sE-selectin and measures of obesity, blood pressure, serum
lipids and insulin. Relationships between sE-selectin and blood pressure appear to
be strongest in young women and may relate to their menopausal status. Soluble
intracellular adhesion molecule 1 levels are also increased in obese individuals

---

Key terms are defined at the end of the chapter.

and those with MetS. Ethnic differences in circulating adhesion molecules and the relationship with MetS have been reported and need further investigation. Expression of cellular adhesion molecules is increased by activation of the nuclear factor-kappaB (NF-κB) pathway and this may provide a therapeutic target for disease prevention. Activation of cellular adhesion molecules may also play a role in the observed association between sleep and cardiovascular disease.

INTRODUCTION

Ischemic heart disease and cardiovascular disease (CVD) remain the most common causes of death. While 'traditional' risk factors such as blood pressure and serum lipids may account for a large proportion of an individual's risk, these factors do not explain it entirely. Evidence increasingly suggests that inflammatory pathways may provide the common underlying mechanism for the development of these diseases. Adhesion molecules have been implicated in early atherosclerotic lesion development (Krieglstein and Granger 2001) and plasma levels of intracellular adhesion molecule 1 (ICAM-1) and E-selectin may act as markers of atherosclerosis and the development of coronary heart disease (CHD) (Hwang *et al.* 1997). Ethnic differences exist, however, in these inflammatory mediators and the prevalence of CVD and CHD (Miller *et al.* 2003, Miller and Cappuccio 2007a).

Innate Immune System and TLRs

The innate immune system is the system conserved by evolution that allows the body to respond immediately to perceived threats to bodily integrity. It is related to, but completely distinct from, the acquired immune system. Generalized dysfunction of the endothelial cells (which cover the walls of the arteries and veins) may precede the development of atherosclerosis and underlie the development of both CVD and CHD. This process is initiated by the accumulation of macrophages and low-density lipoproteins (LDL), which lead to an activation of innate immune toll-like receptors (TLRs). These combine with the pattern recognition molecule CD14 to form a complex (TLR4-CD14), which activates the nuclear factor-kappaB (NF-κB) pathway. Activation of this pathway mediates adhesion molecule expression, cytokine production and associated local inflammation (Miller and Cappuccio 2007a). Leukocyte adhesion molecules and chemokines promote recruitment of monocytes and T cells. Monocytes differentiate into macrophages and further upregulate pattern recognition receptors that internalize accumulated lipoproteins in the intima to form foam-cells and subsequently atherosclerotic plaques. T cells in lesions recognize local antigens and mount T helper-1 responses with secretion of pro-inflammatory cytokines that contribute to local inflammation and plaque growth. Prolonged inflammatory activation may lead to local proteolysis, plaque rupture and thrombus formation, which causes ischemia, infarction and CHD (Fig. 1).

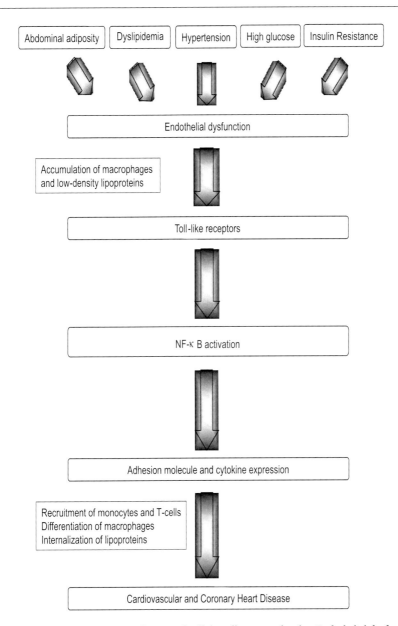

Fig. 1 Obesity, metabolic syndrome and cellular adhesion molecules. Endothelial dysfunction acting through the toll-like receptor pathway may lead to the development of cardiovascular disease and coronary heart disease (M.A. Miller, unpublished).

Adhesion Molecules (ICAM-1, VCAM-1, E-selectin, P-selectin)

During atherosclerotic plaque development the adhesion molecule pathway is activated. Expressed adhesion molecules attract leukocytes to the endothelium (Fig. 2). Cells are attracted to the endothelial adhesion molecule receptors (a). The cells then 'roll' along the endothelium (b), become firmly attached to it (c) and migrate into the subintimal spaces (d), where they take up lipids to become foam cells and fatty plaques (Krieglstein and Granger 2001). The selectin family of adhesion molecules are involved in the adhesion of leukocytes to the activated endothelium and the observed cell 'rolling'. Two different adhesion molecules belonging to the Immunoglobulin (Ig) superfamily, ICAM-1 and vascular cell adhesion molecule 1 (VCAM-1), are involved in the extravasation of leukocytes into the surrounding tissue. The expressed endothelial adhesion molecules bind to their complementary ligands on the leukocytes, many of which belong to the integrin family. P-selectin is stored in specific granules that are present in platelets and endothelial cells from where it is mobilized to the cell surface after stimulation. E-selectin, however, is not stored, but increased surface expression can occur in response to transcription-dependent protein synthesis.

(a) Cells are attracted to endothelial receptors.

(b) The cells 'roll' along the endothelium.

(c) Firm attachment takes place.

(d) Cells migrate into the subintimal spaces. Lipids are taken up and the formation of foam cells and fatty plaques occurs.

Fig. 2 Diagram showing leukocyte attachment to the endothelial lining via endothelial adhesion molecules and their receptors. These images depict the process whereby leukocytes are attracted towards and then bind to the endothelial lining. Unpublished computer generated images: M.A. Miller and S.P. Miller (Poser 7 (c) 1991-2008).

Soluble adhesion molecules lacking cytoplasmic and membrane spanning domains are a reflection of endothelial activation. They can be found in the circulation (Pigott *et al.* 1992) and increased levels have been found in both carotid atherosclerosis and CHD (Hwang *et al.* 1997).

ADHESION MOLECULES AND OBESITY

The definition of obesity varies with ethnicity and gender, but in general an adult who has a body mass index (BMI) between 25 and 29.9 is considered overweight, whereas an adult who has a BMI of 30 or higher is considered obese. The rate of increase in the number of overweight and obese individuals has risen dramatically in the past two decades (Li *et al.* 2007), with the excess body weight being stored as fat. The distribution of the body fat, however, is of utmost importance. In epidemiology studies, despite the plethora of methods used to measure fat (including waist-hip ratio (WHR) and subscapular skinfold measurements), the findings have demonstrated a consistent association between central adiposity and CVD risk, strokes and CHD mortality. This has been observed in both men and women and in most cases is independent of BMI. Those individuals with visceral obesity appear to represent the group of individuals with greatest CVD risk (Després *et al.* 2008).

Adhesion Molecules and WHR and BMI

Evidence relating WHR and BMI to circulating cellular adhesion molecule levels is not clear cut. Some studies indicate a positive correlation between, for example, sICAM-1 and WHR (Demerath *et al.* 2001) and sICAM-1 and BMI (Hwang *et al.* 1997) and others do not (DeSouza *et al.* 1997, Rohde *et al.* 1999, Miller *et al.* 2003, Ponthieux *et al.* 2004). Likewise, sE-selectin has been shown to be related to WHR or BMI in some studies (Hwang *et al.* 1997, Miller *et al.* 2003, Ponthieux *et al.* 2004, Miller and Cappuccio 2006) but not all (DeSouza *et al.* 1997).

In a study to investigate the relationship between adhesion molecules and measures of obesity we found the strongest and most consistent relationships were those between sE-selectin and both WHR and BMI (Miller and Cappuccio 2006). The associations for the total group studied are shown in Table 1. These relationships were maintained following adjustment for multiple confounders including blood pressure and serum lipids and there was no interaction with gender or ethnicity. A 0.01 unit higher WHR and a 1 unit greater BMI would be associated with an approximate 2% higher sE-selectin level (Fig. 3).

In the Health Professional Follow-up Study of 18,225 men, sICAM-1 and sVCAM-1 were found to be directly associated with obesity and CHD risk factors. Following multiple adjustment the relative risk of CHD was 1.69 [1.14-2.51 (95%CI)] for sICAM-1 and 1.34 [0.91-1.96] for sVCAM-1. Furthermore, those

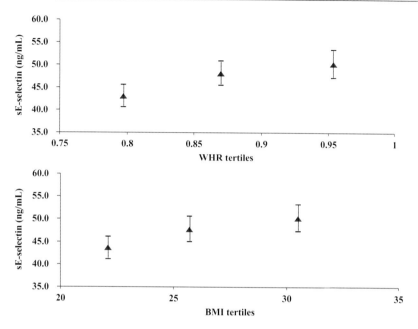

Fig. 3 sE-selectin levels by tertiles of waist-hip ratio (upper panel) and tertiles of body mass index (lower panel) in 664 individuals from the Wandsworth Heart and Stroke Study. Adjusted for age, sex, ethnicity, and smoking. Results are geometric means and 95% CI. P < 0.001 for both. Adapted from Miller and Cappuccio (2006), with permission.

individuals with elevated levels of both molecules had an almost 2.5-fold increased risk of CHD (Shai *et al.* 2006).

sICAM-1 and sE-selectin were significantly correlated with BMI and measures of visceral, but not subcutaneous, fat in morbidly obese individuals both before and one year after bariatric restrictive surgery (Pontiroili *et al.* 2008). By contrast, a study comparing obese and non-obese type 2 diabetes mellitus patients (T2DM) did not find any relationship between sICAM-1 and sVCAM-1 and BMI (Matsumoto *et al.* 2002). Furthermore, while sE-selectin levels were significantly and independently related to BMI, they were not independently related to regional fat distribution. These results are consistent with the idea that obesity may induce endothelial activation or increased shedding of cell surface E-selectin leading to subsequent increase in sE-selectin levels. In another study, in non-diabetic obese women, sICAM-1 and sE-selectin were positively correlated with measures of central obesity. Weight reduction resulted in a decrease in these adhesion molecules that was explained by a change in trunk fat mass (Ito *et al.* 2002). The relationship between adhesion molecules and weight loss are discussed in chapter 14.

Table 1 Partial correlation coefficients of associations between cellular adhesion molecules and risk factors for metabolic syndrome and cardiovascular disease

	Waist-hip ratio	Body mass index	Serum triglycerides #	HDL-cholesterol	Systolic blood pressure	Diastolic blood pressure	Serum fasting glucose	Serum fasting insulin #
sE-selectin#	0.21***	0.19***	0.18***	−0.09*	0.17***	0.19***	0.10*	0.20***
sP-selectin#	0.01	−0.03	0.17***	−0.12**	0.08*	0.03	0.10*	0.07
sICAM-1#	0.09*	−0.02	0.20***	−0.14***	0.04	−0.002	0.04	0.05
sVCAM-1#	0.08*	0.04	0.08*	−0.17***	0.03	−0.01	0.01	0.02

Results from a total group of 664 men and women (261 whites, 215 South Asians and 188 Africans) from the Wandsworth Heart and Stroke Study: Adjusted for age, sex, smoking and ethnicity. *p < 0.05, **p < 0.01, ***p < 0.001. #Analysis performed on $\log_e$ transformed data. Adapted from Miller and Capuccio (2006), with permission.

ADHESION MOLECULES AND METABOLIC SYNDROME

Metabolic syndrome (MetS) is the name given to a group of risk factors that increase an individual's risk of CHD and other related problems including diabetes and stroke. These include: abdominal adiposity (as defined by a large WHR); a higher than normal triglyceride level; a lower than normal level of high density lipoprotein cholesterol (HDL-cholesterol); a higher than normal blood pressure; and a higher than normal fasting glucose level or insulin resistance. While the World Health Organization and the National Cholesterol Education Program Adult Treatment Panel III (NCEP-ATPIII) have slightly different cut-off values and diagnostic criteria (Haffner 2007), in principle a diagnosis of MetS is made when three or more of the five previously listed factors are present. Individuals with MetS have an aggressive form of vascular disease that is characterized by accelerated atherosclerosis and pro-inflammatory changes (Haffner 2007). The exact pathophysiology of MetS is unknown but is thought to be mediated through insulin resistance and visceral obesity. Men with MetS, but not diabetes, have a twofold increased CVD risk, which is increased to threefold in the presence of diabetes. This increased risk in the presence of diabetes is even more apparent in women. In the absence of diabetes, women with MetS have a twofold greater risk of CVD than women without MetS but in the presence of diabetes this risk increases dramatically to eightfold (Haffner 2007), highlighting the therapeutic importance of addressing gender-specific multiple risk factors.

The relationship between WHR and adhesion molecules has been previously examined and in the following section the relationship between adhesion molecule levels and the other risk factors for MetS will be examined before looking at the effect of combinations of these factors.

Adhesion Molecules and Serum Triglycerides

High serum levels of triglyceride-rich lipoproteins, which result either from the diet or from an individual's genetic background, are linked to the development of atherosclerosis. Unstable plaques are characterized by the presence of an abundance of macrophages and other inflammatory cells together with a paucity of smooth muscle cells and a thin fibrous cap. Triglyceride-rich lipoproteins, their remnants and small LDL particles have a pro-inflammatory action and, while this appears to be mainly a result of their very high susceptibility to oxidation, other studies have also indicated that they may have a direct effect on NF-κB, or the effect may be mediated by tissue necrosis factor alpha (TNF-α) (Libby 2007).

Studies have demonstrated that adhesion molecules are related to serum triglyceride levels (Rohde *et al.* 1999, Miller *et al.* 2003, Miller and Cappuccio 2006). Following a meal there is an increase in serum lipids and competition for the endothelium-bound lipoprotein lipase (LPL) enzyme, which is responsible for hydrolysis of triacylglycerols into glycerols and free fatty acids. This competition is particularly evident in individuals with the MetS in whom fasting hypertrigicideaemia is present (Lewis *et al.* 1991). During this post-prandial hyperlipidaemic state there is a concomitant increase in neutrophil count, expression of adhesion molecules, production of pro-inflammatory cytokines and subsequent penetration on the endothelium by chylomicrons and their remnants, leading to the formation of foam cells and thus increased atherosclerotic risk (Alipour *et al.* 2007).

Adhesion Molecules and HDL-Cholesterol

Epidemiological and clinical studies have provided evidence to show that low levels of HDL-cholesterol are a risk factor for CVD and CHD. Furthermore, serum HDL-cholesterol is significantly and inversely related to sE-selectin, sP-selectin, sICAM-1 and sVCAM-1 levels (Miller and Cappuccio 2006) (see Table 1).

Adhesion Molecules and Blood Pressure

Hypertension is recognized as being a multi-factorial disease arising from a combination of both genetic and environmental factors that may also involve inflammatory processes. The relationship between adhesion molecules and blood pressure varies according to the adhesion molecule examined and is dependent on both sex and age. The exact mechanisms relating blood pressure to adhesion molecules are unknown but may, in women, be modified by menopausal status (Miller *et al.* 2004). Studies on this topic have been few to date, with many inconsistencies, but a number of recent studies now report an association between sE-selectin and blood pressure. One study reported a relationship between both systolic and diastolic blood pressure and sE-selectin (Demerath *et al.* 2001). Likewise, we found, in a study based on an untreated population, the strongest relationship between adhesion molecules and both systolic and diastolic blood pressure (SBP and DBP respectively) was with sE-selectin, a marker of endothelial activation (Fig. 4) (Miller *et al.* 2004). This relationship, which is consistent across different ethnic groups, is independent of BMI and smoking and found mainly in women younger than 50 years of age. A more recent study has demonstrated that the level of both sE-selectin and sP-selectin but not sICAM-1 or sVCAM-1 is raised in hypertensive compared to normotensive individuals and that treatment with Benidipine, a long-acting calcium channel blocker, led to a decrease in blood pressure and a time-dependent reduction in adhesion molecule levels (Sanada *et al.* 2005).

Discrepancies between studies may arise when different confounders are adjusted for and when hypertensive status, as opposed to actual blood pressure, is examined. Each adhesion molecule has a specific role in the adhesion pathway and

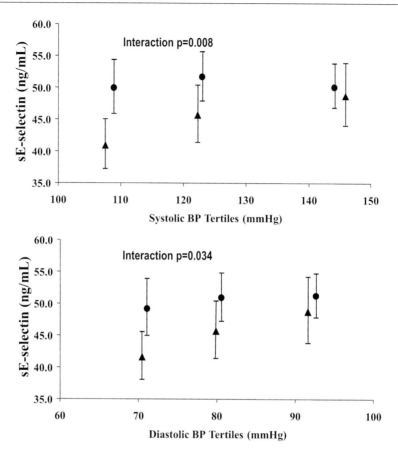

Fig. 4   sE-selectin concentrations in women and men by tertiles of blood pressure (BP). Adjusted for age, ethnicity, body mass index and smoking (geometric means and 95% confidence intervals). Women (triangles); Men (circles). In women < 50 years a 10 mmHg increase in systolic blood pressure is associated with a 7.4% increase in sE-selectin. Produced from author's original data (see also Miller *et al.* 2004).

may therefore have a different role in the development of CHD. Given the clear gender differences observed and the interaction with age, it is also possible that different mechanisms may operate in men and women. This needs to be examined more carefully.

Adhesion Molecules, Glucose Intolerance, Insulin Resistance

Studies have reported that adhesion molecule levels are increased in individuals with T2DM and these relationships are discussed in a separate chapter. A positive relationship between both sE-selectin and sP-selectin and glucose has been reported in the Stanislas study (Ponthieux *et al.* 2004). A similar observation was reported in the Wandsworth Heart and Stroke Study, which also observed an association between sE-selectin and fasting insulin (Miller *et al.* 2003, Miller and Cappuccio 2006) (Table 1).

Adhesion Molecules in Individuals with the Combined Risk Factors of Metabolic Syndrome

Previously, the relationship between the individual factors for MetS and adhesion molecule levels was examined. In this section, although the number of studies to date is few, the relationship between combinations of these factors (as in MetS) and adhesion molecule levels is examined.

In the Bruneck study, levels of sE-selectin, sP-selectin, sICAM-1 and sVCAM-1 were higher in individuals with MetS, as defined by the NCEP-ATPIII criteria, than in controls (Bonora *et al.* 2003). Increased levels were shown to be particularly associated with increased levels of insulin resistance.

Low HDL-cholesterol leads to an increase in inflammation and, in the presence of increasing numbers of risk factors of the MetS, there is an aggravated increase in cellular adhesion molecule levels (Soro-Paavonen *et al.* 2006).

In a community-based sample of elderly male and female individuals, sICAM, sVCAM-1 and sE-selectin (but not sP-selectin) were found to be higher in individuals with MetS. Following multivariate logistic regression, sVCAM-1 and sE-selectin remained strongly associated with MetS and insulin resistance (Ingelsson *et al.* 2008). Furthermore, a recent study in adult males found that, as the number of components of the MetS present in an individual is increased, so there is an incremental increase in soluble adhesion molecule levels (Gomez Rosso *et al.* 2008). Increased levels of sICAM-1 and sE-selectin have also been reported in Caucasians but not in African-American children with MetS (Lee *et al.* 2008). In a population-based study of normal individuals, ethnic differences in cellular adhesion molecules have been previously reported (Miller *et al.* 2003, Miller and Cappuccio 2007a). Likewise, ethnic differences in the strength of the association between the individual components of the MetS and adhesion molecules (Miller and Cappuccio 2006) have been observed. It will be interesting to see whether ethnic differences are observed in adults with MetS and whether these are translated into the observed differences in long-term CVD and CHD risk.

APPLICATIONS TO OTHER AREAS OF HEALTH AND DISEASE

Novel Therapeutic Interventions

Understanding the mechanism by which cellular adhesion molecules are related to MetS and its individual components is important for the development of new diagnostic and treatment regimes. Novel therapies currently being investigated include the use of fish oil supplements and hormone replacement therapy. Lipid-lowering statins may also be important as not only do they reduce cholesterol, but they also have other pleiotrophic effects that may lead to a reduction in adhesion molecule levels. Statins may reduce adhesion molecule expression by increasing nitric oxide production, decreasing the susceptibility of LDL to oxidation, and through their antioxidant effects inhibit the migration of macrophages and smooth muscle cell proliferation (Tsiara *et al.* 2003). Peroxisome proliferator-activated receptor-α (PPAR-α) agonists, used to regulate lipid and glucose homeostasis, may reduce adhesion molecule levels by preventing the activation of NF-κB.

Diagnosis of Disease

Circulating adhesion molecule levels may be useful tools for diagnosis of disease and for stratifying individuals according to disease severity and prognosis. They may also provide insights into disease mechanisms.

Role of Adhesion Molecules in Sleep-related CVD

Emerging evidence suggests that disturbances in sleep and sleep disorders, including obstructive sleep apnoea, play a role in the development and morbidity of chronic conditions including CVD. This in part may be mediated by inflammatory mechanisms (Miller and Cappuccio 2007b). The observed periods of hypoxia that are associated with obstructive sleep apnoea may lead to activation of NF-κB, TNF-α and ultimately cellular adhesion molecule expression (Fig. 5).

SUMMARY

1. Adhesion molecules are important in atherosclerotic plaque development.
2. Soluble adhesion molecules can be found in the circulation.
3. Soluble adhesion molecule levels are related to risk factors for, and the development of, cardiovascular disease (CVD) and coronary heart disease (CHD).
4. The prevalence of CVD is more common in individuals with metabolic syndrome (MetS).
5. Relationships between soluble adhesion molecules and measures of obesity and other risk factors for MetS are adhesion molecule- and gender-specific.
6. Consistent associations have been observed between sE-selectin and measures of obesity, as well as blood pressure, serum lipids and insulin.

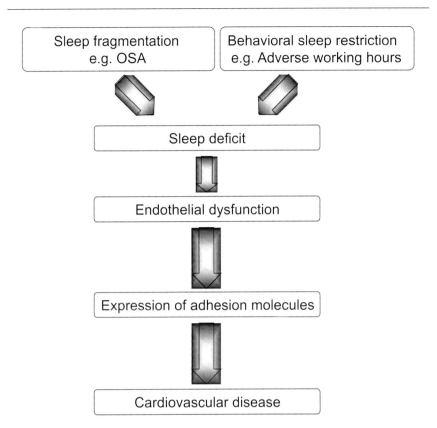

Fig. 5 Sleep loss, adhesion molecule expression and cardiovascular disease. The diagram shows the possible relationship between sleep deprivation and cardiovascular disease (M.A. Miller, unpublished). OSA, obstructive sleep apnoea.

7. Relationships between sE-selectin and blood pressure appear to be strongest in young women and may relate to menopausal status.

8. Ethnic differences in circulating adhesion molecules and the relationship with MetS have been reported and need further investigation.

9. Expression of cellular adhesion molecules is increased by activation of the nuclear factor-kappaB (NF-κB) pathway.

10. Adhesion molecule activation may play a role in the observed association between short sleep duration and CVD.

Acknowledgements

With thanks to Professor F.P. Cappuccio for his collaborative support and S.P. Miller for the computer-generated images used in Fig. 2.

Abbreviations

BMI	body mass index
CHD	coronary heart disease
CVD	cardiovascular disease
DBP	diastolic blood pressure
HDL	high-density lipoprotein
*ICAM-1	intracellular adhesion molecule 1
Ig	immunoglobulin
LDL	low-density lipoproteins
LPL	lipoprotein lipase
MetS	metabolic syndrome
NCEP-ATPIII	national Cholesterol Education Program Adult Treatment Panel III
NF-κB	nuclear factor-kappaB
OSA	obstructive sleep apnoea
PPAR-α	peroxisome proliferator-activated receptor-α
SBP	systolic blood pressure
T2DM	type 2 diabetes mellitus
TLRs	toll-like receptors
TNF-α	tissue necrosis factor-alpha
*VCAM-1	vascular cell adhesion molecule 1
WHR	waist-hip ratio

Key Facts about Obesity and Metabolic Syndrome

1. MetS is the name for a group of risk factors that increase an individual's chance of developing CHD, CVD, diabetes and stroke.
2. The diagnosis of MetS is made when an individual has any three out of the five following risk factors: a large waistline, a higher than normal triglyceride level, a lower than normal level of HDL-cholesterol, a higher than normal blood pressure and a higher than normal fasting blood glucose.
3. An adult with a BMI of 30 or higher is considered to be obese.
4. A number of underlying mechanisms have been proposed for the development of MetS and include insulin resistance and inflammatory mechanisms.

---

* soluble form of adhesion molecules

5. Increased activation of the NF-κB pathway and expression of cellular adhesion molecules are found in individuals with MetS.

6. sICAM-1 and sE-selectin are consistently found to be increased in obese individuals and those with MetS.

7. Healthy lifestyle changes, dietary modifications and therapeutic intervention can help to reverse or reduce the risk of CVD and CHD associated with obesity and MetS.

Definition of Terms

Atherosclerotic plaque development: Endothelial dysfunction leads to the accumulation of macrophages, lipids and smooth muscle cells in the endothelial cells and the subsequent development of plaque, which may result in the formation of a blood clot and CHD.

Endothelial cells: Cells that provide a selective barrier between the blood flowing in the lumen and the inner layers of the vessel wall; they have a number of other roles including regulating blood flow and controlling blood pressure and inflammatory and immune function.

Endothelial dysfunction: Endothelial dysfunction occurs if the endothelium is damaged and the cells lose their normal function. It may result from disease processes such as hypertension and it is a key event in the development of atherosclerosis.

Endothelium: The thin layer of endothelial cells that line the inner walls of blood vessels.

Metabolic syndrome (MetS): The name given to a group of risk factors that increase an individual's risk of CHD and associated problems.

Nuclear Factor-kappaB (NF-κB) signaling pathway: NF-κB is a protein complex that acts as a transcription factor. It is involved in the cellular responses to stimuli such as stress, cytokines, oxidized LDL, and bacterial or viral antigens. It is found in many cell types and plays key roles in regulating the immune response to infection and controlling inflammatory genes. It is important in the development of atherosclerosis and CHD.

Obesity: An adult with a BMI between 25 and 29.9 is considered overweight and one with a BMI of 30 or higher is considered obese.

References

Alipour, A. and J.W. Elte, H.C. van Zaanen, A.P. Rietveld, and M.C. Cabezas. 2007. Postprandial inflammation and endothelial dysfunction. Biochem. Soc. Trans. 35: 466-469.

Bonora, E. and S. Kiechl, J. Willeit, F. Oberhollenzer, G. Egger, R.C. Bonadonna, and M. Muggeo. 2003. Metabolic Syndrome: epidemiology and more extensive phenotypic description. Cross-sectional data from the Bruneck Study. Int. J. Obesity 27: 1283-1289.

Demerath, E. and B. Towne, J. Blangero, and R.M. Siervogel. 2001. The relationship of soluble ICAM-1, VCAM-1, P-selectin and E-selectin to cardiovascular disease risk factors in healthy men and women. Ann. Human Biol. 28: 664-678.

DeSouza, C.A. and D.R. Dengel, R.F. Macko, K. Cox, and D.R. Seals. 1997. Elevated levels of circulating cell adhesion molecules in uncomplicated essential hypertension. Am. J. Hypertens. 10: 1335-1341.

Després, J.P. and B.J. Arsenault, M. Côté, A. Cartier, and I. Lemieux. 2008. Abdominal obesity: the cholesterol of the 21st century? Can. J. Cardiol.. 24 Suppl D: 7D-12D.

Gómez Rosso, L. and M.B. Benítez, M.C. Fornari, V. Berardi, S. Lynch, L. Schreier, R. Wikinski, L. Cuniberti, and F. Brites. 2008. Alterations in cell adhesion molecules and other biomarkers of cardiovascular disease in patients with metabolic syndrome. Atherosclerosis 199: 415-423.

Haffner, S.M. 2007. Abdominal adiposity and cardiometabolic risk: do we have all the answers? Am. J. Med. 120 (9 Suppl 1): S10-6; discussion S16-7.

Hwang, S.J. and C.M. Ballantyne, A.R. Sharrett, L.C. Smith, C.E. Davis, A.M. Gotto Jr, and E. Boerwinkle. 1997. Circulating adhesion molecules VCAM-1, ICAM-1, and E-selectin in carotid atherosclerosis and incident coronary heart disease cases: the Atherosclerosis Risk in Communities (ARIC) study. Circulation 96: 4219-4225.

Ingelsson, E. and J. Hulthe, and L. Lind. 2008. Inflammatory markers in relation to insulin resistance and the metabolic syndrome. Eur. J. Clin. Invest. 38: 502-509.

Ito, H. and A. Ohshima, M. Inoue, N. Ohto, K. Nakasuga, Y. Kaji, T. Maruyama, and K. Nishioka. 2002. Weight reduction decreases soluble cellular adhesion molecules in obese women. Clin. Exp. Pharmacol. Physiol. 29: 399-404.

Krieglstein, C.F. and D.N. Granger. 2001. Adhesion molecules and their role in vascular disease. Am. J. Hypertens. 14: 44S-54S.

Lee, S. and F. Bacha, N. Gungor, and S. Arslanian. 2008. Comparison of different definitions of pediatric metabolic syndrome: relation to abdominal adiposity, insulin resistance, adiponectin, and inflammatory biomarkers. J. Pediatr. 152: 177-184.

Lewis, G.F. and N.M. O'Meara, P.A. Soltys, J.D. Blackman, P.H. Iverius, W.L. Pugh, G.S. Getz, and K.S. Polonsky. 1991. Fasting hypertriglyceridemia in noninsulin-dependent diabetes mellitus is an important predictor of postprandial lipid and lipoprotein abnormalities. J. Clin. Endocrinol. Metab. 72: 934-944.

Li, C. and E.S. Ford, L.C. McGuire, and A.H. Mokdad. 2007. Increasing trends in waist circumference and abdominal obesity among US adults. Obesity (Silver Spring) 15: 216-224.

Libby, P. 2007. Fat fuels the flame: triglyceride-rich lipoproteins and arterial inflammation. Circ. Res. 100: 299-301.

Matsumoto, K. and Y. Sera, Y. Abe, T. Tominaga, K. Horikami, K. Hirao, Y. Ueki, and S. Miyake. 2002. High serum concentrations of soluble E-selectin correlate with obesity but not fat distribution in patients with type 2 diabetes mellitus. Metabolism 51: 932-934.

Miller, M.A. and G.A. Sagnella, S.M. Kerry, P. Strazzullo, D.G. Cook, and F.P. Cappuccio. 2003. Ethnic differences in circulating soluble adhesion molecules. The Wandsworth Heart and Stroke Study. Clin. Sci. 104: 591-598.

Miller, M.A. and F.P. Cappuccio. 2006. Cellular adhesion molecules and their relationship with measures of obesity and metabolic syndrome in a multiethnic population. Int. J. Obes. 30: 1176-1182.

Miller, M.A and F.P. Cappuccio. 2007a. Ethnicity and inflammatory pathways - implications for vascular disease, vascular risk and therapeutic intervention. Curr. Med. Chem. 14: 1409-1425.

Miller, M.A. and F.P. Cappuccio. 2007b. Inflammation, sleep, obesity and cardiovascular disease. Curr. Vasc. Pharmacol. 5: 93-102.

Miller, M.A. and S.M. Kerry, D.G. Cook, and F.P. Cappuccio. 2004. Cellular adhesion molecules and blood pressure: interaction with sex in a multi-ethnic population. J. Hypertens. 22: 705-711.

Pigott, R. and L.P. Dillon, I.H. Hemingway, and A.J. Gearing. 1992. Soluble forms of E-selectin, ICAM-1 and VCAM-1 are present in the supernatants of cytokine activated cultured endothelial cells. Biochem. Biophys. Res. Commun. 187: 584-589.

Ponthieux, A. and B. Herbeth, S. Droesch, N. Haddy, D. Lambert, and S. Visvikis. 2004. Biological determinants of serum ICAM-1, E-selectin, P-selectin and L-selectin levels in healthy subjects: the Stanislas study. Atherosclerosis 172: 299-308.

Pontiroli, A.E. and F. Frigè, M. Paganelli, and F. Folli. 2008. In Morbid obesity, metabolic abnormalities and adhesion molecules correlate with visceral fat, not with subcutaneous fat: effect of weight loss through surgery. Obes. Surg. Jul. 16 [Epub ahead of print].

Rohde, L.E. and C.H. Hennekens, and P.M. Ridker. 1999. Cross-sectional study of soluble intercellular adhesion molecule-1 and cardiovascular risk factors in apparently healthy men. Arterioscler. Thromb. Vasc. Biol. 19: 1595-1599.

Sanada, H. and S. Midorikawa, J. Yatabe, M.S. Yatabe, T. Katoh, T. Baba, S. Hashimoto, and T. Watanabe. 2005. Elevation of serum soluble E- and P-selectin in patients with hypertension is reversed by benidipine, a long-acting calcium channel blocker. Hypertens. Res. 28: 871-878.

Shai, I. and T. Pischon, F.B. Hu, A. Ascherio, N. Rifai, and E.B. Rimm. 2006. Soluble intercellular adhesion molecules, soluble vascular cell adhesion molecules, and risk of coronary heart disease. Obesity (Silver Spring) 14: 2099-2106.

Soro-Paavonen, A. and J. Westerbacka, C. Ehnholm, and M.R. Taskinen. 2006. Metabolic syndrome aggravates the increased endothelial activation and low-grade inflammation in subjects with familial low HDL. Ann. Med. 38: 229-238.

Tsiara, S. and M. Elisaf, and D.P. Mikhailidis. 2003. Early vascular benefits of statin therapy. Curr. Med. Res. Opin. 19: 540-556.

Software

Poser 7 (c). 1991-2008. Smith Micro Software, Inc., California, USA.

CHAPTER 14

Weight Loss and Adhesion Molecules

Jennifer B. Keogh[1, 2,*] and Peter M. Clifton[1, 2,*]

[1]Preventative Health National Research Flagship, Commonwealth Scientific and Industrial Research Organization – Human Nutrition, Adelaide, South Australia, Australia

[2]Centre of Clinical Research Excellence in Nutritional Physiology, Interventions and Outcomes, University of Adelaide, Adelaide, South Australia

Alternative contact: Professor Peter M. Clifton, CSIRO Human Nutrition, Gate 13 Kintore Ave, Adelaide, South Australia 5000, E-mail: peter.clifton@csiro.au

Postal address: CSIRO Human Nutrition, PO Box 10041, Adelaide BC, SA 5000

ABSTRACT

Concentrations of adhesion molecules are raised in obese individuals but the mechanisms for this are unclear. The effect of obesity on cellular adhesion molecules (CAMs) is relatively modest, with those in the highest quintile of body mass index having median levels 14% higher than those in the lowest quintile for intracellular adhesion molecule 1 (ICAM-1). Weight loss of 7-18% following either lifestyle programs or bariatric surgery reduces ICAM-1, vascular cell adhesion molecule 1 (VCAM-1) and E-selectin by 15-28%, 15% and 20-35% respectively. Therefore, weight loss has a relatively modest and highly variable effect on CAMs but levels remain raised as study participants frequently remain overweight or obese. Dietary composition during weight loss appears to have little impact on reducing levels of adhesion molecules. There are a small number of studies exploring the effect of dietary composition on CAMs.

Corresponding author: Dr Jennifer B. Keogh, CSIRO Human Nutrition, Gate 13 Kintore Ave., Adelaide, South Australia 5000, E-mail: jennifer.keogh@csiro.au
Key terms are defined at the end of the chapter.

INTRODUCTION

Adhesion molecules are expressed and shed from endothelial cells, platelets neutrophils and monocytes. It is not clear what the major source of circulating adhesion molecules is and, when the plasma level is increased, whether this is due to increased cellular expression or to increased shedding from cell surfaces. Although plasma-soluble E-selectin is 50% higher in obese women than in lean women, baseline gene expression for E-selectin and vascular cell adhesion molecule 1 (VCAM-1) and intracellular adhesion molecule 1 (ICAM-1) in fat is lower than in lean women and actually increases with weight loss induced by a very low calorie diet (VLCD) (Bosanská *et al.* 2008). However, overexpression of pro-inflammatory factors is seen in subcutaneous adipose tissue of obese subjects in cells of the monocyte/macrophage lineage and weight loss with VLCD lowered pro-inflammatory factors and increased anti-inflammatory factors (Viguerie *et al.* 2005).

ICAM-1 is certainly expressed on endothelial cells in culture (Dustin 1986) and can be observed on the surface of atherosclerotic plaques (De Graba 1997). Cellular adhesion molecules (CAMs) can be induced by the transcription factor NF kappa beta. The production and release of CAMs from endothelial cells has been found to be inducible by interleukin-1, tumour necrosis factor alpha (Cartwright *et al.* 1995), angiotensin II (Pastore *et al.* 1999) and non-oxidized low density lipoprotein (Maeno *et al.* 2000). Leeuwenberg *et al.* (1992) found that sICAM1 and sE-selectin correlated with surface expression on cultured endothelial cells, as did Noutsias *et al.* (2003) in dilated cardiomyopathy. This suggests that there is no specific control mechanism for shedding but there is evidence in neutrophils of control of the process of shedding. In neutrophils p38 activation (via FMLP, LPS and hypertonicity) and protein kinase C (PKC) activation lead to shedding of L-selectin through a protease or sheddase. One example of this is the ADAM family of proteases (A Disintegrin and Metalloprotease) (Edwards *et al.* 2008). The archetypal ADAM is ADAM17. This protease shed the TNF-alpha precursor from the surface to produce the active cytokine and is also referred to as TNF-alpha converting enzyme (TACE) (Bell *et al.* 2007). While a number of ADAMs have been identified in mammalian tissues, this was the first to have a known function. TACE is also involved in VCAM-1 shedding (plus 20 other proteins) (Garton *et al.* 2003). In systemic inflammatory conditions, for example, juvenile rheumatoid arthritis and sepsis particularly with organ failure (Whalen *et al.* 2000), CAMs are elevated but it is not clear whether the source is white cells or the endothelium. It would appear that platelets are the major source of circulating P-selectin in healthy individuals. Endothelial cell activation from sepsis (as assessed by increased levels of circulating E-selectin and ED1-fibronectin) is associated with an increased sP-selectin concentration per platelet (Fijnheer *et al.* 1997).

ADHESION MOLECULES IN OBESITY

Obesity is associated with increased levels of CAMs (Mora *et al.* 2006). The effect of obesity on CAMs is relatively modest, with those in the highest quintile of body mass index (BMI) having median levels 14% higher than those in the lowest quintile for ICAM-1 in a cross-sectional study of 27,158 women in the United States (Mora *et al.* 2006). A similar association was also seen in obese children compared with controls but this was not independent of other risk factors for obesity (Glowinska *et al.* 2005, Meyer *et al.* 2006).

ADHESION MOLECULES AND WEIGHT LOSS

Weight loss of 7-18% following either lifestyle programs or bariatric surgery reduces ICAM-1, VCAM-1 and E-selectin by 15-28%, 15% and 20-35% respectively (Ferri *et al.* 1999, Ito *et al.* 2002, Pontiroli *et al.* 2004, Vazquez *et al.* 2005). However, the effect of weight loss on adhesion molecules is very variable. In a study by Pontiroli *et al.* (2008), BMI was reduced from 44.3 ± 0.43 to 36.4 ± 0.39 kg/m^2 with a corresponding reduction in ICAM-1 concentrations from 313.1 ± 9.69 to 275.2 ± 6.16 ng/m (a 12% reduction). E-selectin was reduced from 64.1 ± 4.04 to 44.1 ± 1.96 ng/ml (a 31% reduction).

Vazquez *et al.* (2005) observed that E- and P-selectin improved after weight loss of 24 kg, but ICAM-1 levels did not change and VCAM-1 increased, PAI-1 and CRP fell but TNF-alpha and IL6 did not change. Plat *et al.* (2007) in a study evaluating the effects of fish oil and moderate weight loss in 11 obese men (BMI 30-35 kg/m^2) found that fish oil did not affect s-ICAM-1, whereas weight loss of 9.4 kg reduced fasting levels (p = 0.009) and post-prandial s-ICAM-1 responses (p < 0.001).

A number of other studies have observed similarly mixed results (Ziccardi *et al.* 2002, Hamdy *et al.* 2003, Hanusch-Enserer *et al.* 2004, Pontiroli *et al.* 2004, Sharman and Volek 2004, Trøseid *et al.* 2005, McLaughlin *et al.* 2006, Konukoglu *et al.* 2007, Swarbrick *et al.* 2008). In a study of 18 morbidly obese patients who underwent gastric banding, subjects were evaluated before surgery and 6 and 12 mon after surgery. BMI dropped from 45.22 ± 5.62 to 36.99 ± 4.34 kg/m^2 after 6 mon and to 33.72 ± 5.55 kg/m^2 after 12 mon (both p < 0.0001). E-selectin levels decreased significantly after 6 and 12 mon (p = 0.05), whereas significantly lower levels of ICAM-1 and PAI-1 were seen after 6 mon but were not sustained at 12 mon. No changes were observed in VCAM-1 (Hanusch-Enserer *et al.* 2004). This was a small study and despite large weight loss these patients remain obese, which may explain the variability in the effects on adhesion molecules after maintenance of weight loss.

Pontiroli *et al.* (2004) found that ICAM-1, E-selectin, and endothelin-1 (ET-1) were higher in 96 obese patients compared with 30 lean controls. In those 26 patients with type 2 diabetes, these variables were higher than in patients with normal (n = 43) or impaired (n = 27) glucose tolerance. Sixty-eight obese patients had significant weight loss following bariatric surgery, with a decrease in these molecules to levels comparable with the lean controls. In 13 patients with a small weight loss achieved by diet the changes were not significant.

McLaughlin *et al.* (2006) observed that subjects (n = 57) who underwent 16 wk of an energy-restricted diet, 15% of energy as protein and either 60% and 25% or 40% and 45% of energy as carbohydrate and fat, showed weight loss of 5.7 ± 0.7 versus 6.9 ± 0.7 kg, respectively; subjects on the lower-carbohydrate (LC) diet (40% of energy) had greater decrease in plasma E-selectin than subjects on the higher-carbohydrate (HC) diet (60% of energy). ICAM-1 decreased after weight loss by 11% with no difference between diets.

Hamdy *et al.* (2003) performed an uncontrolled study in 24 obese subjects in whom weight loss was 6.6 ± 1% and observed that sICAM decreased (251.3 ± 7.7 vs. 265.6 ± 9.3 ng/ml, p = 0.018).

Sharman *et al.* (2004) in a study of 15 obese men who were allocated to two weight-loss diets for two consecutive 6 wk periods, either a very-low-carbohydrate diet (< 10% energy) or a low-fat diet (< 30% energy), observed that both diets resulted in significant decreases in sICAM-1 with no change in sP-selectin concentrations.

In 18 obese patients 1 mon after bariatric surgery, Konukoglu *et al.* (2007) found that weight loss was 13.2 kg and, while plasma sCRP and ADMA concentrations were significantly decreased, they did not return to the same level as age-matched lean controls. VCAM-1 levels did not change after weight loss. These patients remained obese despite the weight loss.

In a study of 56 obese and 40 age-matched lean women compared with non-obese women, obese women had increased concentrations of TNF-alpha, IL-6, P-selectin, ICAM-1 and VCAM-1. TNF-alpha and IL-6 were related to visceral obesity. Weight loss (10%, 9.8 kg) was associated with reductions in TNF-alpha (30%), IL-6 (46%), P-selectin (30%), ICAM-1 (26%) and VCAM-1 (17%) (Ziccardi *et al.* 2002). Swarbrick *et al.* (2008), in a study of 19 obese women (BMI 45.6 ± 1.6 kg/m^2), assessed weight before and 1 yr after Roux-en-Y gastric bypass surgery. Weight loss was 32% (40.5 kg) and IL6 fell by 49%, CRP by 77% and sICAM1 by 11%.

Finally, the effect of physical exercise and pravastatin on levels of CAMs was evaluated in a study of 32 subjects (Trøseid *et al.* 2005). Subjects in the exercise group lost 0.7 BMI units and 7.5 cm in waist circumference. Changes in serum E-selectin were correlated with changes in BMI (r = 0.48, p = 0.006) and waist circumference (r = 0.48, p = 0.006) but not to changes in visceral or subcutaneous fat.

ADHESION MOLECULES, WEIGHT LOSS AND DIETARY COMPOSITION

There are several studies from our own group examining the effects of weight loss and dietary composition on CAMs (Brinkworth *et al.* 2004, Clifton *et al.* 2005, Keogh *et al.* 2007, 2008) (Table 1).

In a study by Clifton *et al.* (2005), 55 overweight and obese subjects were randomized to one of two low-fat, HC weight loss diets (both < 6000 kJ). After 3 mon, weight loss was 6.3 ± 3.7 kg and sICAM1 fell by 8% (p < 0.001) with no difference between diet groups, but no changes in sVCAM1 or IL-6 were noted. In an 8 wk weight loss study, a diet of LC (4% of energy) and high saturated fat (20% of energy) was contrasted with a diet of HC (46% of energy) and low fat in 70 subjects (LC 37, HC 33) with features of metabolic syndrome. E- and P-selectin decreased by approximately 32% and 7% respectively and ICAM-1 by 15% (all p < 0.001) after weight loss (7.5 kg LC and 6.4 kg HC) with no effect of dietary composition (Keogh *et al.* 2008). We observed a small but statistically significant rise in VCAM-1 of 4% in this study. Similarly, in a randomized parallel design of two weight loss diets, LC (33% of energy) and HC (60% of energy), both low in saturated fat, in 36 obese subjects, weight loss of 8.7% was achieved after 12 wk and 5.6% after 52 wk. Adhesion molecules decreased at 12 wk, VCAM-1 by 6% and ICAM-1 by 14% (both p < 0.05), E-selectin also decreased by 14% (p < 0.01) after weight loss with no difference between diets. sICAM-1 remained reduced after 52 wk (p < 0.05), whereas the change in sVCAM-1 did not reach statistical significance (p = 0.08) and E-selectin remained reduced at the end of the study (p < 0.05). P-selectin did not change after 12 wk of weight loss but was reduced at the end of the study (p < 0.05) (Keogh *et al.* 2007). In a long-term study, weight loss was 2.9% on a standard protein diet and 4.1% on a high protein diet after 52 wk of follow-up, and sICAM-1 decreased significantly on both diets (p < 0.05) (Brinkworth *et al.* 2004).

ADHESION MOLECULES AND DIETARY COMPOSITION IN WEIGHT STABILITY

There are a small number of studies exploring the effect of dietary composition on CAMs (Lewis *et al.* 1999, Bemelmans *et al.* 2002, Keogh 2005). Bemelmans *et al.* (2002) found that a reduction in saturated fat intake over 2 yr in a dietary intervention study was associated with reduced levels of sICAM-1, whereas Lewis *et al.* (1999) found there was no significant difference in sICAM-1 or sVCAM-1 concentrations between a diet high in saturated fat and a Mediterranean diet, high in monounsaturated fat, after 4 wk in men with raised cholesterol. In a study of 40 healthy subjects randomly crossed over to four 3 wk isocaloric diets high in polyunsaturated fat, high in monounsaturated fat, or high in saturated fat, or a

Table I Effects of weight loss and dietary composition on adhesion molecules

Author	n	Diet	Wt loss	Effect
Clifton *et al.* 2005	55	Two low-fat diets 1) Meal replacements (MR) or 2) Conventional low-fat diet (C)	After 12 wk −6.0 ± 4.2 kg MR −6.63 ± 3.35 kg C Mean ± SD	sICAM1 ↓ by 8% with no effect of diet composition
Keogh *et al.* 2008	70	1) Low carbohydrate, high saturated fat (LC) or 2) High carbohydrate, low fat (HC)	After 8 wk −7.5 ± 2.6 kg LC −6.2 ± 2.9 kg HC Mean ± SD	E- and P-selectin ↓ by 32% and 7% respectively ICAM-1 ↓15% VCAM-1 ↑4% with no effect of diet composition
Keogh *et al.* 2007	36	1) Low carbohydrate, low saturated fat (LC) or 2) High-carbohydrate, low saturated fat (HC)	After 12 wk −8.5 ± 4.5 kg LC −7.9 ± 5.5 kg HC After 52 wk −4.6 ± 7.6 kg LC −4.0 ± 5.8 kg HC Mean ± SD	After 12 wk VCAM-1 ↓ 6% ICAM-1 ↓ 14% E-selectin ↓ ↓ 14% After 52 wk sICAM-1 ↓ 7% after 52 wk (P < 0.05) E-selectin ↓ 10% P-selectin ↓ 15% with no effect of diet composition
Brinkworth *et al.* 2004	58	1) High protein, moderate carbohydrate (HP) or 2) Standard protein, high carbohydrate, moderate fat (SP)	After 68 wk HP −4.1 ± 1.3% SP −2.9 ± 0.8% Mean ± SEM	sICAM-1 ↓ −27.6 ± 4.6% HP −30.8 ± 4.3% SP with no effect of diet composition

Weight loss achieved a moderate decrease in the majority of adhesion molecules (all P < 0.05). Dietary composition during weight loss had no effect on the reductions in adhesion molecules seen.

low-fat, high carbohydrate diet, P-selectin was highest after the high saturated fat diet (Keogh 2005).

SUMMARY AND CONCLUSION

It is apparent from the literature that levels of adhesion molecules are raised in obese individuals but that the mechanisms for this are unclear. Weight loss generally reduces these levels, although by a relatively small amount. Dietary composition appears to have little role to play, although a diet high in saturated fat may have adverse effects on these molecules.

Acknowledgements

Dr. Keogh and Prof. Clifton contributed equally to this work and have no conflicts of interest to declare.

Abbreviations

ADMA	Asymmetric dimethylarginine
BMI	body mass index
CAMs	cellular adhesion molecules
ET-1	endothelin-1
IL-6	interleukin-6
ICAM-1	intracellular adhesion molecule 1
PKC	protein kinase C
TNF-alpha	tumour necrosis factor-alpha
TACE	TNF-alpha converting enzyme
VCAM-1	vascular cell adhesion molecule 1
VLCD	very low calorie diet
WHR	waist hip ratio

Key Factors about Weight Loss and Adhesion Molecules

1. Blood levels of adhesion molecules are raised in obese individuals.
2. The effect of obesity on adhesion molecules is relatively modest.
3. Weight loss generally decreases levels of adhesion molecules.
4. Dietary composition appears to have little effect on reducing levels of adhesion molecules either during weight loss or weight stability.

Definition of Terms

Bariatric surgery: Name given to a variety of surgical procedures used to treat obesity. The aim of the surgery is to reduce nutrient intake and achieve weight loss.

Endothelial cells: Cells lining the blood vessels and forming a barrier between blood and the wall or structure of the blood vessels.

Obesity: A body mass index of ≥ 30 kg/m^2.

Weight loss: A loss in body weight achieved by reducing energy intake relative to intake.

References

Arkin, J.M. and R. Alsdorf, C.M. Apovian, and N. Gokce. 2008. Relation of cumulative weight burden to vascular endothelial dysfunction in obesity. Am. J. Cardiol. 101: 98-101.

Bemelmans, W.J. and J.D. Lefrandt, J.F. May, and A.J. Smit. 2002. Change in saturated fat intake is associated with progression of carotid and femoral intima-media thickness, and with levels of soluble intercellular adhesion molecule-1. Atherosclerosis 163: 113-120.

Bell, J.H. and A.H. Herrera, Y. Li, and B. Walcheck. 2007. Role of ADAM17 in the ectodomain shedding of TNF-alpha and its receptors by neutrophils and macrophages. J. Leukoc. Biol. 82: 173-176.

Bosanská, L. and Z. Lacinová, M. Matoulek, and M. Haluzík. 2008. The influence of very-low-calorie diet on soluble adhesion molecules and their gene expression in adipose tissue of obese women. Cas. Lek. Cesk. 147: 32-37.

Brinkworth, G.D. and M. Noakes, G.A .Wittert, and P.M. Clifton. 2004. Long-term effects of a high-protein, low-carbohydrate diet on weight control and cardiovascular risk markers in obese hyperinsulinemic subjects. Int. J. Obes. Relat. Metab. Disord. 28: 661-670.

Cartwright, J.E. and G.S. Whitley, and A.P. Johnstone. 1995. The expression and release of adhesion molecules by human endothelial cell lines and their consequent binding of lymphocytes. Exp. Cell Res. 217: 329-335.

Clifton, P.M. and J.B. Keogh, P.R. Foster, and M. Noakes. 2005. Effect of weight loss on inflammatory and endothelial markers and FMD using two low-fat diets. Int. J. Obes. (Lond.). 29: 1445-1451.

De Benedetti, F. and M. Vivarelli, A. Pistorio, and A. Martini. 2000. Circulating levels of soluble E-selectin, P-selectin and intercellular adhesion molecule-1 in patients with juvenile idiopathic arthritis. J. Rheumatol. 27: 2246-2250.

De Graba, T.J. 1997. Expression of inflammatory mediators and adhesion molecules in human atherosclerotic plaque, Neurology 49: 15-19.

Edwards, D.R. and M.M. Handsley, and C.J. Pennington. 2008. The ADAM metalloproteinases. Mol. Aspects Med. 29: 258-289.

Fijnheer, R. and C.J. Frijns, J.J. Sixma, and H.K. Nieuwenhuis. 1997. The origin of P-selectin as a circulating plasma protein. Thromb. Haemost. 77: 1081-1085.

Garton, K.J. and P.J. Gough, P.J. Dempsey, and E.W. Raines. 2003. Stimulated shedding of vascular cell adhesion molecule 1 (VCAM-1) is mediated by tumor necrosis factor-alpha-converting enzyme (ADAM17). J. Biol. Chem. 278: 37459-37464.

Glowinska, B. and M. Urban, J. Peczynska, and B. Florys. 2005. Soluble adhesion molecules (sICAM-1, sVCAM-1) and selectins (sE-selectin, sP-selectin, sL-selectin) levels in children and adolescents with obesity, hypertension, and diabetes. Metabolism 54: 1020-1026.

Herder, C. and M. Peltonen, H. Kolb, and J. Tuomilehto. 2006. Systemic immune mediators and lifestyle changes in the prevention of type 2 diabetes: results from the Finnish Diabetes Prevention Study. Diabetes 55(8): 2340-2346.

Hamdy, O. and S. Ledbury, A. Veves, and E.S. Horton. 2003. Lifestyle modification improves endothelial function in obese subjects with the insulin resistance syndrome. Diabetes Care 26: 2119-2125.

Hanusch-Enserer, U. and E. Cauza, G. Pacini, and R. Prager. 2004. Improvement of insulin resistance and early atherosclerosis in patients after gastric banding. Obes. Res. 12: 284-291.

Ferri, C. and G. Desideri, A. Santucci, and G. De Mattia. 1999. Early upregulation of endothelial adhesion molecules in obese hypertensive men. Hypertension 34: 568-573.

Ito, H. and A. Ohshima, T. Maruyama, and K. Nishioka. 2002. Weight reduction decreases soluble cellular adhesion molecules in obese women. Clin. Exp. Pharmacol. Physiol. 29: 399-404.

Keogh, J.B. and G.D. Brinkworth, J.D. Buckley, and P.M. Clifton. 2008. Effects of weight loss from a very-low-carbohydrate diet on endothelial function and markers of cardiovascular disease risk in subjects with abdominal obesity. Am. J. Clin. Nutr. 87: 567-576.

Keogh, J.B. and G.D. Brinkworth, and P.M. Clifton. 2007. Effects of weight loss on a low-carbohydrate diet on flow-mediated dilatation, adhesion molecules and adiponectin. Br. J. Nutr. 98: 852-859.

Konukoglu, D. and H. Uzun, A. Kocael, and M. Taskin. 2007. Plasma adhesion and inflammation markers: asymmetrical dimethyl-L-arginine and secretory phospholipase A2 concentrations before and after laparoscopic gastric banding in morbidly obese patients. Obes. Surg. 17: 672-678.

Leeuwenberg, J.F. and E.F. Smeets, T.J. Ahern, and W.A. Buurman. 1992. E-selectin and intercellular adhesion molecule-1 are released by activated human endothelial cells in vitro. Immunology 77: 543-549.

Lewis, T.V. and A.M. Dart, and J.P. Chin-Dusting. 1999. Endothelium-dependent relaxation by acetylcholine is impaired in hypertriglyceridemic humans with normal levels of plasma LDL cholesterol. J. Am. Coll. Cardiol. 33(3): 805-812.

Maeno, Y. and A. Kashiwagi, N. Takahara, and R. Kikkawa. 2000. IDL can stimulate atherogenic gene expression in cultured human vascular endothelial cells. Diabetes Res. Clin. Pract. 48: 127-138.

McLaughlin, T. and S. Carter, M. Basina, and G. Reaven. 2006. Effects of moderate variations in macronutrient composition on weight loss and reduction in cardiovascular disease risk in obese, insulin-resistant adults. Am. J. Clin. Nutr. 84: 813-821.

Meyer, A.A. and G. Kundt, P. Schuff-Werner, and W. Kienast. 2006. Impaired flow-mediated vasodilation, carotid artery intima-media thickening, and elevated endothelial plasma markers in obese children: the impact of cardiovascular risk factors. Pediatrics 117: 1560-1567.

Mora, S. and I.M. Lee, J.E. Buring, and P.M. Ridker. 2006. Association of physical activity and body mass index with novel and traditional cardiovascular biomarkers in women. JAMA 295(12): 1412-1419.

Noutsias, M. and C. Hohmann, U. Kühl, and H.P. Schultheiss. 2003. sICAM-1 correlates with myocardial ICAM-1 expression in dilated cardiomyopathy. Int. J. Cardiol. 91: 153-161.

Pastore, L. and A. Tessitore, A. Gulino, and A. Santucci. 1999. Angiotensin II stimulates intercellular adhesion molecule-1 (ICAM-1) expression by human vascular endothelial cells and increases soluble ICAM-1 release in vivo. Circulation 100: 1646-1652.

Plat, J. and A. Jellema, J. Ramakers, and R.P. Mensink. 2007. Weight loss, but not fish oil consumption, improves fasting and postprandial serum lipids, markers of endothelial function, and inflammatory signatures in moderately obese men. J. Nutr. 137: 2635-2640.

Pontiroli, A.E. and P. Pizzocri, K. Esposito, and D. Giugliano. 2004. Body weight and glucose metabolism have a different effect on circulating levels of ICAM-1, E-selectin, and endothelin-1 in humans. Eur. J. Endocrinol. 150: 195-200.

Pontiroli, A.E. and F. Frigè, M. Paganelli, and F. Folli. 2008. In morbid obesity, metabolic abnormalities and adhesion molecules correlate with visceral fat, not with subcutaneous fat: effect of weight loss through surgery. Obes. Surg. Jul. 16. [Epub ahead of print].

Sharman, M.J. and J.S. Volek. 2004. Weight loss leads to reductions in inflammatory biomarkers after a very-low-carbohydrate diet and a low-fat diet in overweight men. Clin. Sci. (Lond.) 107: 365-369.

Swarbrick, M.M. and K.L. Stanhope, B.M. Wolfe and P.J. Havel. 2008. Longitudinal changes in pancreatic and adipocyte hormones following Roux-en-Y gastric bypass surgery. Diabetologia 51: 1901-1911.

Trøseid, M. and K.T. Lappegård, H. Arnesen, and I. Seljeflot. 2005. Changes in serum levels of E-selectin correlate to improved glycaemic control and reduced obesity in subjects with the metabolic syndrome. Scand. J. Clin. Lab. Invest. 65: 283-290.

Vázquez, L.A. and F. Pazos, J. Freijanes, and J.A. Amado. 2005. Effects of changes in body weight and insulin resistance on inflammation and endothelial function in morbid obesity after bariatric surgery. J. Clin. Endocrinol. Metab. 90: 316-322.

Viguerie, N. and C. Poitou, K. Clément, and D. Langin. 2005. Transcriptomics applied to obesity and caloric restriction. Biochimie 87: 117-123.

Whalen, M.J. and L.A. Doughty, P.M. Kochanek, and J.A. Carcillo. 2000. Intercellular adhesion molecule-1 and vascular cell adhesion molecule-1 are increased in the plasma of children with sepsis-induced multiple organ failure. Crit. Care Med. 28: 2600-2607.

Ziccardi, P. and F. Nappo, A.M. Molinari, and D. Giugliano. 2002. Reduction of inflammatory cytokine concentrations and improvement of endothelial functions in obese women after weight loss over one year. Circulation 105: 804-809.

CHAPTER 15

Hypoxia Reperfusion Injury and Adhesion Molecules

Kamran Ghori

Consultant Anaesthetist, Department of Anaesthesia, Bon SECOURS Hospital, College Rd. Cork, Republic of Ireland, E-mail: kamrang@hotmail.com

ABSTRACT

Hypoxia reperfusion (HR) injury has been recognized to play a key role in the pathogenesis of many kinds of organ dysfunction. Ischaemia occurs in various clinical conditions such as myocardial infarction, stroke, peripheral vascular disease and hypovolumeic shock. It is important to restore the blood supply of the ischaemic organ, but sometimes reperfusion itself can cause tissue injury in excess of that caused by ischaemia alone. Reperfusion of ischaemic tissues is associated with microvascular dysfunction, manifested by enhanced leukocyte plugging in capillaries, and the migration of leukocytes into intrestisuum. Activated endothelial cells and leukocytes in all segments lead to the production and release of inflammatory mediators (e.g., platelet-activating factor, tumour necrosis factor) and upregulate the expression of adhesion molecules that promote leukocyte-endothelial cell adhesion. Once leukocytes reach the extravascular space, they exacerbate tissue injury by releasing oxygen free radicals and other destructive enzymes. The production of adhesion molecules in endothelium and leukocytes is regulated by a family of protein kinases, which are important signalling pathways during HR injury. The protein kinases initiate several interconnected intracellular enzyme reactions. The inflammatory mediators released as a consequence of reperfusion also appear to activate endothelial cells in remote organs that are not exposed to the initial ischaemic insult. This distant response to HR can result in severe generalized inflammatory response and can result in multiple organ dysfunction syndrome.

INTRODUCTION

The role of hypoxia reperfusion (HR) injury has been well recognized in the pathogenesis of many kinds of organ dysfunction. Despite the complexity of the mechanism of HR injury, the investigations in last couple of decades have made our understanding more clear about these phenomena. The pathogenesis of HR represents a complex interaction between biochemical, cellular, vascular endothelial and tissue-specific factors, inflammation being a common feature. In recent years there has been a specific focus on describing the mechanism of HR injury. With advances in technology and a better understanding of organization of cellular structures, it is now possible to describe the details of this convoluted phenomenon, to which is attributed the pathogenesis of many kinds of organ dysfunction. This chapter focuses on the cascade of events that occur during HR injury and the factors responsible in the expression and upregulation of adhesion molecules on neutrophil, platelet and vascular cells and their mechanism of tissue injury.

Ischaemia itself can cause direct cell death, in which case damage may lead to activation of many degenerative systems in a rapid and uncontrolled fashion. This form of cell death is called ischaemic or necrotic cell death. It is essential to restore the blood supply of an ishcaemic organ, but reperfusion itself can augment tissue injury. Temporary ischaemic damage causes cells to undergo reperfusion injury when blood flow is re-established. The tissue insult caused by this injury is not restricted locally but also causes damage to distant organs. Thus, HR injury may extend beyond the ischaemic area at risk to include injury of a remote, non-ischaemic organ. Of much clinical and scientific interest is the minimizing of additional loss of tissue that occurs during reperfusion following ischaemia.

MECHANISM OF HR INJURY

Prolonged ischaemia causes a variety of cellular metabolic and structural changes. Ischaemia causes decreased cellular oxidative phosphorylation and results in a failure to re-synthesize adenosine triphosphate. Alteration in membrane ionic pump function also leads to disturbance of ion equilibrium across the cell membrane (Na^+, K^+, Ca^{++}) and intracellular accumulation of harmful metabolites, leading in turn to devastating cellular events (acidosis, inhibition of protein synthesis, oedema), which result in cellular death. During ischaemia, cellular adenosine triphosphate is degraded to form hypoxanthine. Normally, hypoxanthine is oxidized by xanthine dehydrogenase to xanthine. However, during ischaemia, xanthine dehydrogenase is converted to xanthine oxidase. Unlike xanthine dehydrogenase, which uses nicotinamide adenine dinucleotide as its substrate, xanthine oxidase uses oxygen and therefore, during ischaemia, is unable to catalyse the conversion of hypoxanthine to xanthine, resulting in a build-up of excess tissue levels of hypoxanthine. When oxygen is reintroduced during reperfusion, conversion of the excess hypoxanthine by xanthine oxidase results in the formation of toxic reactive oxygen species (ROS). In addition to causing direct cell injury, ROS increase leukocyte activation, chemotaxis, and leukocyte-endothelial adherence after HR (Carden *et al.* 2000, Collard *et al.* 2001).

The ischaemia produces expression of certain pro-inflammatory gene products such as cytokines, endothelial, neutrophil and platelets adhesion molecules. This helps in recruitment of neutrophil at the ischaemic tissue. The infiltration of neutrophil in ischaemic area cause further release of pro-inflammatory cytokines and enhance tissue damage. Thus, ischaemia induces pro-inflammatory state in tissue and increases tissue vulnerability to further damage when oxygen supply is restored. This type of tissue insult, unfortunately, is not restricted to local tissue. If severe enough, it can cause damage to distant organs and leads into systemic inflammatory response syndrome (Ley 1991, Bevilacqa 1993).

ROLE OF ENDOTHELIUM-NEUTROPHIL INTERACTION

The ischaemia promotes expression of certain pro-inflammatory gene products (e.g., adhesion molecules, cytokines) and bioactive agents (endothelium thromboxane) on the endothelium. Thus, ishcaemic insult makes the tissue vulnerable to further injury.

Following ischaemia reperfusion–induced leukocyte activation, leukocytes and endothelium interact in a series of distinct steps characterized by leukocyte rolling on the endothelium and firm adherence of leukocytes to the endothelium and endothelial transmigration.

Within the endothelium, ischaemia promotes the expression of pro-inflammatory gene products (e.g., leukocyte adhesion molecules) and bioactive agents (e.g., endothelin, thromboxane A2), while repressing other 'protective' gene products (e.g., constitutive nitric oxide synthase, thrombomodulin) and bioactive agents (e.g. prostacyclin, nitric oxide). Thus, ischaemia induces a pro-inflammatory state that increases tissue vulnerability to further injury on reperfusion (Bevilacqua 1993, Watson 1998).

The recruitment of neutrophil at the site of injury is an initial important step in pathophysiology of HR injury. Following tissue inflammation and trauma, a portion of the rolling leukocytes are observed to flatten and spread on the endothelium and then to stick firmly. Some of the adherent leukocytes crawl over the endothelial surface, seeming to probe for an opening and then diapedes (crawl) between endothelial cells. Once in the extravascular tissue, the extravasated leukocytes continue to migrate toward the inflammatory site. In the last decade, *in vitro* and *in vivo* studies have identified many of the adhesion molecules on neutrophil and endothelium, and locally generated inflammatory mediators that are involved in the adhesive interaction.

EXPRESSION OF ADHESION MOLECULE IN ISCHAEMIA REPERFUSION INJURY

Selectins

The selectin family comprises three proteins designated by the prefixes E (endothelial), P (platelets), and L (leukocyte). E-selectin and P-selectin are expressed by endothelial cells, and L-selectin is expressed only on leukocytes.

E-selectin (CD62E) was initially described as an antigen that was induced in endothelium after stimulation by interleukin-1 (IL-1) and was involved in the adhesion of neutrophil and several leukemic cell lines. P-selectin (CD62P) is an adhesion protein that was initially characterized in platelet and was subsequently shown to be present also in endothelial cells. The soluble P-selectin has been detected in the plasma of normal individuals in a very small quantity (0.15 μg/mL). However, it has not been shown that this is the secreted form rather than protein shed from platelets or endothelial cells. This raises the possibility that soluble plasma P-selectin may prime leukocyte adhesion molecule to P-selectin expressed on endothelium (Dunlop *et al.* 1992, Hogg *et al.* 1991).

L-selectin (CD62L) is found only on leukocytes. It is shown to be expressed on most other peripheral blood leukocytes and is involved in leukocyte traffic in the systemic microcirculation.

Endothelial Cell Surface Protein

The Ig gene superfamily consists of cell surface proteins that are involved in antigen recognition (C1-type) or complement-binding or cellular (C2-type). Four important members of this family expressed by endothelial cells are involved in leukocyte adhesion: intercellular adhesion molecules 1 and 2 (ICAM-1), (ICAM-2), vascular cell adhesion molecule 1 (VCAM-1), and platelet-endothelial cells adhesion molecule 1 (PECAM) (William *et al.* 1988, Katz *et al.* 1985).

ICAM-1 (CD54) is the primary site for their counterpart leukocyte integrins (CDa/CD18 and CD11b/Cd18), and ICAM-2 (CD102) is a second endothelial ligand for this leukocyte integrin. VCAM (CD106) also has a functional role in leukocyte adhesion to endothelium similar to ICAM-1, and ICAM-2. PECAM-1 (CD31) may play a role in leukocyte adhesion and, especially, in transmigration of leukocytes (Stauton 1989, Arnaout 1996, de Foujerolles 1992).

Leukocyte Integrins

Integrins are transmembrane cell surface proteins that bind to cytoskeletal proteins and communicate extracellular signals. Each integrin has α and β subunits, which are further divided into subtype and represent by numbers. Within the integrin family of adhesion receptors, only five members have so far been shown to be involved in leukocyte adhesion to endothelium: the β_2 leukocyte integrins (CD11a/CD18, CD11b/CD18 and CD11c/CD18, the β_1 integrin VLA-4 (CD49d/CD29) and $\alpha_4\beta_7$.

The expression of the β_2 integrin is confined to leukocytes, but among subtypes of leukocytes the distribution on CD11/CD18 differs. For example, peripheral blood lymphocytes express primarily CD11a/CD18, whereas neutrophil, monocytes and natural killer cells express all three β_2 integrins. Surface expression of CD11b/CD18 and CD11c/CD18 is increased by a variety of stimulus, FMLP, GM-CSF, C5a, TNF-α, and LTB4. Ligands for the leukocyte integrin include proteins expressed by cells (ICAM-1CD11a/CD18, CD11b/CD18; ICAM-2 CD11a/CD18; ICAM-3 for CD11a/CD18) as well as soluble proteins (fibrinogen and factor X for CD11b/CD18 and complement fragment for CD11b/Cd18 and CD11c/CD18). Neutrophil and monocyte adhesion to endothelium relies primarily on the CD11a/CD18 and CD11b/CD18 leukocyte integrins with only a minor role for CD11c/CD18 (Hynes *et al.* 1992, William 1988, Arnaout 1990).

Other Adhesion Pathways

CD11b/CD18 was initially identified as a receptor for iC3b. CD11b/CD18-dependent neutrophil adhesion to endothelium was shown to be rapidly induced by fixation of complement on the endothelial surface (Mark *et al.* 1989, Collard 1999).

Fibrinogen is a soluble ligand for CD11b/CD18. Recently, fibrinogen was shown to promote leukocyte adherence to endothelium by binding both leukocyte and endothelial cells.

REGULATION OF ENDOTHELIAL ADHESION MOLECULE EXPRESSION

Among the endothelial adhesion molecules there are both similarities and differences regarding the stimuli that induce them and the temporal relation of their expression. For example, a triad of agents, IL-1, TNF-α, and lipopolysaccharide, stimulate the expression of ICAM-1, VCAM-1 and E-selectin, but the kinetics of the induced surface expression in vitro differs, with E-selectin having a shorter half-life than ICAM-1 or VCAM-1. Stimulated surface expression of ICAM-1, E-selectin, and VCAM-1 appears to result in large part from increased transcriptional regulation. However, surface expression of P-selectin may also involve a rapid mobilization of cytoplasmic granules induce by non-cytokine agents, and surface expression of E-selectin is in part regulated by rapid internalization. Therefore, important differences in regulatory mechanisms exist among these proteins that may help to explain the recruitment of subsets of leukocytes to specific sites of endothelium during an inflammatory or immune response (Ellist 1990, Smith 1993, McEver *et al.* 1989).

P-selectin is constitutively synthesized by endothelial cells and platelets. With endothelial activation by thrombin, histamine, phorbol ester, complement proteins, cytoplasmic storage granules fuse with the cell membrane, externalizing their contents. Such stimulation of endothelial cells leads to P-selectin-dependent neutrophil adhesion lasting 1.5 to 4 hr. In contrast to ICAM-2 and P-selectin, there is abundant evidence that E-selectin, ICAM-1 and VCAM-1 are transcriptionally regulated by cytokines. Furthermore, there appears to be a difference between endothelium in large vessels versus the microcirculation in the ability to express these proteins. Again, all of these factors may contribute to the selective recruitment of subsets of leukocytes during an inflammatory or immune response.

ADHESION CASCADE

Elegant studies by interracial microscopy have identified a sequence of adhesive interactions involved in leukocyte emigration from the bloodstream to extravascular site of inflammation. Because of lack of cilia, leukocytes cannot swim to the vessel wall in response to extravascular chemotactic stimuli. Initial contact with the vessel wall then is a random event, perhaps enhanced by local alteration in flow characterstics. Although 39% of leukocytes were observed to roll along the endothelium of rat mesenteric venules, only 0.6% rolled along the endothelium of arterioles. With sufficient tissue trauma or inflammation, a portion of the leukocytes are observed to flatten and spread on the endothelium and then to stick firmly. Occasionally, other leukocytes will stick to the leukocyte adherent to the vessel wall, forming small aggregates of leukocytes attached to the endothelium. Some of the adherent leukocytes crawl over the endothelial surface, seeming to probe for an opening, and then diapedes or crawl between endothelial cells; once in subendothelial tissue, the extravasated leukocytes continue to migrate toward the inflammatory site. In the final steps of the cascade, some of the adherent leukocytes migrate between the inter-endothelial cell junctions, and then through the subendothelial extracellular matrix to accumulate finally at the site of inflammatory or immune reaction.

SELECTIVE RECRUITMENT

Experimental models of inflammation have shown that the recruitment of leukocyte subtypes follows a characteristic temporal sequence. There is often selective recruitment of a leukocyte subtype in inflammatory or immune reaction. The most striking example is the marked accumulation of eosinophils at extravascular site of allergic rhinitis or alveolar spaces in asthma, although eosinophils represent only a small percentage of circulating leukocytes. Also, in synovial tissue in rheumatoid arthritis and in skin involved with inflammatory dermatoses there is a preponderance of memory T-cells that does not reflect their frequency in peripheral blood. One possible mechanism to account for the selective recruitment of leukocyte subtype is the expression of specific combination of endothelial adhesion molecules that will preferentially bind certain leukocytes, an endothelial area code in the inflamed systemic vasculature analogous to an address in lymphoid tissue (Marky *et al.* 1990, Elliot *et al.* 1990).

INTRACELLULAR SIGNALLING MECHANISM OF HR INJURY

Acute and chronic inflammations are thought to be central to the pathogenesis of many diseases. The process of tissue injury is complex and requires intercellular communication between infiltrating leukocytes and endothelium. Migration and activation of leukocytes is initiated by physical injury, infection, HR or a local immune response, and require a series of intracellular signals. One of the many signalling pathways used is the mutagen-activated protein kinase (MAPK) pathway. The MAPK signalling pathway is one of the four major signalling systems used by eukaryotic cells to transduce extracellular signals to intracellular response. There is ample evidence of the role that protein kinases play in the signalling pathways secondary to HR injury. The protein kinases initiate several interconnected downstream cascades regulated by phosphorylation and dephosphorylation reactions. The signalling transduction pathways ultimately initiate the nuclear transcription of the inflammatory and anti-inflammatory genes that plays a pivot role in HR injury (Obata 2000, Herlaar 1999).

MAPK SIGNALLING PATHWAY

Three major MAPK pathways are known at present: extracellular-regulated protein kinase [ERK (p42/44)], c-JunNH-terminal kinase [JNK (p46/54)] and P38 mitogen-activated kinase. To date more than 12 MAPKs have been cloned. Their products form a complex network of signalling routes that upon stimulation lead to a variety of cellular responses. Activation of MAPK system produces various responses against a variety of stimuli. For example, the activation of ERK causes cell proliferation, transformation and cell differentiation, while activation of JNK and p38 system is responsible for apoptosis, stress response and inflammation. The recruitment of neutrophil is mediated via the generation of pro-inflammatory cytokines resulting in the production of chemokines by endothelial and infiltrating cells. Neutrophil and macrophage stimulation regulates expression of the TNF-α IL-8 via MAPK p38. The β_2 integrins, which are normally present in the leukocyte cell membrane, require MAPK pathway for their upregulation. The p38 MAPK signalling cascade regulates TNF-α-induced expression of VCAM-1 in endothelial cells, but ICAM-1 is not regulated by p38 but is thought to be regulated through NFκB. The mechanism involved in the generation of ROS in leukocyte endothelial cells is the oxidative burst, which also requires p38 MAPK.

MECHANISM OF TISSUE DAMAGE BY HR INJURY

Once in the tissue, activated leukocytes can subsequently enhance the localized inflammatory response by releasing toxic metabolites such as proteases (e.g., elastase and methyloproteinases) and ROS resulting in damage to the surrounding tissue. The toxic ROS include superoxide anions (O_2^-), hydroxyl radicals (OH^-), hypochlorous acid ($HOCl^-$), and hydrogen peroxide (H_2O_2). ROS are potent oxidizing and reducing agents that directly damage cellular membranes by lipid peroxidation. The released proteases degrade the subendothelial extracellular matrix, enabling more leukocytes to migrate to the site of injury while ROS degrade matrix components, resulting in loss of structural integrity of the affected tissue. ROS have long been recognized as important components in host defence as well as contributors to the pathogenesis of inflammatory disease. In addition, ROS stimulate leukocyte activation and chemotaxis by activating plasma membrane phospholipase A2 to form arachidonic acid, an important precursor for eicosanoid synthesis (e.g., thromboxane A2 and leukotriene B4).

ROS also stimulate leukocyte adhesion molecule and cytokine gene expression via activation of transcription factors such as nuclear factor-κB. In addition to causing direct cell injury, ROS thus increases leukocyte activation, chemotaxis, and leukocyte-endothelial adherence after HR.

Ischaemia reperfusion results in complement activation and the formation of several pro-inflammatory mediators that alter vascular homeostasis. Particularly important are C3a and C5a, and complement components iC3b and C5b-9. In addition to stimulating leukocyte activation and chemotaxis, C5a may further amplify the inflammatory response by inducing production of the cytokines monocyte chemoattractant protein 1, TNFα, IL-1, and IL-6. C3b is a specific ligand for leukocyte adhesion to the vascular endothelium via the β_2 integrin, CD11b/CD18. In addition, C5b-9 may activate endothelial NF-κB to increase leukocyte adhesion molecule transcription and expression. C5b-9 also promotes leukocyte activation and chemotaxis by inducing endothelial IL8 and monocyte chemoattractant protein 1 secretion. C5b-9 may also alter vascular tone by inhibiting endothelium-dependent relaxation and decreasing endothelial cyclic guanosine monophosphate (Mark *et al.* 1989, Collard *et al.* 1999). Thus, complement may compromise blood flow to an ischaemic organ by altering vascular homeostasis and increasing leukocyte-endothelial adherence.

SUMMARY

- Hypoxia reperfusion injury has been recognized to play a key role in the pathogenesis of many kinds of organ dysfunction.
- Reperfusion of ischaemic tissues is associated with microvascular dysfunction that causes leukocyte migration into intrestisuum.

- The endothelial cells and leukocytes in all segments lead to the production and release of inflammatory mediators and upregulate the expression of adhesion molecules in a highly organized fashion that promotes leukocyte-endothelial cell adhesion.
- Many intracellular enzymes in the leukocytes are activated during the process of HR injury.
- On reaching the extravascular space, leukocytes exacerbate tissue injury by releasing these destructive enzymes.
- The tissue insult caused by this injury is not restricted locally but also causes damage to distant organs.

Abbreviations

ERK	extracellular signal-regulated kinase
fMLP	N-formyl-methionyl-leucyl-phenylalanine
ICAM-1	intracellular adhesion molecule 1
ICAM-2	intracellular adhesion molecule 2
IL1	interleukin-1
ILβa	interleukin β
MAPK	mitogen-activated protein kinase
ROS	reactive oxygen species
TNFα	tumour necrosis factor α
VCAM-1	vascular cellular adhesion molecule 1

References

Carden, D.L. and D.N. Granger. 2000. Pathophysiology of ischemia-reperfusion injury. J. Pathol. 190: 255.

Collard, D.C. and S. Gelman. 2001. Pathophysiology, Clinical Manifestations, and Prevention of Ischemia-Reperfusion Injury. Anesthesiology 94: 1133.

Fiebig, E. and G. Ley et al. 1991. Rapid leukocyte accumulation by "spontaneous" rolling and adhesion in the exteriorised rabbit mesentery. J. Microcirc. Clin. Exp. 10: 10127.

Ley, K. and Gaehtgens, P. 1991. Endothelial, not hemodynamic, differences are responsible for preferential leukocyte rolling in rat mesenteric venules. Circ. Res. 69: 1034.

Bevilacqa, M.P. and R.M. Nelson. 1993. Selectins. J. Clin. Invest. 91: 379.

Watson, M.L. et al. 1990. Genomic organization of the selectin family of leukocyte adhesion molecules on human and mouse chromosome-1. J. Exp. Med. 172: 263.

Bevilacqa, M.P. 1987. Recombinant tumor necrosis factor induces procagulant activity in cultured human vascular endothelium: characterization and comparison with the action of interleukin 1. Proc. Natl. Acad. Sci. USA 84: 9238.

McEver, R.P. *et al.* 1989. GMP-140 a platelet a-granule membrane protein, is also synthesized by vascular endothelial cells and is localized in Weibel-Palade bodies. J. Clin. Invest. 84: 92.

Bonfanti, R. *et al.* 1989. PADGEM (GMP-140) is a component of Weibel-Palade bodies of human endothelial cell. Blood 73: 1109.

Dunlop, L.C. *et al.* 1992. Characterization of GMP-140 (p-selectin) as a circulating plasma. J. Exp. Med. 175: 1147.

William, A.F. and A.N. Barclay. 1988. The immunoglobulin superfamily. Annu. Rev. Immunol. 16: 381.

Katz, E.F. *et al.* 1985. Chromosome mapping of cell membrane antigens expressed on activated B cells. Euro. J. Immunol. 15: 103.

Hogg, N. *et al.* 1991. Structure and function of intercellular adhesion molecule-1. Chem. Immunol. 50: 98.

Staunton, D.E. *et al.* 1989. Functional cloning of ICAM-2, a cell adhesion ligand for LFA-I homologous to ICAM-1. Nature 339: 61.

Hynes, R.O. 1992. Integrins: Versatility, modulation, and signaling in cell adhesion. Cell 69: 11.

Arnaout, M.A. 1990. Structure and function of the leukocyte adhesion molecule CD11/CD18. Blood 75: 1037.

de Fougerolles, A.R. and T.A. Springer. 1992. Intercellular adhesion molecule-3, a third counter-receptor for lymphocyte function-associated molecule on resting lymphocytes. J. Exp. Med. 175: 185.

Mark, R. *et al.* 1989. Rapid induction of neutrophil endothelial adhesion by endothelial complement fixation. Nature 339: 314.

Patel, K.D. 1991. Oxygen radicals induce human endothelial cells to express GMP-140 and bind neutrophils. J. Cell Biol. 112: 749.

Smith, C.W. 1993. Endothelial adhesion molecule and their role in inflammation. Can. J. Pharmacol. 71: 76.

Elliot, M.J. *et al.* 1990. IL-3 and granulocyte-macrophage colony-stimulating factor stimulate two distinct phases of adhesion in human monocytes. J. Immunol. 145: 167.

Markey, A.C. *et al.* 1990. T-cell inducer population in cutaneous inflammation: a predominance of T helper-inducer lymphocytes (THI) in the inflammatory dermatoses. Br. J. Dermatol. 112: 325.

Obata, T. and G. Brown, and M. Yaffe. 2000. MAP kinase pathways activated by stress: the p38 MAPK pathway. Crit. Care Med. 28: 67.

Herlaar, E. and Z. Brown. 1999. P38 MAPK signalling cascades in inflammatory disease. Mol. Med. Today 5: 439.

Collard, C.D. *et al.* 1999. Complement-activation following oxidative stress. Mol. Immunol. 36: 941.

CHAPTER **16**

Adhesion Molecules in Kidney Diseases

Maurizio Li Vecchi[1] **and Giuseppe Montalto**[2, *]

[1]Chair of Nephrology, Department of Internal Medicine, Cardiovascular and Renal Disease, Università di Palermo, via del Vespro 141, 90127 Palermo, Italy, E-mail: livecchi@unipa.it

[2]Chair of Internal Medicine, Department of Clinical Medicine and Emerging Pathologies, Università di Palermo, via del Vespro 141, 90127 Palermo, Italy, E-mail: gmontal@unipa.it

ABSTRACT

Cell adhesion molecules are crucial to many biological processes, such as cell-cell and cell-substrate adhesion, and in some cases they act as regulators of intracellular signaling cascades. Adhesion molecules mediate leukocyte adhesion to endothelial, epithelial and mesangial cells and facilitate communication between cells and extracellular matrix proteins.

In the kidney, the specific permeability and transport properties of the nephron segments are determined by the cyto-architecture of the epithelial cells and by the cell-cell and cell-matrix interactions which involve specialized junctional complexes composed of specific cell adhesion molecules. Disorders in the functioning of these molecules may affect the normal absorption/excretion of fluid and solutes. Abundant experimental and clinical evidence indicates that adhesion molecules play a critical role in mediating the inflammatory disease process characterized by tissue leukocyte infiltration. Injury to the kidney triggers a cascade of events leading to changes in the expression of these molecules, conditioning the anatomical and clinical evolution of kidney disease. This chapter summarizes the families of adhesion molecules, including claudins, integrins, the immunoglobulin superfamily, and cadherins; their expression in normal kidney tissue; and the pathophysiological role they play in some types of kidney disease including membranous glomerulonephritis, IgA nephropathy, crescentic glomerulonephritis, glomerulosclerosis, renal cyst, acute renal failure, chronic renal failure, hemodialysis and renal transplant rejection.

*Corresponding author

INTRODUCTION

In the kidney, the specific permeability and transport properties of the nephron segments are determined by the cyto-architecture of the epithelial cells and also by the cell-cell and cell-matrix interactions which involve specialized junctional complexes composed of specific cell adhesion molecules. Disorders in the functioning of these molecules can affect renal function.

The glomerular filtration process involves the passage of solutes, electrolytes and low molecular proteins through the porous glomerular endothelium and through the glomerular basement membrane (GBM). In the final step of the filtration process they pass through the slit diaphragm that bridges adjacent foot processes deriving from different podocytes. The slit diaphragm has been described as having an hourglass shape with high expression levels of P-cadherin and α- and β-catenin (Reiser *et al.* 2000), although some authors have reported no P-cadherin and α- and β-catenin localization in this site (Yaoita *et al.* 2002). In addition, nephrin, an associated protein NEPH1 and the transmembrane protein Fat1 (a member of the Fat subclass of cadherins) are located in the slit diaphragm region (Chugh *et al.* 2003).

The specific barrier and permeability characteristics of the tubular segments of the nephron are determined by the functional state of the apical junctional complexes. These junctional complexes are necessary to restrict permeability, establish epithelial polarity, and direct the traffic of membrane proteins to either the apical or the basolateral cell surface, and for the normal transport of solutes and electrolytes across the tubular barrier (Crean *et al.* 2004).

This chapter summarizes the families of adhesion molecules expressed in normal renal tissue and the changes that occur in the different types of kidney disease.

Adhesion Molecules in Normal Kidney

The expression of adhesion molecules in the normal human kidney is summarized in Table 1.

Claudins

The apical junctional complex has been considered to consist of two primary units. Occludin and the claudins are the structural components of the tight junction (TJ) and are the main determinants of paracellular permeability in the tubular segments of the nephron (Gonzalez-Mariscal *et al.* 2003). Occludin and zonula occludens-1 (ZO), ZO-2 and ZO-3 are integral membrane proteins that interact with each other and anchor the junctional complex to the cytoskeletal elements of adjacent cells (Van Itallie *et al.* 2006). Claudins, but not occludin, are thought to constitute the backbone of TJ. In the kidney, the claudin expression pattern

Table I Major families of adhesion molecules and the main sites of expression

Family	Adhesion molecule	Distribution
Selectins	E-selectin	Endothelium
	L-selectin	Leukocytes
	P-selectin	Platelets, endothelium
Integrins		
β1 integrins	VLA-4	Leukocytes
β2 integrins	CD11a/CD18	Leukocytes
	CD11b/CD18	Monocytes, neutrophils
	CD11c/CD18	Monocytes, neutrophils
	CD11d/CD18	Leukocytes
β7 integrins	α4 β7	B and T cells
Ig superfamily	ICAM-1	Leukocytes, endothelium, epithelium, mesangium
	ICAM-2	Endothelium
	ICAM-3	Lymphocytes
	VICAM-1	Endothelium, epithelium, mesangial cells, glomerular parietal cells
	LFA-2	T cells
	LFA-3	Erythrocytes
Cadherins	E-cadherin	Adult epithelial cells
	N-cadherin	Adult nerve and muscle
	P-cadherin	Adult placenta and epithelial cells
Claudins	claudin-1	Glomerulus
	claudin-2	Proximal tubule
	claudin-4	Collecting tubule
	claudin-5	Endothelial cells
	claudin-6	Glomerular capillary wall
	claudin-7, 8	Distal nephron
	claudin-10a	Medulla
	claudin-10b	Cortex
	claudin-16	Thick ascending limb of the loop of Henle

The table shows the cellular and renal distribution of adhesion molecules in healthy subjects.

is tissue- and segment-specific: the glomerulus expresses claudin-1, the proximal tubule expresses relatively high levels of claudin 2, the thick ascending limb of the loop of Henle expresses claudin-16 (paracellin-1), the distal nephron expresses claudin-7 and -8, the collecting tubule contains claudin-4, and claudin-5 is localized in endothelial cells. Claudin-6 in the glomeruli is distributed along the glomerular capillary wall and co-localized with ZO-1. Claudin-6 is a transmembrane protein of TJ in podocytes present during development and under pathological conditions. Two isoforms of claudin-10 (a and b) are expressed in many tissues, but claudin-10a is unique to the kidney. It is difficult to assign the expression of claudin-10a

Table 2 Expression of major adhesion molecules in renal diseases

Renal diseases	Adhesion molecules
Membranous glomerulonephritis	VLA-3 ↓, VCAM-1 and PECAM-1 ↓, E-selectin ↑, α-integrin ↑↓
IgA nephropathy	ICAM-1 ↑, VCAM-1 ↑, α1β1 and α5β1 integrin ↑
Crescentic glomerulonephritis	ICAM-1 ↑, VCAM-1 ↑, α2β1, α3β1, α5β1, α5β3 integrins ↑
Glomerulosclerosis	Integrin α1β1 ↑
Renal transplant rejection	ICAM-1 ↑, VCAM-1 ↑, β6 integrin ↑, VE-cadherin ↓

Up- or down-regulations of adhesion molecules are expressed by the symbol ↑ or ↓, respectively.

and -10b to specific nephron segments, while their mRNAs are preferentially expressed in either the medulla or cortex, respectively (Table 1).

Integrins

Integrins are heterodimeric transmembrane proteins made up of an α and β subunit. They are ubiquitously distributed and mediate diverse biological functions, including cell-cell and cell-matrix interactions, cell polarity, cell migration and angiogenesis. They serve as receptors for a number of morphogenetic extracellular matrix (ECM) proteins, including laminins, collagens, fibronectin, osteopontin, nephronectin, vitronectin and tenascin. Podocytes predominantly express the α3 integrin subunit, proximal tubules express the α6 subunit, and distal tubules express α2, α3 and α6 subunits.

The β1 integrins have been identified on the cell membranes of lymphocytes and glomerular, endothelial, epithelial and mesangial cells. β1 integrins are also referred to as the VLA family. The β1 integrin VLA-4 (α4β1) is expressed by lymphocytes, basophils and eosinophils, and binds to vascular cell adhesion molecule 1 (VCAM-1). Epithelial and mesangial cells express VCAM-1, facilitating the interaction of these cells with leukocytes bearing VLA-4.

The β2 integrins include several CD11/CD18 subtypes expressed exclusively on leukocytes.

CD11a/CD18 (LFA-1) binds primarily to the inducible receptor intercellular adhesion molecule 1 (ICAM-1) on endothelial cells, epithelial cells, glomerular mesangial cells and leukocytes. CD11a/CD18 also binds to the receptor ICAM-2 on endothelial cells, as well as to ICAM-3 on leukocytes. CD11b/CD18 also binds ICAM-1 and, in addition, engages the complement fragment C3b, factor X and fibrinogen.

The β3 integrin subfamily includes the platelet receptor GPIIb/IIIa (CD41/CD61), which binds fibrinogen and Willebrand's factor. This integrin is a potential target for modulating platelet-mediated kidney diseases.

Immunoglobulin Superfamily

The immunoglobulin superfamily includes ICAM-1 identified on leukocytes, endothelial and epithelial cells, mesangial cells; ICAM-2 on the endothelium; and ICAM-3 on lymphocytes. VCAM-1 is constitutively expressed by glomerular parietal epithelial cells and is also detected on a wide variety of cells, including endothelial, tubular epithelial and mesangial cells following stimulation with cytokines.

Selectins

The selectins are classified according to the cell type on which they were originally identified: E-selectin (endothelial cells), P-selectin (platelets) and L-selectin (leukocytes). They are members of the group of carbohydrate-containing binding proteins known as lectins. Adhesive interactions with the vascular endothelium initiate the migration of leukocytes to sites of inflammation. In this cascade of events (rolling, attachment, spreading and transendothelial migration), the selectin family plays a role mainly in rolling and attachment.

Cadherins

Cadherins are a large family of cell-cell adhesion molecules acting in a homotypic, homophilic manner and playing an important role in the maintenance of tissue integrity. Classical cadherins can be classified into two subfamilies, type I (E-, N-, P- and R-cadherin) and type II (cadherin-5 to -12, -14 and -15). Cadherin-16, also called kidney-specific cadherin, is exclusively expressed in epithelial cells of the adult kidney. E-cadherin expression is restricted to the distal tubules and collecting ducts of the human kidney, whereas N-cadherin and cadherin-6 expression are found on proximal tubules. Cadherin-8 can only be detected on developing tubular; cadherins E and P have been demonstrated in glomeruli (Thedieck *et al.* 2005).

Adhesion Molecules in Kidney Diseases

Glomerular leukocyte infiltration is a common feature of many types of glomerulonephritis (GN). Leukocyte infiltration requires leukocytes to adhere to the vascular endothelium before emigrating across the endothelium and basement membranes to reach the tissue inflammation site. Adherence to leukocytes of the endothelium involves adhesion molecules, present on both endothelial cells and leukocytes. Once the site of inflammation is reached, they release cytokines and chemoattractants to enhance leukocyte infiltration, then release lysosomal enzymes and reactive oxygen molecules causing tissue injury. Leukocytes attach to parenchymal cells of the kidney through specific adhesion molecules (Fig. 1). Numerous studies have focused on the altered glomerular expression of adhesion molecules in human biopsy specimens from different GN.

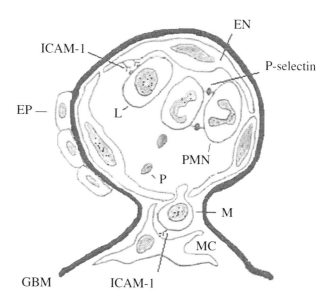

Fig. 1 Leukocytes and adhesion molecules in glomerular inflammation. EN, endothelial cell; EP, epithelial cell; GBM, glomerular basement membrane; L, lymphocyte; MC, mesangial cell; M, macrophage; P, platelet; PMN, polymorphonuclear leukocyte. Modified from Rabb *et al.* (2001).

Membranous Glomerulonephritis

MG is a glomerular disease characterized by diffuse and uniform thickening of the GBM due to the deposition of subepithelial immune complexes. It is often associated with interstitial cellular inflammation and fibrotic changes in the interstitium as well as tubular atrophy. Active glomerular disease results in the release of cytokines capable of enhancing the expression of adhesion molecules that increase the traffic of immunocompetent cells into the interstitium. The disease may slowly progress to renal failure.

A study of integrin VLA-3 expression in renal biopsy specimens from patients with nephrotic syndrome due to MG showed that the normal linear distribution of VLA-3 along the glomerular capillary loop was altered, indicating that changes in MG were not an aspecific effect of proteinuria. In stages I and II of MG, VLA-3 distribution showed an irregular, trabecular pattern. In addition, in stage III a segmental loss of VLA-3 was detected. These observations suggest that in human MG there is a disruption of the normal interaction between VLA-3 and its GBM ligand. In another series, investigating the expression of adhesion molecules ICAM-1, VCAM-1, platelet-endothelial cell adhesion molecule 1 (PECAM-1), E-selectin and two members of the β2-integrin family (LFA-1 and Mac-1), glomerular expression was found to be largely unchanged apart from a reduced expression of VCAM-1 and PECAM-1. No tubular expression of ICAM-1 and VCAM-1 was observed. The sample patients showed significantly elevated E-selectin expression in the peritubular capillaries, suggesting an activated capillary endothelium. Expression of most adhesion molecules significantly increased in the interstitial areas. There was a significant correlation between the number of interstitial cells expressing LFA-1 and E-selectin expression in peritubular capillaries as well as interstitial ICAM-1 expression. This is in agreement with the concept that the activated capillary endothelium promotes leukocyte adhesion and extravasation into the interstitial tissue (Honkanen *et al.* 1998). White *et al.* showed α5-integrin staining in the tubular epithelium in areas of tubulo-interstitial disease, which occurred primarily on the basolateral cell surface in renal biopsies from patients with MG. No tubular epithelial cell staining was seen in biopsies where the tubules and interstitium were normal, suggesting that α5-integrin is expressed in the epithelium only during tubular injury. These data show that α5-integrin is induced in the tubular epithelium of patients with kidney disease and tubulo-interstitial injury and provide new insight into the role of integrins in renal injury and fibrosis (White *et al.* 2007) (Table 2).

IgA Nephropathy

IgAN is the most common GN in patients undergoing renal biopsy worldwide. Diagnosis is based on the occurrence of mesangial deposits of IgA in the glomeruli in the presence of recurrent episodes of intra-infectious macroscopic hematuria or persistent microscopic hematuria and/or proteinuria, frequently followed by hypertension and progressive reduction of the glomerular filtration rate. ICAM-1 expression in glomeruli from patients with IgAN has been reported significantly higher in patients with advanced stage than in patients with the mild stage. Its expression was closely linked to glomerular cell proliferation and to lymphocyte and monocyte infiltration (Tomino *et al.* 1994). Indirect immunofluorescence confirmed ICAM-1 expression in glomeruli from patients with IgAN. A marked expression of ICAM-1 was observed in the glomerular capillary walls and mesangial areas in advanced stage, but not in mild IgAN patients. ICAM-1 expression in the glomeruli may be useful in assessing the degree of renal lesions in these patients. In contrast, as there were no significant changes in ICAM-1 serum levels in the mild and advanced stages of disease, serum ICAM-1 assay cannot be considered useful in evaluating the degree of kidney injury (Nguyen *et al.* 1999).

Staining for the α1β1 subunit and α5β1 integrin in moderate or severe IgAN was increased along the capillary walls and in the mesangial region. Wagrowska-Danilewicz and Danilewicz (2004). reported an upregulation of α5β1 integrins on endothelial cells within glomeruli from IgAN patients with severe proteinuria, reflecting the role of this integrin in the mechanism of glomerular endothelial cell injury. The positive association between the interstitial expression of α5β1 and the degree of interstitial fibrosis, independent of the degree of proteinuria, suggests that this molecule may play a role in the pathogenesis of chronic progressive renal disease in IgAN (Wagrowska-Danilewicz and Danilewicz 2004) (Table 2).

Crescentic Glomerulonephritis

CGN and its clinical corollary, rapidly progressive glomerulonephritis (RPGN), are potentially fatal diseases. The histopathological hallmark of RPGN is the crescent formation resulting from the disruption of glomerular capillaries that allows inflammatory mediators and leukocytes to enter the Bowman space, where they induce epithelial cell proliferation and macrophage maturation, which together produce the cellular crescent. There are three major subgroups of diseases: (1) anti-GBM disease and Goodpasture syndrome; (2) immunocomplex-mediated processes, where there is immune deposition usually resulting from processes such as infections, cryoglobulinemic GN, systemic lupus erythematosus and (3) pauci immune diseases most often associated with antineutrophil cytoplasmic antibodies (ANCA).

Vasculitis affecting the small blood vessels may cause focal necrotizing GN, often with crescents, which presents as RPGN characterized by a rapid loss of renal function, often accompanied by oliguria or anuria and other features of GN, including dysmorphic erythrocyturia, cylindruria and glomerular proteinuria.

Upregulated renal expression of ICAM-1 and VCAM-1 has been reported in ANCA-associated RPGN and related to histological activity. De novo VCAM-1 expression on glomerular capillary walls and de novo ICAM-1 and VCAM-1 expression on tubular cells suggest that endothelial and epithelial cells play a role in adhesive interactions in RPGN. De novo VCAM-1 expression at the glomerular tuft in PR3-ANCA-positive patients seems to be greater than in MPO-ANCA-positive patients, which suggests that immune activation mechanisms play a role in ANCA-associated GN (Arrizabalaga *et al.* 2008). As regards the distribution pattern of integrins, ICAM-1 and VCAM-1 in renal samples from patients with RPGN of different etiologies, a marked upregulation of $\alpha 2\beta 1$, $\alpha 3\beta 1$, $\alpha 5\beta 1$, $\alpha 5\beta 3$ integrins and VCAM-1 on tubular cells of the renal cortex was observed, with as many as 60-90% of tubular cross-sections labeled, while a strong ICAM-1 reactivity was limited to the luminal surface. The same adhesion molecules were also uniformly expressed on crescentic cells. In the glomeruli, integrin upregulation occurred only on apparently preserved capillary tufts in an early stage of lesion, while collapsed and sclerotic tufts showed a reduced integrin expression. The upregulation of $\alpha 5\beta 3$ on podocytes might play a role in the adhesion of crescentic cells (Baraldi *et al.* 1995). Increased serum levels of E-selectin, ICAM-1 and VCAM-1 together with decreased levels of L-selectin in active ANCA-associated vasculitis and the normalization of E-selectin, ICAM-1, and VCAM-1 during the remission phase have been reported, suggesting that the concentration of soluble levels of these adhesion molecules reflects disease activity (Ara *et al.* 2001) (Table 2).

Glomerulosclerosis

Glomerulosclerosis, the process by which glomerular tissue is replaced by ECM, is the final common pathway in the loss of functioning glomeruli. All the three major cell types constituting the glomerulus contribute to this process. Podocytes and endothelial cells are probably critical for the onset of sclerosis. However, mesangial cells are the major contributors to progression. ECM metabolism is, in part, regulated by integrins. Integrin $\alpha1\beta1$, a major collagen receptor, is expressed in all cell types in the glomerulus. This integrin has been associated with kidney disease and is overexpressed in the proliferating mesangium in GN. Lack of integrin $\alpha1$ has been demonstrated to predispose mice to severe acute damage and excessive sclerosis after non-immunologically-induced renal injury. This response to the insult is mediated by the loss of the direct interaction of integrin $\alpha1\beta1$ with its ligand collagen IV, as well as by increased production of reactive oxygen species. Integrin $\alpha1\beta1$ may be regarded as a potent modifier for injury-induced glomerulosclerosis in particular, and possibly for organ fibrosis in general (Chen *et al.* 2004).

Acute Renal Failure

ARF is characterized by a rapid decline in the glomerular filtration rate, accompanied by disturbances in solute and water transport and in acid-base homeostasis. Inflammation plays a central role in ARF; it is classically characterized by the margination of neutrophils to the vascular endothelium and the activated polymorphonuclear leukocytes induce cell injury through a wide range of secreted products including reactive oxygen species, proteolytic and other degradation enzymes, cytokines and numerous other pro-inflammatory substances. Tethering interactions between selectins and their ligands initially slows the neutrophils, allowing firmer adhesion and transmigration by integrins and their ligands. P-selectin has been shown to be the main determinant of P-selectin-mediated renal injury. Blockade of the shared ligand to all three selectins (E-, P-, and L-selectin) significantly protected rats from both renal ischemic reperfusion injury and associated mortality (Nemoto *et al.* 2001). Both in rats and in mice, the selectin ligand blockade, initially targeted to abrogate neutrophil infiltration, resulted in renal protection while neutrophils continued to infiltrate the post-ischemic kidney. Thus, it appears that selectin pathway modulation can alter the outcome of ARF via neutrophil-independent ways. After the slowing of leukocytes at the site of injury by selectins, firm adhesion occurs by the interactions of integrins with ICAM-1. Mice deficient in ICAM-1 were also found to have relative protection from ischemic renal injury. Interestingly, neutrophil depletion in the rat model did not lead to protection, whereas it did do so in the mouse.

Renal Cysts

Renal cyst formation derives from abnormal tubular cell proliferation and is accompanied by abnormalities in the matrix synthesis by cyst-lining cells and changes in epithelial polarity. Tubule cell matrix interactions are mediated by adhesion receptors, mainly of the integrin family. Autosomal dominant polycystic kidney disease (ADPKD) is due to germ-line and somatic PKD1 or PKD2 gene mutations. The majority of ADPKD cases are caused by mutations in the PKD1 gene, which codes for polycystin-1. Huan *et al.* (1999) showed that polycystin-1 co-localized with the cell adhesion molecules E-cadherin and α-, β-, and γ-catenin and co-precipitated and co-migrated with them on sucrose density gradients. They concluded that polycystin-1 is in a complex containing E-cadherin and α-, β-, and γ-catenin. These observations raise the question of whether the defects in cell proliferation and cell polarity observed in ADPKD are mediated by E-cadherin or the catenins (Huan *et al.* 1999). Analysis of cDNA array identified an aberrant β4 integrin expression in ADPKD cells. Furthermore, laminin 5 (Ln-5), the main α6β4 integrin ligand, was also abnormally expressed in ADPKD. Studies performed with ADPKD cyst-lining epithelial cells indicate that integrin α6β4-Ln-5 interactions are involved in cellular events of potential importance for cystogenesis. These studies highlight the role of Ln-5 and α6β4 integrin in the adhesive and motility properties of cyst-lining epithelial cells and further suggest that integrins and ECM modifications may be of general relevance to kidney epithelial cell cyst formation and cyst enlargement in ADPKD (Joly *et al.* 2003).

Renal Transplant Rejection

Renal transplant rejection is a complex continuum of immunological and non-immunological processes that result in graft dysfunction and eventual loss. The cellular and molecular events lead to a group of well-characterized pathological and clinical patterns of rejection. Cells of the immune system infiltrate the graft from nearby lymphoid organs and the bloodstream by a multi-step process. They roll along the vessel wall through interactions between selectins on the endothelium and receptors on immune cells and adhere to the vessel endothelium following chemokine release. Adhesion molecules and chemokines are important rejection regulators and appear to be targets for immunotherapy. Adhesion molecules on T cells include LFA-1, which interacts with ICAM-1 and -2; CD2, which interacts with CD58 (LFA-3); and VLA-4 (α4, β1 integrin CDw49d, CD29), which interacts with VCAM-1, CD106.

Graft rejection is accompanied by a de novo expression of ICAM-1 on the renal tubular cells which, however, is not specific to rejection. Increased ICAM-1 expression on the renal vascular endothelium has occasionally been described and significant de novo ICAM-1 expression on tubules supports the contact between tubular cells and infiltrating leukocytes. In acute rejection, de novo VCAM-1 expression is induced on the renal vascular endothelium and is upregulated in the tubules. VCAM-1 plays a pathogenic role in rejection by potentiating the interaction between tubular cells and effector cells and by promoting leukocyte migration into the renal tissue. Upregulation of β6 integrin, which is indispensable for the activation of TGF-β1, has been investigated in chronic renal allograft dysfunction. TGF-β1 and -β6 integrin reactivity were observed in the distal tubules in acute rejection, and even greater reactivity was observed in the distal tubules. The upregulation of β6 integrin as well as TGF-β1 in CAD may serve as an alternative target for the treatment of CAD (Sawada *et al.* 2004).

Reduced vascular endothelial (VE)-cadherin expression in peritubular endothelial cells in acute rejection after human renal transplantation has been reported. It was postulated that the downregulation of VE-cadherin could be responsible for the lymphocyte transmigration into interstitial tissues leading to graft dysfunction (Roussoulières *et al.* 2007) (Table 2).

Chronic Renal Failure and Hemodialysis

Chronic renal failure is a slow, progressive loss of kidney function resulting in permanent kidney failure, and hemodialysis (HD) is the most common method used to treat advanced and permanent kidney failure. However, HD induces several changes in the circulating blood cells and protein systems. These changes range from activation of the coagulation pathway, the complement and kallikrein-kinin systems to alterations in adhesion molecule expression and cytokine synthesis by the different leukocyte subtypes. Adhesion molecules are involved in the extravasation of leukocytes from the vessel lumen into the surrounding tissue as well as in the binding of leukocytes to other circulating cells. Granulocyte rolling is predominantly mediated by the selectin family and their counterligands. P- and E-selectin (CD62P and CD62E, respectively) are expressed by endothelial cells upon activation.

The current literature on soluble adhesion molecules in HD patients is in part contradictory. L-selectin (CD62L) expression on granulocytes decreases during HD with cellulose dialyzers, while during HD with polysulfone dialyzers it does not. Some authors have demonstrated that soluble P-selectin is released during HD and that this is dependent on the type of dialyzer used. Nevertheless, some studies have shown different plasma level results when using the same type of dialyzer. Some of them found soluble P-selectin unchanged during HD with a polysulfone dialyzer, but P-selectin plasma levels increased (Musial *et al.* 2004) or decreased (Bonomini *et al.* 1998) with cuprophane and cellulose dialyzers.

Suliman *et al.* (2006) investigated concentrations of soluble ICAM-1 and VCAM-1 in relation to all-cause and cardiovascular mortality in pre-dialysis patients. They showed that high levels of ICAM-1 and VCAM-1 are associated with signs of malnutrition, inflammation and cardiovascular disease, and also that high ICAM levels predict all-cause and cardiovascular mortality in patients starting dialysis treatment, even after adjustment for conventional risk factors, suggesting that increased levels of soluble adhesion molecules may be involved in the process of atherosclerosis and increased mortality in pre-dialysis patients (Suliman *et al.* 2006).

CONCLUSION

Adhesion molecules play a critical role in the pathophysiology of a variety of kidney disorders including glomerulonephritis, tubulointerstitial nephritis, renal cysts, acute and chronic renal failure, hemodialysis membrane incompatibility reactions and kidney transplant rejection. Further knowledge of the molecular basis for adhesion molecules may not only shed light on the basic mechanism of inflammation, but also enhance understanding of other biological processes and suggest novel strategies for therapeutic intervention.

Abbreviations

ADPKD	autosomal dominant polycystic kidney disease
ARF	acute renal failure
CAD	chronic allograft dysfunction
CAPD	continuous ambulatory peritoneal dialysis
C-ANCA	cytoplasmic antineutrophil cytoplasmic antibodies
cDNA	complementary deoxyribonucleic acid
CGN	crescentic glomerulonephritis
ECM	extracellular matrix
ELAM	endothelial cell-leukocyte adhesion molecule
GN	glomerulonephritis
GPIIb	platelet glycoprotein
HD	hemodialysis
ICAM	intercellular adhesion molecule
IgAN	IgA nephropathy
LFA-1	leukocyte function-associated molecule 1
MG	membranous glomerulonephritis
MPO	myeloperoxidase
NEPH	nephrin
P-ANCA	perinuclear antineutrophil cytoplasmic antibodies
PECAM	platelet-endothelial cell adhesion molecule
RPGN	rapidly progressive glomerulonephritis
TGF-β1	transforming growth factor-β_1
TNF-α	tumor necrosis factor-α
TJ	tight junction
VCAM	vascular cell adhesion molecule
VE-cadherin	vascular endothelial cadherin
VLA	very late antigens
ZO	zonula occludens

Key Facts about Adhesion Molecules and the Main Kidney Diseases

1. Normal renal function involves specialized junctional complexes that are composed of specific cell adhesion molecules.
2. Disorders in the functioning of these molecules elicit some effects on renal function.
3. MG generally determines downregulation of VLA3, VCAM-1 and PECAM-1 as well as upregulation of E-selectin in peritubular capillaries and α 5-integrin in tubular epithelium.
4. IgAN shows upregulation of ICAM-1 in glomeruli, VCAM-1 in parietal/tubular epithelial cells, α1β1 and α5β1 integrin in capillary walls and mesangial region.
5. CGN involves upregulation of α2β1, α3β1, α5β1, α5β3 integrins, ICAM-1,VCAM-1 in tubular cells and VCAM-1 in glomerular capillary.
6. Renal transplant rejection causes upregulation of ICAM-1, VCAM-1 in vascular endothelium and tubules, β6 integrin in tubules, and downregulation of VE-cadherin in peritubular endothelial cells.

References

Ara, J. and E. Mirapeix, P. Arrizabalaga, R. Rodriguez, C. Ascaso, R. Abellana, J. Font, and A. Darnell. 2001. Circulating soluble adhesion molecules in ANCA-associated vasculitis. Nephrol. Dial. Transplant. 16: 276-285.

Arrizabalaga, P. and M. Solé, R. Abellana, and C. Ascaso. 2008. Renal expression of adhesion molecules in anca-associated disease. J. Clin. Immunol. 28: 411-419.

Baraldi, A. and G. Zambruno, L. Furci, M. Ballestri, A. Tombesi, D. Ottani, L. Lucchi, and E. Lusvarghi. 1995. β1 and β3 integrin up-regulation in rapidly progressive glomerulonephritis. Nephrol. Dial. Transplant. 10: 1259-1267.

Bonomini, M. and M. Reale, P. Santarelli, S. Stuard, N. Settefrati, and A. Albertazzi. 1998. Serum levels of soluble adhesion molecules in chronic renal failure and dialysis patients. Nephron. 79: 399-407.

Chen, X. and G. Moeckel, J.D. Morrow, D. Cosgrove, R.C. Harris, A.B. Fogo, R. Zent, and A. Pozzi. 2004. Lack of Integrin α1β1 leads to severe glomerulosclerosis after glomerular injury. Am. J. Pathol. 165: 617-630.

Chugh, S.S. and B. Kaw, and Y.S. Kanwar. 2003. Molecular structure–function relationship in the slit diaphragm. Semin. Nephrol. 23: 544-555.

Crean, J.K. and F. Furlong, D. Finlay, D. Mitchell, M. Murphy, and B. Conway. 2004. Connective tissue growth factor [CTGF]/CCN2 stimulates mesangial cell migration through integrated dissolution of focal adhesion complexes and activation of cell polarization. FASEB J. 18: 1541-1543.

Gonzalez-Mariscal, L. and A. Betanzos, P. Nava, and B.E. Jaramillo. 2003. Tight junction proteins. Prog. Biophys. Mol. Biol. 81: 1-44.

Honkanen, E. and E. von Willebrand, A.M. Teppo, T. Tornroth, and C. Gronhagen-Riska. 1998. Adhesion molecules and urinary tumor necrosis factor-alpha in idiopathic membranous glomerulonephritis. Kidney Int. 53: 909-917.

Huan, Y. and J. van Adelsberg. 1999. Polycystin-1, the PKD1 gene product, is in a complex containing E-cadherin and the catenins. J. Clin. Invest. 104: 1459-1468.

Joly, D. and V. Morel, A. Hummel, A. Ruello, P. Nusbaum, N. Patey, L.H. Noel, P. Rousselle, and B. Knebelmann. 2003. β4 Integrin and Laminin 5 are aberrantly expressed in Polycystic Kidney Disease. Am. J. Pathol. 163: 1791-1800.

Musial, A.K. and D. Zwolinska, D. Polak-Jonkisz, U. Berny, K. Szprynger, and M. Szczepanska. 2004. Soluble adhesion molecules in children and young adults on chronic hemodialysis. Pediatr. Nephrol. 19: 332-336.

Nemoto, T. and M.J. Burne, F. Daniels, M.P. O'Donnell, J. Crosson, K. Berens, A. Issekutz, B.L. Kasiske, W.F. Keane, and H. Rabb. 2001. Small molecule selectin ligand inhibition improves outcome in ischemic acute renal failure. Kidney Int. 60: 2205-2214.

Nguyen, T.T. and I. Shou, K. Funabiki, I. Shirato, K. Kubota, and Y. Tomino. 1999. Correlations among expression of glomerular intercellular adhesion molecule 1 (ICAM-1), levels of serum soluble ICAM-1, and renal histopatology in patients with IgA Nephropathy. Am. J. Nephrol. 19: 495-499.

Rabb, H. and K. Modi, and M. O'Donnel. 2001. Role of leukocytes and leukocytes adhesion molecules in glomerular, tubular and vascular disease. In Massry and Glassock's Textbook of Nephrology. Lippincot Williams & Wilkins, pp. 628-638.

Reiser, J. and W. Kriz, M. Kretzler, and P. Mundel. 2000. The glomerular slit diaphragm is a modified adherens junction. J. Am. Soc. Nephrol. 11: 1-8.

Roussoulières, A. and B. McGregor, L. Chalabreysse, C. Cerutti, J.L. Garnier, P. Pascale Boissonnat, O. Bastien, J.Y. Scoazec, F. Thivolet-Bejui, L. Sebbag, and J.L. McGregor. 2007. Expression of VE-Cadherin in peritubular endothelial cells during acute rejection after human renal transplantation. J. Biomed. Biotechnol. (6): 41705.

Sawada, T. and M. Abe, K. Kai, K. Kubota, S. Fuchinoue, and S. Teraoka. 2004. β6 integrin is up-regulated in chronic renal allograft dysfunction. Clin. Transplant. 18: 525-528.

Suliman, M.E. and A.R. Qureshi, O. Heimburger, B. Lindholm, and P. Stenvinkel. 2006. Soluble adhesion molecules in end-stage renal disease: a predictor of outcome. Nephrol. Dial. Transplant. 21: 1603-1610.

Thedieck, C. and M. Kuczyk, K. Klingel, I. Steiert, C.A. Müller, and G. Klein. 2005. Expression of Ksp-cadherin during kidney development and in renal cell carcinoma. Br. J. Cancer 92: 2010-2017.

Tomino, Y. and H. Ohmuro, T. Kuramoto, I. Shirato, K. Eguchi, H. Sakai, K. Okumara, and H. Koide. 1994. Expression of intercellular adhesion molecule-1 and infiltration of lymphocytes in glomeruli of patients with IgA Nephropathy. Nephron 67: 302-307.

Van Itallie, C.M. and S. Rogan, A. Yu, L. Seminario Vidal, J. Holmes and J.M. Anderson. 2006. Two splice variants of claudin-10 in the kidney create paracellular pores with different ion selectivities. Am. J. Physiol. Renal Physiol. 291: 1288-1299.

Wagrowska-Danilewicz, M. and M. Danilewicz. 2004. Expression of α5β1 and α6β1 integrins in IgA nephropathy (IgAN) with mild and severe proteinuria. An immunohistochemical study. Int. Urol. Nephrol. 36: 81-87.

White, L.R. and J.B. Blanchette, L. Ren, A. Awn, K. Trpkov, and D.A. Muruve. 2007. The characterization of α5-integrin expression on tubular epithelium during renal injury. Am. J. Physiol. Renal. Physiol. 292: 567-576.

Yaoita, E. and N. Sato, Y. Yoshida, M. Nameta, and T. Yamamoto. 2002. Cadherin and catenin staining in podocytes in development and puromycin aminonucleoside nephrosis. Nephrol. Dial. Transplant. 17: 16-19.

CHAPTER 17

Adhesion Molecules and Smoking

Matthew J. Garabedian[1],[*] and Kristine Y. Lain[2]
Department of Obstetrics and Gynecology, University of Kentucky,
800 Rose Street, Room C-358, Lexington, KY 40536-0293,
[1]E-mail: matt.garabedian@uky.edu
[2]E-mail: kylain2@email.uky.edu
Departmental Administrator: Linda Sager, Department of Obstetrics and
Gynecology, University of Kentucky, 800 Rose Street, Room C-358, Lexington,
KY 40536-0293, E-mail: lsage2@uky.edu

ABSTRACT

The association between smoking and disease was identified prior to elucidating the
underlying molecular mechanisms. Cigarette smoke is a complex mixture of many
compounds, of which nicotine is considered the most important biologically active
agent. Exposure to cigarette smoke and nicotine creates a systemic inflammatory
condition increasing the risk of cardiovascular disease (CVD), atherosclerosis,
asthma, and chronic obstructive pulmonary disease (COPD). Cellular adhesion
molecules play a role in these disease states. Systemic inflammation enhances
adhesion molecule expression, which facilitates leukocyte recruitment and
migration into tissue and leads to inflammatory injury.

In endothelial cell cultures, administration of cigarette smoke condensate and
purified nicotine enhances expression of adhesion molecules and inflammatory
cytokines via second messenger signaling pathways. These signaling pathways
trigger altered genetic expression and the upregulation of inflammatory cytokines.
This correlates with the clinical understanding of atherosclerosis, asthma, and
COPD as inflammatory conditions. In human populations, the same adhesion
molecules that are enhanced by cigarette smoke condensate and nicotine

---

*Corresponding author

Key terms are defined at the end of the chapter.

exposure— ICAM-1, VCAM-1, P-selectin, and E-selectin—are associated with these diseases in epidemiological studies. Although the exact mechanisms have yet to be fully characterized, the pathophysiology in part relies on these molecules.

Initially, epidemiological studies linked cigarette smoking to abnormal adhesion molecule expression. The molecular pathways leading to disease have since been examined. Smoking leads to abnormal monocyte expression of pro-inflammatory cytokines, which in turn leads to abnormal adhesion molecule expression. This incites inflammatory injury and disease. Each cellular adhesion molecule plays its own role. In this chapter, we examine the influence of smoking on ICAM-1 and P-selectin expression and the initiation of atherosclerosis and chronic airway inflammation.

INTRODUCTION

Tissue damage caused by cigarette smoking, at a molecular level, results from leukocyte-mediated damage of the endothelium and subendothelium. Circulating leukocytes are recruited from the circulation and adhere to the vascular endothelium. Following cellular adhesion, the leukocytes migrate to the subendothelium and cause inflammatory injury to the underlying tissues.

Adhesion molecules mediate the interactions between inflammatory cells, specifically circulating leukocytes, and the endothelium. Changes in the expression of these molecules alter the balance between homeostasis and disease. Adhesion molecules are present in two forms: those bound to the epithelial cells and those found in a soluble form, circulating within the blood or other body fluids. Soluble adhesion molecules have been studied as biomarkers for disease activity and for possible usefulness as prognostic indicators.

CARDIOVASCULAR DISEASE

Atherosclerosis

Background

Atherosclerosis is the result of inflammatory damage to the vascular endothelium and an important etiology of CVD. Elevated serum concentrations of soluble adhesion molecules are associated with atherosclerosis and are predictive of future cardiovascular events. Different adhesion molecules play distinct roles in the pathology of CVD (Fig. 1) (Güray *et al.* 2004). While smoking increases circulating concentrations of soluble adhesion molecules, cessation of smoking leads to a decrease in concentrations of soluble adhesion molecules (Blann *et al.* 1997). This corresponds to the clinical phenomenon of decreased risk of atherosclerosis and other ill effects of smoking that is observed following cessation. This phenomenon

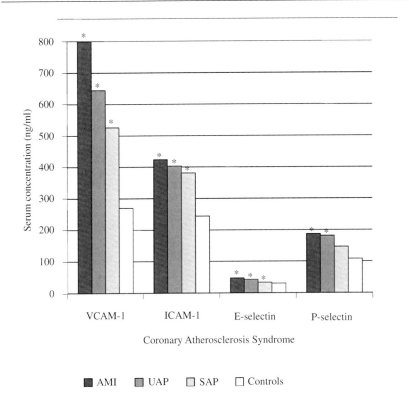

AMI ■ **UAP** ■ **SAP** □ **Controls** □

Fig. 1 Serum cellular adhesion molecules in coronary atherosclerosis syndromes and controls. Different manifestations of coronary atherosclerosis are associated with different patterns of expression of adhesion molecules. AMI, acute myocardial infarction; UAP, unstable angina pectoris; SAP, stable angina pectoris. *Statistically different from controls (p < 0.0001) within grouping. (Adapted from Güray *et al.* (2004), with permission.)

also serves as the basis for adhesion molecules serving as potential biomarkers for disease activity.

Elevation of sICAM-1 is associated with initiation of inflammatory injury, whereas sVCAM-1 is associated with current atherosclerotic burden and acute coronary syndrome. sVCAM-1 concentrations differ by the clinical scenario of coronary atherosclerosis (Fig. 1), suggesting that sVCAM-1 is a marker for ongoing endothelial injury and becomes more elevated with progression to more severe disease. Elevation in sVCAM-1 (but not sICAM-1, sE-selectin, or sP-selectin) in patients with unstable angina or non-Q-wave myocardial infarction is predictive of major cardiovascular ischemic event in the subsequent six months. The predictive potential of sVCAM-1 in this situation is comparable to, and additive with, that of C-reactive protein (Blankenberg *et al.* 2001). The differences seen in Fig. 1 for sICAM-1 between the different presentations of coronary artery disease, however, are not different from each other, suggesting a role for ICAM-1 in the initiation of, rather than progression of, atherosclerosis. Once atherosclerosis is established, the elevation of sICAM-1 becomes stable.

In healthy individuals, sICAM-1 is predictive of future cardiovascular events. Multiple secondary analyses of data from prospective studies (Table 1) have examined the association between sICAM-1 and CVD and atherosclerosis (Table 2). Although these studies have not specifically examined the association of smoking with adhesion molecule expression, the analyses suggest that ICAM-1 mediates the initiation of CVD.

sICAM-I and Healthy Men

Table I   Select epidemiological studies of cardiovascular disease

Study	Population	Study design	Outcomes of interest
The Prospective Epidemiological Study of Myocardial Infarction (PRIME) N=9,758	European (France and Northern Ireland) apparently healthy men, aged 50-59 yr with history of CHD	Prospective cohort	Fatal AMI CD CHD
Physician's Health Study N=22,071	Apparently healthy male U.S. physicians	Prospective, randomized double-blind trial of aspirin and beta-carotene for primary prevention	CHD PAD Cancer
The Atherosclerosis Risk in Communities Study (ARIC) N=15,792	Adults aged 45-64 yr from four U.S. communities (Forsyth County, NC; Jackson, MS; northwestern suburbs of Minneapolis, MN; Washington County, MD)	Prospective cohort	CHD CVA
Women's Health Study N=28,263	Apparently healthy U.S. middle-aged women	Placebo-controlled prevention trial	CVD Cancer

AMI, acute myocardial infarction; CD, cardiac death; CHD, coronary heart disease; PAD, symptomatic peripheral artery disease; CVA, cerebral vascular accident/stroke.

A nested case control study of the data from the PRIME study demonstrated that elevated sICAM-1, but not sVCAM-1, in healthy men was predictive of the development of cardiac disease. This association remained after controlling for traditional risk factors for coronary atherosclerosis including smoking and

Table 2 Epidemiological studies linking sICAM-I to incident cardiovascular and atherosclerotic risk

Study	Findings
PRIME	sICAM-1 (highest third vs. lowest third) associated with 2-fold increase in CHD in crude and multivariate analysis
Physician's Health Study	sICAM-1 (highest quartile vs. lowest quartile) associated with 80% increase in AMI in crude and multivariate analyses and 3-fold increase in PAD
ARIC	sICAM-1 associated with CHD and CAA in crude and multivariate analyses
Women's Health Study	sICAM-1 (and other inflammatory markers) higher in current smokers than non-smokers after adjusting for cardiovascular risk factors
	Inflammatory marker levels higher in current vs. former smokers

CHD, coronary heart disease; AMI, acute myocardial infarction; PAD, symptomatic peripheral artery disease; CAA, carotid artery atherosclerosis.

demonstrates the independent association between sICAM-1 and atherosclerosis. The data also suggests that sICAM-1 is a non-specific biomarker for systemic inflammation (Luc *et al.* 2003).

Similarly, elevated levels of sICAM-1 predicted development of CVD in both smokers and non-smokers in the Physician's Health Study (Ridker *et al.* 1998). Participants were matched on smoking status at enrollment, making direct investigation of the influence of smoking on CVD impossible. Subjects with the highest sICAM-1 concentrations had nearly double the risk of acute myocardial infarction (AMI) that those with the lowest concentrations had, and the risk of AMI increased over time. Smokers had a higher mean concentration of sICAM-1, but the relative risk of AMI by quartile of sICAM-1 did not differ by smoking status. The lack of independence of smoking status and sICAM-1 concentration on the relative risk of AMI in this study suggests the two exposures are in the same causal pathway. Additionally, elevated sICAM-1, but not sVCAM-1, at enrollment was associated with development of peripheral artery disease (Pradhan *et al.* 2002), demonstrating sICAM-1's association with atherosclerosis in general. In a subsequent nested case control analysis of the Physician's Health Study, no association was found between sVCAM-1 and risk of AMI, suggesting different roles for the two adhesion molecules (de Lemos *et al.* 2000).

The association between smoking and sICAM-1, but not sVCAM-1, is consistent with the findings from the Atherosclerosis Risk in Communities (ARIC) Study (Hwang *et al.* 1997). sICAM-1 was elevated among current and former smokers in comparison to subjects who had never smoked. sVCAM-1 concentrations were not altered by smoking status. sE-selectin and sICAM-1 were elevated among subjects with coronary heart disease (CHD) and carotid artery atherosclerosis in comparison to controls. sVCAM-1 levels were not. These findings support different roles for sICAM-1 and sVCAM-1. sE-selectin and sICAM-1 may reflect ongoing atherosclerotic activity, while sVCAM-1 concentration is associated with cumulative atherosclerotic burden.

sICAM-1 and Healthy Women

In the Women's Health Study, a study of healthy post-menopausal women, C-reactive protein, sICAM-1, sE-selectin, and IL-6 were elevated among women with a smoking history when compared to non-smoking controls. These biomarkers remained elevated after controlling for other risk factors for CVD, including dyslipidemia, hypertension, diabetes, hormone replacement therapy, and family history. Concentrations were more elevated in current compared to former smokers. Individuals with any smoking history had higher concentrations of these biomarkers than those with no smoking history (Bermudez *et al.* 2002).

While circulating levels of adhesion molecules are elevated in individuals at increased risk for CVD, they are biomarkers for systemic inflammation and do not directly confer risk. The function of cellular adhesion molecules expressed on endothelial cells provides insight on how circulating leukocytes gain entry into tissues, allowing for inflammatory injury. The upregulation of ICAM-1 expression is mediated via pro-inflammatory cytokines, such as TNF-α and IL-6, and enhances recruitment of circulating leukocytes out of the bloodstream, facilitating migration into the subendothelium and initiating the generation of atherosclerotic plaques.

Predictive Value of P-Selectin

P-selectin facilitates platelet-leukocyte interaction and the initiation of leukocyte rolling along the endothelium. P-selectin concentration is elevated with acute coronary syndrome and CHD (Blankenberg *et al.* 2003, Güray *et al.* 2004). The initial inflammatory injury caused by cigarette smoke leads to elevations in sP-selectin in addition to its influence on sICAM-1 expression.

Data from the Women's Health Study showed a strong correlation between smoking status and sP-selectin. sP-selectin was also correlated with sICAM-1. Healthy women with the highest levels of sP-selectin were at greatest risk of future CVD. While correlated with sICAM-1, controlling for sICAM-1 in multivariate analysis had minimal effect on the association between P-selectin and future cardiovascular events, demonstrating the independence of these two biomarkers (Ridker *et al.* 2001).

CHRONIC AIRWAY DISEASE

Chronic Inflammation

Chronic airway inflammation is an important component of pulmonary diseases such as COPD and asthma. Cigarette smoking increases the risk of, and worsens the course of, these diseases. The enhancement of cellular adhesion molecules is one part of a molecular cascade facilitating inflammatory damage to lung tissue. ICAM-1 is an important mediator for the recruitment, retention, and activation of neutrophils in chronic airway inflammation. Environmental stimuli can activate alveolar macrophages, inducing the expression of IL-1β, TNF-α, and other pro-inflammatory cytokines. Consequently, this stimulates ICAM-1 expression, enhancing leukocyte infiltration and inflammatory injury (Churchill *et al.* 1993).

The inflammation induced by smoke exposure is reversible in the early stages. However, once chronic inflammation has been established, persistent pathological change occurs in the airway. Using a rat model, chronic bronchitis and pathological changes were induced within 7 wk of smoke exposure (Li *et al.* 2007). These inducible pathological changes were associated with increased inflammation. An increase in neutrophils in bronchoalveolar lavage fluid (BALF) is noted as early as 2 wk after onset of smoke exposure. Additionally, increased expression of ICAM-1 was found in the airway tissue. The pathological changes, inflammation, and elevations in ICAM-1 were all reversible, to some degree, with cessation of smoking. This reversal of pathological changes is a function of time; the longer the smoking exposure, the less likely the pathologic changes are to resolve.

Cessation of smoking does not lead to complete reversal of systemic inflammation. One interpretation of this finding is that systemic inflammation from sources other than smoking predisposes individuals to develop chronic inflammatory conditions such as COPD (Gan *et al.* 2004). Smoking then exacerbates this systemic inflammation by altering ICAM-1 expression. This is supported by the observation that COPD often is accompanied by other inflammatory diseases, such as CVD and osteoporosis. Systemic inflammation is associated with both asthma and COPD, and increased concentrations of sICAM-1 are found in patients who develop severe pulmonary disease (Hollander *et al.* 2007). Serum sE-selectin, another biomarker for inflammation, is also elevated in individuals with COPD and chronic bronchitis (Riise *et al.* 1994).

Pathological Changes in the Human Airway

In humans, increased ICAM-1 expression is found in the pulmonary vasculature of current smokers (Schaberg *et al.* 1996). Individuals underwent lung biopsy in evaluation of a peripheral lung tumor. Differences in VCAM-1, E-selectin, and P-selectin were not observed between smokers and non-smokers. ICAM-1 expression increased linearly by pack-year history of smoking, thus showing a duration-response to cigarette smoke exposure. These findings may not be generalizable, as 24 of 26 patients in this study had a bronchial cancer. The authors cite the inability to recruit healthy volunteers for lung biopsy as justification for studying individuals requiring lung biopsy for evaluation of a pulmonary mass. The tumors in this group, however, were not squamous cell carcinomas, not thought to be associated with smoking, and unlikely to alter the interpretation of their findings.

In children with asthma, BALF concentrations of sICAM-1 were higher among those with active disease than those with recent asthma flares. sICAM-1 does not correlate with either neutrophil or eosinophil counts in children with asthma. Also, sICAM-1 in BALF correlated with findings on chest x-ray suggestive of diffuse airway inflammation (Marguet *et al.* 2000). sICAM-1 concentrations are elevated among patients experiencing acute asthma exacerbations in comparison to control patients as well as those with stable disease. Consistent with the association of sICAM-1 with airway inflammation, the production of sICAM-1 is inhibited by corticosteroid administration (Ren-Bin Tang *et al.* 2002). Concentrations of sVCAM-1 and sE-selectin also are elevated among patients with acute asthma flares (Riise *et al.* 1994, Ren-Bin Tang *et al.* 2002). If the airway inflammation of asthma is mediated via cellular adhesion molecules, tobacco smoke may worsen the course via upregulation of expression of these molecules.

BASIC SCIENCE

Altered Leukocyte/Endothelium Interaction

Since the epidemiological association of smoking and disease, insight to the molecular pathways involved has been garnered. The study of these pathways is difficult, as smoke from a burning cigarette is a complex mixture of both gaseous and particulate matter from approximately 4,000 different components (Table 3). The constituent molecules may act singly or in combination to cause disease. Particulate cigarette smoke condensate (CSC) can be collected and used for research purposes. The expression of ICAM-1, ELAM-1, and VCAM-1 on human and bovine endothelial cells is increased when cell cultures are exposed to CSC (Kalra *et al.* 1994).

CSC affects both leukocyte adhesion and leukocyte transmigration. Time- and dose-dependent increases in adhesion molecule expression and monocyte adhesion to endothelial cells is demonstrated with exposure of endothelial cells to CSC at concentrations lower than those obtained from a single cigarette. The monocyte-endothelial cell interaction can be blocked with antibodies against CD11b (an adhesive ligand; ICAM-1 functions as a CD11b receptor), with protein kinase C (PKC) inhibitors, and with anti-inflammatory medications (Kalra *et al.* 1994, Shen *et al.* 1996). Activation of PKC leads to increased binding activity of nuclear

Table 3 Major constituents of cigarette smoke

Gaseous phase	*Particulate phase*
Carbon monoxide	Nicotine
Nitrogen oxides	Phenol
Ammonia	Antracyclic hydrocarbons
Hydrogen cyanide	Nitrosamines
Formaldehyde	Heavy metals (e.g., cadmium)
Acrolein	
Nitroso-compounds	
Benzene	

Cigarette smoke is a complex mixture of over 4,000 compounds (Kalra *et al.* 1994).

transcription factor NF-κB and expression of its products, including ICAM-1, ELAM-1, and VCAM-1 (Fig. 2) (Shen *et al.* 1996).

Furthermore, CSC augments leukocyte migration across the endothelium by its effect on PECAM-1, a platelet-derived adhesion molecule. CSC incites phosphorylation of PECAM-1 via PKC, which alters cell-to-cell adhesion in the endothelium. This allows for increased leukocyte migration to the subendothelium. Treatment with either PKC inhibitors or antibody to PECAM-1 attenuates the affects of CSC on leukocyte transmigration (Shen *et al.* 1996).

CSC-induced Changes of Cytokine Expression

The Role of ICAM-1

The specific components of the CSC that are responsible for the physiological effects are difficult to determine, but nicotine alters leukocyte adhesion and transendothelial migration. ICAM-1 and VCAM-1 are not expressed in HUVEC cultures under physiological conditions, but exposure to nicotine induces cellular expression of these adhesion molecules (Albaugh *et al.* 2004, Ueno *et al.* 2006), mediated via expression of two pro-inflammatory cytokines, TNF-α and IL-1β (Zhang *et al.* 2002, Albaugh *et al.* 2004, Wang *et al.* 2004). ICAM-1 is elevated in the supernatant from HUVEC cultures exposed to IL-1β and to cotinine, a bioactive nicotine metabolite. The ICAM-1 concentration is related to cotinine exposure in a dose-dependent fashion (Lain *et al.* 2006).

Macrophage expression of sICAM-1, sVCAM-1, and sE-selectin is induced with nicotine (Wang *et al.* 2004, Ueno *et al.* 2006) and associated with increased expression of TNF-α and IL-1β. Expression of other inflammatory cytokines (INF-γ and IL-8) is not increased with nicotine administration, demonstrating specificity of the TNF-α and IL-1β response of macrophages to nicotine exposure (Wang *et al.* 2004). The CSC does not directly stimulate expression of sICAM-1; rather,

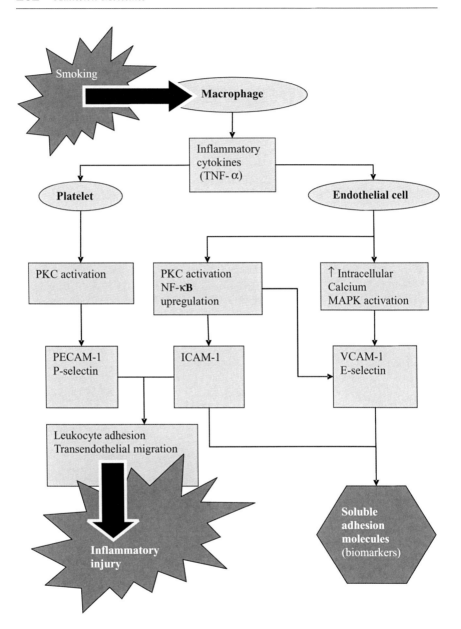

Fig. 2 Simplified cascade triggered by cigarette smoke leading to abnormal soluble adhesion molecule expression and inflammatory injury.

the increase in expression of the adhesion molecule is mediated through CSC stimulation of macrophages to produce TNF-α, which in turn enhances sICAM-1 expression (Zhang *et al.* 2002). This stimulation of TNF-α and IL-1β in turn leads to increased expression of ICAM-1 and VCAM-1 through NF-κB-mediated cellular expression of these adhesion molecules (Fig. 2) (Albaugh *et al.* 2004).

Not all authors have found the same response to nicotine exposure. Nicotine-induced expression of VCAM-1 and E-selectin, but not ICAM-1, is achieved by increasing intracellular calcium levels. Increased calcium concentration enhances ERK1/2 and phosphorylation of nicotinic receptors that activate the MAPK second messenger signaling pathway (Wang *et al.* 2006). NF-κB and ERK1/2 are activated at different points in time, which may explain the different effects on ICAM-1 expression observed. For both pathways, the effect of nicotine is blocked via specific inhibitors of the second signaling pathways.

The Role of P-Selectin

P-selectin is involved in the initiation of fatty-streak formation and atherosclerosis. P-selectin is found in platelets, endothelial cells overlying atherosclerotic lesions, and serum as a soluble biomarker. In mouse models, mice deficient in low-density lipoprotein receptor and P-selectin have attenuated fatty-streak formation when fed high-fat diets. The influence of P-selectin expression diminishes once the fatty streaks mature and develop a fibrinous cap, confirming the importance of P-selectin in the initiation of atherosclerotic disease (Johnson *et al.* 1997). A similar delay in fatty-streak formation is observed in mice deficient in both apolipoprotein-E and P-selectin. Mice with apolipoprotein-E deficiency that were capable of P-selectin expression developed significant atherosclerotic lesions of the aorta when fed a high-fat diet. Mice with P-selectin deficiency, however, developed lesions only after more prolonged exposure. Mice with P-selectin had more monocytes within the atherosclerotic lesion, emphasizing P-selectin's role in monocyte recruitment (Dong *et al.* 2000). The required presence of P-selectin in the initiation of fatty-streak formation and atherosclerosis provides the molecular basis for the association between elevated risk of CHD and elevated concentrations of P-selectin seen in the Women's Health Study (Ridker *et al.* 2001).

CONCLUSION

Much progress has been made in understanding the molecular mechanisms of smoking-related disease. Systemic inflammation causes perturbations of adhesion molecule expression. These changes facilitate leukocyte infiltration of subendothelial tissue. Leukocyte infiltration leads to the inflammatory insults characteristic of diseases such as atherosclerosis and chronic airway inflammation.

Soluble adhesion molecules serve as biomarkers for disease. Their expression correlates with inflammation. Pro-inflammatory cytokines such as TNF-α and IL-1β enhance expression of adhesion molecules, in both bound and soluble forms. Interventions targeted at decreasing inflammation, such as anti-inflammatory medications and cessation of smoking, have been shown to decrease the enhanced expression of adhesion molecules.

Different adhesion molecules are thought to be involved at different points in the molecular pathways leading to disease. Furthermore, for coronary heart disease, elevation of adhesion molecule concentrations has predictive value. Smoking has been shown to increase circulating levels of ICAM-1 and P-selectin. These two molecules are involved in the initiation of disease. Once inflammation is established, the molecular pathways shift and other adhesion molecules become important biomarkers for the burden of disease.

SUMMARY

- Risk of atherosclerosis, coronary heart disease, and other cardiovascular diseases is increased with both smoking and enhanced sICAM-1 expression.
- Risk of chronic pulmonary obstructive disease and asthma is increased with both smoking and enhanced sICAM-1 expression.
- Cigarette smoking creates a systemic inflammatory condition leading to endothelial damage.
- Systemic inflammation leads to altered expression of adhesion molecules, thus facilitating inflammatory damage of the endothelium and underlying tissues.
- sICAM-1 is a soluble adhesion molecule associated with smoking-induced inflammatory damage.
- ICAM-1 is likely involved with the initiation of inflammatory injury, whereas other adhesion molecules are involved with different points in the inflammatory cascade.

Abbreviations

AMI	acute myocardial infarction
BALF	bronchoalveolar lavage fluid
CD	cardiac death
CHD	coronary heart disease
CSC	cigarette smoke condensate
COPD	chronic obstructive pulmonary disease

CVD	cardiovascular disease
ELAM-1	endothelial leukocyte adhesion molecule 1
1ERK1/2	extracellular signal-regulated kinases 1 and 2
HUVEC	human umbilical vein endothelial cell
ICAM-1	intercellular adhesion molecule 1
INF	interferon
IL	interleukin
MAPK	mitogen-activated protein kinase
NF-κB	nuclear factor kappa B
PECAM-1	platelet endothelial adhesion molecule 1
PKC	protein kinase C
TNF-α	tumor necrosis factor alpha
VCAM-1	vascular cellular adhesion molecule 1

Key Facts about Cigarette Smoking and Its Deleterious Health Effects (Giovino 2007)

1. Smoking has caused more than 14 million premature deaths in the United States since 1964.
2. Smoking is the single most preventable cause of death in the United States, with 1 in 5 deaths being caused by smoking.
3. Smoking is a risk factor for:
   - Lung cancer
   - Other cancers (e.g., larynx, mouth, pancreas, cervix)
   - Coronary heart disease
   - Stroke
   - Respiratory disease of adults (e.g., COPD, pneumonia, chronic bronchitis)
   - Childhood diseases (e.g., asthma, otitis media)
   - Reproductive complications (e.g., infertility, low birth weight, sudden infant death syndrome)
4. Exposure to second-hand smoke is an established risk factor for disease, with no known risk-free level of exposure.
5. Nicotine addiction is similar to heroin or cocaine addiction.
6. The epidemiology of cigarette smoking is complex and includes biological, familial, social, cultural, economic, political, and media-based factors.

Definition of Terms

Biomarker: A molecule or trait used to indicate the extent or progress of disease.

Cellular adhesion molecules: Proteins that facilitate interaction between endothelial cells, other cells (e.g., circulating leukocytes), and the extracellular matrix. Found either on the cellular surface or in a soluble form. Soluble form denoted with an 's-' prefix, e.g., sICAM-1.

Endothelial cell: Cells that line tissue surfaces, e.g., the inner layer of cells in a blood vessel.

Leukocyte: White blood cell. Includes monocytes, macrophages, and neutrophils.

Monocyte/Macrophage: Mononuclear leukocyte involved in phagocytosis and production of inflammatory cytokines. Monocytes become macrophages after migrating from circulation to tissue.

Neutrophil: A leukocyte responsible for fighting infection. Activated by pro-inflammatory cytokines and responsible for the damage caused by inflammation.

Pro-inflammatory cytokines: Molecules involved in cell-to-cell communication that promote an inflammatory response.

References

Albaugh, G. and E. Bellavance, L. Strande, S. Heinburger, C.W. Hewitt, and J.B. Alexander. 2004. Nicotine induces mononuclear leukocyte adhesion and expression of adhesion molecules, VCAM and ICAM, in endothelial cells in vitro. Ann. Vasc. Surg. 18: 302-307.

Bermudez, E.A. and N. Rifai, J.E. Buring, J.E. Manson, and P.M. Ridker. 2002. Relation between markers of systemic vascular inflammation and smoking in women. Am. J. Cardiol. 89: 1117-1119.

Blankenberg, S. and S. Barbaux, and L. Tiret. 2003. Adhesion molecules and atherosclerosis. Atherosclerosis 170: 191-203.

Blankenberg, S. and H.J. Rupprecht, C. Bickel, D. Peetz, G. Hafner, L. Tiret, and J. Meyer. 2001. Circulating cell adhesion molecules and death in patients with coronary artery disease. Circulation 104: 1336-1342.

Blann, A.D. and C. Steele, and C.N. McCollum. 1997. The influence of smoking on soluble adhesion molecules and endothelial cell markers. Thromb. Res. 85: 433-438.

Churchill, L. and R.H. Gundel, L.G. Letts, and C.D. Wegner. 1993. Contribution of specific cell-adhesive glycoproteins to airway and alveolar inflammation and dysfunction. Am. Rev. Respir. Dis. 148: S83-87.

de Lemos, J.A. and C.H. Hennekens, and P.M. Ridker. 2000. Plasma concentration of soluble vascular cell adhesion molecule-1 and subsequent cardiovascular risk. J. Am. Coll. Cardiol. 36: 423-426.

Dong, Z.M. and A.A. Brown, and D.D. Wagner. 2000. Prominent role of P-selectin in the development of advanced atherosclerosis in APOe-deficient mice. Circulation 101: 2290-2295.

Gan, W.Q. and S.F.P. Man, A. Senthilselvan, and D.D. Sin. 2004. Association between chronic obstructive pulmonary disease and systemic inflammation: A systematic review and a meta-analysis. Thorax 59: 574-580.

Giovino, G.A. 2007. The tobacco epidemic in the United States. Am. J. Prev. Med. 33: S318-S326.

Güray, Ü. and A.R. Erbay, Y. Güray, M.B. Yilmaz, A.A. BoyacI, H. Sasmaz, S. Korkmaz, and E. Kütük. 2004. Levels of soluble adhesion molecules in various clinical presentations of coronary atherosclerosis. Int. J. Cardiol. 96: 235-240.

Hollander, C. and B. Sitkauskiene, R. Sakalauskas, U. Westin, and S.M. Janciauskiene. 2007. Serum and bronchial lavage fluid concentrations of IL-8, SLPI, SCD14 and sICAM-1 in patients with COPD and asthma. Respir. Med. 101: 1947-1953.

Hwang, S.J. and C.M. Ballantyne, A.R. Sharrett, L.C. Smith, C.E. Davis, A.M. Gotto, Jr., and E. Boerwinkle. 1997. Circulating adhesion molecules VCAM-1, ICAM-1, and E-selectin in carotid atherosclerosis and incident coronary heart disease cases: The atherosclerosis risk in communities (ARIC) study. Circulation 96: 4219-4225.

Johnson, R.C. and S.M. Chapman, Z.M. Dong, J.M. Ordovas, T.N. Mayadas, J. Herz, R.O. Hynes, E.J. Schaefer, and D.D. Wagner. 1997. Absence of P-selectin delays fatty streak formation in mice. J. Clin. Invest. 99: 1037-1043.

Kalra, V.K. and Y. Ying, K. Deemer, R. Natarajan, J.L. Nadler, and T.D. Coates. 1994. Mechanism of cigarette smoke condensate induced adhesion of human monocytes to cultured endothelial cells. J. Cell. Physiol. 160: 154-162.

Lain, K.Y. and P. Luppi, S. McGonigal, J.M. Roberts, and J.A. DeLoia. 2006. Intracellular adhesion molecule concentrations in women who smoke during pregnancy. Obstet. Gynecol. 107: 588-594.

Li, Q.Y. and S.G. Huang, H.Y. Wan, H.C. Wu, T. Zhou, M. Li, and W.W. Deng. 2007. Effect of smoking cessation on airway inflammation of rats with chronic bronchitis. Chin. Med. J. 120: 1511-1516.

Luc, G. and D. Arveiler, A. Evans, P. Amouyel, J. Ferrieres, J.-M. Bard, L. Elkhalil, J.-C. Fruchart, and P. Ducimetiere. 2003. Circulating soluble adhesion molecules ICAM-1 and VCAM-1 and incident coronary heart disease: The PRIME study. Atherosclerosis 170: 169-176.

Marguet, C. and T.P. Dean, and J.O. Warener. 2000. Soluble intercellular adhesion molecule-1 (sICAM-1) and interferon-gamma in bronchoalveolar lavage fluid from children with airway diseases. Am. J. Respir. Crit. Care Med. 162: 1016-1022.

Pradhan, A.D. and N. Rifai, and P.M. Ridker. 2002. Soluble intercellular adhesion molecule-1, soluble vascular adhesion molecule-1, and the development of symptomatic peripheral arterial disease in men. Circulation 106: 820-825.

Ren-Bin Tang, S.-J.C. and W-J. Soong, and R-L. Chung. 2002. Circulating adhesion molecules in sera of asthmatic children. Pediatr. Pulmonol. 33: 249-254.

Ridker, P.M. and J.E. Buring, and N. Rifai. 2001. Soluble P-selectin and the risk of future cardiovascular events. Circulation 103: 491-495.

Ridker, P.M. and C.H. Hennekens, B. Roitman-Johnson, M.J. Stampfer, and J. Allen. 1998. Plasma concentration of soluble intercellular adhesion molecule 1 and risks of future myocardial infarction in apparently healthy men. Lancet 351: 88-92.

Riise, G.C. and S. Larsson, C.G. Lofdahl, and B.A. Andersson. 1994. Circulating cell adhesion molecules in bronchial lavage and serum in COPD patients with chronic bronchitis. Eur. Respir. J. 7: 1673-1677.

Schaberg, T. and M. Rau, R. Oerter, U. Liebers, W. Rahn, D. Kaiser, C. Witt, and H. Lode. 1996. Expression of adhesion molecules in peripheral pulmonary vessels from smokers and nonsmokers. Lung 174: 71-81.

Shen, Y. and V. Rattan, C. Sultana, and V.K. Kalra. 1996. Cigarette smoke condensate-induced adhesion molecule expression and transendothelial migration of monocytes. Am. J. Physiol. 270: H1624-1633.

Ueno, H. and S. Pradhan, D. Schlessel, H. Hirasawa, and B.E. Sumpio. 2006. Nicotine enhances human vascular endothelial cell expression of ICAM-1 and VCAM-1 via protein kinase c, p38 mitogen-activated protein kinase, NF-kappaB, and AP-1. Cardiovasc. Toxicol. 6: 39-50.

Wang, Y. and L. Wang, X. Ai, J. Zhao, X. Hao, Y. Lu, and Z. Qiao. 2004. Nicotine could augment adhesion molecule expression in human endothelial cells through macrophages secreting TNF-alpha, IL-1beta. Int. Immunopharmacol. 4: 1675-1686.

Wang, Y. and Z. Wang, Y. Zhou, L. Liu, Y. Zhao, C. Yao, L. Wang, and Z. Qiao. 2006. Nicotine stimulates adhesion molecular expression via calcium influx and mitogen-activated protein kinases in human endothelial cells. Int. J. Biochem. Cell Biol. 38: 170-182.

Zhang, X. and L. Wang, H. Zhang, D. Guo, J. Zhao, Z. Qiao, and J. Qiao. 2002. The effects of cigarette smoke extract on the endothelial production of soluble intercellular adhesion molecule-1 are mediated through macrophages, possibly by inducing TNF-alpha release. Methods Find. Exp. Clin. Pharmacol. 24: 261-265.

CHAPTER 18

Osteoprotegerin and Adhesion Molecules

Catherine Rush[1] and Jonathan Golledge[2, *]

[1]Vascular Biology Unit, School of Medicine and Dentistry, James Cook University, Douglas, QLD 4811, E-mail: Catherine.Rush@jcu.edu.au
[2]Vascular Biology Unit, School of Medicine and Dentistry, James Cook University, Douglas, QLD 4811, E-mail: jonathan.golledge@jcu.edu.au

ABSTRACT

The glycoprotein osteoprotegerin (OPG) plays an important role in controlling normal bone remodelling by inhibiting osteoclast function. More recently, OPG has been implicated in a number of diseases associated with inflammation. This chapter describes current work being undertaken to try and understand the mechanisms underlying the association between high circulating levels of OPG and inflammatory pathologies. Human *in vitro* work suggests OPG promotes leukocyte adhesion via at least two mechanisms. It stimulates the upregulation of angiopoietin-2 within endothelium, thereby promoting endothelial responsiveness to the pro-inflammatory cytokine tumour necrosis factor alpha. Thus, OPG acts in concert with pro-inflammatory cytokines to promote adhesion molecule upregulation and favour leukocyte recruitment. It also has the ability to cause rapid increases in leukocyte adhesion to the endothelium by acting as a direct bridge between leukocytes and endothelial cells. Despite this convincing data for a pro-inflammatory role of OPG in human cells, *in vitro* studies in knockout mouse models have not confirmed the role of OPG in inflammation. The reason for this disparity is currently unknown. Further studies using different animal models and alternative ways of studying human disease will be required to clarify these disparities.

---

*Corresponding author
Key terms are defined at the end of the chapter.

INTRODUCTION

Osteoprotegerin (OPG) is a soluble glycoprotein originally identified to play a fundamental role in inhibiting osteoclast function and differentiation by blocking the interaction between the ligand receptor activator of NF-κB ligand (RANKL) and receptor activator of NF-κB (RANK) expressed on osteoclasts (Simonet *et al.* 1997). In this chapter, we discuss the putative effects of OPG outside the recognized role of OPG in bone remodelling. We initially discuss human association data linking OPG with human diseases known to involve inflammation. We then centre our discussion on recent investigations carried out in our laboratory looking at the interaction between OPG and leukocyte adhesion molecules.

THE OPG, RANKL AND RANK INTERACTION

No discussion about OPG is possible without mention of its relationship with RANKL and RANK (Schoppet *et al.* 2002, Collin-Osdoby 2004). RANKL, expressed on a variety of cells, such as osteoblasts, stromal and T cells binds RANK present on other cell types, including osteoclasts, endothelial cells, monocytes and dendritic cells. There are very low concentrations of soluble RANKL in the circulation of healthy individuals. In contrast, OPG is a soluble glycoprotein expressed in most human tissues and circulates at ng/ml quantities. OPG has high affinity for RANKL and therefore blocks the interaction between RANKL and RANK, explaining its ability to inhibit osteoclast activation and bone remodelling (Simonet *et al.* 1997). OPG also displays ability to bind other ligands, such as tumour necrosis factor-related apoptosis-inducing ligand (TRAIL), and thereby block their functions (Emery 1998). It is assumed that many of the effects of OPG are due to binding ligands, thereby blocking their function, although in the authors' opinion it is likely that some direct functions of OPG also exist. Given these interactions of OPG we will also include some discussion of the effect of RANKL within this chapter.

OPG AND INFLAMMATION: HUMAN ASSOCIATION STUDIES

A role for OPG in bone remodelling has been convincingly demonstrated (Kearns *et al.* 2008); however, a direct role in other patho-physiological processes has also been suggested (Schoppet *et al.* 2002, Collin-Osdoby 2004). It has been suggested that OPG plays both protective and pathological roles within the vasculature. Several studies have shown that OPG at physiological concentrations influences endothelial cell and vascular smooth muscle cell (VSMC) survival and prevents arterial calcification (Van Campenhout and Golledge 2008). Other studies have shown that expression and release of OPG by endothelial cells and VSMCs is markedly upregulated in response to pro-inflammatory cytokines like tumour necrosis factor alpha (TNF-α) and correlates with increased cardiovascular risk (Van Campenhout and Golledge 2009). A number of the diseases linked with OPG involve inflammation as critical in their pathology. Circulating concentrations of OPG have been associated with the severity of vascular calcification, progression of abdominal aortic aneurysm, complications of atherosclerosis, rheumatoid arthritis progression and Crohn's disease (Golledge *et al.* 2004, Kiechl *et al.* 2004, Bernstein *et al.* 2005, Moran *et al.* 2005, Clancy *et al.* 2006, Geusens *et al.* 2006, Kadoglou *et al.* 2008). Whether OPG and/or RANKL are directly involved in these diseases or their elevated expression is simply a secondary consequence of the diseases is currently controversial (Van Campenhout and Golledge 2009). Expression of OPG within tissue from patients with these diseases has also been related to localized or systemic inflammation in some but not all studies (Smith *et al.* 2003, Golledge *et al.* 2004, Gannage-Yared *et al.* 2008, Vandooren *et al.* 2008). In this chapter, we discuss evidence for a direct role of OPG in leukocyte adhesion and hence the inflammation process.

OPG AND LEUKOCYTE ADHESION

Studies have suggested that OPG promotes leukocyte adhesion, although the mechanisms involved are currently controversial (Mangan *et al.* 2007, Zauli *et al.* 2007). Zauli *et al.* (2007) found that incubating cultured human umbilical vein endothelial cells (HUVECs) and human microvascular endothelial cells (HMVECs) with recombinant OPG for 16 hr promoted the adhesion of primary polymorphonuclear neutrophils (PMNs; drawn from healthy volunteers) and the cell line HL-60 to endothelial cells. Adhesion of leukocytes was maximally stimulated at concentrations between 0.1 and 0.5 ng/mL, which are similar to levels reported in the sera of patients affected by a number of diseases associated with inflammation, such as rheumatoid arthritis (Geusens *et al.* 2006), symptomatic atherosclerosis (Golledge *et al.* 2004, Van Campenhout and Golledge 2009) and abdominal aortic aneurysm (Moran *et al.* 2005). Zauli and colleagues found that incubating endothelial cells with OPG alone had no effect on adhesion molecule expression. These investigators also reported that in the presence of the pro-inflammatory cytokine TNF-α, OPG (0.5 ng/ml) did not promote additional leukocyte adhesion (Zauli *et al.* 2007). Further experiments by Zauli *et al.* suggested that incubation periods as short as 5 min were enough to stimulate increased adhesion of leukocytes to endothelial cells. Furthermore, the effects of OPG could be stimulated by incubation with either the PMNs or the endothelial cells. Further experiments provided additional data suggesting that OPG bound to PMNs via its ligand- (RANKL and TRAIL) binding domain, while endothelial cells bound to OPG via its heparin-binding domain (Zauli *et al.* 2007). This would suggest that circulating OPG acts as a bridge between the PMNs and endothelial cells, facilitating leukocyte rolling and firm adhesion. The investigators also reported that topical administration of OPG to rat mesenteric post-capillary venules increased leukocyte rolling. These studies strongly suggest that elevated levels of OPG contribute to inflammation. The investigators also reported that TNF-α induced production of OPG from endothelial cells, suggesting a link between mechanisms promoting inflammation (Zauli *et al.* 2007).

At approximately the same time as Zauli *et al.* (2007) reported their findings we reported a similar study of the influence of OPG on leukocyte adhesion (Mangan *et al.* 2007). In line with Zauli and colleagues, we reported that incubating HUVECs with OPG alone (0.5-10 ng/ml) had no influence on expression of a range of adhesion molecules. In contrast, OPG stimulated a dose-dependent increase in intercellular adhesion molecule 1 (ICAM-1), vascular cell adhesion molecule 1 (VCAM-1) and E-Selectin (CD62E) messenger ribonucleic acid (mRNA) and surface protein expression in HUVECs activated with TNF-α compared to control cells (Fig. 1) (Mangan *et al.* 2007). The maximal effect was seen at an OPG concentration of approximately 10 ng/ml. In keeping with this finding, we demonstrated that in the presence of TNF-α (50 U/ml), OPG (0.5-10 ng/ml for 12 hr) stimulated a dose-dependent increase in binding of a monocyte cell line (THP-1) to HUVECs (Fig. 2) (Mangan *et al.* 2007). The latter findings are not in agreement to those reported by Zauli and colleagues, who reported no co-stimulatory effect of OPG and TNF-α on adhesion of PMNs to HUVECs (Zauli *et al.* 2007). These differences may relate to cell types used, incubation times or doses of OPG, all of which differed between experiments (Mangan *et al.* 2007, Zauli *et al.* 2007). In contrast to the TNF-α–related effects, we found that OPG did not augment adhesion molecule expression in Interleukin-1 beta (IL-1β) activated endothelial cells. Our overall findings suggested a TNF-α- and OPG-specific cooperation in promoting leukocyte adhesion.

In order to investigate the mechanisms that might be responsible for the upregulation of adhesion molecules in TNF-α-activated HUVECs, we studied gene expression in resting HUVECs using expression arrays and showed that angiopoietin-2 (Ang-2) expression was significantly altered following OPG incubation (Fig. 3). Using quantitative gene expression analysis (MassArray), we found that incubating HUVECs with 10 ng/ml concentrations of OPG for 4 hr induced a two-fold increase in Ang-2 gene expression (162.3 ± 26.9 fM compared to 88.2 ± 8.9 fM in control cells). We found increased Ang-2 protein within Weibel-Palade bodies in OPG-treated resting HUVECs (Fig. 4).

Fiedler *et al.* (2006) have shown that Ang-2 facilitates leukocyte adhesion by potentiating the TNF-α–mediated expression of the adhesion proteins ICAM-1 and VCAM-1. Ang-2-deficient mice have impaired inflammatory responses to some stimuli. Studies in cell culture by Fiedler *et al.* (2006) showed that Ang-2 sensitized cells to the effect of TNF-α. In particular, these investigators showed that Ang-2 augmented the ability of TNF-α to upregulate adhesion molecules.

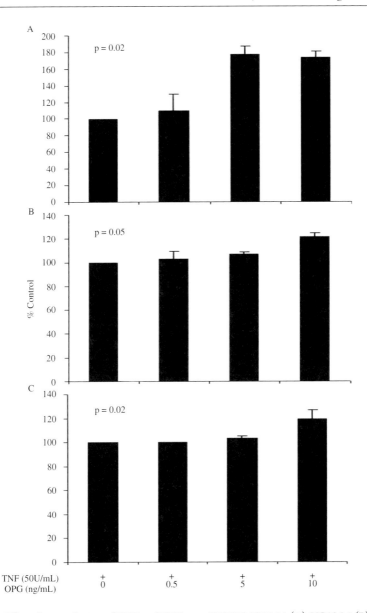

Fig. 1 Effect of co-incubation of OPG and TNF-α on HUVEC ICAM-1 (A), VCAM-1 (B) and E-selectin surface expression (C). Adhesion molecule levels were measured using flow cytometry at 6, 8 and 12 hr for VCAM-1, E-selectin and ICAM-1, respectively. Results are expressed as percentage of control (± standard error) of three independent experiments. Statistical analysis was by ANOVA. Reproduced with permission from Mangan *et al.* (2007).

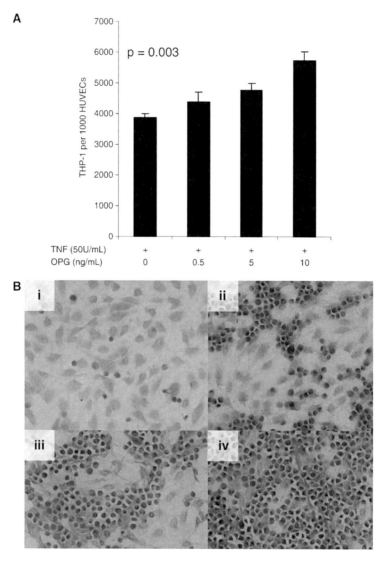

Fig. 2 Effect of incubation of TNF-α-activated HUVEC monolayers with OPG on the binding of THP-1 monocytic cells. HUVEC monolayers were pre-incubated with TNF-α (50 U/mL) and OPG (0-10 ng/mL) prior to exposure to monocytes for 30 min. The ratio of firmly adherent THP-1 cells to HUVECs was measured by flow cytometry. Results are representative of one of three independent experiments carried out in triplicate. Monocyte binding is expressed as the ratio of THP-1 cells bound per 1,000 HUVECs (± standard error) (A). Parallel wells were fixed and a haematoxylin and eosin stain performed (B): 0 treatment (i), TNF-α (50 U/mL) (ii), OPG (5 ng/mL)/TNF-α (50 U/mL) (iii), and OPG (10 ng/mL)/TNF-α (50 U/mL) (iv). Statistical analysis was by ANOVA. Reproduced with permission from Mangan *et al.* (2007).

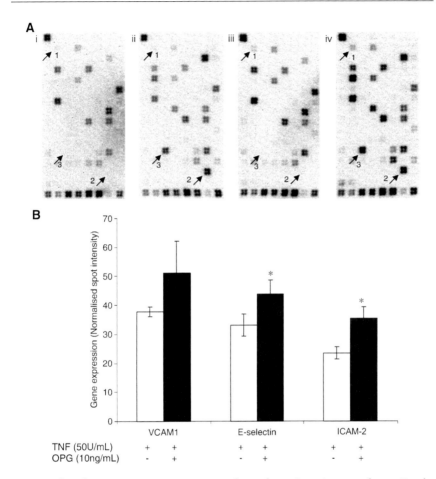

Fig. 3 Effect of OPG on HUVEC gene expression detected using SuperArray membranes. Results are representative of one of three independent experiments. Endothelial cells were incubated with 0 treatment (i), TNF-α(50 U/mL) (ii), OPG (10 ng/mL) (iii), and OPG (10 ng/mL)/TNF-α(50 U/mL) (iv). Differences in the expression of genes such as 1, angiopoietin-2; 2, VCAM-1; 3 E-selectin are noted by an arrowhead. (B) Bars and error bars represent mean ± standard error of intensities for specific genes from three independent membranes. OPG significantly increased expression of E-selectin and ICAM-2 on activated HUVECs. *p < 0.05. Reproduced with permission from Mangan *et al.* (2007).

In view of these published findings, we reasoned that the ability of OPG to upregulate Ang-2 could underlie its TNF-α sensitizing effect on endothelial cells. We used small interfering RNA (siRNA) to reduce Ang-2 expression in cultured endothelial cells. When endothelial cell Ang-2 expression was attenuated, the ability of OPG to upregulate adhesion molecules in TNF-α–activated HUVECs

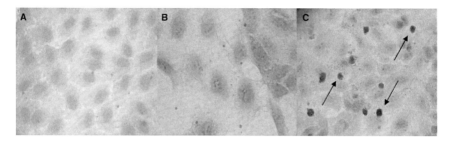

Fig. 4 Effect of OPG on HUVEC angiopoietin-2 expression assessed by immunohistochemistry. Images are representative of four independent experiments. Endothelial cells were incubated with isotype control antibody (negative control) (A), 0 treatment (B) or 10 ng/mL OPG for 10 hr (C). Images were taken at ×10 magnification (B and C). Treatment with OPG is associated with increased immunostaining for angiopoietin-2 demonstrated in a granular pattern (positive granules identified by arrows), as previously described [18]. Reproduced with permission from Mangan *et al.* (2007).

was blocked (Fig. 5). These findings suggest angiopoietin-2 is a key regulator of endothelial cell inflammatory responses. Ang-2 may control endothelial responsiveness in different situations. Thus, in the presence of low levels of Ang-2 the ability of pro-inflammatory cytokines to promote leukocyte accumulation is limited. In contrast, when Ang-2 is expressed at high concentrations, the effects of pro-inflammatory cytokines are augmented. The ability of OPG to promote Ang-2 expression is a potentially important mechanism by which this glycoprotein can promote inflammation-based pathology.

RANK/RANKL AND ADHESION MOLECULES

Binding of RANKL to its receptor RANK on endothelial cells stimulates proliferation, migration and activation of endothelial cells, resulting in promotion of angiogenesis. RANKL also stimulates inflammatory responses by increasing endothelial cell-leukocyte interactions and the expression of cell adhesion molecules ICAM-1 and VCAM-1 on endothelial cells (Min *et al.* 2005). These effects occur through a signalling cascade leading to NF-κB activation. Thus, if OPG was solely a RANK/RANKL inhibitor, then one would expect OPG to attenuate adhesion molecule expression, which is not the case, as we and others have shown that OPG augments adhesion molecule expression. The relative balance of the effects of OPG and RANKL likely vary *in vivo* depending on the other cytokine milieux and relative concentrations of the proteins. Since OPG is found within the circulation at much greater concentrations, however, it is likely its effects dominate.

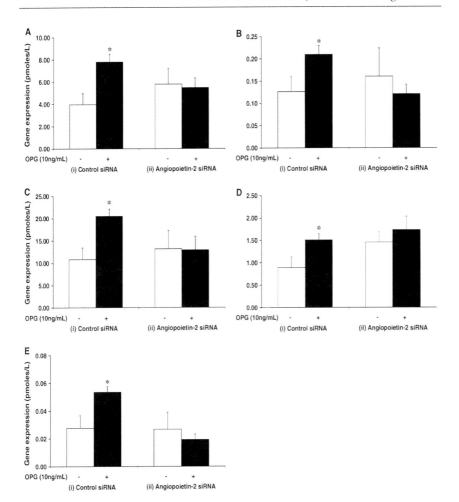

Fig. 5 Effect of OPG on adhesion molecule expression in TNF-α-activated HUVECs transfected with control (i) or angiopoietin-2-targeted (ii) siRNA. HUVECs were transfected with siRNA 48 hr prior to incubation with TNF-α (50 U/mL) and OPG (10 ng/mL) for 4 hr. Gene expression of ICAM-1 (A), VCAM-1 (B), E-selectin (C), ICAM-2 (D) and angiopoietin-2 (E) was measured by MALDI-TOF MS. In control siRNA-transfected HUVECs, OPG induced an upregulation in ICAM-1, VCAM-1, E-selectin, ICAM-2 and angiopoietin-2 expression (*p = 0.05). However, in endothelial cells transfected with angiopoietin-2-targeted siRNA, OPG alone had no effect on adhesion molecule or angiopoietin-2 expression (p = 0.4-0.9). Reproduced with permission from Mangan *et al.* (2007).

STUDIES OF OPG DEFICIENCY AND EXCESS IN ANIMAL MODELS WITH RELATIONSHIP TO ADHESION MOLECULES AND INFLAMMATION

A number of groups have investigated the effect of OPG deficiency or excess in mouse models, although its effects on inflammation have not been central to these studies (Bucay *et al.* 1998, Bennett *et al.* 2006, Morony *et al.* 2008). A number of groups have demonstrated that OPG deficiency promotes aortic calcification to varying degrees depending on the genetic background of the mice (Bucay *et al.* 1998, Bennett *et al.* 2006). Inflammation is commonly associated with vascular calcification and thus this finding is counter-intuitive given the data presented about the pro-inflammatory effects of OPG. Indeed, in one mouse model (the hyperlidaemia mouse deficient in apo lipoprotein e), OPG deficiency appears to promote atherosclerosis, where inflammation is believed to play an important role (Bennett *et al.* 2006). The latter findings are not confirmed in all studies in this area. Thus, in the mouse model of atherosclerosis deficient in low-density lipoprotein receptor, upregulating OPG appears to have no effect on atherosclerosis progression (Morony *et al.* 2008). On the basis of human association data, OPG would be expected to promote atherosclerosis. The findings from mouse models are thus currently difficult to interpret in terms of the large amount of human association data linking high levels of OPG with a number of different inflammation-associated diseases. A number of possibilities exist, such as that the positive association between OPG and inflammation-associated diseases is simply a consequence of the process with no role in the inflammation itself. It is also likely that inflammation pathways within mice do not completely reflect the situation in humans, as has been seen with a number of cytokines.

CONCLUSION

There is a large amount of data linking high levels of circulating OPG to diseases associated with inflammation. Experiments *in vitro* using human cells support a pathology role of OPG in promoting inflammation. Work by our own group suggests that this in part is due to the ability of OPG to promote production of Ang-2, which sensitizes the endothelium to TNF-α, thereby promoting upregulation of leukocyte adhesion molecules. The exact role of the mechanisms *in vivo* is, however, unclear. Mice studies have generally not supported a pro-inflammatory role for OPG. Further work is required to more completely understand the role of OPG in inflammation, such as that employing *ex vivo* human samples.

SUMMARY

- Circulating concentrations of osteoprotegerin are increased in patients with a number of pathologies, many of which are associated with inflammation.
- Human cell culture studies indicate that osteoprotegerin can promote leukocyte adhesion via a number of mechanisms.
- The *in vivo* importance of these *in vitro* findings is currently unclear.

Abbreviations

Ang-2	angiopoietin-2
CD62E	E-selectin
HMVEC	human microvascular endothelial cell
HUVEC	human umbilical vein endothelial cell
ICAM-1	intercellular adhesion molecule 1
IL-1β	interleukin-1 beta
mRNA	messenger ribonucleic acid
OPG	osteoprotegerin
PMN	primary polymorphonuclear neutrophil
RANKL	receptor activator of NF-κB ligand
RANK	receptor activator of NF-κB
siRNA	small interfering RNA
TNF-α	tumour necrosis factor alpha
TRAIL	tumour necrosis factor-related apoptosis-inducing ligand
VCAM-1	vascular cell adhesion molecule 1
VSMC	vascular smooth muscle cells

Key Facts about the Interaction between Osteoprotegerin, Inflammation and Adhesion Molecules

1. In human subjects, high circulating levels of OPG have been associated with a number of diseases where inflammation is integral in the pathology.
2. In human cells *in vitro*, OPG stimulates leukocyte adhesion.
3. The mechanisms by which OPG stimulates leukocyte adhesion in human cells is currently controversial but may involve a number of mechanisms:
   (a) the ability of OPG to sensitize the endothelial to TNF-α-mediated upregulation of adhesion molecules;

(b) the ability of OPG to act as a bridge between endothelial cells and leukocytes.

4. Studies in mouse models did not support a major role of OPG in diseases associated with inflammation and even suggested protective effects. Whether these findings represent species differences from humans or other experimental effects is currently unclear.

Definition of Terms

Angiopoietin-2: A regulator stored in Weibel-Palade bodies within the endothelium that, evidence now suggests, plays an important role in controlling the activation state of the endothelium.

Intercellular adhesion molecule 1 (ICAM-1), vascular cell adhesion molecule 1 (VCAM-1) and E-selectin (CD62E): Endothelial expressed adhesion molecules important in stages of leukocyte adhesion to the endothelium.

Leukocytes: White blood cells that are key components of the immune response. They defend the body against infection and bacteria.

Osteoprotegerin: A soluble glycoprotein originally identified to protect bone from breakdown by inhibiting osteoclasts. Subsequently it has been associated with a wide range of human diseases.

RANKL: A regulator of bone remodelling; RANK is the receptor for this molecule.

References

Bennett, B.J. and M. Scatena, E.A. Kirk, M. Rattazzi, R.M. Varon, M. Averill, S.M. Schwartz, C.M. Giachelli, and M.E. Rosenfeld. 2006. Osteoprotegerin inactivation accelerates advanced atherosclerotic lesion progression and calcification in older ApoE-/- mice. Arterioscler. Thromb. Vasc. Biol. 26: 2117-2124.

Bernstein, C.N. and M. Sargent, and W.D. Leslie. 2005. Serum osteoprotegerin is increased in Crohn's disease: a population-based case control study. Inflamm. Bowel Dis. 11: 325-330.

Bucay, N. and I. Sarosi, C.R. Dunstan, S. Morony, J. Tarpley, C. Capparelli, S. Scully, H.L. Tan, W. Xu, D.L. Lacey, W.J. Boyle, and W.S. Simonet. 1998. Osteoprotegerin-deficient mice develop early onset osteoporosis and arterial calcification. Genes Dev. 12: 1260-1268.

Clancy, P. and L. Oliver, R. Jayalath, P. Buttner, and J. Golledge. 2006. Assessment of a serum assay for quantification of abdominal aortic calcification. Arterioscler. Thromb. Vasc. Biol. 26: 2574-2576.

Collin-Osdoby, P. 2004. Regulation of vascular calcification by osteoclast regulatory factors RANKL and osteoprotegerin. Circ. Res. 95: 1046-1057.

Emery, J.G. and P. McDonnell, M.B. Burke, K.C. Deen, S. Lyn, C. Silverman, *et al.* 1998. Osteoprotegerin is a receptor for the cytotoxic ligand TRAIL. J. Biol. Chem. 273: 14363-14367.

Fiedler, U. and Y. Reiss, M. Scharpfenecker, V. Grunow, S. Koidl, G. Thurston, *et al.* 2006. Angiopoietin-2 sensitizes endothelial cells to TNF-alpha and has a crucial role in the induction of inflammation. Nat. Med. 12: 235-239.

Gannage-Yared, M.H. and C. Yaghi, B. Habre, S. Khalife, R. Noun, M. Germanos-Haddad, *et al.* 2008. Osteoprotegerin in relation to body weight, lipid parameters insulin sensitivity, adipocytokines, and C-reactive protein in obese and non-obese young individuals: results from both cross-sectional and interventional study. Eur. J. Endocrinol. 158: 353-539.

Geusens. P.P. and R.B. Landewe, P. Garnero, D. Chen, C.R. Dunstan, W.F. Lems, *et al.* 2006. The ratio of circulating osteoprotegerin to RANKL in early rheumatoid arthritis predicts later joint destruction. Arthritis Rheum. 54: 1772-1777.

Golledge, J. and M. McCann, S. Mangan, A. Lam, and M. Karan. 2004. Osteoprotegerin and osteopontin are expressed at high concentrations within symptomatic carotid atherosclerosis. Stroke 35: 1636-1641.

Kadoglou. N.P. and T. Gerasimidis, S. Golemati, A. Kapelouzou, P.E. Karayannacos, and C.D. Liapis. 2008. The relationship between serum levels of vascular calcification inhibitors and carotid plaque vulnerability. J. Vasc. Surg. 47: 55-62.

Kearns, A.E. and S. Khosla, and P.J. Kostenuik. 2008. Receptor activator of nuclear factor kappaB ligand and osteoprotegerin regulation of bone remodeling in health and disease. Endocr. Rev. 29: 155-92.

Kiechl, S. and G. Schett, G. Wenning, K. Redlich, M. Oberhollenzer, A. Mayr, *et al.* 2004. Osteoprotegerin is a risk factor for progressive atherosclerosis and cardiovascular disease. Circulation 109: 2175-2180.

Mangan, S.H. and A.V. Campenhout, C. Rush, and J. Golledge. 2007. Osteoprotegerin upregulates endothelial cell adhesion molecule response to tumor necrosis factor-alpha associated with induction of angiopoietin-2. Cardiovasc. Res. 76: 494-505.

Min, J.K. and Y.M. Kim, S.W. Kim, M.C. Kwon, Y.Y. Kong, I.K. Hwang, M.H. Won, J. Rho, and Y.G. Kwon. 2005. TNF-related activation-induced cytokine enhances leukocyte adhesiveness: induction of ICAM-1 and VCAM-1 via TNF receptor-associated factor and protein kinase C-dependent NF-kappaB activation in endothelial cells. J. Immunol. 175: 531-540.

Moran, C.S. and M. McCann, M. Karan, P. Norman, N. Ketheesan, and J. Golledge. 2005. Association of osteoprotegerin with human abdominal aortic aneurysm progression. Circulation 111: 3119-3125.

Morony, S. and Y. Tintut, Z. Zhang, R.C. Cattley, G. Van, D. Dwyer, M. Stolina, P.J. Kostenuik, and L.L. Demer. 2008. Osteoprotegerin inhibits vascular calcification without affecting atherosclerosis in ldlr(-/-) mice. Circulation 117: 411-420.

Schoppet, M. and K.T. Preissner, and L.C. Hofbauer. 2002. RANK ligand and osteoprotegerin: paracrine regulators of bone metabolism and vascular function. Arterioscler. Thromb. Vasc. Biol. 22: 549-553.

Simonet, W.S. and D.L. Lacey, C.R. Dunstan, M. Kelley, M.S. Chang, R. Luthy, *et al.* 1997. Osteoprotegerin: a novel secreted protein involved in the regulation of bone density. Cell 89: 309-319.

Smith, M.D. and E. Barg, H. Weedon, V. Papengelis, T. Smeets, P.P. Tak, M. Kraan, M. Coleman, and M.J. Ahern. 2003. Microarchitecture and protective mechanisms in synovial tissue from clinically and arthroscopically normal knee joints. Ann. Rheum. Dis. 62: 303-307.

Van Campenhout, A. and J. Golledge. 2009. Osteoprotegerin, vascular calcification and atherosclerosis. Atherosclerosis 204: 321-329.

Vandooren, B. and T. Cantaert, T. Noordenbos, P.P. Tak, and D. Baeten. 2008. The abundant synovial expression of the RANK/RANKL/Osteoprotegerin system in peripheral spondylarthritis is partially disconnected from inflammation. Arthritis Rheum. 58: 718-729.

Zauli, G. and F. Corallini, F. Bossi, F. Fischetti, P. Durigutto, C. Celeghini, *et al.* 2007. Osteoprotegerin increases leukocyte adhesion to endothelial cells both in vitro and in vivo. Blood 110: 536-543.

CHAPTER 19

Adhesion Molecules in Skin

Robert C. Fuhlbrigge[1,*] and Ahmed Gehad[2]

[1, 2]Department of Dermatology, Harvard Medical School,
Brigham and Women's Hospital,
221 Longwood Ave., Boston, MA 02115
[1]E-mail: rfuhlbrigge@partners.org
[2]E-mail: agehad@rics.bwh. harvard. edu

ABSTRACT

Selective recruitment of lymphocyte populations to the skin and other tissues is a crucial element of both immune surveillance and immune response to specific stimuli. Dysregulation of leukocyte recruitment plays a key role in the pathogenesis of various skin diseases, including inflammatory disorders such as atopic dermatitis or psoriasis, and in the evasion of immune defenses by malignancies, such as squamous cell carcinoma and melanoma. Leukocyte homing depends on the stepwise interaction of blood cells in flow with the vascular wall. Each step in this cascade is necessary, but not sufficient, to direct recruitment and retention of individual cells at a defined site of interest. Together they provide specificity of site and selectivity in leukocytes recruited, as well as protection against inappropriate accumulation of effector leukocytes. Therapeutic strategies targeting key molecules in the recruitment paradigm have proved effective and promise to provide even more options for tissue- and/or cell-selective manipulation of leukocyte homing. In this chapter, we discuss the current model of tissue-specific leukocyte recruitment, focusing on skin, and the role of adhesion receptors in regulating primary and memory immune responses.

---

*Corresponding author
Key terms are defined at the end of the chapter.

INTRODUCTION

The recruitment paradigm describes the movement of specific populations of leukocytes from the pool of cells circulating in blood across the vascular wall and into the underlying tissue. This process can be divided into discreet and independently regulated steps. The process and the primary molecules involved in each step are summarized in Fig. 1 and Tables 1 and 2.

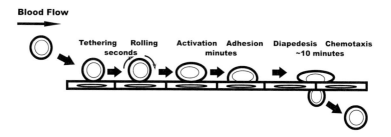

Fig. 1 The multi-step paradigm for circulating leukocyte recruitment to skin. Extravasation of leukocytes depends on the specific interactions of multiple receptor-ligand pairs on both the migrating leukocyte (see Table 1) and on the local endothelium (see Table 2). The sequence and kinetics of events and the qualities of the involved molecules confer a specific address code for the skin homing leukocyte. (Source: RCF, unpublished.)

Table I Leukocyte molecules involved in endothelial adhesion and transmigration

Tethering	Rolling	Activation	Adhesion	Diapedesis	Chemotaxis
Selectin ligands	*Selectin ligands*	*CC Chemokine receptors*	*Integrins*	*Integrins*	*CC Chemokine receptors*
PSGL-1	**PSGL-1**	CCR2, 3, **4**, 8, **10**	**LFA-1**	LFA-1	CCR1, 2, 3, **4**, 5,
CD43	CD43		MAC-1	MAC-1	6, 8, **10**
CD44	CD44		VLA-4		
			IgSF		
			JAM C		*CXC Chemokine*
Selectin	*Selectin*				*receptor*
CD62L	CD62L			*C-type lectin*	CXCR3, 4
				CD209	
Integrin	*Integrin*				*CX₃C chemokine*
VLA-4	VLA-4				*receptors*
					CX₃CR1
	C-type lectin				
	CD209				

Leukocytes subsets express variable adhesion receptors that will direct their accumulation at specific endothelial sites. The molecules primarily involved in skin homing are listed in bold type. (Source: RCF, unpublished.)

Table 2 Endothelial molecules involved in leukocyte adhesion and transmigration

Tethering	Rolling	Activation	Adhesion	Diapedesis	Chemotaxis
Selectins	*Selectins*	*CC chemokines*	*IgSF*	*IgSF*	*CC chemokines*
E-selectin	**E-selectin**	CCL1, 2, 5, 11,	**ICAM-1**	PECAM-1	CCL1, 2, 5, 11, 13,
P-selectin	P-selectin	13, 26,	ICAM-2	ICAM-2, -3	20, 26
		17 (TARC),	VCAM-1	JAM A, B, C	**17 (TARC),**
		27 (CTACK)			**27 (CTACK)**
s-Le^x	*s-Le^x*			*Cadherins*	
PNAd	PNAd			VE-cadherin	*CC chemokines*
					CXCL10, 11, 12
IgSF	*IgSF*			*Type II*	
VCAM-1	VCAM-1			*membrane*	
	ICAM-3			*protein*	*CX$_3$C chemokine*
				CD99	CX$_3$CL1

Endothelial sites express variable adhesion receptors and chemoattractant compounds that will direct recruitment of specific leukocyte subsets. The molecules primarily involved in T cell homing to skin are listed in bold type. (Source: RCF, unpublished.)

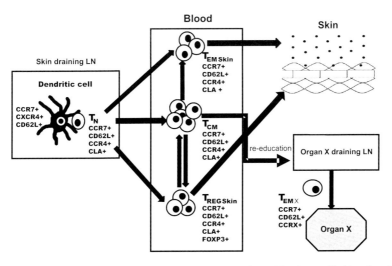

Fig. 2 Immune surveillance and T cell education. Mature dermal dendritic cells (dDC) and/or Langerhans cells (LC) migrate into skin draining lymph nodes, where they prime and imprint a skin homing phenotype on antigen-specific naive T cells (T$_N$). Skin effector memory T cells (T$_{EM}$ skin) and skin regulatory T cells (T$_{REG}$ skin) enter the circulation after leaving the lymph node, and home specifically to the skin to attack the invading pathogen (T$_{EM}$) or restrain the immune response (T$_{REG}$) in the skin. Central memory T cells (T$_{CM}$) leave skin draining lymph nodes and enter lymph nodes of other organs (X) where they can become re-educated to become effector cells of other organs (T$_{EM}$-X). (Source: RCF, unpublished.)

LEUKOCYTE-ENDOTHELIAL INTERACTIONS

Adhesion Molecules in Tethering and Rolling

In the first step of leukocyte recruitment, cells in blood flow tether to the endothelial cells in specific areas of post-capillary venules and roll at reduced velocity across the luminal surface (Fuhlbrigge and Weishaupt 2007, Ley *et al.* 2007). The primary molecules mediating tethering and rolling are vascular E-selectin (CD62E) and P-selectin (CD62P). P-selectin is produced by all endothelia and stored in organelle bodies for rapid transport to the luminal surface in response to inflammatory stimuli. E-selectin, in contrast, is expressed constitutively on post-capillary venules in skin but not in other tissues, although it can be induced in all vessels by inflammation (Chong *et al.* 2004). While P-selectin glycoprotein ligand-1 (PSGL-1, CD162) is the sole major ligand on leukocytes for P-selectin, carbohydrate epitope(s) recognized by E-selectin can be expressed on at least three distinct glycoproteins, PSGL-1, CD43, and CD44. These scaffold glycoproteins are differentially expressed and decorated with relevant carbohydrate(s) at distinct stages of development in the various populations of circulating leukocytes (Fuhlbrigge *et al.* 1997, 2006, Dimitroff *et al.* 2001). We believe, however, that the full repertoire of E-selectin ligands is yet to be discovered. Creation of selectin ligands requires the glycosylation of these carrier or scaffold glycoproteins with sialylated, fucosylated carbohydrate epitopes similar or identical to sialyl-Lewis^x. As a group, these functional ligand structures are recognized by a specific monoclonal antibody called HECA-452 and termed cutaneous lymphocyte-associated antigen (CLA) (Fuhlbrigge *et al.* 1997). CLA is a distinctive marker for skin homing leukocytes expressed by the vast majority of T cells in uninflamed skin and at sites of cutaneous inflammation (Clark *et al.* 2006). Production of CLA and related carbohydrate epitopes is regulated by key glycosyltransferases including α1,3-fucosyltransferase-IV (FucT-IV) and FucT-VII. In addition to E- and P-selectin, very late antigen 4 (VLA-4, integrin $α_4β_1$) can also mediate tethering and rolling through interaction with its ligand, vascular cell adhesion molecule 1 (VCAM-1) expressed on endothelia. The interplay between integrin and selectin-mediated rolling on dermal endothelium at baseline and in disease remains to be defined.

Adhesion Molecules in Activation

To enter a target tissue, leukocytes need to adhere firmly to vessel walls in the area. Details of the molecules involved in this process are discussed in a recent review (Ley *et al.* 2007). Briefly, cells moving at the hydrodynamic velocity of blood pass through the vessels too quickly to sense local environmental signals or form stable interactions with endothelial cells. Cells that have tethered and are rolling move at an order of magnitude slower velocity and maintain close contact with the endothelial surface. This allows them to sample the local endothelial surface for chemoattractant compounds and to form firm adhesions if activated. The chemokines are a multigene family of chemoattractant compounds that bind to transmembrane receptors expressed on leukocytes. Chemokines and their receptors are the most heterogeneous group of molecules involved in leukocyte recruitment, with more than 50 chemokines and 19 chemokine receptors described. The specific pattern of chemokines and chemoattractant molecules produced at an anatomical site is influenced by the local cytokine milieu, which is in turn determined by the type of tissue and the mechanism of activation (e.g., pathogen invasion, thermal burn, irritant exposure). Chemokines are transported to the endothelial luminal surface, where they bind to proteoglycans for presentation to rolling cells. Chemokine binding to a suitable receptor activates an intracellular signaling cascade that results in upregulation of leukocyte integrin avidity and promotes binding to endothelia. Chemokine receptors on circulating leukocytes are variably expressed, influenced by cell lineage, developmental state and degree of activation. In addition, individual chemokines may bind multiple receptors and individual receptors may bind multiple chemokines. This complexity allows for subtle effects resulting from summed signals received from varied sources, reminiscent of neural networks.

With regard to T cell recruitment to skin, recent investigations have emphasized the roles of CC-chemokine receptor 4 (CCR4) and CCR10 and their ligands CCL17 and CCL27, respectively (Campbell *et al.* 2007, Sigmundsdottir *et al.* 2007). Virtually all CLA+ T cells observed in normal or inflamed skin express CCR4, and CCL17 is constitutively expressed on post-capillary venules in skin (Chong *et al.* 2004, Clark *et al.* 2006). CCR10 is expressed by a smaller subset of skin homing T cells than CCR4 and has been suggested to have a role in epidermal positioning of T cells (Soler *et al.* 2003). CCR6 and CCR8 are also expressed on populations of skin homing leukocytes and are implicated in cell recruitment to skin (Fitzhugh *et al.* 2000, Ebert *et al.* 2006). As with rolling ligands, the precise interplay between the various chemoattractant compounds and their receptors in leukocyte recruitment to resting and inflamed skin is an area of intense investigation with many details yet to be elucidated.

Adhesion Molecules in Firm Adhesion

Chemokines and chemoattractant compounds upregulate the function of leukocyte integrins and promote their binding to counter ligands expressed on the vascular endothelium (Ley *et al.* 2007). Integrins are relatively unique in that their activity can be regulated independent of their expression level. These interactions result in firm (non-rolling) adhesion and are involved in the subsequent transmigration across the endothelial layer and movement through the surrounding tissue. Integrins of the β_2 family, including lymphocyte function associated antigen 1

(LFA-1, integrin $\alpha_L\beta_2$, CD11a/CD18) and macrophage-1 antigen (Mac-1, integrin $\alpha_M\beta_2$, CD11b/CD18), and the β_1 family, including VLA-4 (integrin $\alpha_4\beta_1$, CD49d/ CD29), have been shown to be crucial elements of the leukocyte recruitment pathway in skin and other tissues.

Adhesion Molecules in Transmigration and Diapedesis

Transmigration across the endothelium and basement membrane and movement within the tissue are the final steps in the process of leukocyte recruitment into the skin. A detailed description of the variety of molecules involved in endothelial transmigration is available in recent reviews (Ley *et al.* 2007, Vestweber 2007). Briefly, a complex interplay of endothelial junctional proteins and leukocyte surface glycoproteins, including PECAM-1, junctional adhesion molecules, CD99, LFA-1/ICAM-1 complexes, and VE-cadherin, supports and directs leukocyte transmigration through and around endothelial cells and into the underlying extravascular space. Adding to this complexity, transmigrating leukocytes express surface proteases that are selective for basement membrane proteins, thereby exposing new or hidden binding sites for adhesion molecules, and/or generating protein fragments that can be chemotactic and recruit additional cells into the involved tissue.

IMMUNE SURVEILLANCE IN SKIN

In this section, we discuss the ways in which selective leukocyte recruitment drives immune surveillance and response in peripheral tissues.

Innate Immune Surveillance

Effective host defense against invading microorganisms requires detection of foreign invaders. In higher vertebrates, the immune system has developed rapid non-specific (innate), and sustained specific (adaptive) mechanisms to defend the host against pathogen invasion. Central to our understanding of cutaneous innate immune surveillance is the observation that cells resident in skin serve as sentinels for danger signals, including pathogen invasion and physical injury. We have reviewed these concepts in detail elsewhere and will discuss them only briefly here (Kupper and Fuhlbrigge 2004, Clark and Kupper 2005). Keratinocytes and Langerhans cells in the epidermis, as well as dermal dendritic cells, mast cells and macrophages, express a variety of pattern recognition receptors (PRR) that recognize conserved molecular patterns found on microbial pathogens and trigger downstream activation cascades and release of antimicrobial peptides, chemotactic proteins, and inflammatory cytokines. An important subset of these PRR belong to the toll-like receptor (TLR) family, which recognize pathogen-associated molecules such as lipopolysaccharide, bacterial lipoproteins, flagellin, yeast mannans and DNA containing unmethylated CpG motifs. Activation via TLRs results in increased production of pro-inflammatory cytokines and prostanoid, induction and release of chemoattractants, and activation of the JNK and NF-κB signaling pathways. These compounds are seen as key links between innate and adaptive immune systems, regulating the expression of numerous genes involved in the initiation of the inflammatory response including adhesion receptors, chemokines and cytokines involved in leukocyte trafficking. Keratinocytes are important and often under-appreciated participants in cutaneous immunity. They produce large quantities of IL-1α, TNF, β-defensins and cathelicidins in response to a variety of stimuli, including kinetic and thermal trauma, and UV radiation. Keratinocytes also produce chemokines and immunoregulatory cytokines that promote recruitment of leukocytes and modulate their functions, providing a link between innate and adaptive immunity in skin. Ultimately, skin-resident dendritic cells take up antigen, enter the local lymphatic channels and travel to the draining lymph node(s) where they present antigen to T cells to promote adaptive immune responses.

Adaptive Immune Surveillance

The adaptive immune system enhances responses to specific pathogens and provides a means for maintaining memory of past encounters. Creating this response requires the interaction of antigen-presenting cells with T or B cells bearing cognate receptors for those antigens and the direction of those cells to their required sites of function. This process operates at three levels, which we have termed primary, secondary and tertiary immune surveillance (Kupper and Fuhlbrigge 2004). For skin, primary immune surveillance involves the mechanisms for bringing antigens, antigen presenting cells (APC), and naïve T cells together in the specialized microenvironment of the skin-draining lymph nodes. Secondary immune surveillance involves the production and distribution of antigen-specific effector T cells back to the skin. Tertiary immune surveillance relates to the production of 'central' memory T cells that circulate through lymph nodes draining non-skin tissues and effector memory T cells directed to tissues other than the skin.

Primary Immune Surveillance

Epidermal Langerhans cells (LC) and dermal dendritic cells (dDC) are professional APCs that, when activated, develop the capacity to present antigens efficiently and to drive the maturation of naïve T cells to a memory/effector phenotype. APCs become activated at sites of injury or pathogen invasion in the skin through innate mechanisms, including PRRs such as TLRs and exposure to pro-inflammatory cytokines, rapidly engulf antigen and emigrate through the afferent lymphatics to the local skin-draining lymph node (LN). Skin-draining LNs are hubs where mature DCs carrying antigens can interact with T cells expressing cognate receptors for those antigens, thus connecting the innate and adaptive immune systems. T cells that encounter their cognate antigen will undergo clonal expansion, produce autocrine growth factors, and differentiate into effector memory (T_{EM}) or central memory (T_{CM}) cells. Both unactivated and activated/expanded cell populations return to the blood via the lymphatics. Naïve and T_{CM} cells express L-selectin, CCR7, and LFA-1, which mediate interaction with PNAd, CCL21, and ICAM-1, respectively. These ligands are expressed by high endothelial venules in peripheral LNs and promote traffic of these cell populations into LNs throughout the body. Activation of T cells in skin-draining LNs will not only produce T_{EM} cells, but will 'imprint' them with a skin homing phenotype (combination of adhesion molecules and chemokine receptors), allowing them to traffic to skin. In this way, the immune response is preferentially targeted back to the site of the inciting infection or inflammation. Imprinting of homing phenotypes involves the specialized microenvironment of tissue LNs, tissue-specific DCs, and soluble factors (Edele *et al.* 2008, Sigmundsdottir and Butcher 2008). Recent reports have highlighted an interesting role for vitamin D in T cell homing to the skin (Sigmundsdottir *et al.* 2007).

Secondary Immune Surveillance

Activation of secondary surveillance occurs in the generic context of injury that leads to the release of inflammatory cytokines in the epidermis, increased local expression of endothelial adhesion molecules and recruitment of skin-homing cells from the circulation. The T cells that accumulate at this site will bear multiple distinct antigenic specific receptors comprising, in effect, a representative sample of the circulating library of T cells specific for all antigens previously encountered in the skin. Thus, there is immunological memory not only for antigen, but also for the anatomical context in which that antigen was encountered. This is an elegant evolutionary strategy, ensuring that skin injury is considered infectious until proven otherwise. A non-specific stimulus in skin leads to accumulation of a polyclonal population of skin-homing memory T cells that, in the absence of appropriately presented antigen, eventually exit the skin via the draining lymphatics. If the stimulus is accompanied by antigen, then the subset of recruited T cells with the appropriate specific receptor will become activated by local DCs to perform their effector functions. While circulating CLA+ T cells can be recruited rapidly to sites of infection or inflammation, there is also evidence for constitutive homing of T cells to skin. E-selectin, CCL17, and ICAM-1 are constitutively expressed on cutaneous microvessels, providing a molecular basis for steady state homing of CLA+ cells to the skin. Furthermore, T cells recovered from non-inflamed skin express high levels of CLA and CCR4 (Chong *et al.* 2004, Clark *et al.* 2006). In fact, calculations based on histological examination indicate that the majority of CLA+ T_{EM} cells actually reside in skin (Clark *et al.* 2006). In this fashion, the immune system maintains surveillance by directing the accumulation of cells expanded in response to a select antigen to the site where that antigen is most likely to be encountered again. Recently, it has become apparent that effector T cell functions are modulated by the actions of regulatory T cells (T_{REG}). T_{REG} recovered from skin display a receptor homing phenotype similar to effector T cells from skin, suggesting they use the same recruitment paradigm for migration to skin. The balance of effector and regulatory functions in normal and disease states, and the role of adhesion receptors in determining these ratios, are an area of active research.

Tertiary Immune Surveillance

Since pathogens are not always encountered at the same tissue interface, the immune system has evolved mechanisms to disperse memory to other sites. T_{CM}, produced along with T_{EM} in the response to specific antigen stimulation, retain expression of CD62L and CCR7 and the ability to circulate through LNs. Antigen-specific T_{CM} produced in skin-draining LNs enter LNs draining other tissue sites and acquire the homing phenotype of that tissue (Liu *et al.* 2006). In this way, the immune system ensures a rapid and effective response even if the next encounter with an antigen happens at a different interface.

ADHESION RECEPTORS IN SKIN DISEASE

Tissue-selective recruitment of lymphocytes to the skin and other organs is crucial for immune surveillance under normal conditions. Dysregulation of leukocyte recruitment also plays a key role in the pathogenesis of various diseases including skin inflammatory disorders such as atopic dermatitis or psoriasis (Zollner *et al.* 2007), and skin malignancies such as squamous cell carcinoma and melanoma (Weishaupt *et al.* 2007). The selective targeting of key molecules involved in the recruitment paradigm promises new therapeutic options for treating inflammation and malignancies in the skin.

CUTANEOUS TUMORS

In individuals with intact immune function, immune surveillance mechanisms rapidly and efficiently eliminate abnormal cells before they have a chance to develop into tumors. Conversely, tumors that persist develop mechanisms to evade immune surveillance, including lack of specific surface antigens and production of soluble inhibitory factors that activate regulatory T cells and suppress anti-tumor immune responses. We have studied tumor evasion of immune surveillance via the dysregulation of adhesion molecules on the tumor vasculature (Weishaupt *et al.* 2007). Although patients often produce tumor-specific T_{EM} cells, these cells are unable to efficiently enter the tumor tissue and kill tumor cells (Appay *et al.* 2006). Endothelial cells in various tumors express lower levels of adhesion molecules essential in leukocyte recruitment, including ICAM-1, VCAM-1, and E-selectin (Dirkx *et al.* 2003, Weishaupt *et al.* 2007, Clark *et al.* 2008). Tumor endothelial cells are also unresponsive to inflammatory cytokines that induce adhesion molecule expression (Griffioen 2008). Interestingly, treatment of squamous cell carcinomas with a topical TLR agonist can enhance expression of E-selectin on tumor vessels, resulting in recruitment of CLA+ CD8 T cells and tumor regression (Clark *et al.* 2008). Excessive angiogenesis will also result in abnormal tumor vessel architecture that can block recruitment of T cells and even drug delivery to the tumor (Manzur *et al.* 2008). Combining immunotherapy with angiostatic and proinflammatory agents has led to better prognoses in solid tumor therapy (Cirone *et al.* 2004, Ganss *et al.* 2004). These findings have important implications for the design of future immunotherapy protocols, which will depend on the identification of methods to upregulate the expression of homing molecules on tumor vasculature or otherwise bypass this block to effective T cell function.

SUMMARY

- Leukocyte homing is a crucial element of immune surveillance and immune response, playing an important role in both health and disease.
- Leukocyte homing depends on a stepwise process of tethering, rolling, activation, firm adhesion and transmigration.
- Every step in the homing cascade is necessary, but individually insufficient, to direct leukocyte recruitment and retention, thus providing a mechanism for specificity and selectivity.
- Keratinocytes are important and often under-appreciated participants in cutaneous immunity.
- Immune memory encompasses not only specificity for antigen, but also a homing bias to the anatomical site of first exposure.
- Therapeutic strategies targeting key molecules in the recruitment paradigm have led to effective treatments for skin inflammation and malignancies.

Abbreviations

APC	antigen presenting cells
CCR	CC chemokine receptor
CCL	CC chemokine ligand
CD	cluster of differentiation
dDC	dermal dendritic cells
ERK	extracellular signal-regulated kinase

HECA	human endothelial cutaneous antigen
IFN-γ	interferon gamma
IgSF	immunoglobulin superfamily
IL	interleukin
JNK	c-Jun N-terminal kinase
LC	epidermal Langerhans cells
LFA	lymphocyte function associated antigen
LN	lymph node
NF-κB	nuclear factor kappa B
PECAM	platelet endothelial cell adhesion molecule
PNAd	peripheral node addressin
PRR	pattern recognition receptors
PSGL-1	platelet selectin glycoprotein ligand-1
T_{CM}	central memory T cells
T_{EM}	effector memory T cells
T_{REG}	regulatory T cells
TCR	t cell receptor

Key Facts about Immunosurveillance and Leukocyte Migration to Cutaneous Tissues

1. Leukocytes that bear selectin ligands can tether and roll across cutaneous capillary endothelial surfaces that express selectins. Cells that do not have an appropriate ligand will move through the vessel too quickly to interact with the vascular surface and are thus not eligible for recruitment.

2. Rolling cells sample the vessel surface for activation molecules. If the cells encounter a chemokine for which they have the appropriate counter receptor, they become activated, upregulate integrin avidity and form firm attachments.

3. Firmly attached cells, in turn, must successfully negotiate interaction with the endothelial cells, junctional proteins and extracellular matrix components to effectively transmigrate and move into the tissue.

4. Each step in this cascade is thus required, but individually insufficient, providing multiple checkpoints for regulation.

5. All the major ligands necessary for recruitment of circulating leukocytes to the skin are expressed constitutively on the endothelium of cutaneous capillaries to ensure continuous surveillance even under non-inflammatory conditions.

Definition of Terms

Angiogenesis: The physiological process involving the growth of new blood vessels from pre-existing vessels.

Antigen presenting cells: Any cell that can process and present antigenic peptides as well as deliver stimulatory signals necessary to activate T cells.

Chemokines: A family of small proteins secreted by cells, characterized by the presence of cysteine residues in conserved locations. Their name is derived from their ability to induce directed movement (chemotaxis) in nearby responsive cells.

Dendritic cells: Leukocytes that specialize in antigen presentation.

Dermis: The layer of skin between the epidermis and subcutaneous tissues. The dermis contains blood and lymph vessels, hair follicles, and sweat glands. The dermis also contains T cells, dendritic cells, macrophages, and mast cells.

Diapedesis: Passage of cells through the intact wall of a blood vessel to the inside of the tissue.

Endothelium: Layer of cells that line the interior surfaces of blood vessels, forming an interface between circulating blood and the rest of the surrounding tissue.

Epidermis: Outermost layer of the skin, avascular and nourished by diffusion from the dermis. Keratinocytes are the major cellular constituent of the epidermis.

Integrins: A family of heterodimeric cell surface receptors that bind to adhesion molecules on the endothelium allowing cells to adhere to the endothelium. There are many types of integrin, and many cells have multiple types on their surface.

Junctional adhesion molecules (JAMs): Proteins expressed at cell junctions in epithelial and endothelial cells as well as on the surface of leukocytes, platelets, and erythrocytes. JAMs are important for a variety of cellular processes, including tight junction assembly, leukocyte transmigration, and angiogenesis.

Selectins: Single chain transmembrane adhesion molecules bearing a lectin domain that bind to carbohydrate ligands displayed on leukocytes.

Sialyl Lewis^X: Tetrasaccharide carbohydrate, usually attached to O-glycans on the surface of the cells, and known to play a vital role in cell-cell recognition and attachment.

References

Appay, V. and C. Jandus, V. Voelter, S. Reynard, S.E. Coupland, D. Rimoldi, D. Lienard, P. Guillaume, A.M. Krieg, J.C. Cerottini, P. Romero, S. Leyvraz, N. Rufer, and D.E. Speiser. 2006. New generation vaccine induces effective melanoma-specific CD8+ T cells in the circulation but not in the tumor site. J. Immunol. 177(3): 1670-1678.

Campbell, J.J. and D.J. O'Connell, and M.A. Wurbel. 2007. Cutting Edge: Chemokine receptor CCR4 is necessary for antigen-driven cutaneous accumulation of CD4 T cells under physiological conditions. J. Immunol. 178(6): 3358-3362.

Chong, B.F. and J.E. Murphy, T.S. Kupper, and R.C. Fuhlbrigge. 2004. E-selectin, thymus- and activation-regulated chemokine/CCL17, and intercellular adhesion molecule-1 are constitutively coexpressed in dermal microvessels: a foundation for a cutaneous immunosurveillance system. J. Immunol. 172(3): 1575-1581.

Cirone, P. and J.M. Bourgeois, F. Shen, and P.L. Chang. 2004. Combined immunotherapy and antiangiogenic therapy of cancer with microencapsulated cells. Hum. Gene Ther. 15(10): 945-959.

Clark, R. and T. Kupper. 2005. Old meets new: the interaction between innate and adaptive immunity. J. Invest. Dermatol. 125(4): 629-637.

Clark, R.A. and B. Chong, N. Mirchandani, N.K. Brinster, K. Yamanaka, R.K. Dowgiert, and T.S. Kupper. 2006. The vast majority of CLA+ T cells are resident in normal skin. J. Immunol. 176(7): 4431-4439.

Clark, R.A. and S.J. Huang, G.F. Murphy, I.G. Mollet, D. Hijnen, M. Muthukuru, C.F. Schanbacher, V. Edwards, D.M. Miller, J.E. Kim, J. Lambert, and T.S. Kupper. 2008. Human squamous cell carcinomas evade the immune response by down-regulation of vascular E-selectin and recruitment of regulatory T cells. J. Exp. Med. 205(10): 2221-2234.

Dimitroff, C.J. and J.Y. Lee, S. Rafii, R.C. Fuhlbrigge, and R. Sackstein. 2001. CD44 is a major E-selectin ligand on human hematopoietic progenitor cells. J. Cell Biol. 153(6): 1277-1286.

Dirkx, A.E. and M.G. Oude Egbrink, M.J. Kuijpers, S.T. van der Niet, V.V. Heijnen, J.C. Bouma-ter Steege, J. Wagstaff, and A.W. Griffioen. 2003. Tumor angiogenesis modulates leukocyte-vessel wall interactions in vivo by reducing endothelial adhesion molecule expression. Cancer Res. 63(9): 2322-2329.

Ebert, L.M. and S. Meuter, and B. Moser. 2006. Homing and function of human skin gammadelta T cells and NK cells: relevance for tumor surveillance. J. Immunol. 176(7): 4331-4336.

Edele, F. and R. Molenaar, D. Gutle, J.C. Dudda, T. Jakob, B. Homey, R. Mebius, M. Hornef, and S.F. Martin. 2008. Cutting edge: instructive role of peripheral tissue cells in the imprinting of T cell homing receptor patterns. J. Immunol. 181(6): 3745-3749.

Fitzhugh, D.J. and S. Naik, S.W. Caughman, and S.T. Hwang. 2000. Cutting edge: C-C chemokine receptor 6 is essential for arrest of a subset of memory T cells on activated dermal microvascular endothelial cells under physiologic flow conditions in vitro. J. Immunol. 165(12): 6677-6681.

Fuhlbrigge, R.C. and J.D. Kieffer, D. Armerding, and T.S. Kupper. 1997. Cutaneous lymphocyte antigen is a specialized form of PSGL-1 expressed on skin-homing T cells. Nature 389(6654): 978-981.

Fuhlbrigge, R.C. and S.L. King, R. Sackstein, and T.S. Kupper. 2006. CD43 is a ligand for E-selectin on CLA+ human T cells. Blood 107(4): 1421-1426.

Fuhlbrigge, R.C. and C. Weishaupt. 2007. Adhesion molecules in cutaneous immunity. Seminars in Immunopathology 29(1): 45-57.

Ganss, R. and B. Arnold, and G.J. Hammerling. 2004. Mini-review: overcoming tumor-intrinsic resistance to immune effector function. Eur. J. Immunol. 34(10): 2635-2641.

Griffioen, A.W. 2008. Anti-angiogenesis: making the tumor vulnerable to the immune system. Cancer Immunol. Immunother. 57(10): 1553-1558.

Kupper, T.S. and R.C. Fuhlbrigge. 2004. Immune surveillance in the skin: mechanisms and clinical consequences. Nat. Rev. Immunol. 4(3): 211-222.

Ley, K. and C. Laudanna, M.I. Cybulsky, and S. Nourshargh. 2007. Getting to the site of inflammation: the leukocyte adhesion cascade updated. Nat. Rev. Immunol. 7(9): 678-689.

Liu, L. and R.C. Fuhlbrigge, K. Karibian, T. Tian, and T.S. Kupper. 2006. Dynamic programming of CD8+ T cell trafficking after live viral immunization. Immunity 25(3): 511-520.

Manzur, M. and J. Hamzah, and R. Ganss. 2008. Modulation of the "blood-tumor" barrier improves immunotherapy. Cell Cycle (Georgetown, Tex) 7(16): 2452-2455.

Sigmundsdottir, H. and E.C. Butcher. 2008. Environmental cues, dendritic cells and the programming of tissue-selective lymphocyte trafficking. Nat. Immunol. 9(9): 981-987.

Sigmundsdottir, H. and J. Pan, G.F. Debes, C. Alt, A. Habtezion, D. Soler, and E.C. Butcher. 2007. DCs metabolize sunlight-induced vitamin D3 to 'program' T cell attraction to the epidermal chemokine CCL27. Nat. Immunol. 8(3): 229-230.

Soler, D. and T.L. Humphreys, S.M. Spinola, and J.J. Campbell. 2003. CCR4 versus CCR10 in human cutaneous TH lymphocyte trafficking. Blood 101(5): 1677-1682.

Vestweber, D. 2007. Adhesion and signaling molecules controlling the transmigration of leukocytes through endothelium. Immunol. Rev. 218: 178-196.

Weishaupt, C. and K.N. Munoz, E. Buzney, T.S. Kupper, and R.C. Fuhlbrigge. 2007. T-cell distribution and adhesion receptor expression in metastatic melanoma. Clin. Cancer Res. 13(9): 2549-5256.

Zollner, T.M. and K. Asadullah, and M.P. Schon. 2007. Targeting leukocyte trafficking to inflamed skin: still an attractive therapeutic approach? Exp. Dermatol. 16(1): 1-12.

CHAPTER 20

Expression of Cell–Cell Adhesion Molecules in Matrical Tumors of Hairs, Nails, Teeth and Pituitary Gland

Alejandro Peralta Soler[1,*] and **James L. Burchette**[2]

[1]The Richfield Laboratory of Dermatopathology, 9844 Redhill Drive, Cincinnati, OH 45242, E-mail: AlexPSoler@hotmail.com

[2]Immunopathology Lab, Department of Pathology, Duke University Medical Center, Trent Drive, Room 4338 South Clinic Building, Durham, NC 27710 E-mail: jburch01@aol.com

ABSTRACT

In mammals, hard adnexal structures comprise hairs, nails and teeth. These structures can give rise to tumors with phenotypical features of the matrix cells that originate them. Although the pituitary gland is not considered a hard structure, matrical tumors of the pituitary gland will be included in this group, because of their embryological and phenotypical similitude with tumors of the teeth. Although matrical tumors are not usually grouped together, the purpose of this chapter is to classify them as a unique type of tumors that share morphological features and β-catenin oncogenic signal pathways.

Matrical tumors of the hair follicles include the pilomatricoma, pilomatrix carcinoma and melanocytic matricoma. Other follicular tumors that may exhibit partial matrical features include trichoepithelioma, trichoblastoma, basal cell carcinoma and panfolliculoma. The matrical tumors of the nail are rare and include the onychomatricoma and its variants, the unguioblastoma and unguioblastic fibroma. Matrical tumors of the teeth include the calcifying odontogenic cysts, and ameloblastoma and its variants. The matrical tumors of the pituitary gland include craniopharyngioma and its variants, originated from remnants of the Rathke's pouch, which shares embryological origin with the buccopharynx.

---

*Corresponding author

Cadherins are calcium-dependent cell-cell adhesion proteins involved in the sorting of cells and tissue morphogenesis. The most abundant and best characterized are the classical cadherins, a group that includes E(epithelial)-, P(placental)- and N(nerve)-cadherin. Cadherins are linked to a heterogeneous group of cytoskeletal proteins termed catenins. Among the catenins, β-catenin is a multifunctional protein with adhesion and signaling properties that plays a crucial role in the development of many tumors.

The purpose of this chapter is to review the data on the expression of cadherins and catenins in matrical tumors of the hair, nails, teeth and pituitary gland. We propose that matrical tumors of these structures can be classified together as a group based on similar phenotypical features and common activating cadherin/catenin-derived oncogenic signaling.

GENERAL CONCEPTS

The matrix tissue of hairs, teeth, and nails and from the Rathke's pouch of the pituitary gland comprises pluripotential fibroblast-like cells, which upon activation by transcriptional signals induce the growth of epithelial cells. This critical interaction of mesenchymal and epithelial tissues results in the development of hairs, nails and teeth and the Rathke's pouch of the pituitary gland. There are tumors from these structures that exhibit morphological characteristics of matrical tissues. However, these tumors are not usually grouped together, and in pathology books they are described separately under the chapters on each tissue from which they originate (Rosai 2004). The classification of tumors into distinct categories is important for pathologists in outlining differential diagnoses, for clinicians in determining patient management and treatment, and for biologists in understanding mechanisms of oncogenesis and designing better therapeutic strategies. Most tumors categories are defined by pathologists on the basis of phenotypical similarities. However, new knowledge of tumor markers and oncogenic signals can result in the grouping of tumors into categories that would make their management more feasible and more biologically accurate.

CADHERINS AND CATENINS

Cadherins are calcium-dependent cell-cell adhesion proteins involved in cell sorting and tissue morphogenesis (Takeichi 1990). The best characterized are the type I or classical cadherins (Kemler 1992), a group that includes E-(epithelial)-, P(placental)- and N(nerve)-cadherin. E- and P-cadherin (Takeichi 1990) are essential in promoting and maintaining morphogenesis of the skin and appendages (Hirai *et al.* 1989). Cadherins are linked to a family of cytoskeletal proteins termed catenins (Gumbiner and McCrea 1993). Among the catenins, β-catenin, a multifunctional protein with cell-cell adhesion and signaling properties (Behrens *et al.* 1996) plays a crucial role in the development of hair and hair tumors

(Gat *et al.* 1998). Activated nuclear β-catenin is also involved in the development of tumors of the teeth (Sekine *et al.* 2003, Kumamoto and Ooya 2005), and pituitary craniopharyngiomas (Sekine *et al.* 2002).

MATRICAL TUMORS: PHENOTYPE AND NOMENCLATURE

Matrical tumors of the hair, nails, teeth and pituitary gland are usually described separately in pathology books (Rosai 2004). However, because those tumors have similar histological features, they are grouped together in this chapter (Table 1). We propose that classification of some tumors on the basis of phenotype and oncogenic signaling pathways (Table 2) rather than tissue of origin may be important for the understanding of their biology and management.

Table 1 Nomenclature of matrical tumors from hairs, nails, teeth and pituitary gland

Hair	*Nails*	*Teeth*	*Pituitary gland*
Pilomatricoma	Onychomatricoma	Calcifying odontogenic cyst	Craniopharyngioma
Matricoma	Unguioblastoma	Ameloblastoma	
Pigmented matricoma	Unguioblastic fibroma	Extraosseous ameloblastoma	
Melanocytic matricoma			
Pilomatrix carcinoma			

The nomenclature of matrical tumors of hair (Ackerman *et al.* 1993), nails (Baran and Kint 1992, Ko *et al.* 2004), teeth (Neville *et al.* 2002) and pituitary gland (Rosai 2004) is based on their histological appearance.

Matrical Hair Tumors

Matrical hair tumors comprise pilomatricoma, matricoma, pilomatrix carcinoma and melanocytic matricoma. Pilomatricomas are by far the most common of the group. Pilomatricomas are composed of basaloid cells and 'shadow' fully differentiated anucleated cells. The lesions have a tendency to become cystic and to calcify and ossify (Ackerman *et al.* 1993). An architectural variant with a more pronounced differentiation towards the hair follicle inner sheath has been named matricoma (Ackerman *et al.* 1993). Aggressive and confirmed malignant variants have been described under the name of pilomatrix carcinoma (Ackerman *et al.* 1993). Melanocytic matricoma is a more rare neoplasm thought to recapitulate the bulb of the hair follicle in anagen (Ackerman *et al.* 1993). In addition to the epithelial hair matrix cells and the terminally differentiated 'shadow' cells, there is a conspicuous population of S-100, HMB45, and vimentin-positive pigmented dendritic melanocytes, which differentiate it from pilomatricomas, matricomas (Ackerman *et al.* 1993) and pigmented matricoma variants (Peralta Soler *et al.* 2007).

Table 2   β-catenin mutations and nuclear expression in matrical tumors of the hair, teeth, nails and pituitary gland

Tissue	*Matrical tumor*	*β-catenin mutations*	*Nuclear β-cat (IHC)*	*Reference*
Hair	Pilomatricoma	Frequent	In basaloid cells	Chan *et al.* 1999
	Matricoma	Not known	Not known	
	Pilomatrix carcinoma	Frequent	In basaloid cells	Hassanein and Glanz 2004 Lazar *et al.* 2005
	Melanocytic matricoma	Not known	In basaloid cells	Peralta Soler *et al.* 2007
Teeth	Calcifying odontogenic cyst	Frequent	In basaloid cells	Sekine *et al.* 2003
	Ameloblastoma	Infrequent	In basaloid cells	Sekine *et al.* 2003 Kumamoto and Ooya 2005
Nails	Onychomatricoma	Not known	Negative	Burchette *et al.* 2008
Pituitary gland	Craniopharyingioma	Frequent	In basaloid cells	Sekine *et al.* 2001

The majority of matrical tumors from hard structures exhibit β-catenin mutations and/or nuclear translocation of β-catenin. However, there are differences among tumors, and subcellular distribution, mutations and signaling activation of β-catenin have not been studied in many of this group of matrical tumors.

Matrical Nail Tumors

Onychomatricoma is an exceedingly rare nail tumor with distinctive biphasic epithelial and stromal tissues. The architectural characteristics include proximal cavitated invaginations of nail matrix epithelium into the stroma, and distal dermal protrusions perforating the nail plate (Baran and Kint 1992). Electron microscopy and immunohistochemistry (Perrin *et al.* 2002) studies support the origin of the tumor from the nail matrix, a structure containing keratinocytes, Langerhans cells, melanocytes and Merkel cells. Although very few cases of onychomatricoma have been reported, there is a proposal to subclassify them into predominantly epithelial unguioblastomas, most closely associated with trichoblastomas and ameloblastomas, and predominantly stromal tumors called unguioblastic fibromas (Ko *et al.* 2004).

Matrical Tooth Tumors

Tumors of the teeth with matrical features include the calcifying odontogenic cyst (Gorlin cyst, dentinogenic ghost cell tumor, calcifying ghost cell odontogenic cyst) and the ameloblastoma and its variants. Calcifying odontogenic cyst is a predominantly intraosseous neoplasm of the jaw that includes benign, purely cystic forms and more aggressive solid variants (Neville *et al.* 2002). They exhibit 'ghost cells' similar to matrical tumors from the hair follicle. Ameloblastomas are tumors originating from the odontogenic epithelium and include phenotypical variants with follicular, plexiform, desmoplastic, granular, basaloid and acanthomatous histological patterns (Neville *et al.* 2002). Less frequently, ameloblastomas can be extraosseous, probably arising from misplaced odontogenic remmants, and frankly malignant forms with metastatic potential (Rosai 2004). Ameloblastomas are biphasic tumors with epithelial tissues containing peripheral cylindrical cells, a central loose network of stellate epithelial cells, and surrounding stroma with various degrees of maturation, resembling the developing tooth (Rosai 2004).

Matrical Pituitary Gland Tumors

The matrical tumor of the pituitary gland is the craniopharyngioma. Craniopharyngiomas most likely originate from odontogenic embryonic remnants of the Rathke's pouch, which exhibits histological features similar to the enamel tissue of the teeth. Moreover, the microscopic appearance of craniopharyngioma is highly similar to that of calcifying odontogenic cysts and ameloblastomas (Rosai 2004). Craniopharyngiomas are frequently cystic tumors. They exhibit a biphasic architecture consisting of anastomosing epithelial strands with peripheral cylindrical cells aligned in palisade, a center of loose stellate epithelial cells, and a fibroblastic stroma with variable degrees of cellularity (Rosai 2004). Foci of squamous differentiation, calcification and inflammatory reaction and clearly discernible embryonic tooth structures are often seen (Rosai 2004).

CADHERIN AND CATENIN EXPRESSION AND SIGNALING IN MATRICAL TUMORS

The study of signaling pathways is crucial for the understanding of tumor development, as well as for the clarification of the cellular progeny of the tumors. Tumors with a phenotype of matrix cells from the hair, nails, teeth and pituitary gland share morphological features. There are also immunohistochemical evidence and mutational studies that support the role of β-catenin as a transcriptional activator in the oncogenesis of these tumors. There are, however, differences in expression of nuclear-cytoplasmic and mutation rates of β-catenin within tumors of each group (Table 2). These differences indicate that the molecular mimicry of tumors with the tissues of origin is more complex than previously thought, and there is still much work to do in understanding the different stimuli that lead to the formation and maintenance of these tumors.

Matrical Hair Tumors

E- and P-cadherin are crucial in determining the formation and maintenance of the epidermis and the skin adnexae, including the hair follicles (Hirai *et al.* 1989). The hair cycles during the lifespan of mammals, although there are specific time and regulatory differences in different parts of the body and between species (Camacho *et al.* 2000). The maximal growth phase of the hair is called the anagen stage. During this period, E-cadherin is expressed in the outer and inner root sheaths and in the outer portion of the hair matrix, and P-cadherin is expressed in the innermost portion of the hair matrix near the dermal papilla (Hardy and Vielkind 1996, Muller-Rover *et al.* 1999). Functional P-cadherin induces the segregation of proliferating hair matrix cells to form a follicular epithelial unit (Muller-Rover *et al.* 1999), and its absence interferes with the normal hair cycle. Mutations in the CDH3 gene encoding P-cadherin cause the autosomal recessive disorder called ectodermal dysplasia-ecterodactyly-macular dystrophy, manifested by sparse scalp hair and blindness (Sprecher *et al.* 2001).

During the formation of the hair follicle, primitive mesenchymal cells aggregate to form the dermal papilla and release Wnt signals, which in turn prevent the degradation of β-catenin. Stable β-catenin acts as a transcription factor and activates LEF1 (lymphocyte enhancing factor 1) and downregulates E-cadherin. The process can be reversed by forced elevation of E-cadherin, which results in interruption of the development of normal hair follicles (Jamora *et al.* 2003). In the adult hair follicle, β-catenin is expressed inside the nucleus of inner hair matrix cells and in the cell membranes of the outer and inner root sheaths (Jamora *et al.* 2003). This differential pattern of expression of β-catenin suggests a switch from a transcriptional factor role (Behrens *et al.* 1996) in hair matrix cells during development to a structural role as part of the cadherin-catenin cell-cell adhesion complex in the differentiated cells of the adult hair follicle. This differential distribution of cadherins and β-catenin is also seen in the different portions of hair matrical tumors such as pilomatricoma, melanocytic matricoma (Peralta Soler *et al.* 2006) and pilomatrix carcinoma (Hassanein and Glanz 2004). In melanocytic matricoma, the less differentiated basaloid cells express nuclear and cytoplasmic β-catenin (Fig. 1A). This is consistent with the transcriptional role of β-catenin in these cells and reflects the frequent activating mutations in exon 3 of CTNNB1, the gene encoding for β-catenin, found in both benign (Chan *et al.* 1999) and malignant (Lazar *et al.* 2005) pilomatricomas. In contrast, the more differentiated squamoid cells express cell membrane β-catenin together with E-cadherin (Fig. 1B). P-cadherin is expressed at the cell membrane but restricted to the basaloid cells and is absent from the anucleated fully differentiated keratinized shadow cells (Fig. 1C). The tumor is colonized by numerous HMB-45 (a marker of melanosomes) positive dendritic cells (Fig. 1D), which gives it its name.

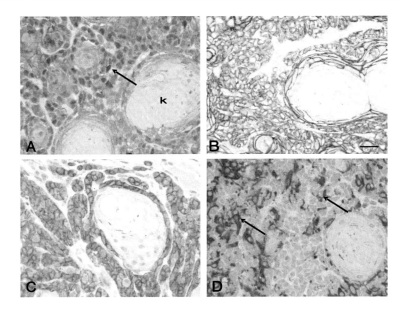

Fig. 1 Immunohistochemical expression of β-catenin, cadherins and a marker of melanosomes, the hair matrical tumor called melanocytic matricoma. Expression of β-catenin (A), E-cadherin (B), P-cadherin (C) and HMB45, a marker of melanosomes (D) in a melanocytic matricoma. Nuclear β-catenin (arrow) is seen in the basal cells but not in the more differentiated keratinized cells (k). E- and P-cadherin show cell membrane distribution typical of their cell-cell adhesion function. The expression of HMB45 (arrows) highlights the presence of numerous dendritic cells that characterize this tumor. Scale bar = 200 μm.

Matrical Nail Tumors

In contrast to the signals mediating the development of hairs and teeth, the signals regulating the development of the nail are still poorly understood, although it is known that the interaction between epithelium and mesenchymal tissue plays a significant role in nail development (Bergmann *et al.* 2006). Furthermore, there are differences between the mechanisms of hair development and nail development. Hair growth is a cyclic process that alternates between growth and rest phases. In contrast, the nail grows continuously. The growth rate of nails is the result of cell proliferation of the matrix and mechanisms of terminal differentiation. The growth and development of hairs is known to depend on the presence of activated β-catenin in the matrix cells (Gat *et al.* 1998). In contrast, growth of nails appears to depend on the activation of signals in different portions of the nail. On one hand, nail growth and increased cell proliferation can be triggered by activation of Notch1 in post-mitotic keratinized cells. This results in activation of Wnt signaling but no β-catenin in nail matrix cells (Lin and Kopan 2003).

Moreover, overexpression of Notch1 in transgenic mice causes an increase of proliferating cells of the nail matrix (Lin and Kopan 2003), forming structures similar to those seen in onychomatricomas. However, mutations in a gene called R-sponding4 (RSPO4), an activator of the Wnt/β-catenin signaling pathway, results in anonychia (autosomal recessive absence of nails) as a result of disruption of the mesenchymal-epithelial interactions that are crucial in nail development (Bergmann *et al.* 2006). These findings suggest that the Wnt/β-catenin signaling pathway may be involved in the primary matrix nail development, whereas Notch1 signals may promote the terminal differentiation steps in a tissue with continuous growth. In a case of onychomatricoma we found a distribution of β-catenin similar to E-cadherin, along the cell membranes as in a cell-cell adhesion role (Burchette *et al.* 2008). However, we studied a recent case of onychomatricoma and found focal intranuclear β-catenin within the less differentiated basal layer of the tumor (Fig. 2A), suggesting an activated transcriptional role (Peralta Soler and Burchette, unpublished data). E-cadherin is expressed at the cell membranes in most epithelial cells of the tumor (Fig. 2B), whereas P-cadherin, although similarly distributed at the cell membranes, is more restricted to the epithelial basal layer (Fig. 2C). The finding of variable nuclear expression of β-catenin in different onychomatricomas may support the concept that, although there are very few cases reported of this rare tumor, there may be different variants, reflecting various stages of developmental mimicry.

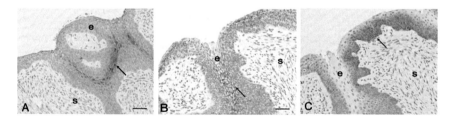

Fig. 2 Immunohistochemical expression of β-catenin and cadherins in the nail matrical tumor called onychomatrichoma. Expression of β-catenin (A), E-cadherin (B), P-cadherin (C) in an onychomatricoma. Nuclear β-catenin (arrow) is seen mostly in the basal cells. E- and P-cadherin show cell membrane distribution (arrows) typical of their cell-cell adhesion function, although P-cadherin is more restricted to the basal layers. Scale bar = 800 μm in A and 400 μm in B and C.

Matrical Tooth Tumors

The β-catenin/Wnt signal transduction pathway is crucial for the development of teeth (Sarkar and Sharpe 1999). Mice deficient in the β-catenin-interacting LEF1 or overexpressing Dickkopf1 (diffusible inhibitor of Wnt action) exhibit early arrest in tooth development (van Genderen *et al.* 1994, Andl *et al.* 2002). Somatic β-catenin mutations were found in most calcifying odontogenic cysts. Serine/threonine residues substitution of the GSK-3β-dependent phosphorylation sites causes inhibition of β-catenin phosphorylation and results in protein stabilization (Sekine *et al.* 2003). In contrast to calcifying odontogenic cysts, mutations in β-catenin were rare in ameloblastomas (Sekine *et al.* 2003). However, intracytoplasmic and intranuclear β-catenin was found in ameloblastomas by immunohistochemistry (Sekine *et al.* 2003, Kumamoto and Ooya 2005), suggesting epigenetic activation of β-catenin-dependent signal pathways.

Matrical Pituitary Gland Tumors

During the development of the pituitary gland, remnants of the Rathke's pouch may give rise to craniopharyngiomas, tumors with a similar phenotype to calcifying odontogenic cysts and ameloblastomas (Rosai 2004). Moreover, like most calcifying odontogenic cysts, craniopharyngiomas have been found to harbor β-catenin mutations, which resulted in stabilization and intranuclear accumulation of β-catenin (Sekine *et al.* 2002).

Proposal for a Reclassification of Matrical Tumors of Hard Adnexal Structures

We propose the grouping of matrical tumors from hair, nails, teeth and pituitary glands into a single category of neoplasms, regardless of their tissue of origin. This is based on phenotypical similarities and the sharing of β-catenin signaling oncogenic pathways. Most of the matrical tumors from hard tissues and pituitary gland express nuclear β-catenin at least in the more proliferating, less differentiated basaloid cell compartment (Fig. 3). This indicates increased protein stabilization, upregulation of Wnt signals, disassociation from the cadherin cell-cell adhesion complex, and activation of its transcriptional factor functions. There are, however, differences in the mutation rates of β-catenin between different tumors, indicating that nuclear β-catenin translocation in some of these tumors may be the result of upstream signal activation. In the case of onychomatricomas, although the number of cases is very limited, the finding of both nuclear and cell membrane β-catenin and the activation of two different pathways involved in nail development suggest a spectrum of tumors reflecting different stages of development mimicry. More studies are necessary to better define the mechanisms of nail matrix tumor oncogenesis. Similarly, more detailed dissection of β-catenin signaling in matrical tumors of the hair, teeth and pituitary gland can provide helpful targets for the design of better therapeutic strategies.

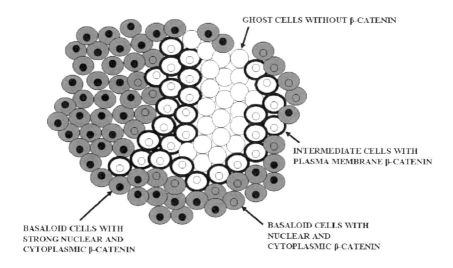

Fig. 3 Schematic representation of the distribution of β-catenin in different cell populations of a matrical tumor. The expression of nuclear β-catenin is either very strong or similar to its cytoplasmic expression in basaloid cells, indicating a signaling transduction and oncogenic activating role. In the more differentiated cells, β-catenin is mostly at the plasma membrane, similar to the distribution of cadherins (not shown), indicating a cell-cell adhesion role. β-catenin is mostly absent in the terminally differentiated anuclear ghost cells.

Abbreviations

Dickkopf1	diffusible inhibitor of Wnt action
HMB45	the clone identifying a monoclonal antibody against melanosomes
IHC	immunohistochemistry
LEF1	lymphocyte enhancing factor 1
Notch1	notch homolog 1, translocation-associated (Drosophila)
RSPO4	R-sponding4
Wnt	combined homologous Wg (wingless) and Int genes

References

Ackerman, A.B. and P.A. DeViragh, and N. Chongchtnant. 1993. Pilomatricoma and Matricoma, pp. 477-506. *In:* Neoplasms with Follicular Differentiation, Lea & Febiger, Philadelphia.

Andl, T. and S.T. Reddy, T. Gaddapara, and S.E. Millar. 2002. WNT signals are required for the initiation of hair follicle development. Dev. Cell 2: 643-653.

Baran, R. and A. Kint. 1992. Onychomatrixoma. Filamentous tufted tumour in the matrix of a funnel-shaped nail: a new entity (report of three cases). Br. J. Dermatol. 126: 510-515.

Bergmann, C. and J. Senderek, D. Anhuf, C.T. Thiel, A.B. Ekici, P. Poblete-Gutierrez, M. van Steensel, D. Seelow, G. Nürnberg, H.H. Schild, P. Nürnberg, A. Reis, J. Frank, and K. Zerres. 2006. Mutations in the gene encoding the Wnt-signaling component R-spondin 4 (RSPO4) cause autosomal recessive anonychia. Am. J. Hum. Genet. 79: 1105-1109.

Behrens, J. and J.P. von Kries, M. Kuhl, L. Bruhn, D. Wedlich, R. Grosschedl, and W. Birchmeier. 1996. Functional interaction of beta-catenin with the transcription factor LEF-1. Nature 382: 638-642.

Burchette, J.L. and T.T. Pham, S.P. Higgins, J.L. Cook, and A. Peralta Soler. 2008. Expression of cadherin/catenin cell-cell adhesion molecules in a onychomatricoma. Int. J. Surg. Pathol. 16: 349-353.

Camacho, F.M. and V.A. Randall, and V.H. Price. 2000. Hair and Its Disorders. Biology, Pathology and Management. MartinDunitz, London.

Chan, E.F. and U. Gat, J.M. McNiff, and E. Fuchs. 1999. A common human skin tumour is caused by activating mutations in beta-catenin. Nature Genet. 21: 410-413.

Gat, U. and R. DasGupta, L. Degenstein, and E. Fuchs. 1998. De novo hair follicle morphogenesis and hair tumors in mice expressing a truncated beta-catenin in skin. Cell 95: 605-614.

Gumbiner, B.M. and P.D. McCrea. 1993. Catenins as mediators of the cytoplasmic functions of cadherins. J. Cell. Sci. Suppl. 17: 155-158.

Hardy, M.H. and U. Vielkind. 1996. Changing patterns of cell adhesion molecules during mouse pelage hair follicle development. 1. Follicle morphogenesis in wild-type mice. Acta Anat. (Basel) 157: 169-182.

Hassanein, A.M. and S.M. Glanz. 2004. Beta-catenin expression in benign and malignant pilomatrix neoplasms. Br. J. Dermatol. 150: 511-516.

Hirai, Y. and A. Nose, S. Kobayashi and M. Takeichi. 1989. Expression and role of E- and P-cadherin adhesion molecules in embryonic histogenesis. II. Skin morphogenesis. Development 105: 271-277.

Jamora, C. and R. DasGupta, P. Kocieniewski, and E. Fuchs. 2003. Links between signal transduction, transcription and adhesion in epithelial bud development. Nature 422: 317-322.

Kemler, R. 1992. Classical cadherins. Semin. Cell. Biol. 3: 149-155.

Ko, C.J. and L. Shi, R.J. Barr, L. Mölne, A. Ternesten-Bratel, and J.T. Headington. 2004. Unguioblastoma and unguioblastic fibroma. An expanded spectrum of onychomatricoma. J. Cutan. Pathol. 31: 307-311.

Kumamoto, H. and K. Ooya. 2005. Immunohistochemical detection of beta-catenin and adenomatous polyposis coli in ameloblastomas. J. Oral. Pathol. Med. 34: 401-406.

Lazar, A.J. and E. Calonje, W. Grayson, A.P. Dei Tos, M.C. Mihm Jr, M. Redston, and P.H. McKee. 2005. Pilomatrix carcinomas contain mutations in CTNNB1, the gene encoding beta-catenin. J. Cutan. Pathol. 32: 148-157.

Lin, M.H. and R. Kopan. 2003. Long-range, nonautonomous effects of activated Notch1 on tissue homeostasis in the nail. Dev. Biol. 263: 343-359.

Muller-Rover, S. and Y. Tokura, P. Welker, F. Furukawa, H. Wakita, M. Takigawa, and R. Paus. 1999. E- and P-cadherin expression during murine hair follicle morphogenesis and cycling. Exp. Dermatol. 8: 237-246.

Neville, B.W. and D.D. Damm, C.M. Allen and J.E. Bouquot. 2002. Odontogenic cysts and tumors, pp. 589-642. *In:* Oral and Maxillofacial Pathology, Saunders, Philadelphia.

Peralta Soler, A. and J.L. Burchette, J.S. Bellet, and J.A. Olson. 2007. Cell adhesion protein expression in melanocytic matricoma. J. Cutan. Pathol. 34: 456-460.

Perrin, C. and R. Baran, A. Pisani, J.P. Ortonne, and J.F. Michiels. 2002. The onychomatricoma: Additional histologic criteria and immunohistochemical study. Am. J. Dermatopathol. 24: 199-203.

Rosai, J. 2004. Rosai and Ackerman's Surgical Pathology. Mosby, Philadelphia.

Sarkar, L. and P.T. Sharpe. 1999. Expression of Wnt signalling pathway genes during tooth development. Mech. Dev. 85: 197-200.

Sekine, S. and S. Sato, T. Takata, Y. Fukuda, T. Ishida, M. Kishino, T. Shibata, Y. Kanai, and S. Hirohashi. 2003. Beta-catenin mutations are frequent in calcifying odontogenic cysts, but rare in ameloblastomas. Am. J. Pathol. 163: 1707-1712.

Sekine, S. and T. Shibata, A. Kokubu, Y. Morishita, M. Noguchi, Y. Nakanishi, M. Sakamoto, and S. Hirohashi. 2002. Craniopharyngiomas of adamantinomatous type harbor beta-catenin gene mutations. Am. J. Pathol. 161: 1997-2001.

Sprecher, E. and R. Bergman, G. Richard, R. Lurie, S. Shalev, D. Petronius, A. Shalata, Y. Anbinder, R. Leibu, I. Perlman, N. Cohen, and R. Szargel. 2001. Hypotrichosis with juvenile macular dystrophy is caused by a mutation in CDH3, encoding P-cadherin. Nat. Genet. 29: 134-136.

Takeichi, M. 1990. Cadherins: A molecular family important in selective cell-cell adhesion. Annu. Rev. Biochem. 59: 237-252.

van Genderen, C. and R.M. Okamura, I. Fariñas, R.G. Quo, T.G. Parslow, L. Bruhn, and R. Grosschedl. 1994. Development of several organs that require inductive epithelial-mesenchymal interactions is impaired in LEF-1-deficient mice. Genes Dev. 8: 2691-2703.

CHAPTER 21

Bioinformatics Analysis to Identify Cell Adhesion Molecules in Cancer

Anguraj Sadanandam[1,2], William J. Gibb[1] and Rakesh K. Singh[3,*]

[1]Life Sciences Division, Ernest O. Lawrence Berkeley National Laboratory,
Berkeley, CA
[3]Department of Pathology and Microbiology, University of Nebraska Medical
Center, Omaha, NE
[2]E-mail: Asadanandam@lbl.gov

ABSTRACT

Cancer is one of the leading causes of death in humans. Dysregulation of cell-cell or cell-extracellular matrix adhesion is the prominent feature of cancer cells during tumor formation, progression and metastasis. As regulators of tumor cell and extracellular matrix interactions, cell adhesion molecules (CAMs) play a significant role in these processes. And in the case of organ-specific metastasis, certain primary tumor cells display frequent metastasis to specific organ of preference through unique interaction between particular CAMs that appear on both the surface of tumor cells and the vascular endothelium of the preferred organs. Less is known, however, about the differential expression of CAMs between various tumor types or between vascular endothelium preferred and non-preferred host tissues of metastasis. Therefore, understanding the molecular diversity of CAMs associated with cell surfaces of different types of tumors or host tissues becomes

---

Corresponding author: Department of Pathology and Microbiology, University of Nebraska, Medical Center, 985845 Nebraska Medical Center, Omaha, NE 68198-5845; E-mail: rsingh@unmc.edu
[2]*Alternative communication/corresponding author*
Key terms are defined at the end of the chapter.

important for the development of diagnostic markers and anticancer therapies. In the post-genomic era, biomarkers can be screened using whole genome approaches using high-throughput technologies and by interpreting the voluminous data generated from these technologies. However, our limited ability to interpret the voluminous data is complemented by bioinformatics approaches. We summarize recently developed bioinformatics methods that are being used to identify and predict the functional CAMs involved in various processes of tumor progression and metastasis.

INTRODUCTION

Tumor formation and metastasis are multistep processes during which malignant cells acquire the potential for unlimited replication, increased proliferation and enhanced angiogenesis leading to tissue invasion and metastasis. They acquire these abilities by becoming sensitive to growth signals, insensitive to anti-growth signals and resistant to apoptosis (Hanahan and Weinberg 2000). In addition to dysregulated proliferation, apoptosis and angiogenesis, an essential process during tumor formation and metastasis is the change in the ability of cancer cells to adhere to one another, to neighboring stromal cells or to the extracellular matrix. This selective process of cell-cell or cell-matrix interactions is typically mediated by high affinity binding of cell adhesion molecules (CAMs) to neighboring cells or the extracellular matrix (Lukas and Dvorak 2004, Stein *et al.* 2005)[*]. The central role of CAMs in a wide variety of processes during tumor progression and metastasis makes their study applicable in a diverse array of tumor types.

Both gene expression and the functional attributes of CAMs differ depending on cell type. Primary tumors with the capability to metastasize express unique CAMs that bind to counterpart CAMs expressed on the surface of endothelial cells present in the target organ's vasculature. Through this mechanism, the interacting CAMs anchor the metastatic cells to the target organ (Sadanandam *et al.* 2007). Recent evidence suggests that CAMs play a major role in the sensitivity of tumor cells to various drugs, and the sensitivity is mediated through suppression of apoptotic pathways in cancer cells (Stein *et al.* 2005). While high-throughput technologies such as phage-display peptide library screening and microarrays allow one to study a wide range of CAMs, bioinformatics techniques are needed to distill large volumes of data so that those CAMs with a functional role in tumor progression and metastasis can be identified. Once identified, the key players can help guide the development of clinical diagnostic tools and anticancer therapies. Here, we review the current high-throughput molecular techniques and the pertinent bioinformatics methods.

---

[*]CAMs are a subset of cell surface molecules. Hence, methods used to identify cell surface molecules were discussed in the context of CAM identification.

APPLICATION OF KNOWLEDGE

Cell-cell and cell-matrix adhesion are critical for various biological processes, and CAMs mediate these processes. During tumor formation, progression and metastasis, the tumor cell CAMs change their ability to bind to their neighboring cells and/or the extracellular matrix. The breaking and forming of bonds between CAMs on the tumor and endothelial cells of the secondary organs is important for the tumor cells to undergo local or distant metastasis. These cells express unique CAMs on their surface for functional tumor progression and metastasis. In contrast, relatively few CAMs have been identified as targets for clinical diagnosis or the treatment of different types of cancers. Therefore, the molecular diversity of CAMs associated with different types of tumor cell surfaces needs to be understood. This can be achieved using high-throughput technologies and analyzing data using bioinformatics approaches. This approach using appropriate models for cancer might provide CAM targets on the surface of tumor cells that could be used as diagnostic and therapeutic targets for tumors.

PHAGE DISPLAY LIBRARY

Organ-specific homing of tumor cells suggests that the vasculature in different tissues is different and indicates that organ microenvironments carry unique CAMs accessible to circulating tumor cells (Pasqualini and Ruoslahti 1996, Pasqualini 1999). In addition, tumor tissues express unique adhesion molecules important for homing and organ-specific colonization of tumor cells (Fidler and Hart 1982, Sadanandam *et al.* 2007). Up until now, identification of tumor- and organ-specific adhesion molecules has progressed slowly, but several recent reports have demonstrated the use of phage display libraries to identify unique CAMs in the tumor and organ microenvironments (Arap *et al.* 1998, Sato *et al.* 2007).

Since its introduction in 1985 (Smith 1985), the phage display library has developed into a very useful technique for studying protein-protein interactions. Specifically, ligand-binding regions are identified using the random peptide library displayed on the surface of the coat proteins of genetically engineered phage (Burritt *et al.* 1996, Cwirla *et al.* 1990). The desired peptides are affinity-selected on the basis of their binding to target molecule by injecting a pool of phages intravenously into mice (Figs. 1 and 2) or alternatively by binding to an immobilized target on a solid surface (Koivunen *et al.* 1994). Multiple rounds of affinity selection (biopanning) enrich the highly specific peptides that bind to their targets. This type of approach can yield tissue-specific homing peptides (Pasqualini and Ruoslahti 1996, Sadanandam *et al.* 2007). The strength of this technology is its ability to identify interactive regions of peptides and other molecules without pre-existing notions about the nature of interactions. In this way, billions of peptides can be effectively surveyed for tight binding to a given protein

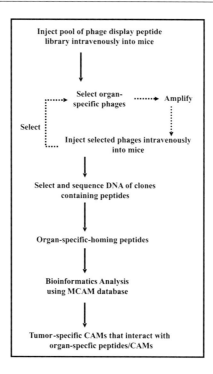

Fig. 1 A representation of *in vivo* phage display peptide library screening and bioinformatics analysis to find tumor-associated CAMs. A pool of phages containing seven amino acid peptides was injected intravenously into mice. Different organs were harvested after 5 min to isolate, purify and count organ-specific phage. Selected phages were reinjected two times to avoid non-specific phage. The DNA of the phage were cloned and sequenced to identify nucleotides corresponding to organ-specific peptides. Bioinformatics analysis of peptide sequences using MCAM database resulted in tumor-specific CAMs that bind to organ-specific peptides or CAMs (Sadanandam *et al.* 2007, 2008a).

using simple microbiological and recombinant DNA procedures. The number of peptides that can be accommodated with this technology far exceeds that which is achievable using conventional expression systems. Phage display peptide libraries not only allow selective exploration of cell surface features of the tumor or organ-specific microenvironments, but also provide ligand-directed therapeutics to the tumor surface (Kolonin *et al.* 2006). To this end, peptides identified using phage display libraries are being used to identify tumor- or organ-specific CAMs by bioinformatics strategies.

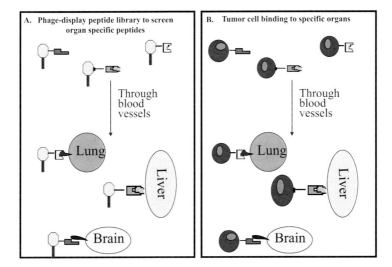

Fig. 2 A schematic representation demonstrating the similarities between organ-specific homing of phages and malignant cells. A. Unique peptides displayed on the surface of phages have affinity for receptors expressed on specific organs and help the phages to home to that organ. B. Similarly, receptors expressed on the surface of malignant cells help them home to organ-specific microenvironment during metastasis.

Bioinformatics and Phage-display Assay

The screened peptide sequences from a phage-display assay can be used to identify CAMs that have residues in their conserved ligand-binding regions. After successful identification of the CAMs expressed on the surface of the tumor cells, they can be targeted using the specific peptides. Bioinformatics tools and databases can help achieve the CAM identification process (Sadanandam *et al.* 2007). At present, phage-display technology based CAM identification can be divided into four steps (Fig. 3): (1) identification of consensus binding motifs using phage-display peptide sequences, (2) screening the tumor-associated proteins that have these peptides in their ligand-binding regions using sequence database searches, (3) understanding the structural significance of the peptide sequence at the ligand-binding regions of tumor-associated CAMs, and (4) screening the gene expression of tumor-associated CAMs to understand their role in cancer. In this section, we review each of these steps in detail.

Consensus-binding Motifs

One can detect phage-displayed peptides as those that occur much more frequently than one would expect by random chance. Several reports have demonstrated that

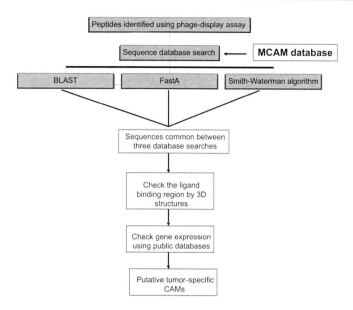

Fig. 3 Bioinformatics strategy used to identify and validate tumor-specific CAMs using *in vivo* phage display peptides. The description is available in the text as well as in Sadanandam *et al.* (2007).

three amino acids are sufficient for two proteins to physically interact with one another (Arap *et al.* 2002, Kolonin *et al.* 2006, Sadanandam *et al.* 2007). Indeed, one can detect tumor-associated proteins by first finding conserved tripeptide motif(s) from peptides identified using single phage-display experiment. With the tripeptide motif(s) in hand, one can then use sequence alignment tools to identify specific CAMs enriched for tripeptide motifs. Because traditional sequence alignment tools such as basic local alignment search tool (BLAST), FASTA, Clustal W and Smith-Waterman algorithms (Arap *et al.* 2002, Sadanandam *et al.* 2007) are not optimized for small peptides, Arap *et al.* applied high-throughput pattern recognition software to identify the consensus-binding motifs by comparing the frequencies of the phage-selected tripeptide motif(s) against motifs from the unselected library (Arap *et al.* 2002).

Another tool, called Receptor Ligand Contacts (RELIC, see Table 1), is a suite of motif-hunting programs that considers different combinations of motif properties and scoring systems. This helps to detect even the weaker sequence conservation among selected peptides. A program has also been developed to measure the sequence bias in individual selected peptides as opposed to groups of peptides (Mandava *et al.* 2004). These measures not only will identify conserved sequence motifs, but also will provide a clear indication whether the affinity selection process was effective in cases where the phage-selected peptides exhibit poor sequence conservation.

Table I Bioinformatics resources for identification of tumor-associated CAMs

Resources	Web address
Receptor Ligand Contacts (RELIC): set of programs to process and manipulate the experimental data of affinity selected peptides	http://relic.bio.anl.gov/programs.aspx
Mammalian Cell Adhesion Molecule database: a consolidated database of mammalian CAMs	http://app1.unmc.edu/mcam/index.cfm
Similar enRiched Parikh Vector Searching Algorithm (SRPVS): new bio-sequence comparison method	http://biigserver.ist.unomaha.edu:8080/x/SRPVS.html
Gene Expression Omnibus (GEO): a gene expression/molecular abundance repository	http://www.ncbi.nlm.nih.gov/geo/
ARRAY EXPRESS: a public repository for transcriptomics data	http://www.ebi.ac.uk/microarray-as/ae/
SAGE Genie: visual displays of human and mouse gene expression	http://cgap.nci.nih.gov/SAGE
CPHmodels 2.0: a homology modeling server	http://www.cbs.dtu.dk/services/CPHmodels/
NCI Molecular Target Database	http://www.dtp.nci.nih.gov

Identifying Tumor-associated CAMs Using Peptides

The final objective of a phage-display peptide library is to identify the functional tumor-associated CAMs that contain the small peptides in their ligand-binding region (Sadanandam *et al.* 2007). However, the identification of the tumor-associated CAMs from the consensus-binding sequences is a challenging task because of short peptide sequences from phage-display assay. In addition, the phage-display peptide sequences could be non-contiguous in the natural proteins. Usually the sequence alignment tools, BLAST, FASTA and Smith-Waterman algorithm, are used to identify the tumor-associated CAMs (Arap *et al.* 2002, Pillutla *et al.* 2001). However, conventional sequence alignment tools will result in many non-CAM protein hits along with CAMs when querying with such small peptide sequences using whole genome or proteome databases. In order to circumvent this problem, we used a CAM-specific database called mammalian cell adhesion molecule (MCAM) database (Sadanandam *et al.* 2008a) developed in our lab to perform conventional sequence database searches, as shown in Fig. 3 (Sadanandam *et al.* 2007, 2008a). This database contains information about adhesion molecules from mouse, human and rat collected mainly from the Gene Ontology and NCBI Entrez-Gene databases and cross-referenced to NCBI Reference Sequence (RefSeq) database and Swiss Institute of Bioinformatics Swiss-Prot Protein Knowledgebase. The MCAM database can be downloaded to set up sequence database searches locally and is freely available online (see Table 1) (Sadanandam *et al.* 2008a). Using this approach, we identified less than 25 DNA or protein hits for each phage-display peptide selected *in vivo*, whereas using non-redundant databases of all proteins resulted in more than 150 hits (Sadanandam *et al.* 2007).

As a robust approach to finding tumor-associated CAMs, we developed a new motif search program called Similar enRiched Parikh Vector Searching (SRPVS, see Table 1) (Huang *et al.* 2005) by assuming that when two sequences are homologous, they share biological structure, irrespective of sequence order. In this method, a protein sequence under analysis is split into a pre-defined word size and its sequence similarity to another protein is detected in a manner that is flexible with respect to local sequence order. The order difference between the two sequences is scored. This method will also analyze proteins with reverse or shuffled order of their ligand-binding sequences (Huang *et al.* 2005). In addition, the RELIC software (described above) applies the bioinformatics summation of multiple binding sequences to produce collective ligand-binding signatures that can be used to search for the tumor-associated proteins (Mandava *et al.* 2004).

Structure and Gene Expression of Tumor-associated Proteins

Once the putative tumor-associated proteins are identified using the preceding step, the next step is to identify the structural significance of the peptide sequences at the ligand-binding regions of tumor-associated CAMs and then to screen the gene expression of tumor-associated CAMs to understand their role in cancer (Fig. 3). To our knowledge, not many phage-display-based bioinformatics analyses have considered these two steps. But it is essential to understand that without these two steps the whole process of genome-based identification of biomarkers is not complete. In order to confirm that the identified phage-display peptides are part of the ligand-binding region, we successfully used the protein modeling server CPHModels (Table 1) to perform three-dimensional molecular modeling based on the known or predicted structures of the tumor-associated CAMs (Fig. 3) (Sadanandam *et al.* 2007). RELIC software also has a set of programs that allows the user to compare peptides to sequence of known structures (Mandava *et al.* 2004).

To analyze gene expression of putative tumor-associated binding proteins in tumor tissues, gene expression databases such as gene expression atlas, Serial Analysis of Gene Expression data repository (SAGEmap) and TisssueInfo (database of tissue-specific gene expression) can be used (Boon *et al.* 2002, Scherf *et al.* 2000) (Fig. 3). The current version of the MCAM 3.0 database is an interactive Web-based database and provides the resources needed to search function and gene expression in normal and tumor tissues for a given gene (Sadanandam *et al.* 2008a). We have performed such analysis and have identified 30 putative CAMs.

A few selected CAMs and their phage-display peptides are shown in Table 2. Among these, Semaphorin 5A (SEMA5A) and Plexin B3 were predicted as tumor markers for pancreatic and prostate cancers, respectively. To test whether these findings might be applicable to humans, we verified the expression of SEMA5A and Plexin B3 in human pancreatic and prostate cancer cell lines, respectively, using RT-PCR analysis (Sadanandam *et al.* 2007).

Table 2 Selected CAMs identified by phage display peptide library and bioinformatics approaches

Phage display peptide	GenBank accession no.	Protein
NAFTPDY	NP_033180	SEMA5A
	NP_062533.2	Plexin B3
	NP_035069.1	NRP2
YQDSANT	NP_034039.1	CXCR2
TPLQPTA	NP_001034239.1	CD44
KSWKVYV	XP_903659.2	SEMA4C

The peptides were identified using *in vivo* phage display peptide library. These peptides were used as queries for bioinformatics approach to identify CAMs that have tumor-specific expression leading to organ-specific metastasis (Sadanandam *et al.* 2007).

Identification of CAMs by Protein-Protein Interactions

Protein-protein interactions between CAMs expressed on the surface of tumor cells and the vasculature of secondary organs play a major role during metastasis. Computational methods can be used to predict interactions between CAMs that are expressed on the surface of primary tumors and secondary organs. We used a unique and integrated approach to identify interacting partner(s) of a CAM, SEMA5A, by beginning with seven CAMs as putative binding partners of SEMA5A (Sadanandam *et al.* 2008b). We chose SEMA5A because we identified it as one of the candidates from the phage display analysis and observed its expression in aggressive pancreatic cancer cell lines (Sadanandam *et al.* 2007). The putative ligand interacting residues are seven (NAFTPDY) amino acids that were screened to bind specifically to endothelia of liver, brain and bone marrow using a phage display peptide library. In keeping with Dwyer and Root-Bernstein/Dillon theories of protein evolution, we chose eight proteins, including SEMA5A, for protein interaction analysis based on similarities in their putative ligand-binding residues. To achieve the goal of identifying SEMA5A interacting partners, we used integrated bioinformatics approaches such as hydrophobic complementarity of protein structure, functional patterns, information on domain-domain interactions, co-expression of genes and protein evolution. Among the set of seven proteins selected as putative SEMA5A interacting partners, we found the functions of Plexin B3 to be associated with SEMA5A. We modeled the structure of the semaphorin domain of Plexin B3 and found that it shares similarity in the ligand-binding region with that of SEMA5A. Furthermore, we observed co-expression of SEMA5A and Plexin B3 in aggressive pancreatic cancer cell lines. Interestingly, phylogenetic analysis has shown that SEMA5A and Plexin B3 co-evolved. We confirmed the interaction of SEMA5A and Plexin B3 in normal tissue samples using co-immunoprecipitation (Sadanandam *et al.* 2008b). This demonstrates that integrated methods can be used at the genome level to discover protein interactions of many unknown CAMs expressed on the surface tumor and secondary tissues using ligand-binding information of CAMs.

GENE EXPRESSION MICROARRAY ANALYSIS AND BIOINFORMATICS

The advent of DNA microarrays revolutionized many aspects of biology by allowing investigators to quantitatively study gene expression on a genome-wide scale. The applications of microarrays are many, but for this review we will be describing microarray technologies that measure gene expression in tissues or cells. There are two primary types of gene expression microarrays, cDNA and oligonucleotides arrays. Both of these arrays work on the principle of hybridization between nucleic acids to measure the mRNA abundance of tissues or cells. In the case of cDNA arrays, polynucleotides corresponding to 3′ end of RNA transcripts are immobilized on a solid support, whereas the probes on oligonucleotide arrays are built up one base at time using, for example, photo-masking technologies. Although experimental objectives and resources typically determine the most suitable platform, some investigators have suggested that the choice of platform can significantly influence experimental outcomes. In response to this criticism, more recent comparisons of microarray technologies have found good cross-platform agreement, but in order to achieve this level of consistency, experimental and analytical protocols must be carefully observed (Kuo *et al.* 2006). All of the microarray platforms have at least one thing in common: they generate large volumes of data that are best analyzed using appropriate statistical and bioinformatics methods.

Gene expression microarrays can be used to identify CAMs responsible for organ-specific metastasis of tumor cells (Kakiuchi *et al.* 2003). In this vein, Kakiuchi *et al.* used a metastatic model by injecting SBC-5 lung cancer cell line intravenously into mice lacking tumor-associated natural killer cells. By performing permutation tests of the metastatic foci of the lung cancer cells developed in lung, liver, kidney and bone, they identified 435 genes that are differentially expressed between different organ-specific metastatic foci. They observed that the differentially expressed genes were enriched for CAMs. In particular, they observed that lectin family of proteins such as LGALS1 in pulmonary metastases and LGALS9 in renal metastases are highly expressed. Lectin, a family of β-galactoside-binding proteins associated in modulating cell-cell and cell-matrix interactions, may play an important role in organ-specific metastasis. They also observed multiple cell adhesion molecules that were differentially regulated between micrometastases and macrometastases (Kakiuchi *et al.* 2003).

IDENTIFICATION OF NON-LINEAGE-SPECIFIC TUMOR CAMs

High-throughput techniques and bioinformatics application is not limited to the identification of markers specific to one type of tumor tissue (lineage-specific), for example, CAMs specific to breast or lung cancer. Kolonin *et al.* provide evidence that phage display peptide library and bioinformatics analysis can be used to select common peptides against the vasculature of multiple tumor types (Kolonin *et al.* 2006). The US National Cancer Institute (NCI) established a panel of 60 cell lines (NCI60) from multiple tumors in the late 1980s with an objective to develop a model system for *in vitro* drug discovery (Shoemaker 2006). This cell line panel is being widely used as a rich source of information to understand the mechanisms of action of various existing and experimental drugs (Shoemaker 2006). Kolonin *et al.* used the NCI60 panel to screen a spectrum of peptides using a phage display library that targeted the surface of NCI60 cells. They profiled an enriched spectrum of peptides that preferentially bind to non-integrin cell surface molecules expressed in NCI60 cell lines by selecting against an excess of a competing Arg-Gly-Asp (RGD) synthetic integrin-binding peptide. Since tripeptide motifs were shown to capture the protein-peptide interaction in the phage display library, the authors surveyed the spectrum of tripeptides that resulted from the phage-display screen using statistical and bioinformatics approaches to compare the differential binding of phage-displayed peptides across the NCI60 panel. They found a unique set of tripeptides with differential binding affinity across the NCI60 cell lines (Kolonin *et al.* 2006).

In order to infer the variation in peptide-binding specificity across the NCI60 cell lines, Kolonin *et al.* performed hierarchical cluster analysis of relative tripeptide frequencies found among 7-mer peptides binding to each cell. Thus, cell lines within a given cluster share similar peptide frequencies. By associating these frequencies with the distribution of tripeptides among the NCI-60 cell lines, the authors suggested the existence of shared targeted surface receptor(s) that are common to multiple tumor types. Later, they successfully predicted the lead targeted receptors common to multiple tumor types correlating frequencies of the peptides and the protein expression patterns available from NCI Molecular Target Database (Table 1) for the NCI60 cell lines. Thus, their approach presents a simple way to search drug-accessible tumor cell surface receptors (including CAMs) and to find peptide ligands that can serve as ligands to target the receptors (Kolonin *et al.* 2006).

Along similar lines, a study by Stull *et al.* performed gene expression microarray and bioinformatics analysis using xenograft tumor samples generated by injecting human cancer cell lines derived from different organs to identify common cell surface molecules (including CAMs) as therapeutic targets (Stull *et al.* 2005). Their analytical approaches included principal component analysis (PCA) and multivariate analysis of variance (MANOVA) on all tumor data and on individual tumors. Using PCA, it is possible to identify genes that account for the majority of the gene expression variance across tumor types. In contrast, MANOVA can be used to identify combinations of genes that are most differentially expressed between tumor types. By isolating genes that contribute both to high variance across tumor types (using PCA) and differential expression between tumor types (using MANOVA), the authors identified 12 informative genes (Stull *et al.* 2005).

IDENTIFICATION OF MULTIPLE TUMOR-SPECIFIC CAMs AS TARGETS FOR MULTIMERIC LIGANDS

The methods discussed earlier were focused on identifying one or more CAMs involved in tumor progression and metastasis. However, they were all focused on targeting single tumor- or organ-specific CAMs using monomeric ligands such as phage-display peptides or antibodies. It is well known that tumors are heterogeneous. It is difficult to find an individual CAM expressed in every cell of an individual patient tumor or all the patients suffering from same tumor. To cast a wider net for CAMS with therapeutic potential, Balagurunathan *et al.* (2008) searched for CAMS that could be targets of multimeric ligands by comparing 28 different unique tissues/cell types to pancreatic tumors. They chose to study cell surface molecules by restricting their analysis to cell surface molecules present on the Agilent Human 1A array. As a way of focusing on post-translational targets, the authors combined stringent gene expression criteria with tissue microarrays at the protein level. To identify potentially high avidity, tumor-specific CAMs, their algorithm screens for combinations of receptors that appear together in tumors but are either absent or only partially represented in normal tissues. Based on binding avidity arguments, they estimated that drug target sites with therapeutic potential could be identified as those heteromultimeric binding sites that express at least two additional CAM receptors in pancreatic tumor tissue relative to normal tissue. Furthermore, they demonstrated by tissue microarrays that in their study the target combinations of cell surface molecules covered 82% of patient samples. Hence they demonstrate that multiple CAM targets can be identified using gene expression and bioinformatics and, further, suggest that multiple combinations of CAM targets will be needed to treat the broadest possible spectrum of pancreatic cancer patients (Balagurunathan *et al.* 2008).

IDENTIFICATION OF TUMOR-SPECIFIC CAMs RESPONSIBLE FOR DRUG RESISTANCE

Different tumor types can exhibit differential responses to a given drug, irrespective of their tissue of origin, partly because of differential expression of CAMs (Stein *et al.* 2004, 2005). Stein *et al.* used an approach called 'intractability measurement' that quantitatively defines how tumors from different tissues are differentially sensitive to currently used chemotherapies. They used a bioinformatics approach to illustrate how CAMs are involved in tumor intractability. To measure the intractability, they used survival data for different types of tumors from the Survival Epidemiology and End Results (SEER) project (Ries *et al.* 1983). Treatment success was gauged by response rates of different tumor types to various drugs as surveyed from literature (Stein *et al.* 2004, 2005). Their analysis showed pancreas, liver, lung and colon as the most intractable cancers, breast, ovary and prostate as intermediately intractable cancers, and testis as the least intractable tumor. Based on the evidence, they performed bioinformatics analysis using serial analysis of gene expression (SAGE) databases and different tumor types to identify molecules that could predict the intractability of tumors from different tissues. They found numerous genes that were either overexpressed or underexpressed in intractable tumors compared to tractable tumors. Later, for each tissue, they performed correlation analysis of each gene to the SEER 5-year 'distant tumors' survival numbers (Stein *et al.* 2004). They performed similar analysis using cDNA gene expression microarray data and SEER survival data. Based on the analysis, they found that most of the genes that correlate negatively with survival in intractable tumors were CAMs and cytoskeletal genes (Stein *et al.* 2005). The survival outcome and intractability measures from this study suggest that CAMs are responsible for drug resistance found in poor survival tumors, irrespective of their tissue of origin.

CONCLUSION

The identification of novel CAMs involved in tumor progression and metastasis becomes important because of their role in various processes of malignancy. Identification of tumor-associated CAMs by experimental biology techniques such as RT-PCR and chromatography is time consuming and low-throughput. However, the search for CAMs involved in various processes including tumorigenicity, metastasis and organ-specific homing can be abetted by bioinformatics analysis of high-throughput data generated from various models of cancer. Nevertheless, one must bear in mind that the results obtained from the high-throughput techniques and bioinformatics analysis are limited to the tumor models used. Moreover, predictions made by bioinformatic techniques need to be validated using proper experimental techniques. Because each high-throughput technique has its own limitations, we suggest an integrated bioinformatics approach using different sources of high-throughput data to identify CAMs with high therapeutic or diagnostic potential. A few of the bioinformatics resources related to phage-display assays and microarrays are provided in Table 1. Taken together, high-throughput techniques and bioinformatics approaches to identify tumor-associated CAMs could help guide the development of new drugs against different types and stages of cancer.

Abbreviations

BLAST	basic local alignment search tool
CAM	cell adhesion molecules
MANOVA	multivariate analysis of variance
MCAM	mammalian cell adhesion molecule
NCBI	National Center for Biological Information
NCI	National Cancer Institute
PCA	principal component analysis
RefSeq	Reference Sequence database
RELIC	Receptor Ligand Contacts
RGD	Arg-Gly-Asp peptides
SAGE	serial analysis of gene expression
SAGEmap	Serial Analysis of Gene Expression data repository
SEER	Survival Epidemiology and End Results
SEMA 5A	Semaphorin 5A
SRPVS	Similar enRiched Parikh Vector Searching

Key Facts

1. Cancer is one of the leading causes of death in humans.
2. Because most cancer deaths occur as a result of metastasis, developing therapeutic options to inhibit metastasis would likely have major clinical benefits.
3. Inasmuch as cell adhesion molecules (CAMs) are involved in the release and trafficking of tumor cells, it is important to elucidate the diverse patterns by which CAMs are displayed on the surface of tumor cells and/ or host tissues, and to understand how those patterns guide tumor cells from origin to destination.
4. Bioinformatics methods can be used to refine CAMs that play major roles during various steps of tumor progression and metastasis.

Definition of Terms

Binding motif: A short conserved region in a DNA or protein sequence that acts as a binding substrate.

Bioinformatics: The application of statistical, mathematical, and computer methods to analyze and interpret voluminous data generated from various fields of biology.

Ligand-binding region: A ligand-binding region of a protein is a region that contains amino acids that help the protein to physically interact with its binding partner.

Metastasis: The process by which tumor cells segregate from their place of origin and migrate to a different place in the same organ or different organ through blood or lymphatic vessels.

Microarray: A technique used to quantitatively measure the expression of genes in cells or tissues. The technique uses certain nucleic acid sequences as probes that are immobilized on a solid support. These probes bind specific fluorescence-labeled mRNA expressed in cells or tissues. After binding, the abundance of bound mRNA is measured as the intensity of the fluorescence and interpreted as a measure of gene expression.

Organ-specific metastasis: A process of metastasis whereby cell adhesion molecules cause migrating tumor cells to preferentially bind to the cell surfaces of specific organs.

Phage display peptide library: A method to study protein-protein and protein-peptide interactions using bacteriophages that are genetically engineered to display the entire protein or peptides to proteins of interest.

References

Arap, W. and R. Pasqualini, and E. Ruoslahti. 1998. Chemotherapy targeted to tumor vasculature. Curr. Opin. Oncol. 10(6): 560-565.

Arap, W. and M.G. Kolonin, M. Trepel, J. Lahdenranta, M. Cardo-Vila, R.J. Giordano, P.J. Mintz, P.U. Ardelt, V.J. Yao, C.I. Vidal, L. Chen, A. Flamm, H. Valtanen, L.M. Weavind, M.E. Hicks, R.E. Pollock, G.H. Botz, C.D. Bucana, E. Koivunen, D. Cahill, P. Troncoso, K.A. Baggerly, R.D. Pentz, K.A. Do, C.J. Logothetis, and R. Pasqualini. 2002. Steps toward mapping the human vasculature by phage display. Nat. Med. 8(2): 121-127.

Balagurunathan, Y. and D.L. Morse, G. Hostetter, V. Shanmugam, P. Stafford, S. Shack, J. Pearson, M. Trissal, M.J. Demeure, D.D. Von Hoff, V.J. Hruby, R.J. Gillies, and H. Han. 2008. Gene expression profiling-based identification of cell-surface targets for developing multimeric ligands in pancreatic cancer. Mol. Cancer. Ther. 7(9): 3071-3080.

Boon, K. and E.C. Osorio, S.F. Greenhut, C.F. Schaefer, J. Shoemaker, K. Polyak, P.J. Morin, K.H. Buetow, R.L. Strausberg, S.J. De Souza, and G.J. Riggins. 2002. An anatomy of normal and malignant gene expression. Proc. Natl. Acad. Sci. U.S.A. 99(17): 11287-11292.

Burritt, J.B. and C.W. Bond, K.W. Doss, and A.J. Jesaitis. 1996. Filamentous phage display of oligopeptide libraries. Anal. Biochem. 238(1): 1-13.

Cwirla, S.E. and E.A. Peters, R.W. Barrett, and W.J. Dower. 1990. Peptides on phage: a vast library of peptides for identifying ligands. Proc. Natl. Acad. Sci. U.S.A. 87(16): 6378-6382.

Fidler, I.J. and I.R. Hart. 1982. Biological diversity in metastatic neoplasms: origins and implications. Science 217(4564): 998-1003.

Hanahan, D. and R.A. Weinberg. 2000. The hallmarks of cancer. Cell 100(1): 57-70.

Huang, H. and H. Ali, A. Sadanandam, and R.K. Singh. 2005. Protein Motif Searching Through Similar Enriched Parikh Vector Identification. Proceedings of the Fifth IEEE Symposium on Bioinformatics and Bioengineering. pp. 285-289.

Kakiuchi, S. and Y. Daigo, T. Tsunoda, S. Yano, S. Sone, and Y. Nakamura. 2003. Genome-wide analysis of organ-preferential metastasis of human small cell lung cancer in mice. Mol. Cancer Res. 1(7): 485-499.

Koivunen, E. and B. Wang, and E. Ruoslahti. 1994. Isolation of a highly specific ligand for the alpha 5 beta 1 integrin from a phage display library. J. Cell Biol. 124(3): 373-380.

Kolonin, M.G. and L. Bover, J. Sun, A.J. Zurita, K.A. Do, J. Lahdenranta, M. Cardo-Vila, R.J. Giordano, D.E. Jaalouk, M.G. Ozawa, C.A. Moya, G.R. Souza, F.I. Staquicini, A. Kunyiasu, D.A. Scudiero, S.L. Holbeck, E.A. Sausville, W. Arap, and R. Pasqualini. 2006. Ligand-directed surface profiling of human cancer cells with combinatorial peptide libraries. Cancer Res. 66(1): 34-40.

Kuo, W.P. and F. Liu, J. Trimarchi, C. Punzo, M. Lombardi, J. Sarang, M.E. Whipple, M. Maysuria, K. Serikawa, S.Y. Lee, D. McCrann, J. Kang, J.R. Shearstone, J. Burke, D.J. Park, X. Wang, T.L. Rector, P. Ricciardi-Castagnoli, S. Perrin, S. Choi, R. Bumgarner, J.H. Kim, G.F. Short, 3rd, M.W. Freeman, B. Seed, R. Jensen, G.M. Church, E. Hovig, C.L. Cepko, P. Park, L. Ohno-Machado, and T.K. Jenssen. 2006. A sequence-oriented comparison of gene expression measurements across different hybridization-based technologies. Nat. Biotechnol. 24(7): 832-840.

Lukas, Z. and K. Dvorak. 2004. Adhesion molecules in biology and oncology. Veterinaria Brno. 73(1): 93-104.

Mandava, S. and L. Makowski, S. Devarapalli, J. Uzubell, and D.J. Rodi. 2004. RELIC—a bioinformatics server for combinatorial peptide analysis and identification of protein-ligand interaction sites. Proteomics 4(5): 1439-1460.

Pasqualini, R. and E. Ruoslahti. 1996. Organ targeting in vivo using phage display peptide libraries. Nature 380(6572): 364-366.

Pasqualini, R. 1999. Vascular targeting with phage peptide libraries. Q. J. Nucl. Med. 43(2): 159-162.

Pillutla, R.C. and K. Hsiao, R. Brissette, P.S. Eder, T. Giordano, P.W. Fletcher, M. Lennick, A.J. Blume, and N.I. Goldstein. 2001. A surrogate-based approach for post-genomic partner identification. BMC Biotechnol. 16.

Ries, L.G. and E.S. Pollack, and J.L. Young, Jr. 1983. Cancer patient survival: Surveillance, Epidemiology, and End Results Program, 1973-79. J. Natl. Cancer Inst. 70(4): 693-707.

Sadanandam, A. and M.L. Varney, L. Kinarsky, H. Ali, R.L. Mosley, and R.K. Singh. 2007. Identification of functional cell adhesion molecules with a potential role in metastasis by a combination of in vivo phage display and in silico analysis. Omics 11(1): 41-57.

Sadanandam, A. and S.N. Pal, J. Ziskovsky, P. Hegde, and R.K. Singh. 2008a. MCAM: A Database to Accelerate the Identification of Functional Cell Adhesion Molecules. Cancer Informatics 6(47-50): 47.

Sadanandam, A. and M. Varney, and R. Singh. 2008b. Identification of Semaphorin 5A interacting protein by applying *a priori* knowledge and peptide complementarity related to protein evolution and structure. Geno. Prot. Bioinfo. 6(3-4): 1-12.

Sato, M. and W. Arap, and R. Pasqualini. 2007. Molecular targets on blood vessels for cancer therapies in clinical trials. Oncology (Williston Park). 21(11): 1346-1352.

Scherf, U. and D.T. Ross, M. Waltham, L.H. Smith, J.K. Lee, L. Tanabe, K.W. Kohn, W.C. Reinhold, T.G. Myers, D.T. Andrews, D.A. Scudiero, M.B. Eisen, E.A. Sausville, Y. Pommier, D. Botstein, P.O. Brown, and J.N. Weinstein. 2000. A gene expression database for the molecular pharmacology of cancer. Nat. Genet. 24(3): 236-244.

Shoemaker, R.H. 2006. The NCI60 human tumour cell line anticancer drug screen. Nat. Rev. Cancer 6(10): 813-823.

Smith, G.P. 1985. Filamentous fusion phage: novel expression vectors that display cloned antigens on the virion surface. Science 228(4705): 1315-1317.

Stein, W.D. and T. Litman, T. Fojo, and S.E. Bates. 2004. A Serial Analysis of Gene Expression (SAGE) database analysis of chemosensitivity: comparing solid tumors with cell lines and comparing solid tumors from different tissue origins. Cancer Res. 64(8): 2805-2816.

Stein, W.D. and T. Litman, T. Fojo, and S.E. Bates. 2005. Differential expression of cell adhesion genes: implications for drug resistance. Int. J. Cancer 113(6): 861-865.

Stull, R.A. and R. Tavassoli, S. Kennedy, S. Osborn, R. Harte, Y. Lu, C. Napier, A. Abo, and D.J. Chin. 2005. Expression analysis of secreted and cell surface genes of five transformed human cell lines and derivative xenograft tumors. BMC Genomics 6(1): 55.

CHAPTER 22

Endothelial Cell Adhesion Molecules and Cancer Progression

Kimberly C. Boelte[1], Hanako Kobayashi[2] and
P. Charles Lin[1, 2, 3, 4,*]

Department of Cancer Biology[1], Department of Radiation Oncology[2],
Department of Cell and Developmental Biology[3], Vanderbilt-Ingram Cancer
Center[4], Vanderbilt University Medical Center, Nashville, TN, 37232, USA
Kimberly C. Boelte, 771 PRB, Cancer Biology, Nashville, TN 37232,
E-mail: kimberly.c.boelte@vanderbilt. edu
Hanako Kobayashi, 315 PRB, 2220 Pierce Avenue, Vanderbilt University Medical
Center, Nashville, TN 37232, E-mail: hanako.kobayashi@vanderbilt. edu
P. Charles Lin, Ph.D, 315 PRB, 2220 Pierce Avenue, Vanderbilt University
Medical Center, Nashville, TN 37232, E-mail: charles.lin@vanderbilt. edu

ABSTRACT

Endothelial cell adhesion molecules play an integral role in adhesion and
infiltration of leukocytes into tissues, which is important to immune function.
They also play an important role in cancer progression. Cancer progression
involves many processes, beginning with proliferation of the tumor cells at the
primary site. Leukocytes are recruited into the primary tumors by tumor-secreted
factors, mediating pro- and anti-tumor responses, as well as secreting factors
themselves that lead to more recruitment. Leukocytes are also known to promote
vascularization of the tumor, another key step in cancer progression. Later in
cancer progression, the tumor cells intravasate and circulate, eventually adhering
to the endothelium at distant organ sites, where they can extravasate and form

---

*Corresponding author
Key terms are defined at the end of the chapter.

metastases. Tumor cells are known to express adhesion molecules similar to those found on leukocytes, expressing molecules containing sialyl Lewis determinants, implicating the reciprocal endothelial cell adhesion molecules, E-selectin and P-selectin, in this process. Indeed, E-selectin and P-selectin levels are increased in cancer patients. The endothelial cell adhesion molecules intercellular adhesion molecule 1 (ICAM-1) and vascular cell adhesion molecule 1 (VCAM-1) are found to be upregulated in cancer patients, and tumor cells express many integrins, the ligands for these adhesion molecules. Clinical and research data indicate that endothelial cell adhesion molecules are major players in metastasis. This chapter will outline the roles that endothelial cell adhesion molecules play in cancer progression.

INTRODUCTION

Cancer progression involves many different processes and is regulated on many different levels (Gassmann and Haier 2008, Brooks *et al.* 2009). At the primary site, this involves tumor initiation and growth. Tissue reorganization occurs, mediated by processes such as angiogenesis and immune cell infiltration. As the tumor progresses, the tumor cells will invade the surrounding tissues and eventually intravasate into the circulation. From there, these cells can arrest/adhere at distant sites, extravasate, and form secondary tumors, or metastases (Fig. 1, see Key Features). This process is very inefficient, with the majority of tumor cells dying before a metastatic growth can be established.

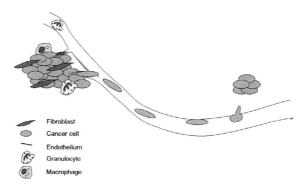

Fibroblast
Cancer cell
Endothelium
Granulocyte
Macrophage

Fig. 1 Scheme of cancer progression. At the primary site, the tumor cells proliferate and secrete factors that lead to leukocyte infiltration. As the tumor progresses, cells intravasate, either as individual cells or a cluster of cells, and enter the circulation. At a distant site, the tumor cell(s) adheres to the endothelium and extravasates, migrating into the tissue, where it can now proliferate, creating a metastatic tumor.

Tumor progression is intimately linked with host response, being mediated by interactions of various cell types in the stroma. As such, endothelial cells play an important role in cancer progression. As with other inflammatory processes, infiltrating cells must first cross endothelial cells to get to the tumor. Leukocytes are recruited to the primary tumor site and extravasate to the tissue using a rolling mechanism (Fig. 2). The circulating leukocytes first interact loosely with selectins and start rolling along the endothelium. As the cells roll, firm interaction occurs through cell adhesion molecules, and subsequently, transendothelial migration ensues. Similarly, tumor cells must intravasate from the primary site and then cross endothelial cells at distant sites to create metastases, or secondary tumor growths. Though the extravasation of tumor cells is not as well characterized as that of leukocytes, the mechanisms seem to have some characteristics in common (Fig. 3A). Some studies indicate that tumor cells do not roll like leukocytes, but tumor cells express adhesion molecules similar to leukocytes, leading to the hypothesis that these cells interact directly with the endothelium (Fig. 3B) (Gassmann and Haier 2008). However, there are some studies indicating that tumor cells are clustered with leukocytes in circulation, and that the leukocytes mediate extravasation through their own cell adhesion molecules (Gassmann and Haier 2008) (Fig. 3C). Other studies suggest that tumor cell interactions with platelets mediate adhesion (Gassmann and Haier 2008).

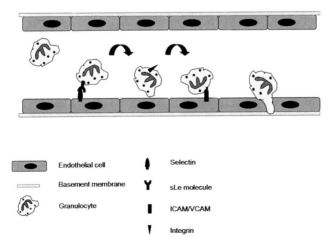

Fig. 2 Leukocyte adhesion to endothelium. Leukocytes adhere to endothelium through a mechanism described as rolling adhesion. As the leukocyte moves with blood flow, mucin molecules on its surface interact with E- or P-selectin on the endothelium, allowing loose adhesion. This interaction leads to cell rolling along the endothelium, exposing integrins on the leukocyte to cell adhesion molecules, such as ICAM and VCAM, on the endothelium, mediating strong adhesion. The leukocyte is then able to undergo transendothelial migration and move into the tissue.

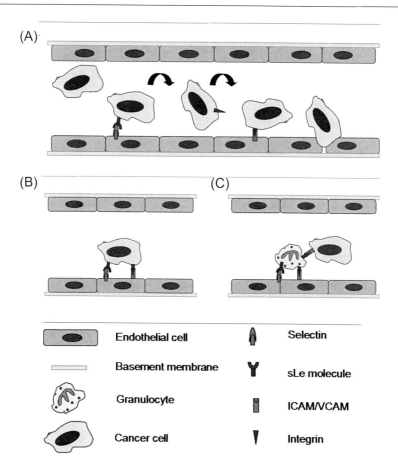

Fig. 3 Tumor cell adhesion to endothelium. (A) Studies indicate that tumor cells express adhesion molecules similar to those found on leukocytes, suggesting that tumor cells may adhere by means of the rolling mechanism. (B) The tumor cells may directly interact with endothelial cells, or (C) they may interact indirectly by adhering to leukocytes, which then adhere to endothelial cells.

This chapter will focus on the role of endothelial adhesion molecules in cancer progression. We will discuss the role of endothelial cell adhesion molecules in the host response to the primary tumor. Infiltrating leukocytes have been shown in many studies to function as both pro- and anti-tumor, indicating that they have a major role in tumor progression. Infiltrating leukocytes have also been shown to be pro-metastatic through production of proteinases and cytokines, mediating intravasation of tumor cells, or clusters of tumor cells and leukocytes. We will then discuss the role of endothelial cell adhesion molecules in extravasation of tumor cells at secondary sites. Lastly, we will focus on specific endothelial cell adhesion molecules, highlighting clinical data and findings from *in vivo* and *in vitro* experiments.

ROLE AT THE PRIMARY TUMOR SITE

Tumor cells secrete many inflammatory factors, which lead to increased leukocyte recruitment. Cancer progression is affected by the infiltrating immune cells. Over the years, it has become clear that immune cells can function in both anti- and pro-tumorigenic roles. Myeloid-derived suppressor cells and T regulatory cells, through their ability to suppress immune function, augment tumor progression. Infiltrating myeloid cells can secrete factors that enhance angiogenesis and promote tumor growth. In contrast, some infiltrating immune cells kill the tumor cells, acting in an anti-tumorigenic fashion. The endothelium in and around the tumor is important in the recruitment process.

As with other inflammatory reactions, inflammatory factors secreted by the tumor, such as interleukin 1 beta (IL-1β) and tumor necrosis factor alpha (TNF-α), increase adhesion molecule expression on the endothelium leading to an activated state (Tedder *et al.* 1995, Goldsby *et al.* 2003, Castermans and Griffioen 2007, Dittmer *et al.* 2008, Brooks *et al.* 2009). Circulating leukocytes adhere to this activated endothelium and, using the rolling mechanism described above, move into the tumor (Castermans and Griffioen 2007). As leukocyte-endothelial interactions are described in detail in a previous chapter, our main focus will be the role of endothelial cell adhesion molecules in metastasis.

ROLE IN METASTASIS

Many cancers have been shown to metastasize in a manner specific to tumor type and location. For example, breast cancer most often metastasizes to the bone. There are many theories on what drives the locations of metastases, such as 'seed and soil' or mechanical forces (Dittmar *et al.* 2008, Gassmann and Haier 2008, Brooks *et al.* 2009). It is likely that a combination of the two is occurring. Some *in vivo* studies show tumor cells in the circulation occluding capillaries, while others show tumor cells adhering without blocking blood flow (Gassmann and Haier 2008). Chemokine axes are known to be important in homing of tumor cells to metastatic locations (Kucia *et al.* 2005, Dittmar *et al.* 2008, Gassmann and Haier 2008). Many tumor cells, such as breast cancer cells, express CXCR4 and therefore migrate to areas where the ligand, SDF-1, is expressed, such as bone (Kucia *et al.* 2005). Once at the site, the CXCR4-expressing tumor cells can adhere through interactions of cell adhesion molecules and transmigrate through the endothelium into the tissue. As tumor cells circulate, they have to travel through small capillaries, are forced to change shapes to fit through, and, at times, become lodged in the capillary. This puts the surface of the tumor cell in close contact with the surface of the endothelium, increasing the likelihood that the molecules on the cell will interact with adhesion molecules of the endothelium (Gassmann and Haier 2008). Regardless of the mechanism involved in localization of tumor cells to specific sites, endothelial cell adhesion molecules play an important role.

ENDOTHELIAL CELL ADHESION MOLECULES IN CANCER PROGRESSION

There are several molecules involved in adhesion to endothelial cells. The first interactions in the rolling mechanism involve selectins. Selectins are a family of glycoproteins, each containing a lectin domain for ligand interactions (Fig. 4). Among the family members, E-selectin is expressed on endothelial cells, P-selectin is expressed on endothelial cells and platelets, and L-selectin is expressed on leukocytes. The selectin molecules each contain an amino terminal lectin domain, an epidermal growth factor (EGF)-like domain, and several short consensus repeat (SCR) domains in the extracellular region (Tedder *et al.* 1995). The selectin family members recognize carbohydrate moieties, binding to oligosaccharides and glycoproteins, often called mucins (Tedder *et al.* 1995). For example, L-selectin can interact with the glycoprotein CD34 expressed by endothelial cells. As leukocyte-endothelial interactions are discussed elsewhere in this text, we will focus mainly on E-selectin and P-selectin. Stronger adhesion to the endothelium is mediated

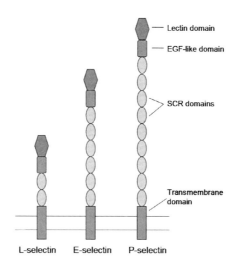

Fig. 4 The structures of the selectin family. The three members of the selectin family are L-selectin, E-selectin, and P-selectin. L-selectin is expressed on leukocytes, E-selectin is expressed on endothelial cells, and P-selectin is expressed on endothelial cells and platelets. Each contains in the extracellular region an amino terminal lectin domain, an epidermal growth factor (EGF)-like domain, and several short consensus repeat (SCR) domains.

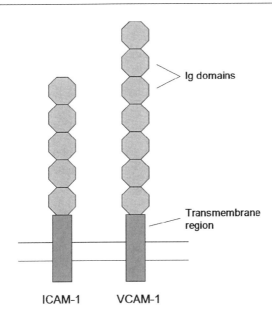

Fig. 5 The structures of ICAM-1 and VCAM-1. ICAM-1 and VCAM-1 are members of the immunoglobulin (Ig) gene superfamily. As such, they contain several extracellular Ig domains.

by adhesion molecules of the Immunoglobulin (Ig) gene superfamily. Members of the Ig gene superfamily contain several Ig-like domains and, within this family, vascular cell adhesion molecule 1 (VCAM-1) and intercellular adhesion molecule 1 (ICAM-1) are expressed on activated endothelium (Fig. 5) (Kobayashi *et al.* 2007, Brooks *et al.* 2009). There are other members of this family, such as platelet endothelial cell adhesion molecule 1 (PECAM-1), but there are few studies as to their roles in cancer progression. The ligands for these molecules are integrins expressed by leukocytes, and also by cancer cells (Dittmer *et al.* 2008). There are studies indicating that cancer cells express adhesion molecules similar to those on endothelium, indicating that the interactions discussed could occur in the opposite orientation, i.e., ICAM-1 on tumor cells interacting with integrins on endothelial cells (Kobayashi *et al.* 2007). However, this has not been well characterized. More likely, expression of these molecules facilitates adhesion of tumor cells to other tumor cells or to leukocytes, allowing the cells to migrate in a group.

E-Selectin in Cancer Progression

E-selectin is expressed mainly by endothelial cells and plays a major role in leukocyte rolling and adhesion. The molecules that act as ligands for E-selectin, as well as P-selectin, belong to a group of heavily glycosylated molecules often referred to as mucins. These mucins often contain carbohydrate moieties called Lewis determinants. These moieties can be sialofucosylated and are then referred to as sialyl Lewis determinants, denoted as sLe. The ones most commonly studied in cancer progression are sLe^a and sLe^x, which differ in their fucosylation pattern (Kobayashi *et al.* 2007, Dittmer *et al.* 2008). Various studies have implicated E-selectin in tumor cell metastasis, indicating a role for this molecule in tumor cell adhesion (Kobayashi *et al.* 2007).

Clinical evidence points to a major role for E-selectin interactions in cancer progression. Increased levels of E-selectin have been reported in malignant breast and colon cancer versus normal and benign tissue, suggesting that, as these tumors progress, they express more of this molecule (Table 1) (Kobayashi *et al.* 2007). In addition, the soluble form of E-selectin, sE-selectin, is present at higher levels in serum of patients with several types of cancer compared to healthy individuals (Table 1) (Kobayashi *et al.* 2007). Several studies have indicated that levels of serum sE-selectin correlate with cancer progression. Sialyl Lewis determinant expression has also been examined in cancer patients. As with sE-selectin, serum levels of sLe^x increase with metastases. Levels of sLe^x in tissue follow a similar pattern, with upregulated expression in colon cancer versus normal tissue, and even higher expression in metastases (Kobayashi *et al.* 2007). In gastric cancer, sLe^a levels also correlate with tumor progression (Kobayashi *et al.* 2007). These studies together suggest that levels of these molecules may aid clinicians in determining the prognosis for a patient.

Table I   Levels of E-selectin and P-selectin detected in cancer patient samples

	E-selectin	*sE-selectin*	*P-selectin*	*sP-selectin*
Breast	↑	↑	↑	↑
Colon	↑		↓	↑
Lung		↑		↑
Gastrointestinal		↑	↑	
Ovary		↑		
Bladder				↑
Lymphoma				↑
Myeloma		↑		

Summary from the literature of levels of E-selectin and P-selectin in samples from cancer patients compared to normal tissues. ↑, increased in samples, ↓, decreased in samples.

The involvement of E-selectin and its ligands in cancer progression have also been examined with *in vivo* and *in vitro* studies. In animal models, the correlations observed clinically between levels of sLe^a and sLe^x and metastasis seem to be upheld, but further study is warranted (Kobayashi *et al.* 2007). Chemical inhibitors and blocking antibodies have been used to study the role of E-selectin interactions in metastasis. Macrosphelide B, an inhibitor of sLe^x, was able to inhibit adhesion, and subsequent metastasis, of cancer cells expressing sLe^x, while not affecting cells that expressed low levels of sLe^x (Fukami *et al.* 2002). Cimetidine, a drug that can act as a ligand for sialyl Lewis determinants, was able to decrease adhesion of a colon cancer cell line to endothelial cells *in vitro* (Kobayashi *et al.* 2000). Blocking, or neutralizing, antibodies targeted against E-selectin, sLe^a, or sLe^x, could reduced adhesion of several cancer cell types *in vitro* and metastasis in animal models *in vivo* (Kobayashi *et al.* 2007). In addition, a soluble E-selectin fusion protein and a synthetic sLe^x ligand were separately able to attenuate metastasis, likely by binding to their counterparts on tumor cells and endothelial cells, respectively, and therefore blocking interaction of the two cell types (Kobayashi *et al.* 2007).

The cell adhesion molecule CD44 has been studied as a ligand for E-selectin in cancer progression (Kobayashi *et al.* 2007). CD44 is a glycoprotein that can be expressed in different isoforms. The common isoform may actually negatively affect metastasis. However, the sialofucosylated isoform correlates with tumor progression. In addition, the sialofucosylated isoform binds much more strongly to all of the selectin family members than the common isoform (Kobayashi *et al.* 2007). These studies, along with those described above, suggest that E-selectin and its ligands play a major, and perhaps integral, role in metastases of some tumor types, and that therapeutic programs aimed at obstructing the interaction of E-selectin and its ligands would be beneficial in preventing metastasis.

P-Selectin in Cancer Progression

P-selectin is also expressed on the endothelium. The main ligand for P-selectin is P-selectin glycoprotein ligand-1, or PSGL-1. While this interaction is commonly found in leukocyte-endothelial cell adhesion, it does not seem to play a major role in adhesion of tumor cells (Kobayashi *et al.* 2007). As such, P-selectin has not received as close scrutiny as E-selectin. However, P-selectin has other ligands. Similar to E-selectin, P-selectin can bind the sialyl Lewis determinants, suggesting a multitude of potential binding partners. CD24 has been implicated as a ligand for P-selectin (Kobayashi *et al.* 2007). Studies have shown that both the sLe^x determinant and CD24 are important in adhesion of tumor cells and subsequent metastasis (Kobayashi *et al.* 2007). Further, animal studies show that lack or inhibition of P-selectin leads to a reduction in metastases, indicating P-selectin may play an important role in cancer progression (Kobayashi *et al.* 2007).

Clinical studies of P-selectin levels in various cancers, including invasive tumors and metastatic tumors, do not point to a clear trend (Table 1) (Kobayashi *et al.* 2007). Invasive breast and gastric cancers have increased levels of P-selectin, and this was associated with increased leukocyte infiltration in gastric cancer (Kobayashi *et al.* 2007). On the contrary, decreased levels were observed with melanoma and colon cancer progression, correlating with a reduced influx of leukocytes (Kobayashi *et al.* 2007). While the decrease in P-selectin could be due to downregulation in endothelial cells, it could alternatively be due to processing into soluble P-selectin (sP-selectin). Whereas studies do not point to a clear role for levels of P-selectin in cancer progression, it is a different story for sP-selectin. Several studies have shown that high levels of sP-selectin correlate with tumor progression and indicate a poor prognosis for patient survival (Table 1) (Kobayashi *et al.* 2007). What role is sP-selectin playing in tumor progression? One could argue that sP-selectin functions as an inhibitor for P-selectin interactions, thereby inhibiting leukocyte infiltration and metastasis. However, a recent study demonstrated that sP-selectin actually increases leukocyte adhesion in an integrin-dependent manner (Woollard *et al.* 2008). It will be interesting to determine in future studies whether sP-selectin enhances tumor cell adhesion, leading to increased metastases. These observations all together point to a complex role for P-selectin, and additional study will lead to a better understanding of its contribution to cancer progression.

VCAM-1 and ICAM-1 in Cancer Progression

In the rolling process, the weak interactions involving selectins are followed by stronger interactions of cell adhesion molecules on the endothelium such as VCAM-1 and ICAM-1 with integrins on the adhering cells. While the role of these adhesion molecules in infiltration of leukocytes at sites of inflammation is well established, their role in metastasis is still being determined. As with the selectins, soluble forms of VCAM-1 and ICAM-1 exist. In some types of cancers, VCAM-1 and ICAM-1 are upregulated, but the soluble forms are upregulated in many more types of cancers (Table 2) (Kobayashi *et al.* 2007). Unlike what has been suggested with P-selectin, soluble ICAM-1 (sICAM-1) has been indicated to reduce adhesion, perhaps providing a mechanism to reduce infiltration of potentially anti-tumor leukocytes (Kobayashi *et al.* 2007). Studies have shown that VCAM-1 mediates tumor cell adhesion and that blocking the molecule can inhibit this adhesion, indicating a possible role in metastasis (Kobayashi *et al.* 2007).

Table 2   Levels of VCAM-1 and ICAM-1 detected in cancer patient samples

	VCAM-1	sVCAM-1	ICAM-1	sICAM-1
Breast	↑	↑	↑	↑
Colon				
Lung				↑
Liver		↑	↑	↑
Gastrointestinal		↑		↑
Kidney				↑
Ovary		↑		↑
Bladder		↑		↑
Lymphoma		↑		↑
Myeloma		↑		↑

Summary from the literature of levels of VCAM-1 and ICAM-1 in samples from cancer patients compared to normal tissues. ↑, increased in samples, ↓, decreased in samples.

So far, clinical data do not lead to a clear conclusion on the correlation between these molecules and cancer progression. Several studies suggest that these molecules, both the cell surface and the soluble forms, increase concurrently with progression and metastasis, but not consistently between different cancer types, suggesting the involvement of particular cell adhesion molecules may be dependent on site and tumor type (Kobayashi *et al.* 2007). This is already known to occur in leukocyte trafficking (Goldsby *et al.* 2003), with studies demonstrating leukocytes trafficking to various lymphoid organs and inflammatory regions due to homing receptor interactions and upregulation of particular adhesion molecules at these sites. This mechanism relies upon distinct sets of adhesion molecules being expressed by both the leukocyte and the endothelial cells. A similar mechanism is likely used by circulating tumor cells.

What makes the potential metastatic site attractive to the circulating cancer cells? Many studies suggest the primary tumor secretes factors that act upon tissues at distant sites. For example, carcinoembryonic antigen (CEA) produced by colon cancer cells leads to production of inflammatory factors by certain cells in the liver, which then leads to an upregulation of ICAM-1 in the surrounding endothelial cells, creating an environment amenable to adhesion of circulating colon cancer cells (Gangopadhyay *et al.* 1998). Along these lines, studies show that blocking inflammatory molecules inhibits metastases, while addition of inflammatory molecules increases metastases. Other studies show that certain organs constitutively express specific adhesion molecules, making it more likely that cells with the counterpart molecules will adhere at those locations (Dittmer *et al.* 2008). Future studies aimed at determining the mechanisms of site-directed metastasis may uncover major roles for endothelial adhesion molecules, and perhaps suggest targets for therapeutic intervention.

CONCLUSIONS

As the data discussed above have shown, endothelial cell adhesion molecules play an important role in cancer progression. Tumor cells, similar to an inflammatory reaction, secrete factors such as IL-1β and TNF-α, which mediate upregulation of adhesion molecules on the surrounding endothelial cells. These endothelial cell adhesion molecules allow leukocytes to infiltrate the tumor, where they function in multiple ways. Later in cancer progression, cancer cells themselves adhere to the endothelium via endothelial cell adhesion molecules, and subsequently extravasate and develop metastases. Clinical data indicate that many endothelial adhesion molecules are upregulated in several cancer types as progression occurs. Perhaps levels of these molecules may be used in determining patient prognosis. Studies show that inhibiting adherence of tumor cells to the endothelium using chemical inhibitors or blocking antibodies can reduce metastases. These data suggest that chemotherapy regimens would benefit from including such inhibitors.

All together, the clinical and experimental data indicate that endothelial adhesion molecules are important players in cancer progression.

Acknowledgement

Supported in part by grants from NIH (CA108856, NS45888 and AR053718) to PCL, and training grants from NCI to HK (T32-CA093240) and KCB (T32CA009582).

SUMMARY POINTS

- Cancer progression includes many steps that involve endothelial cell adhesion molecules.
- Tumor cells secrete factors that increase levels of endothelial cell adhesion molecules, allowing leukocytes to adhere and infiltrate the tumor.
- Tumor cells express adhesion molecules similar to leukocytes and therefore may use a similar mechanism for adhesion and extravasation.
- E-selectin and P-selectin levels correlate with cancer progression in patients, especially the soluble forms.
- The most studied ligands for selectins with regard to cancer contain the sialyl Lewis determinants, sLe^a or sLe^x, which are detected at increased levels in cancer patients.
- Blocking selectin interactions has been shown to reduce tumor cell adhesion and subsequent metastasis.
- The soluble forms of VCAM-1 and ICAM-1 are detected at increased levels in cancer patients.
- Location of metastatic sites is likely influenced by specific tissue expression of endothelial cell adhesion molecules.

Abbreviations

CEA	carcinoembryonic antigen
CXCR4	chemokine (CXC motif) receptor 4
EGF	epidermal growth factor
ICAM	intercellular adhesion molecule
Ig	immunoglobulin
IL-1β	interleukin 1 beta
PECAM	platelet endothelial cell adhesion molecule
PSGL-1	P-selectin glycoprotein ligand 1

SCR	short consensus repeat
SDF-1	stromal derived factor 1
sE-selectin	soluble E-selectin
sP-selectin	soluble P-selectin
TNF-α	tumor necrosis factor alpha
VCAM	vascular cell adhesion molecule

Key Features of Cancer Progression

1. Cancer progression involves several steps.
   a. Growth at primary tumor site
   b. Angiogenesis
   c. Invasion of tumor cells into surrounding tissues
   d. Intravasation, adherence, and subsequent extravasation of tumor cells
   e. Growth at a secondary site
2. A tumor is not considered malignant until it has become invasive.
3. Intravasation, adherence, extravasation, and growth at a secondary site are all steps of metastasis.
4. Metastasis is a very inefficient process, with most tumor cells unable to complete all steps.
5. The majority of cancer-related deaths are due to metastasis.

Definition of Terms

Angiogenesis: New blood vessel formation from existing vessels.

Benign: Adjective to describe non-cancerous growth.

Chemokine: Cytokine that directs chemotaxis, the process by which cells migrate toward an attractant.

Extravasate: To move out of a blood vessel into a tissue.

Intravasate: To move from a tissue into a blood vessel.

Lewis determinants: A set of carbohydrate moieties recognized by selectins, often contained within mucin molecules.

Malignant: Adjective to describe cancerous growth, used once tumor cells have become invasive.

Metastasis: The process by which tumor cells move from the primary tumor site, migrate and adhere to endothelium at a distant site, extravasate, and proliferate into a new tumor.

Sialyl Lewis determinants: Lewis determinants that have been altered by sialofucosylation, which alters the recognition by selectins.

Transendothelial migration: Migration across the endothelium.

Tumorigenic: Capable of leading to tumor formation.

References

Brooks, S.A. and H.J. Lomax-Browne, T.M. Carter, C.E. Kinch, and D.M.S. Hall. 2009. Molecular interactions in cancer cell metastasis. Acta Histochem. (in press).

Castermans, K. and A.W. Griffioen. 2007. Tumor blood vessels, a difficult hurdle for infiltrating leukocytes. Biochim. Biophys. Acta 1776: 160-174.

Dittmar, T. and C. Heyder, E. Gloria-Maercker, W. Hatzmann, and K.S. Zanker. 2008. Adhesion molecules and chemokines: the navigation system for circulating tumor (stem) cells to metastasize in an organ-specific manner. Clin. Exp. Metastasis 25: 11-32.

Fukami, A. and K. Iijima, M. Hayashi, K. Komiyama, and S. Omura. 2002. Macrosphelide B suppressed metastasis through inhibition of adhesion of sLe(x)/E-selectin molecules. Biochem. Biophys. Res. Commun. 291: 1065-1070.

Gassmann, P. and J. Haier. 2008. The tumor cell-host organ interface in the early onset of metastatic organ colonization. Clin. Exp. Metastasis 25: 171-181.

Gangopadhyay, A. and D.A. Lazure, and P. Thomas. 1998. Adhesion of colorectal carcinoma cells to the endothelium is mediated by cytokines from CEA stimulated Kupffer cells. Clin. Exp. Metastasis 16: 703-712.

Goldsby, R.A. and T.J. Kindt, B.A. Osborne, and J. Kuby. 2003. Leukocyte migration and inflammation. pp. 338-360. *In*: Immunology, 5th Ed. W.H. Freeman and Company, New York.

Kobayashi, H. and K.C. Boelte, and P.C. Lin. 2007. Endothelial cell adhesion molecules and cancer progression. Curr. Med. Chem. 14: 377-386.

Kobayashi, K. and S. Matsumoto, T. Morishima, T. Kawabe, and T. Okamoto. 2000. Cimetidine inhibits cancer cell adhesion to endothelial cells and prevents metastasis by blocking E-selectin expression. Cancer Res. 60: 3978-3984.

Kucia, M. and R. Reca, K. Miekus, J. Wanzeck, W. Wojakowski, A. Janowska-Wieczorek, J. Ratajczak, and M.Z. Ratajczak. 2005. Trafficking of normal stem cells and metastasis of cancer stem cells involve similar mechanisms: pivotal role of the SDF-1-CXCR4 axis. Stem Cells 23: 879-894.

Tedder, T.F. and D.A. Steeber, A. Chen, and P. Engel. 1995. The selectins: vascular adhesion molecules. FASEB J. 9: 866-873.

Woollard, K.J. and A. Suhartoyo, E.E. Harris, S.U. Eisenhardt, S.P. Jackson, K. Peter, A.M. Dart, M.J. Hickey, and J.P.F. Chin-Dusting. 2008. Pathophysiological levels of soluble P-selectin mediate adhesion of leukocytes to the endothelium through Mac-1 activation. Circ. Res. 103: 1128-1138.

CHAPTER 23

Adhesion Molecule Phosphorylation in Cancer Chemotherapy

Geraldine M. O'Neill

Focal Adhesion Biology Group, Oncology Research Unit, The Children's Hospital at Westmead, NSW Australia 2145 and Discipline of Paediatrics and Child Health, University of Sydney
Contact details: Oncology Research Unit, The Children's Hospital at Westmead, Locked Bag 4001,Westmead NSW 2145, Australia, Email: Geraldio@chw.edu.au
Departmental contact: Ms Janett Clarkson,Research and Development Manager, Oncology Research Unit, The Children's Hospital at Westmead, Locked Bag 4001, Westmead, NSW, 2145, Australia, E-mail: JanettC@chw.edu.au

ABSTRACT

Transmembrane integrin receptor adhesion to extracellular matrix stimulates the activation of phosphorylation-dependent signalling cascades that control cell proliferation, survival and migration: each of these processes is characteristically altered in cancer cells. Due to increasing understanding of the contribution of adhesion to cancer progression, adhesion-dependent signalling cascades are emerging as targets both for novel anti-cancer chemotherapeutics and as potential biomarkers for cancer therapies. The integrin receptors have no intrinsic enzymatic activity but instead recruit non-receptor protein tyrosine kinases, such as focal adhesion kinase (FAK) and Src family kinases, that catalyse tyrosine phosphorylation downstream from integrin receptor stimulation. The phosphorylation of target proteins such as paxillin and p130Cas then creates protein binding sites and together the web of protein-protein interactions

---

Key terms are defined at the end of the chapter.

established forms the signalling network. In this first decade of the 21st century, we now have an extensive (although not yet comprehensive) view of the signalling networks that are established by integrin receptor stimulation. While there are still important questions to be answered, we now have sufficient knowledge to begin targeting adhesion signalling as a new approach to the treatment of cancer. This chapter describes current understanding of the phospho-regulation of the adhesion proteins, focusing on FAK, Src, paxillin and p130Cas. We discuss the antibodies that have been created to study the phosphorylation modifications of these proteins, approaches that are being employed to target these proteins in novel anti-cancer therapies and the potential for these phospho-proteins as biomarkers for chemotherapeutic activity.

INTRODUCTION

Hanahan and Weinberg (2000) described six essential changes to normal cellular function that determine progression to malignancy: 'self-sufficiency in growth signals, evasion of programmed cell death, limitless replicative potential, sustained angiogenesis, and tissue invasion and metastasis'. Transmembrane integrin receptor interaction with the extracellular matrix (ECM) stimulates the activation of phosphorylation-dependent signalling cascades that play a critical role in each of these six hallmarks of cancer. Protein phosphorylation creates binding sites that facilitates protein-protein interaction, thereby creating networks of interacting proteins. Kinases catalyse the transfer of the γ-phosphoryl group of ATP to serine, threonine or tyrosine amino acids in the recipient protein. Based on the growing realization that kinases are essential regulators of adhesion-dependent signalling in cancer, they—and their phosphorylated substrates—have become important new targets for treating malignancies and for monitoring activity of chemotherapeutic agents. The following sections provide information regarding the phospho-regulation of focal adhesion kinase (FAK), Src, paxillin and p130Cas and discuss how studies of these proteins have led to novel anti-cancer therapies and biomarkers for chemotherapy efficacy.

ADHESION SIGNALLING AND CANCER

Integrin Receptors

The integrin receptor family of proteins are the major receptors that mediate cellular adhesion to the ECM. In mammals, 8 β- and 18 α-integrin subunits heterodimerize to form at least 24 αβ receptors with both overlapping and distinct specificities for ECM components including fibronectin, laminin, fibrinogen, tenascin and collagen. The repertoire of receptor expression is tissue and cell-type specific, is regulated during development and alters during cancer progression.

As the cytoplasmic tails of the integrin receptors directly link to polymerized actin filaments that critically determine cell structure, it was initially believed that integrin receptors simply regulated cell structure. The realization that a vast array of signalling molecules are recruited to the receptors made it clear, however, that integrins are important sites of signal transduction. Under light microscopy, ligated and clustered integrin receptor complexes are visualized in cultured cells as macromolecular structures with a dash-like morphology and these structures are known as focal adhesions/contacts (Fig. 1). The integrin 'adhesome' has 156 components networked by up to 690 interactions (Zaidel-Bar *et al.* 2007).

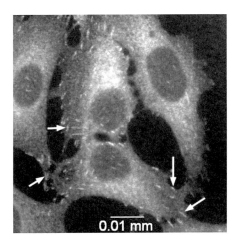

Fig. 1   Paxillin-positive focal adhesions in cultured neuroblastoma cells. SHEP neuroblastoma cells were processed for immunofluorescence analysis as previously described (Cowell *et al.* 2006) and immunostained with anti-paxillin antibodies. Arrows indicate the large paxillin-positive focal adhesions. Note also the punctate cytoplasmic staining, absent from the nucleus. Scale bar 10 μm. (G. O. unpublished data.)

Integrin Receptor Stimulated Phosphorylation

Integrin receptors have no intrinsic enzymatic activity but instead recruit non-receptor protein tyrosine kinases, such as FAK and Src family kinases, that catalyse tyrosine phosphorylation (Fig. 2). Phosphorylation of consensus tyrosine sites creates binding sites for partner molecules, the most prevalent of which are those containing Src-Homology 2 protein domains (SH2). Different SH2 domains show preference for binding to conserved phospho-peptide sequences, thus providing specificity in the assemblage of signalling networks. SH2 domains are present not only in proteins with intrinsic enzymatic activity (e.g., FAK and Src) but also in adaptor proteins (e.g., paxillin and p130Cas) that interact with, and are

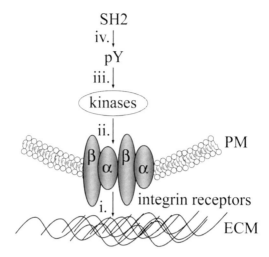

SH2
iv. |
pY
iii. |
kinases
ii. |
PM
β α β α
integrin receptors
i. |
ECM

Fig. 2 Integrin receptor stimulated phosphorylation. The integrin receptors consisting of αβ heterodimers localize to the plasma membrane (PM). (i) Following integrin receptor engagement with the extracellular matrix (ECM), (ii) tyrosine kinases such as FAK and Src become activated. (iii) These kinases catalyse phosphorylation of substrate proteins, (iv) creating binding sites for Src homology 2 (SH2)-domain containing proteins. This establishes extensive protein-protein interaction networks downstream of integrin receptor stimulation. In addition to these tyrosine phosphorylation events, other enzymatic reactions are stimulated, including serine/threonine phosphorylation. The full complement of the integrin adhesome can be viewed in Zaidel-Bar *et al.* (2007).

phosphorylated by, these kinases. Thus, the activation of tyrosine kinases associated with integrin receptors creates multi-protein complexes through protein-protein interaction domains.

Adhesion in Cancer Growth and Metastasis

Unlike their normal counterparts, cancer cells continue to survive when deprived of attachment to the ECM. In the quest to understand this, in the late 1970s researchers investigated the properties of viral proteins that caused cellular transformation to anchorage-independent growth. Src kinase was the first described virally encoded protein that caused oncogenic transformation characterized by anchorage-independence. The viral Src (v-Src) protein transpired to have a matching cellular Src (c-Src) counterpart with the regulatory region of c-Src truncated in v-Src resulting in a constitutively active enzyme. Constitutive activation of Src and phosphorylation of its targets FAK, paxillin and p130Cas are implicated both in cell proliferation and in cell migration and invasion that underlies metastasis.

Metastasis describes the process of cancer dissemination away from the primary tumour and establishment of secondary tumours at distal sites. Integrin-based interaction with the ECM is required for mesenchymal-type cell motility that is observed in many metastasizing cancer cells. Very recently, it has become apparent that migrating cancer cells can switch between integrin-dependent mesenchymal motility and non–integrin-dependent motility (Friedl 2004); successful approaches to block cancer metastasis may thus necessitate blocking of both migration modes. Few therapies currently directly target metastasis, yet given the high burden of mortality resulting from progression to metastatic disease there is considerable scope for therapeutic improvements in this area.

ADHESION-ACTIVATED PHOSPHORYLATION SIGNALLING CASCADES

Src Kinases

Enhanced Src kinase activity and elevated expression levels have been reported in a wide variety of tumour types (Brunton *et al.* 2008). This enzyme has potent kinase activity and of all the molecules in the integrin adhesome displays the greatest connectivity with other molecules in the network (Zaidel-Bar *et al.* 2007).

Phosphorylation of residue Y^{530} in human c-Src carboxyl-terminal creates a binding site for the amino-terminal Src SH2 domain, thereby forming an intra-molecular interaction holding the molecule in a closed and inhibited conformation (Fig. 3). Src pY^{530} de-phosphorylation releases this repression permitting auto-phosphorylation of Y^{418} in the kinase domain activation loop resulting in increased kinase activity. Importantly, there are a number of commercially available robust antibodies that are specific for Src phosphorylation sites. The most commonly employed is anti-pY^{418} as this site positively correlates with enzyme activity and is generally accepted to provide a readout for increased Src activation. The phospho-specific antibody efficiently detects phosphorylated Src kinase on immunoblots of protein extracts that have been separated by sodium-dodecyl-sulphate polyacrylamide gel electrophoresis (SDS-PAGE) and this is a standard technique for measuring Src activity in extracts from cultured cells (e.g., Cowell *et al.* 2006). Increasingly, the phospho-specific antibody is used for immunohistochemical detection of phosphorylated and activated c-Src in patient tissue samples (e.g., Campbell *et al.* 2008).

FAK

Similar to Src, FAK expression levels and activity are elevated in a wide variety of cancers, while elevated FAK expression correlates with increased metastatic behaviour in colon, breast and ovarian tumours (Li *et al.* 2008). The essential

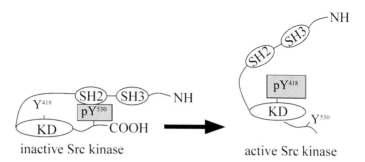

Fig. 3   Key tyrosine residues for Src kinase activity. The protein domains of Src kinase protein include an amino-(NH) terminal Src Homology 3 (SH3) domain, followed by a Src Homology 2 domain (SH2), the kinase domain (KD) and finally the carboxyl-terminus (COOH). Phosphorylation of Y^{530} in the COOH-terminal end of the molecule creates a binding site for the Src SH2 domain, causing an intra-molecular interaction and inhibition of the molecule's kinase activity. Following Y^{530} dephosphorylation, the intra-molecular interaction is lost, exposing Y^{418}. Auto-phosphorylation of this site then activates the kinase. Note that the amino acid designations are based on the sequence of human c-Src. The schematics (not to scale) are based on information from Brunton and Frame (2008) and references therein.

role that FAK plays in cancer was revealed by studies demonstrating that FAK expression is required for tumour formation and malignant progression (Brunton *et al.* 2008). Thus increasingly this protein is a target for novel therapeutic approaches to cancer.

In response to integrin receptor stimulation, FAK undergoes auto-phosphorylation at Y^{397} (Ruest *et al.* 2001) creating a binding site for the Src SH2 domain and other SH2 domain-containing proteins (Fig. 4). An additional 5 tyrosine residues in FAK (Y^{407}, Y^{576}, Y^{577}, Y^{861}, Y^{925}) are subsequently phosphorylated by Src. These phosphorylation events have been elegantly linked to discrete signalling consequences. PY^{407}, pY^{576}, pY^{577} significantly increase *in vitro* kinase activity of FAK, while pY^{407}, pY^{861} or pY^{925} create binding sites for SH2 domain-containing proteins such as the adaptor molecule Grb2. Grb2 in turn can activate cell proliferation via the Ras/mitogen-activated protein kinase (MAPK) pathway (reviewed in Li *et al.* 2008). While FAK pY^{397} can therefore be used as a readout for FAK activity, FAK pY^{576} and pY^{577} levels are regularly used as a measure of Src function. There are well-established phospho-specific antibodies to detect FAK phospho-modifications on immunoblots of protein extracts separated by SDS-PAGE (e.g., Cowell *et al.* 2006). Moreover, the FAK phospho-specific antibodies successfully stain formalin-fixed, paraffin-embedded tissue samples (e.g., Matkowskyj *et al.* 2003, Halder *et al.* 2005). When used for immunofluorescence analysis, the phospho-specific FAK pY^{397} antibodies almost exclusively localize to the focal adhesions (see example in Fig. 6).

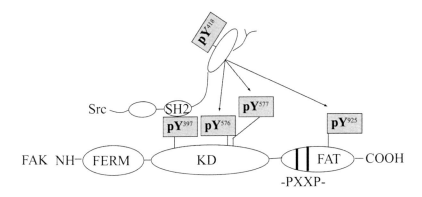

Fig. 4 Interaction between FAK and Src. The protein domains of FAK include the amino-(NH) terminal FERM (band four point one, ezrin, radixin, and moesin homology domain), followed by the kinase domain (KD) and the focal adhesion targeting (FAT) domain at the carboxyl-(COOH) terminus. The FAT domain includes two poly-proline (-PXXP-) motifs that mediate interaction with Src Homology 3(SH3) domain-containing proteins. Following integrin receptor stimulation, FAK undergoes auto-phosphorylation of Y^{530}. This creates a binding site for the Src Homology 2 (SH2) domain of Src (and other proteins). Docking of Src to FAK pY^{397} then promotes Src-mediated phosphorylation of the FAK tyrosine residues Y^{576}, Y^{577} and Y^{925} creating new docking sites. The schematics (not to scale) are based on information in Brunton and Frame (2008) and references therein.

Paxillin

In contrast to Src and FAK, paxillin contains no intrinsic kinase activity (reviewed in Deakin *et al.* 2008). Instead, paxillin is a multi-domain scaffold that organizes complexes of signalling proteins at adhesion sites. Paxillin interaction with partner proteins such as either the anti-apoptotic protein Bcl-2 or the actin-binding protein vinculun can stimulate anti-apoptosis signalling and thus paxillin may promote cancer cell resistance to apoptosis (Deakin *et al.* 2008). Recent *in vitro* data suggests that paxillin phosphorylation by Src may be required for adhesion turnover that is essential for mesenchymal cell motility (Bach *et al.* 2009).

Paxilllin binds to the focal adhesion targeting (FAT) domain in the FAK c-terminus. Together, FAK and Src stimulate the phosphorylation of two main paxillin tyrosine residues Y^{31} and Y^{118} (Deakin *et al.* 2008) (Fig. 5A). This creates binding sites for SH2 domain-containing proteins including the adaptor protein Crk—a critical step in the regulation of cell motility. Available phospho-antibodies for paxillin pY^{31} and pY^{118} are highly successful for immunoblotting (e.g., Chang *et al.* 2008, Bach *et al.* 2009) and immunofluorescence (e.g., Bach *et al.* 2009).

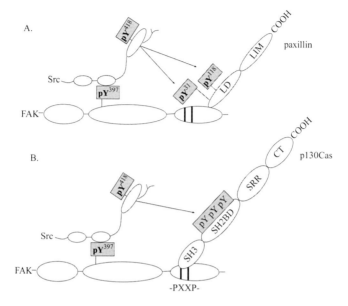

Fig. 5   FAK and Src work in concert to phosphorylate paxillin and p130Cas. (A) The amino terminal half of paxillin contains five leucine-rich sequences (LD motifs). The second half of the protein contains four zinc finger LIM (Lin11, Isl-1 & Mec-3) domains. Paxillin binds to the FAK carboxyl-terminal focal adhesion targeting (FAT) domain and together FAK and Src catalyse the phosphorylation of paxillin Y^{31} and Y^{118}. (B) p130Cas has a multi protein-protein interaction domain structure that includes an amino terminal Src Homology 3 (SH3) domain, followed by a Src Homology 2 (SH2) binding domain that contains 15 consensus tyrosine phosphorylation sites. This is followed by a serine-rich region (SRR) and finally the C-terminal (CT) domain. The p130Cas SH3 domain binds directly to the poly-proline (-PXXP-) domains found in the FAK FAT domain. Src and p130Cas docking to FAK significantly enhances phosphorylation of the p130Cas SH2 substrate binding domain (pY pY pY). The schematics (not to scale) are based on information from Ruest *et al.* (2001) and references therein.

p130Cas

The mounting evidence supporting a role for p130Cas and related Cas family proteins in proliferation and tumour invasion and metastasis (Defilippi *et al.* 2006) has given increasing urgency to an accurate survey of the expression of this family of proteins in human tumour samples. To date, p130Cas expression has been measured in a cohort of breast cancer tumours, confirming that elevated p130Cas expression is correlated with enhanced tumorigenesis (van der Flier *et al.* 2000). Moreover, transgenic overexpression of p130Cas in a mouse mammary carcinoma model caused hyper-proliferation and tumour promotion (Defilippi *et al.* 2006).

The substrate binding domains of p130Cas encompasses over a dozen consensus tyrosine phosphorylation sites and Src is the major kinase that phosphorylates these sites. The currently favoured model suggests that, by directly binding both p130Cas and Src, FAK may serve as a scaffold bringing the enzyme and its substrate into close proximity and thereby drive p130Cas protein phosphorylation by Src (Ruest *et al.* 2001) (Fig. 5B). A number of phospho-specific antibodies have been generated against tyrosine residues in the p130Cas substrate binding domain. One caveat in the use of these antibodies is the potential for cross-reaction between p130Cas and the related family member NEDD9. These two proteins exhibit very high sequence conservation and this must be taken into consideration when using the p130Cas phospho-antibodies. Antibodies to p130Cas pY^{410} have been used in a few studies performed to date (Buettner *et al.* 2008, Chang *et al.* 2008), although the rationale for choosing this particular phospho-antibody is unclear. Recent reports investigating the role of individual p130Cas phosphorylation sites suggest that a comprehensive picture will soon be obtained.

CHEMOTHERAPEUTIC TARGETING OF ADHESION-DEPENDENT SIGNALLING IN CANCER

Therapeutic Targeting of Src

At the present time there are 4 Src inhibitors reported to be in clinical development (Brunton *et al.* 2008). Three of these agents, dasatinib, AZD0530 and bosutinib, target the Src ATP binding site, while the fourth, KXO1, targets the substrate binding site. Although it was hoped that the Src kinase inhibitors would block cell proliferation, evidence is emerging to suggest that Src inhibitors are potent inhibitors of migration and invasion and therefore hold promise for treating metastatic cancers. Dasatinib reduced melanoma migration and invasion (Buettner *et al.* 2008), while AZD0530 is entering phase II clinical trials and has anti-migratory effects in breast cancer cells (Hiscox *et al.* 2006) and growth inhibitory effects in an orthotopic xenograft mouse model (Chang *et al.* 2008). Reduced primary tumour burden is a common endpoint for most trials of new chemotherapeutics, but this is problematic for agents that reduce invasion and metastasis rather than tumour size. In this respect, phosphorylation of the adhesion proteins FAK, Cas and paxillin (and of Src kinase itself) represent potentially useful biomarkers for monitoring inhibition of invasion and metastasis.

Therapeutic Targeting of FAK

Current FAK inhibitors under development include PF-562,271, an ATP-competitive reversible inhibitor, TAE226 (Brunton *et al.* 2008) and PF-573,228 (Slack-Davis *et al.* 2007). TAE226 showed efficacy in orthotopic mouse models of ovarian cancer and when administered in combination with docetaxel, an agent currently in clinical use, tumour burden was significantly reduced (Halder *et al.*

2005). TAE226 also showed efficacy against a range of cultured breast cancer cell lines (Golubovskaya *et al.* 2008). Similarly, PF-562,271 stimulated increased apoptosis and decreased vascularization in multiple mouse tumour models of human sub-cutaneous xenografts (Roberts *et al.* 2008).

PHOSPHORYLATION OF ADHESION PROTEINS AS POTENTIAL BIOMARKERS

Phosphorylation Response to Anti-FAK and Anti-Src Therapeutics

Phospho-specific antibodies to FAK pY^{397} and Src pY^{418} are proving to be excellent markers for the efficacy of therapies targeting FAK and Src, respectively. Indeed, a number of studies confirm loss of FAK and Src phosphorylation in a range of cancers following application of anti-FAK and anti-Src drugs (Table 1). The report that Src phosphorylation was reduced following dasitinib treatment in dasatinib-sensitive but not -resistant cell lines (Eustace *et al.* 2008) suggests that Src phosphorylation status may serve as a biomarker to potentially discriminate responsive from non-responsive tumours. One study suggested Src phosphorylation status prior to therapy did not predict response to Src inhibitors (Johnson *et al.* 2005); however, the potential for Src phosphorylation status as a predictor of therapeutic response warrants further investigation with large patient cohorts of different tumour types.

Table I Studies of FAK and Src phosphorylation following anti-FAK and anti-Src treatments

Chemotherapeutic	Phospho-target	Tumour type	References
TAE226	FAK Y397	neuroblastoma	(Beierle *et al.* 2008)
		breast	(Golubovskaya *et al.* 2008)
		ovarian	(Halder *et al.* 2005)
		glioma	(Shi *et al.* 2007)
PF-562,271	FAK Y397	glioblastoma*	(Roberts *et al.* 2008)
Dasatinib	Src Y418	melanoma	(Eustace *et al.* 2008)
		head and neck	(Johnson *et al.* 2005)
		prostate*	(Luo *et al.* 2008)
		colon*	(Serrels *et al.* 2006)

In vivo cancer models. The table shows the anti-FAK and anti-Src therapies that have been demonstrated to specifically reduce phosphorylation of their target protein (phospho-target) in the indicated tumour cell types.

Phosphorylation of paxillin, FAK and p130Cas has also been used to monitor Src kinase inhibition. A number of studies report reduced phosphorylation of the Src phosphorylation target sites in FAK, while FAK pY^{397} is maintained (Johnson *et al.* 2005, Buettner *et al.* 2008, Chang *et al.* 2008, Eustace *et al.* 2008). Similarly, loss of paxillin pY[118] (Serrels *et al.* 2006) and p130Cas pY[249] (Johnson *et al.* 2005) and pY[410] (Buettner *et al.* 2008, Chang *et al.* 2008) is observed following Src inhibition.

Adhesion Protein Phosphorylation in Response to Anti-integrin Therapies

In recent times, there has been a major effort to develop anti-integrin therapies to block angiogenesis that is required for tumour survival. Since integrins regulate FAK, Src, paxillin and p130Cas phosphorylation, it is likely that the phosphorylation of these proteins may also reflect the efficacy of the anti-integrin treatments. One such promising anti-integrin therapy is the agent cilengitide that targets αVβ3 integrin receptors. Cilengitide caused FAK dephosphorylation and apoptosis in both endothelial cells (required for the formation of new blood vessels) and glioblastoma cell lines that express elevated levels of αVβ3 (Oliveira-Ferrer *et al.* 2008). Although a recent pharmacodynamics investigation of cilengitide administered to patients with a range of solid tumours failed to identify statistically significant differences in soluble angiogenic molecule expression and tumour microvessel density (Hariharan *et al.* 2007), the authors concluded that the phosphorylation of FAK and paxillin may serve as superior biomarkers for cilengitide action in future studies.

Adhesion Phospho-proteins as Biomarkers for Drug Activity

Mounting data suggest that the phospho-adhesion proteins may be excellent candidates as biomarkers for drug activity. Some interesting recent studies suggest that it is feasible to assess adhesion protein phosphorylation in tissue samples that can be easily obtained from patients undergoing therapy. Following dasatinib treatment of mice with experimentally stimulated cancer, both Src pY[418] and paxillin pY[118] (Serrels *et al.* 2006, Luo *et al.* 2008) were reduced in peripheral blood mononuclear cells (PBMCs) isolated from the treated mice. These data suggest that PBMCs may provide a useful surrogate tissue for biomarker analysis of anti-Src therapeutics.

Circulating micro-metastatic populations of cancer cells can also be detected at low frequency in preparations of PBMCs by staining cell preparations for the epithelial cell marker cytokeratin, to distinguish the circulating metastatic tumour cells. Kallergi and colleagues used this approach to investigate PBMC preparations from breast cancer patients and examined the levels of FAK pY[397] in the circulating metastatic tumour cells by confocal microscopy (Kallergi *et al.* 2007). In future studies, if elevated FAK pY[397] can be demonstrated to predict response to anti-FAK targeted therapies, the examination of FAK pY[397] in circulating micrometastases may provide a method to select patients for treatment with anti-FAK therapies.

ADHESION PROTEIN PHOSPHORYLATION AND CHEMO-RESISTANCE

Cell Adhesion-Mediated Drug Resistance

It appears almost inevitable that following identification of a novel therapeutic agent, shortly thereafter reports of resistance to the agent will appear in the literature. Among the mechanisms described to mediate drug resistance, cell adhesion-mediated drug resistance (CAM-DR) has emerged as an important contributor (reviewed in Hehlgans *et al.* 2007). Thus, cancer cells deprived of attachment to the ECM can be sensitized to the apoptosis-inducing effects of chemotherapeutic agents. Conversely, integrin stimulation can cause drug resistance. The influence of integrin interaction with the ECM in promoting resistance to chemotherapeutics and radiotherapy has been documented in a broad variety of cancer types (reviewed in Hehlgans *et al.* 2007). Importantly, elevated levels of FAK expression and/or activity are correlated with CAM-DR, while in contrast suppression of FAK expression can sensitize cells to drug and radiotherapy treatments.

Tamoxifen Resistance

The anti-estrogen tamoxifen has had great success as an adjuvant therapy for treating women with estrogen-receptor (ER)-positive breast cancer; however, up to 40% of cancer patients receiving adjuvant tamoxifen therapy acquire resistance, whereas 50% of ER-positive tumours are intrinsically resistant to tamoxifen (Cowell *et al.* 2006 and references therein). Both p130Cas and Src kinase have been implicated as playing important roles in the resistance to tamoxifen. Elevated p130Cas expression in primary breast tumours correlates with the failure to respond to tamoxifen and can drive cellular proliferation in the presence of tamoxifen (van der Flier *et al.* 2000). Moreover, p130Cas, Src and FAK phosphorylation are increased following exposure to tamoxifen (Fig. 6) and p130Cas phosphorylation appears to be required for survival in the presence of tamoxifen (Cowell *et al.* 2006). To date, p130Cas phosphorylation has not been assessed in patient tumours; however, the availability of p130Cas phospho-specific antibodies means that this should now be achievable.

AZD0530 inhibits Src activation in tamoxifen-resistant breast cancer cells (Hiscox *et al.* 2006), raising the possibility that anti-Src directed therapies may be beneficial for breast cancer patients with tamoxifen-resistant disease. It is therefore surprising that nuclear localized phosphorylated Src Y418 was positively correlated with improved overall survival in a cohort of breast cancer patients with known ER status biopsied prior to tamoxifen therapy (Campbell *et al.* 2008). It is difficult to reconcile with the role for Src kinase in the development of tamoxifen resistance with this finding. Although active Src was detected, sequestration in the

control tamoxifen

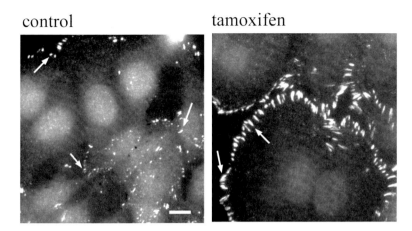

Fig. 6 FAK pY^{397} positive focal adhesions increase following exposure to tamoxifen. MCF-7 ER-positive breast cancer cells cultured in the anti-estrogen tamoxifen as previously described (Cowell *et al.* 2006). Immunofluorescence analysis with antibodies to FAK pY^{397} shows remarkably increased FAK pY^{397}-positive adhesions in the tamoxifen-treated cultures (compare control versus tamoxifen-treated cultures, arrows). Scale bar 10 μm. (G. O. unpublished data.)

nucleus may block integrin-dependent Src activity. Alternatively the progesterone receptor status of the tumours may play a role in regulating Src function in the cytoplasm (Campbell *et al.* 2008). Before Src and FAK inhibitors can fulfil their promise as anti-cancer therapies, these questions highlight the importance for using the well-validated adhesion protein phospho-antibodies to probe extensive cohorts of patient-derived tumour material to relate adhesion protein phospho-status with patient outcomes.

APPLICATION TO OTHER AREAS OF HEALTH AND DISEASE

The critical role for integrin adhesion signalling in cell proliferation, survival and migration means that these receptors are implicated in a number of different pathologies. An area of particular interest is the contribution that integrins play in cell migration. Cell migration not only is critical in cancer metastasis, but also underlies chronic inflammation conditions, is vital to wound healing and when deregulated can lead to congenital defects during embryogenesis. Each of the molecules discussed in this chapter is a strong candidate for regulating migratory processes in other pathological conditions. Worldwide there are extensive resources and research hours devoted to understanding the cell biology of migration, exemplified by the Cell Migration Consortium (http://www.cellmigration.org). Thus, there is the potential for advances in understanding the phospho-regulation of adhesion proteins during cancer chemotherapy to inform novel treatments in other pathologies caused by aberrantly regulated cell migration.

CONCLUSION

Since the concept of integrins and focal adhesions as signalling centres was first realized in the 1970s and 1980s, researchers have amassed increasingly more comprehensive pictures of the signalling networks that are established following integrin receptor stimulation. These basic biology studies of integrin signalling pathways have now progressed to the investigation of these molecules in cancer progression and in cancer therapy. Remarkable progress has been made in recent years towards applying our understanding of fundamental biological processes to improved approaches to treating cancer. This progress suggests that monitoring and manipulating adhesion molecule phosphorylation will become an important part of the arsenal of approaches to improved therapies for cancer patients.

SUMMARY

- Transmembrane integrin receptor interaction with the extracellular matrix is a critical component in cellular processes that are characteristically altered during cancer progression.
- Stimulation of integrin receptors activates non-receptor tyrosine kinases and creates multi-protein phosphorylation-dependent signalling complexes.
- Integrin stimulation activates focal adhesion kinase (FAK) and Src family kinases. These enzymes then catalyse tyrosine phosphorylation of substrate proteins creating binding sites for Src-Homology 2 protein domains.
- Both Src and FAK are elevated in a variety of cancers and therefore are now being targeted in novel therapeutic approaches.
- Phospho-specific antibodies that bind to activated Src and FAK, and their phosphorylated substrates paxillin and p130Cas, provide unique reagents for measuring the efficacy of anti-Src, anti-FAK and anti-integrin therapies.

Acknowledgements

The author acknowledges financial support from the National Health and Medical Research Council of Australia, the New South Wales Cancer Council and kind donations from the Oncology Children's Foundation. The author also acknowledges the excellent technical assistance of Sana Ajaj and Sharmilla Bargon in the preparation of the immunofluorescence micrographs.

Abbreviations

ATP	adenosine tri-phosphate
ECM	extracellular matrix
ER	estrogen receptor
FAK	focal adhesion kinase
FAT	focal adhesion targeting domain
MAPK	mitogen-activated protein kinase
SDS-PAGE	sodium-dodecyl-sulphate polyacrylamide gel electrophoresis
SH2	Src-homology 2 domain
Y	tyrosine

Key Facts about the Detection of Phosphorylated Adhesion Proteins

1. Phospho-specific antibodies can be used to measure the activation of adhesion signalling pathways.
2. The available phospho-specific antibodies for the adhesion proteins Src, FAK, paxillin and p130Cas work well for immunblot and immunofluorescence.
3. Both Src and FAK phospho-specific antibodies have been used successfully on patient-derived tissues including tumour material.
4. Phospho-specific Src and FAK antibodies that detect Src pY418 and FAK pY397 respectively provide a readout for enzymatic activity.
5. Phospho-specific antibodies may be used to monitor adhesion protein phosphorylation as a biomarker for the efficacy of anti-Src, anti-FAK and anti-integrin therapeutics.

Definition of Terms

Angiogenesis: The formation of new blood vessels.

Apoptosis: A regulated series of programmed biochemical cellular events in the cells, culminating in death. Most cancer cells become resistant to apoptosis.

Focal adhesions: Macromolecular structures comprising clusters of ligated and cross-linked integrin receptors together with their associated signalling partners.

Integrin receptor: Transmembrane receptors composed of one α and one β subunit. The receptors bind to the extracellular matrix on the external surface and internally bind to the actin cytoskeleton and a broad array of signalling molecules.

Kinases: Enzymes that catalyse phosphorylation.

Metastasis: The dissemination of tumour cells away from the primary tumour site and subsequent formation of secondary tumours at distal sites.

Protein phosphorylation: A post-translational modification of proteins where the γ-phosphoryl group of ATP is transferred to serine, threonine or tyrosine amino acids in the recipient protein.

References

Bach, C.T. and S. Creed, J. Zhong, M. Mahmassani, G. Schevzov, J. Stehn, L.N. Cowell, P. Naumanen, P. Lappalainen, P.W. Gunning, and G.M. O'Neill. 2009. Tropomyosin isoform expression regulates the transition of adhesions to determine cell speed and direction. Mol. Cell Biol. 29(6): 1506-1514.

Beierle, E.A. and A. Trujillo, A. Nagaram, V.M. Golubovskaya, W.G. Cance, and E.V. Kurenova. 2008. TAE226 inhibits human neuroblastoma cell survival. Cancer Invest. 26(2): 145-151.

Brunton, V.G. and M.C. Frame. 2008. Src and focal adhesion kinase as therapeutic targets in cancer. Curr. Opin. Pharmacol. 8(4): 427-432.

Buettner, R. and T. Mesa, A. Vultur, F. Lee, and R. Jove. 2008. Inhibition of Src family kinases with dasatinib blocks migration and invasion of human melanoma cells. Mol. Cancer Res. 6(11): 1766-1774.

Campbell, E.J. and E. McDuff, O. Tatarov, S. Tovey, V, Brunton, T.G. Cooke, and J. Edwards. 2008. Phosphorylated c-Src in the nucleus is associated with improved patient outcome in ER-positive breast cancer. Br. J Cancer. 99(11): 1769-1774.

Chang, Y.M. and L. Bai, S. Liu, J.C. Yang, H.J. Kung, and C.P. Evans. 2008. Src family kinase oncogenic potential and pathways in prostate cancer as revealed by AZD0530. Oncogene 27(49): 6365-6375.

Cowell, L.N. and J.D. Graham, A.H. Bouton, C.L. Clarke, and G.M. O'Neill. 2006. Tamoxifen treatment promotes phosphorylation of the adhesion molecules, p130Cas/BCAR1, FAK and Src, via an adhesion-dependent pathway. Oncogene 25(58): 7597-7607.

Deakin, N.O. and C.E. Turner. 2008. Paxillin comes of age. J. Cell Sci. 121(Pt15): 2435-2444.

Defilippi, P. and P. Di Stefano, and S. Cabodi, S. 2006. p130Cas: a versatile scaffold in signaling networks. Trends Cell Biol. 16(5): 257-263.

Eustace, A.J. and J. Crown, M. Clynes, and N. O'Donovan. 2008. Preclinical evaluation of dasatinib, a potent Src kinase inhibitor, in melanoma cell lines. J. Transl. Med. 6:53.

Friedl, P. 2004. Prespecification and plasticity: shifting mechanisms of cell migration. Curr. Opin. Cell Biol. 16(1): 14-23.

Golubovskaya, V.M. and C. Virnig, and W.G. Cance. 2008. TAE226-induced apoptosis in breast cancer cells with overexpressed Src or EGFR. Mol. Carcinog. 47(3): 222-234.

Halder, J. and C.N. Landen, Jr., S.K. Lutgendorf, Y. Li, N.B. Jennings, D. Fan, G.M. Nelkin, R. Schmandt, M.D. Schaller, and A.K. Sood. 2005. Focal adhesion kinase silencing augments docetaxel-mediated apoptosis in ovarian cancer cells. Clin. Cancer Res. 11(24 Pt 1): 8829-8836.

Hanahan, D. and R.A. Weinberg. 2000. The hallmarks of cancer. Cell 100(1): 57-70.

Hariharan, S. and D. Gustafson, S. Holden, D. McConkey, D. Davis, M. Morrow, M. Basche, L. Gore, C. Zang, C.L. O'Bryant, A. Baron, D. Gallemann, D. Colevas, and S.G. Eckhardt.

2007. Assessment of the biological and pharmacological effects of the alpha nu beta3 and alpha nu beta5 integrin receptor antagonist, cilengitide (EMD 121974), in patients with advanced solid tumors. Ann. Oncol. 18(8): 1400-1407.

Hehlgans, S. and M. Haase, and N. Cordes. 2007. Signalling via integrins: implications for cell survival and anticancer strategies. Biochim. Biophys. Acta 1775(1): 163-180.

Hiscox, S. and L. Morgan, T.P. Green, D. Barrow, J. Gee, and R.I. Nicholson. 2006. Elevated Src activity promotes cellular invasion and motility in tamoxifen resistant breast cancer cells. Breast Cancer Res. Treat. 97(3): 263-274.

Johnson, F.M. and B. Saigal, M. Talpaz, and N.J. Donato. 2005. Dasatinib (BMS-354825) tyrosine kinase inhibitor suppresses invasion and induces cell cycle arrest and apoptosis of head and neck squamous cell carcinoma and non-small cell lung cancer cells. Clin. Cancer Res. 11(19 Pt 1): 6924-6932.

Kallergi, G. and D. Mavroudis, V. Georgoulias, and C. Stournaras. 2007. Phosphorylation of FAK, PI-3K, and impaired actin organization in CK-positive micrometastatic breast cancer cells. Mol. Med. 13(1-2): 79-88.

Li, S. and Z.C. Hua. 2008. FAK expression regulation and therapeutic potential. Adv. Cancer Res. 101: 45-61.

Luo, F.R. and Y.C. Barrett, Z. Yang, A. Camuso, K. McGlinchey, M.L. Wen, R. Smykla, K. Fager, R. Wild, H. Palme, S. Galbraith, A. Blackwood-Chirchir, and F.Y. Lee. 2008. Identification and validation of phospho-SRC, a novel and potential pharmacodynamic biomarker for dasatinib (SPRYCEL), a multi-targeted kinase inhibitor. Cancer Chemother. Pharmacol. 62(6): 1065-1074.

Matkowskyj, K.A. and K. Keller, S. Glover, L. Kornberg, R. Tran-Son-Tay, and R.V. Benya. 2003. Expression of GRP and its receptor in well-differentiated colon cancer cells correlates with the presence of focal adhesion kinase phosphorylated at tyrosines 397 and 407. J Histochem. Cytochem. 51(8): 1041-1048.

Oliveira-Ferrer, L. and J. Hauschild, W. Fiedler, C. Bokemeyer, J. Nippgen, I. Celik, and G. Schuch. 2008. Cilengitide induces cellular detachment and apoptosis in endothelial and glioma cells mediated by inhibition of FAK/Src/AKT pathway. J. Exp. Clin. Cancer Res. 27(1):86.

Roberts, W.G. and E. Ung, P. Whalen, B. Cooper, C. Hulford, C. Autry, D. Richter, E. Emerson, J. Lin, J. Kath, K. Coleman, L. Yao, L. Martinez-Alsina, M. Lorenzen, M. Berliner, M. Luzzio, N. Patel, E. Schmitt, S. LaGreca, J. Jani, M. Wessel, E. Marr, M. Griffor, and F. Vajdos. 2008. Antitumor activity and pharmacology of a selective focal adhesion kinase inhibitor, PF-562,271. Cancer Res. 68(6): 1935-1944.

Ruest, P.J. and N.Y. Shin, T.R. Polte, X. Zhang, and S.K. Hanks. 2001. Mechanisms of CAS substrate domain tyrosine phosphorylation by FAK and Src. Mol. Cell Biol. 21(22): 7641-7652.

Serrels, A. and I.R. Macpherson, T.R. Evans, F.Y. Lee, E.A. Clark, O.J. Sansom, G.H. Ashton, M.C. Frame, and V.G. Brunton. 2006. Identification of potential biomarkers for measuring inhibition of Src kinase activity in colon cancer cells following treatment with dasatinib. Mol. Cancer Ther. 5(12): 3014-3022.

Shi, Q. and A.B. Hjelmeland, S.T. Keir, L. Song, S. Wickman, D. Jackson, O. Ohmori, D.D. Bigner, H.S. Friedman, and J.N. Rich. 2007. A novel low-molecular weight inhibitor of focal adhesion kinase, TAE226, inhibits glioma growth. Mol. Carcinog. 46(6): 488-496.

Slack-Davis, J.K. and K.H. Martin, R.W. Tilghman, M. Iwanicki, E.J. Ung, A. Autry, M.J. Luzzio, B. Cooper, J.C. Kath, W.G. Roberts, and J.T. Parsons. 2007. Cellular characterization of a novel focal adhesion kinase inhibitor. J. Biol. Chem. 282(20): 14845-14852.

van der Flier, S. and A. Brinkman, M.P. Look, E.M. Kok, M.E. Meijer-van Gelder, J.G. Klijn, L.C. Dorssers, and J.A. Foekens. 2000. Bcar1/p130Cas protein and primary breast cancer: prognosis and response to tamoxifen treatment. J. Nat. Cancer Inst. 92(2): 120-127.

Zaidel-Bar, R. and S. Itzkovitz, A. Ma'ayan, R. Iyengar, and B. Geiger. 2007. Functional atlas of the integrin adhesome. Nat. Cell Biol. 9(8): 858-867.

CHAPTER 24

Adhesion Molecules in Normal Hematopoiesis and Leukemia

Mirela de Barros Tamarozzi[1] and Eduardo Magalhães Rego[2]

[1,2]Hematology Division, Department of Internal Medicine, and National Institute of Science and Technology on Cell Therapy, Medical School of Ribeirão Preto, University of São Paulo, Brazil

Additional author: [1]Ms.C. Hematology Division, Department of Internal Medicine, and National Institute of Science and Technology on Cell Therapy, Medical School of Ribeirão Preto, University of São Paulo, Brazil, E-mail: tamarozzi@usp.br

[2]M.D., Ph.D, Hematology Division, Department of Internal Medicine, Medical School of Ribeirão Preto, and National Institute of Science and Technology on Cell Therapy, Av Bandeirantes 3900, CEP 14049-900, Ribeirão Preto, SP, Brazil, E-mail: emrego@hcrp.fmrp.usp.br

Institution communication: Divisão de Hematologia, Departamento de Clínica Médica, Faculdade de Medicina de Ribeirão Preto, Av Bandeirantes 3900, CEP 14049-900, Ribeirão Preto, SP, Brazil, E-mail: dalvinha@hemocentro.fmrp.usp.br.

Alternative contact: Ms Dalva Cato

ABSTRACT

The fate of hematopoietic stem cells is largely regulated by interactions with the bone marrow environment. This cross-talk mainly occurs in two niches: along the endosteal surface of trabecular bone in close proximity to osteoblasts (osteoblastic niche) and the endothelial cells that line blood vessels (vascular niche). Adhesion molecules are expressed by stromal cells, extracellular matrix, hematopoietic stem cells and committed hematopoietic progenitors and play a pivotal role in the distribution of hematopoietic stem cells in the bone marrow, regulation of

---

*Corresponding author
Key terms are defined at the end of the chapter.

quiescence and determination of their fate (self-renewal or differentiation). In addition, adhesion molecules also participate in the processes of migration and homing of hematopoietic stem cells.

Leukemic stem cells share several characteristics with normal hematopoietic stem cells, including the need for interactions with the microenvironment. For example, the cross-talk between CXCR4 on leukemic cells and SDF-1 at the niche is necessary for homing and *in vivo* growth. Moreover, VLA-4 integrins, along with CXCR4 chemokine receptors, are essential for protection of acute myelogenous leukemia cells from spontaneous or drug-induced apoptosis. More differentiated leukemic cells, mainly those of acute promyelocytic leukemia, express high levels of the β1 integrins VLA-4 and VLA-5, express sLe^x, the ligand of the endothelial E-selectin and may adhere to the endothelium and extravasate from circulation to the tissue. One of the most feared complications of acute promyelocytic leukemia treatment with the all-*trans* retinoic acid (ATRA) and/or arsenic trioxide is differentiation syndrome, characterized by extravascular migration of myeloid cells and clinically associated with fever, weight gain, dyspnea, pleural effusion, and pulmonary infiltrates on chest radiograph. Different authors had demonstrated that the treatment with ATRA modulates the expression of adhesive molecules in acute promyelocytic leukemia cells such as CD11b, CD 18, CDw65, VLA-4, LFA-1, ICAM-1 (CD54) and Mac-1. Moreover, in contrast with wild-type controls, CD54 or CD18 knockout mice did not present the characteristic increase of myeloid cells in lung vasculature when injected with acute promyelocytic leukemia cells and treated with ATRA.

Finally, the expression of adhesion molecules has been associated with treatment outcome in acute leukemias. CD56 expression was significantly associated with a reduced probability of achieving complete remission as well as with a shorter survival in acute myelogenous leukemia and with higher frequency of extramedullary disease in acute lymphoblastic leukemia.

The accumulating evidence of the role of adhesion molecules in normal and leukemic hematopoiesis suggests that they are important biomarkers and may be explored for therapeutic purposes.

INTRODUCTION

The role of adhesion molecules in migration and activation of mature cells from peripheral blood has been extensively studied. Nevertheless, only recently is there accumulating evidence that they play a key role in the physiological control of the fate, migration and homing of hematopoietic stem cells. In addition, compared to their normal counterparts, leukemic stem cells and progenitors have a distinct expression profile of adhesion molecules, which mediate the interaction with the bone marrow microenvironment, thus contributing to survival advantage. In the distinct Acute Myelogenous Leukemia subtype termed Acute Promyelocytic

Leukemia, the modulation of adhesion molecules by differentiating agents such as the all-*trans* retinoic acid (ATRA) and arsenic trioxide constitute the molecular basis of differentiation syndrome, a life-threatening complication of the treatment. In the present chapter, we review some key aspects of the role of adhesion molecules in normal and malignant hematopoiesis.

ADHESION MOLECULES IN NORMAL AND LEUKEMIC STEM CELLS NICHE

Hematopoietic stem cells require a specific microenvironment in order to self-renew or commit to differentiate. The term 'niche' designates this microenvironment, which is composed of stromal cells (macrophages, endothelial cells, adipocytes, reticular cells and T cells) and of extracellular matrix molecules. There are myriad interactions between hematopoietic stem cells and the niche, some of them dependent on the adhesion molecules. From the anatomical point of view, hematopoietic stem cells reside along the endosteal surface of trabecular bone in close proximity to the bone-forming osteoblasts (forming the osteoblastic niche) and the endothelial cells that line blood vessels (vascular niche). Two adhesion molecules exert a pivotal role in the osteoblastic niche function: N-cadherin and β-catenin, which are asymmetrically localized at the interface between hematopoietic stem cells and the osteoblastic niche (Zhang *et al.* 2003) and are thought to facilitate the anchoring of stem cells. However, Haug *et al.* (2008) have demonstrated that bone marrow cells expressing high levels of N-cadherin were unable to reconstitute hematopoietic lineages in irradiated recipient mice in contrast to those expressing low levels of N-cadherin and expressing genes known to prime hematopoietic stem cells to mobilize. In fact, N-cadherin expression at low levels was more frequent in hematopoietic stem cells mobilized from bone marrow to spleen. Therefore, it has been proposed that N-cadherin is required to maintain hematopoietic stem cell adhesion to the niche, to regulate quiescence, to regulate β-catenin signaling, and to maintain hematopoietic stem cells in an undifferentiated state. This concept has been recently challenged by the demonstration that the conditional deletion of N-cadherin from hematopoietic stem cells and other hematopoietic cells in adult mice had no detectable effect on hematopoietic stem cell maintenance or hematopoiesis (Kiel *et al.* 2009).

Hematopoietic stem cells also express the integrins α4β1 (also named very late antigen 4, VLA-4) and α5β1 (VLA-5), which are involved in the adhesion to bone marrow stromal cells through fibronectin. Moreover, α4β1-mediated adhesion promoted proliferation and prevented apoptosis of hematopoietic stem cells (Wang *et al.* 1998) and the hematopoietic stem cells from β1-integrin-deficient failed to colonize the spleen and bone marrow (Potocnik *et al.* 2000). Nevertheless, Brakebusch *et al.* (2002), using a conditional model in which the deletion of β-1 integrin was restricted to the hematopoetic cells, demonstrated that the mutants presented normal hematopoiesis and retained normal numbers of hematopoietic stem cells in the bone marrow.

Sinusoidal endothelial cells in bone marrow have been revealed as an alternative hematopoietic stem cells niche called the 'vascular niche'. Sinusoids are specialized vessels that allow cells to extravasate from venous circulation into hematopoietic tissues. The vascular niche in adult bone marrow is defined as a place for stem cell mobilization or proliferation and differentiation, different from osteoblastic niche, which is thought to maintain hematopoietic stem cell quiescence over the long term. Sequential migration of hematopoietic cells, which begins with stem cells in the osteoblastic niche, progresses through proliferation and differentiation and finally maturation. In close proximity of the vascular niche, Sugiyama *et al.* (2006) found that sinusoids are surrounded by reticular cells in bone marrow that express very high levels of CXCL12 (also known as stroma derived factor 1, SDF-1), a chemokine required for the maintenance of hematopoietic stem cells. They also demonstrated that both hematopoietic stem cells localized around sinusoids and hematopoietic stem cells localized to the endosteum or to other locations were consistently in contact with these CXCL12-expressing reticular cells. The main receptor for CXCL12 is CXCR4, which is expressed by hematopoietic stem cells. CXCL12-CXCR4 signaling is required for the colonization of bone marrow by hematopoietic stem cells during development and regulates the grafting of hematopoietic stem cells after transplantation.

Migration and homing of hematopoietic stem cells are important physiological processes in which adhesion molecules, such as sialomucins, selectins and integrins, play a key role. Migration refers to the leaving of hematopoietic stem cells from the osteoblastic niche, mobilizing to the vascular niche, entering the blood vessel through the endothelial cells, and circulating in the vascular system, where hematopoietic stem cells are in constant contact with endothelial cells. Homing of hematopoietic stem cells is simply the reverse of this process, with hematopoietic stem cells leaving circulation to transendothelial migration reaching the vascular niche, and finally coming back to the osteoblastic niche (Fig. 1). Cytokines such as stem cell factor (SCF), chemokines such as SDF-1 and interleukin 8 (IL-8), and proteolytic enzymes such as the metalloproteinase (MMP) superfamily are also involved in migration and homing. Among the membrane-bound molecules involved in these processes, one of special interest is a sialofucosylated glycoform of CD44 expressed exclusively on human hematopoietic cells and called hematopoietic cell E-/L-selectin ligand (HCELL). CD44 is a ubiquitously expressed glycoprotein that helps cells to adhere to other cells and to matrix proteins, and participates in the recruitment of certain white blood cells to sites of inflammation and in their migration through lymphatic tissues. In hematopoietic stem cells, the isoforms of CD44 bind to its ligand, hyaluronic acid, which is expressed in the bone marrow sinusoids and in the endosteal region. On human hematopoietic

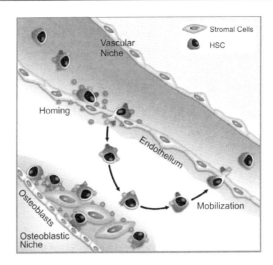

Fig. 1 Homing and mobilization in the hematopoietic stem cells niches. Hematopoietic stem cells express a number of adhesion molecules on their surface that are important to both homing and mobilization. Key molecules involved in these processes are selectins, integrins, CD44 and their ligands. Chemokines are important in guiding the homing of stem cells from the periphery into the bone marrow and ultimately into the stem cell niche.

Color image of this figure appears in the color plate section at the end of the book.

stem cells, HCELL is the most potent E- and L-selectin ligand, mediating these binding hemodynamic flow conditions.

Leukemic stem cells, also named leukemia-initiating cells, are derived from hematopoietic stem cells or proliferating progenitor cells through the acquisition of stem cell properties. In acute myelogenous leukemia, it has been demonstrated that interaction between CXCR4 on leukemic cells and SDF-1 at the niche is necessary for proper homing and *in vivo* growth of leukemic cells, suggesting that leukemic stem cells as the normal hematopoietic stem cells may require niche interactions (Tavor *et al.* 2004). Moreover, it has been demonstrated that the VLA-4 integrins, along with CXCR4 chemokine receptors, are essential for protection of acute myelogenous leukemia cells from spontaneous or drug-induced apoptosis (Delforge *et al.* 2005, Spoo *et al.* 2007). VLA-4-expressing acute myelogenous leukemia cells can better adhere to fibronectin and stromal cells in response to cytokine stimulation, thereby protecting them from apoptosis *in vitro* and possibly leading to minimal residual disease *in vivo* with the potential to trigger leukemic relapse (Matsunaga *et al.* 2003). Other relevant adhesion molecules expressed by acute myelogenous leukemia blasts are PECAM (CD31) and CD38 that mediate the interaction with microenvironmental elements, i.e., CD31 on the surface of marrow endothelial cells (CD31/CD31 and CD38/CD31 interactions) and hyaluronate (CD38/hyaluronate interactions). Excess of CD31 relative to CD38 on the cell membrane of leukemic cells promoted a homotypic interaction with marrow endothelial cells, resulting in higher transendothelial migration. Conversely, excess of CD38 resulted in arrest of the blasts within the bone marrow through hyaluronate adhesion (Gallay *et al.* 2007).

Besides acute myelogenous leukemia, adhesive defects have also been described in acute lymphoblastic leukemia. Bradstock *et al.* (2000) have demonstrated that acute lymphoblastic leukemia blasts migrate into layers of bone marrow fibroblasts *in vitro* by using the α1 integrins VLA-4 and VLA-5. Accordingly, Kollet *et al.* (2001) have shown that homing of leukemic acute lymphoblastic leukemia blasts into the bone marrow of NOD/SCID mice is dependent on VLA-4, VLA-5, and LFA-1.

ADHESION MOLECULES AND ACUTE PROMYELOCYTIC LEUKEMIA

Several acute leukemia subtypes are known to express adhesion molecules. However, acute promyelocytic leukemia represents the paradigm of the clinical importance of these molecules. It is characterized by a balanced reciprocal translocation between chromosomes 15 and 17. As a result, the *Promyelocytic Leukemia* (PML) gene is translocated and fused with the *Retinoic Acid Receptor α* (RARα) generating the PML/RARα hybrid gene. The PML/RARα fusion protein acts as a transcription repressor blocking the differentiation of acute promyelocytic leukemia blasts at the stage of promyelocytes. In immunophenotypic terms, acute promyelocytic leukemia cells express myeloid markers such as CD33 and CD117, while the expression of CD13 is variable. In addition, mature myeloid antigens, such as CD15 and CD14, are expressed weakly or are absent. Furthermore, there appears to be an increased incidence of expression of the neural cell adhesion molecule (NCAM), CD56, which has been associated with poor clinical outcome in acute promyelocytic leukemia. Acute promyelocytic leukemia cells present some differences in the expression of adhesive molecules when compared to neutrophils: they express high levels of the β1 integrins VLA-4 and VLA-5 and lower levels of CD11b and usually express sLe^x, the ligand of E-selectin that probably contributes to adherence to E-selectin on activated endothelium (Di Noto *et al.* 1994).

Currently, the use of ATRA- and anthracycline-based chemotherapy represents the mainstay in treatment of acute promyelocytic leukemia, inducing complete hematological and molecular remission in a high proportion of patients. ATRA induces the malignant cells to differentiate into phenotypically mature myeloid cells, and this differentiation determines the activation of several leukocyte functions including expression of adhesion molecules and secretion of cytokines (Di Noto *et al.* 1994). Acute promyelocytic leukemia cells attach to a layer of endothelial cells in a manner similar to neutrophils. Moreover, acute promyelocytic leukemia cells produce IL-1, which might sustain their growth and induce local activation of adhesion molecules on the endothelium, leading to an increase of adhesiveness of the blasts.

The use of arsenic trioxide in acute promyelocytic leukemia treatment improved the clinical outcome of refractory or relapsed as well as newly diagnosed acute promyelocytic leukemia. Arsenic trioxide at lower doses induces myeloid differentiation, whereas in higher doses it induces apoptosis. Although ATRA and arsenic trioxide are well tolerated, approximately one-fourth of the patients develop differentiation syndrome, formerly known as retinoid acid syndrome. Symptoms occur after 2 to 21 d of treatment and are generally associated with increasing white blood cell count and combined fever, weight gain, dyspnea, pleural effusion, and pulmonary infiltrates on chest radiograph and, in some patients, renal failure, hypotension, and pericardial effusion. The reported incidence of differentiation syndrome ranges from 6% to 27%, with mortality rates varying from 1% to 7% (Santos *et al.* 2004).

Different reports have shown that leukocyte emigration from blood is a key event in the development of differentiation syndrome. It depends on a sequential cascade of leukocyte-endothelium interactions that are regulated by adhesion molecules. Many authors have demonstrated that treatment with ATRA modulates the expression of adhesive molecules in acute promyelocytic leukemia cells such as CD11b, CD 18, CDw65, VLA-4, LFA-1 and ICAM-1 (CD54) (Di Noto *et al.* 1994, Cunha De Santis *et al.* 2007). For NB4 cells, ATRA upregulated the expression of adhesion molecules CD11a, CD11b, CD11c, CD54, CD66c and CD138, consistent with the increased adhesiveness of leukemia cells observed for acute promyelocytic leukemia patients treated with ATRA (Cunha De Santis *et al.* 2007, Barber *et al.* 2008). Interestingly, the β2 integrin LFA-1 is expressed on ATRA-treated NB4 cells in an active form and does not require activation by cytokines or chemokines, as normally required to induce ligand-binding. All these events might explain the increased adhesiveness of the leukemic cells after treatment and how they can contribute to differentiation syndrome development (Fig. 2).

Recently, our group demonstrated that the treatment of acute promyelocytic leukemia primary cells with granulocyte colony-stimulating factor (G-CSF), a cytokine that increase the commitment of hematopoietic progenitors to terminal differentiation, upregulated CD11b expression and potentiated ATRA-induced CD18 and CD11b expression on these cells (Cunha De Santis *et al.* 2007). In the same study, arsenic trioxide upregulated ICAM-1 in NB4 cells. The increase in CD11b, CD18 and CD54 expression was accompanied by a higher adhesion to Matrigel. In addition, the injection of ATRA-treated NB4 cells in Balb/c mice resulted in an increase of the number of myeloid cells adhered to pulmonary endothelium. Both *in vitro* and *in vivo* effects were blocked by pre-incubation with dexamethasone, anti-CD54 or anti-CD18. To test the role of these adhesion molecules *in vivo*, CD54 or CD18 knockout mice and their wild-type controls were injected with NB4 cells and treated intraperitoneally with ATRA. There was an increase in lung cell migration in wild type but not in knockout animals, confirming the results obtained *in vitro*. Thus, these results suggest that both leukocyte and endothelial adhesion molecules are essential for differentiation syndrome development.

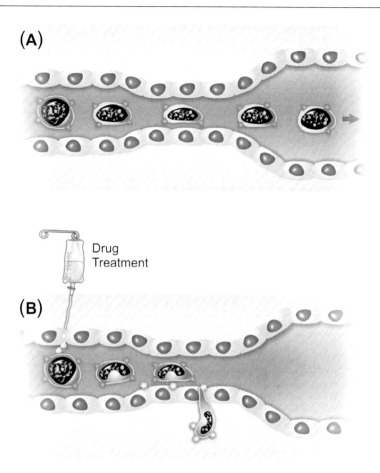

Fig. 2 Effect of ATRA and/or arsenic trioxide induced differentiation on adhesion molecule expression and cell extravasation. Compared to non-treated cells (A), acute promyelocytic leukemia cells treated with ATRA and/or arsenic trioxide (B) express higher levels of adhesion molecules such as CD11b, CD18 and ICAM-1. Similar effects are observed on endothelial cells. Consequently, there is facilitation of the stable arrest and transmigration.

There is a significant association between ICAM-1 polymorphisms and the risk of differentiation syndrome development. Patients harboring the AA genotype at codon 469 of ICAM-1 presented 3.5 times the chances of developing differentiation syndrome as those with GA or GG genotypes. The 469 E/K polymorphism in exon 6 results in a change from glutamic acid to lysine in Ig-like domain 5 of ICAM-1, which is thought to affect interactions with LFA-1 and adhesion of B-cells (Dore *et al.* 2007).

ADHESION MOLECULES AS BIOMARKERS IN ACUTE MYELOGENOUS LEUKEMIA

Many associations have been described between individual antigen expression on myeloid blasts and prognosis; however, few are consistent. Some adhesion molecules have been significantly related with a prognostic association such as CD15, CD11b, CD34 and CD56. CD15 is cell surface glycoprotein that is the ligand of E and P selectins, expressed on maturing cells of monocyte lineage and more weakly on maturing cells of granulocyte lineage. Tien *et al.* (1995) found that CD15 and CD11b were of prognostic value: CD15 with a higher complete remission rate and CD11b with a shorter complete remission duration. CD34 is a monomeric cell adhesion molecule of the syalomucin family that is expressed on leukemic blasts and hematopoietic stem cells. CD34 positivity has been correlated with a poor response to induction chemotherapy in acute myelogenous leukemia patients. Raspadori *et al.* (1997) investigated the expression of CD34 antigen on leukemic cells in 141 adult patients with diagnosis of acute myeloid leukemia. In patients whose blasts expressed CD34 antigen, a significantly lower rate of complete remission was observed than in CD34-negative cases (61% vs 88%), suggesting that a high CD34 intensity of expression should be considered as a reliable poor prognostic factor.

CD56 antigen has been found frequently expressed in several lympho-hematopoietic neoplasms. In fact, it has been reported that the presence of CD56 antigen on the blasts of acute myelogenous leukemia patients with t(8;21) (q22;q22), and in those with acute promyelocytic leukemia, identifies a subgroup of patients with a more unfavorable prognosis. On the basis of these findings, Raspadori *et al.* (2001) evaluated CD56 surface expression in 152 newly diagnosed acute myelogenous leukemia patients and demonstrated that it was significantly associated with a reduced probability of achieving complete remission as well as with a shorter survival. These results were confirmed by other authors, showing that in general the expression of CD56 in acute myelogenous leukemia is a 'negative prognostic marker'. CD56 expression also has been correlated with extramedullary disease. This could be explained by the hypothesis that CD56-expressing blasts can bind to $\beta3$ integrins on endothelial cells and are thus responsible for the higher incidence of extramedullary manifestations in CD56+ leukemias. Another molecule associated with prognosis is the CD44 antigen, which on leukemic blasts from most acute myelogenous leukemia patients is involved in myeloid differentiation. The expression of variant forms of CD44 has been associated with poor prognosis in acute myelogenous leukemia. CD44 displays many variant isoforms (CD44v) generated by alternative splicing of exons 2v to 10v. Legras *et al.* (1998), demonstrated that the expression of CD44-6v correlates with a shorter survival of patients treated with conventional chemotherapy. CD44 is a potential therapeutic target, and treatment with an activating mAb specific to CD44 (H90) was able to eradicate acute myelogenous leukemia leukemic stem cells *in vivo* by blocking leukemic stem cells trafficking to supportive microenvironments and by altering their stem cell fate in a NOD/SCID xenotransplant model (Jin *et al.* 2006).

Taken together, the experimental and clinical data discussed above demonstrate that signaling through adhesion molecules is essential for the regulation of hematopoieses, and their aberrant expression and function provide a survival advantage to leukemic cells. Therefore, an ever-increasing number of studies corroborate the idea that adhesion molecules are important biomarkers and targets for the development of new therapeutic strategies.

SUMMARY

- Hematopoietic stem cells reside in specific niches that regulate their survival, proliferation, self-renewal, or differentiation in the bone marrow. Both osteoblastic and vascular niches may play important roles in regulating hematopoietic stem cell homing and mobilization, processes mediated by adhesion molecules.
- Leukemic stem cells are derived from hematopoietic stem cells or proliferating progenitor cells through the acquisition of stem cell properties. Some adhesion molecules expressed by acute myelogenous leukemia blasts participate in the interaction with the bone marrow microenvironment and give the leukemic cell a survival advantage.
- The expression of adhesion molecules has been associated with the prognosis of patients with acute leukemia.
- Treatment with ATRA induces the differentiation of acute promyelocytic leukemia cells, which is associated with an increase in the expression of adhesion molecules.
- The blockage of adhesion molecule function, such as CD44, may be exploited therapeutically.

Acknowledgments

The authors are grateful to Mrs. Sandra Navarro for her assistance with the illustrations and to Drs. Alexandre Krause and Barbara Santana-Lemos for critical review of the text.

Abbreviations

ATRA	all-*trans* retinoic acid
CD44v	CD44 variant isoform

G-CSF	granulocyte colony-stimulating factor
HCELL	hematopoietic cell E-/L-selectin ligand
ICAM-1	intercellular adhesion molecule 1
IL-1	interleukin 1
IL-8	interleucin 8
LFA-1	lymphocyte function-associated antigen 1
MAC-1	macrophage antigen 1
MMP	metalloproteinase
NCAM	neural adhesion molecules
NOD/SCID mice	Non-obese diabetic mice
PECAM	platelet endothelial cell adhesion molecule
RARα	retinoic acid receptor α
SCF	stem cell factor
SDF-1	stroma derived factor 1
sLe^x	sialyl-Lewis^x-like
VLA	very late antigen

Key Facts about Hematopoiesis

1. Hematopoiesis is the process by which blood cells are formed.
2. All mature blood cells derive from a pool of progenitors known as hematopoietic stem cells, which, after birth, reside in the bone marrow and may circulate.
3. Hematopoietic stem cells have the capacity for unlimited or prolonged self-renewal that can give rise by asymmetric division to a progenitor cell characterized by loss of self-renewal capacity.
4. There are two main hematopoietic lineages that are differentiated from immature progenitors which derive from the hematopoietic stem cells: (a) the lymphoid progenitors that ultimately give rise to B-cells and T-cells and (b) the myeloid progenitors that eventually give rise to monocytes, platelets, granulocytes/monocytes (neutrophils, eosinophils, basophils), and erythrocytes.

Key Facts about Leukemia

1. Leukemia is a malignant and monoclonal proliferation of hematopoietic cells. In leukemia, bone marrow is infiltrated by a large number of abnormal progenitors leading to a block of the production of normal white blood cells.

2. Leukemia can be classified in terms of lineage and degree of maturation of the predominant malignant cells. Thus, there are lymphoid and myeloid leukemias and each subtype is divided into multiple subtypes based on morphology and immunological or genetic cell markers. Acute leukemia is characterized by the appearance of immature, abnormal cells in the bone marrow and peripheral blood.

3. Acute leukemias are associated with several genetic abnormalities, which may result from chromosomal abnormalities and/or from gene mutations.

4. The genetic abnormalities lead to the deregulation of several molecular pathways and are associated with the response to treatment and prognosis.

Key Facts about Acute Promyelocytic Leukemia

1. Acute promyelocytic leukemia is a subtype of acute myelogenous leukemia characterized by the presence of translocations involving chromosome 17 at the locus of the *Retinoic Acid Receptor* α (RARα). In most patients, the chromosomal translocation between chromosomes 15 and 17 [t(15;17)] is detected, which generates the PML-RARα fusion gene.

2. The leukemic cells of acute promyelocytic leukemia resemble normal promyelocytes, which are one of the steps during myeloid maturation; for this reason the leukemic cells are said to present a 'block of differentiation'.

3. Treatment with ATRA induces the differentiation of the leukemic cells into mature myeloid cells, thus leading to the remission of acute promyelocytic leukemia.

4. Although patients with acute promyelocytic leukemia respond well to treatment with ATRA, they may develop a life-threatening complication called differentiation syndrome, which is characterized by the infiltration of lung and other organs by differentiating myeloid cells.

5. Changes in the expression of adhesion molecules induced by ATRA play an important role in the pathophysiology of differentiation syndrome.

Definition of Terms

Acute myelogenous leukemia: A malignant disease of the bone marrow in which hematopoietic precursors are arrested in an early stage of development. It is characterized by the presence of more than 20% leukemic cells of myeloid origin in the bone marrow or blood.

Acute lymphoblastic leukemia: A malignant (clonal) disease of the bone marrow in which early lymphoid precursors proliferate and replace the normal hematopoietic cells of the marrow. It is characterized by the presence of more than 20% leukemic cells of lymphoid origin in the bone marrow or blood.

Acute promyelocytic leukemia: A subtype of acute myelogenous leukemia characterized by the infiltration of the bone marrow by leukemic promyelocytes that harbor translocations involving the chromosome 17.

Bone marrow: The soft blood-forming tissue that fills the cavities of bones and contains fat and immature and mature blood cells, including white blood cells, red blood cells, and platelets.

Differentiation syndrome: A treatment complication that can occur in acute promyelocytic leukemia patients treated with ATRA or arsenic trioxide, which is characterized by enhanced leukocyte transmigration, endothelium damage with increase in capillary permeability, microcirculation obstruction, and tissue infiltration.

Hematopoietic cell E-/L-selectin ligand: A glycoform of CD44 expressed by hematopoietic stem cells. This molecule is the most potent E-selectin ligand natively expressed on any human cell.

Hematopoietic stem cells: Cells that can renew themselves and generate mature cells of the blood-forming and immune systems. They give rise to common lymphoid progenitors and common myeloid progenitors.

Leukemic stem cells: Rare cells with indefinite proliferation potential that drive tumor formation and growth. Because normal stem cells and leukemic stem cells share the ability to self-renew, as well as various developmental pathways, it has been postulated that leukemic stem cells are hematopoietic stem cells that have become leukemic as the result of accumulated mutations or more restricted progenitors, which have reacquired the stem cell capability of self-renewal.

Promyelocytic leukemia gene: A gene first identified through its fusion to the gene encoding the *Retinoic Acid Receptor* alpha (RARα) in acute promyelocytic leukemia patients. The PML:RARα fusion gene is detected in 98% of patients with acute promyelocytic leukemia.

References

Barber, N. and L. Belov, and R.I. Christopherson. 2008. All-trans retinoic acid induces different immunophenotypic changes on human HL60 and NB4 myeloid leukaemias. Leuk. Res. 32: 315-322.

Bradstock, K.F. and V. Makrynikola, A. Bianchi, W. Shen, J. Hewson, and D.J. Gottlieb. 2000. Effects of the chemokine stromal cell-derived factor-1 on the migration and localization of precursor-B acute lymphoblastic leukemia cells within bone marrow stromal layers. Leukemia 14: 882-888.

Brakebusch, C. and S. Fillatreau, A.J. Potocnik, G. Bungartz, P. Wilhelm, M. Svensson, P. Kearney, H. Körner, D. Gray and R. Fässler. 2002. Beta1 integrin is not essential for hematopoiesis but is necessary for the T cell-dependent IgM antibody response. Immunity 16: 465-477.

Cunha De Santis, G. and M.B. Tamarozzi, R.B. Souza, S.E. Moreno, D. Secco, A.B. Garcia, A.S. Lima, L.H. Faccioli, R.P. Falcão, F.Q. Cunha and E.M. Rego. 2007. Adhesion molecules and Differentiation Syndrome: phenotypic and functional analysis of the effect of ATRA, As2O3, phenylbutyrate, and G-CSF in acute promyelocytic leukemia. Haematologica 92: 1615-1622.

Delforge, M. and V. Raets, V. Van Duppen, P. Vandenberghe, and M. Boogaerts. 2005. CD34+ marrow progenitors from MDS patients with high levels of intramedullary apoptosis have reduced expression of alpha4beta1 and alpha5beta1 integrins. Leukemia 19: 57-63.

Di Noto, R. and E.M. Schiavone, F. Ferrara, C. Manzo, C. Lo Pardo, and L. Del Vecchio. 1994. Expression and ATRA-driven modulation of adhesion molecules in acute promyelocytic leukemia. Leukemia 8: 1900-1905.

Dore A.I. and B.A.A. Santana-Lemos, V.M. Coser, F.L.S. Santos, L.F. Dalmazzo, A.S.G. Lima, R.H. Jacomo, J. Elias, Jr., R.P. Falcão, W.V. Pereira and E.M. Rego. 2007. The association of ICAM-1 Exon 6 (E469K) but not of ICAM-1 Exon 4 (G241R) and PECAM-1 Exon 3 (L125V) polymorphisms with the development of differentiation syndrome in acute promyelocytic leukemia. J. Leukoc. Biol. 82: 1340-1343.

Gallay, N. and L. Anani, A. Lopez, P. Colombat, C. Binet, J. Domenech, B.B. Weksler, F. Malavasi and O. Herault. 2007. The role of platelet/endothelial cell adhesion molecule-1 (CD31) and CD38 antigens in marrow microenvironmental retention of acute myelogenous leukemia cells. Cancer Res. 16: 8624-8632.

Haug, J.S. and X.C. He, J.C. Grindley, J.P. Wunderlich, K. Gaudenz, J.T. Ross, A. Paulson, K.P. Wagner, Y. Xie, R. Zhu, T. Yin, J.M. Perry, M.J. Hembree, E.P. Redenbaugh, G.L. Radice, C. Seidel and L. Li. 2008. N-cadherin expression level distinguishes reserved versus primed states of hematopoietic stem cells. Cancer Stem Cell 2: 367-379.

Jin, L. and K.J. Hope, Q. Zhai, F. Smadja-Joffe, and J.E. Dick. 2006. Targeting of CD44 eradicates human acute myeloid leukemic stem cells. Nat. Med. 12: 1167-1174.

Kiel, M.J. and M. Acar, G.L. Radice, and S.J. Morrison. 2009. Hematopoietic stem cells do not depend on N-cadherin to regulate their maintenance. Cell Stem Cell 4: 170-179.

Kollet, O. and A. Spiegel, A. Peled, I. Petit, T. Byk, R. Hershkoviz, E. Guetta, G. Barkai, A. Nagler and T. Lapidot. 2001. Rapid and efficient homing of human CD34(+)CD38(-/low)CXCR4(+) stem and progenitor cells to the bone marrow and spleen of NOD/SCID and NOD/SCID/B2m(null) mice. Blood 97: 3283-3291.

Legras, S. and U. Günthert, R. Stauder, F. Curt, S. Oliferenko, H.C. Kluin-Nelemans, J.P. Marie, S. Proctor, C. Jasmin and F. Smadja-Joffe. 1998. A strong expression of CD44-6v correlates with shorter survival of patients with acute myeloid leukemia. Blood 91: 3401-3413.

Matsunaga, T. and N. Takemoto, T. Sato, R. Takimoto, I. Tanaka, A. Fujimi, T. Akiyama, H. Kuroda, Y. Kawano, M. Kobune, J. Kato, Y. Hirayama, S. Sakamaki, K. Kohda, K. Miyake and Y. Niitsu. 2003. Interaction between leukemic-cell VLA-4 and stromal fibronectin is a decisive factor for minimal residual disease of acute myelogenous leukemia. Nat. Med. 9: 1158-1165.

Potocnik, A.J. and C. Brakebusch, and R. Fässler. 2000. Fetal and adult hematopoietic stem cells require beta1 integrin function for colonizing fetal liver, spleen, and bone marrow. Immunity 12: 653-663.

Raspadori, D. and F. Lauria, M.A. Ventura, D. Rondelli, G. Visani, A. Vivo, and S. Tura. 1997. Incidence and prognostic relevance of CD34 expression in acute myeloblastic leukemia: Analysis of 141 cases. Leuk. Res. 21: 603-607.

Raspadori, D. and D. Damiani, M. Lenoci, D. Rondelli, N. Testoni, G. Nardi, C. Sestigiani, C. Mariotti, S. Birtolo, M. Tozzi and F. Lauria. 2001. CD56 antigenic expression in acute myeloid leukemia identifies patients with poor clinical prognosis. Leukemia 15: 1161-1164.

Sackstein, R. and J.S. Merzaban, D.W. Cain, N.M. Dagia, J.A. Spencer, C.P. Lin, and R. Wohlgemuth. 2008. Ex vivo glycan engineering of CD44 programs human multipotent mesenchymal stromal cell trafficking to bone. Nat. Med. 14: 181-187.

Santos, F.L.S. and A.I. Dore, A.S.G. Lima, A.B. Garcia, M.A. Zago, E.G. Rizzatti, J. Elias, Jr., R.P. Falcão and E.M. Rego. 2004. Hematological features and expression profile of myeloid antigens of acute promyelocytic leukemia patients. Analysis of prognostic factors for development of the retinoic acid syndrome. Rev. Assoc. Med. Bras. 50: 286-292.

Spoo, A.C. and M. Lübbert, W.G. Wierda, and J.A. Burger. 2007. CXCR4 is a prognostic marker in acute myelogenous leukemia. Blood 109: 786-791.

Sugiyama, T. and H. Kohara, M. Noda, and T. Nagasawa. 2006. Maintenance of the hematopoietic stem cell pool by CXCL12-CXCR4 chemokine signaling in bone marrow stromal cell niches. Immunity 25: 977-988.

Tavor, S. and I. Petit, S. Porozov, A. Avigdor, A. Dar, L. Leider-Trejo, N. Shemtov, V. Deutsch, E. Naparstek, A. Nagler and T. Lapidot. 2004. CXCR4 regulates migration and development of human acute myelogenous leukemia stem cells in transplanted NOD/SCID mice. Cancer Res. 64: 2817-2824.

Tien, H.F. and C.H. Wang, M.T. Lin, F.Y. Lee, M.C. Liu, S.M. Chuang, Y.C. Chen, M.C. Shen, K.H. Lin and D.T. Lin. 1995. Correlation of cytogenetic results with immunophenotype, genotype, clinical features, and RAS mutation in acute myeloid leukemia. A study of 235 Chinese patients in Taiwan. Cancer Genet. Cytogen. 84: 60-68.

Wang, M.W. and U. Consoli, C.M. Lane, A. Durett, M.J. Lauppe, R. Champlin, M. Andreeff and A.B. Deisseroth. 1998. Rescue from apoptosis in early (CD34-selected) versus late (non-CD34-selected) human hematopoietic cells by very late antigen 4- and vascular cell adhesion molecule (VCAM) 1-dependent adhesion to bone marrow stromal cells. Cell Growth Differ. 9: 105-112.

Zhang, J. and C. Niu, L. Ye, H. Huang, X. He, W.G. Tong, J. Ross, J. Haug, T. Johnson, J.Q. Feng, S. Harris, L.M. Wiedemann and Y. Mishina and L. Li. 2003. Identification of the haematopoietic stem cell niche and control of the niche size. Nature 425(6960): 836-841.

Carbamylated Low-density Lipoprotein and Adhesion Molecules

Eugene O. Apostolov[1], Sudhir V. Shah[2,3], Ercan Ok[4]
and Alexei G. Basnakian[1, 3, *]

[1]Department of Pharmacology and Toxicology, University of Arkansas for Medical Sciences, Slot 501, 4301 West Markham St., Little Rock, AR, 72205, USA
[2]Division of Nephrology, Department of Internal Medicine, University of Arkansas for Medical Sciences, Slot 501, 4301 West Markham St., Little Rock, AR, 72205, USA
[3]Renal Medicine Service, Central Arkansas Veterans Healthcare System, 4300 West 7th St., Little Rock, AR, 72205, USA
[4]Division of Nephrology, Department of Internal Medicine, Ege University Medical School, Bornova 35100, Izmir, Turkey
Departmental administrator: Edith Adams, University of Arkansas for Medical Sciences, Department of Pharmacology & Toxicology, 4301 West Markham, # 638, Little Rock, AR 72205, USA, E-mail: adamsedithr@uams.edu
Yevgeniy (Eugene) O. Apostolov, MD, PhD, University of Arkansas for Medical Sciences, Department of Pharmacology & Toxicology, 4301 West Markham, # 638, Little Rock, AR 72205, USA, E-mail: apostolovyevgeniyo@uams.edu
Sudhir V. Shah, MD, University of Arkansas for Medical Sciences, Department of Internal Medicine, Division of Nephrology, 4301 W. Markham St., #501, Little Rock, AR 72205, USA, E-mail: shahsudhirv@uams.edu
Ercan Ok, MD, Professor, Chair, Division of Nephrology, Department of Internal Medicine, Ege University Medical School, Bornova 35100, Izmir, Turkey, E-mail: ercan.ok@ege.edu.tr
Alexei G. Basnakian, MD, PhD, University of Arkansas for Medical Sciences, Department of Pharmacology & Toxicology, 4301 West Markham, # 638, Little Rock, AR 72205, USA, E-mail: basnakianalexeig@uams.edu

Corresponding author

ABSTRACT

Carbamylated LDL (cLDL), the most abundant modified LDL isoform in human blood, has been recently implicated in causing atherosclerosis-prone injuries to endothelial cells *in vitro* and atherosclerosis in patients with chronic kidney disease. cLDL acts by inducing monocyte adhesion to endothelial cells via activation of adhesion molecules responsible for the recruitment of monocytes. Recent data indicate that intercellular adhesion molecule 1 (ICAM-1) in cooperation with vascular cell adhesion molecule 1 (VCAM-1) is essential for monocyte adhesion by cLDL-activated human vascular endothelial cells *in vitro*. Exposure of human coronary artery endothelial cells (HCAECs) with cLDL but not native LDL caused monocyte adhesion and the induction of ICAM-1 and VCAM-1. Silencing of ICAM-1 by siRNA or its inhibition using neutralizing antibody resulted in decreased monocyte adhesion to the endothelial cells. Similar silencing or neutralizing of VCAM-1 alone did not have an effect but was shown to contribute to ICAM-1 when tested simultaneously.

INTRODUCTION

Chronic kidney disease (CKD) is a common disorder that can result from a wide variety of diseases including diabetes, hypertension and glomerulonephritis, and that affects about 10% of the world's population (Levey *et al.* 2003). For unknown reasons, CKD is an independent risk factor for the development of cardiovascular disease (Sarnak *et al.* 2003). After stratification for age, sex, race, and the presence or absence of diabetes, cardiovascular mortality in patients with advanced kidney disease is 10 to 20 times that in the general population (Sarnak *et al.* 2003). The most frequent causes of cardiovascular complications of CKD such as ischemic heart disease, sudden death, peripheral artery diseases, arterial hypertension and congestive heart failure are occlusive lesions due to atherosclerosis.

Monocyte (leukocyte) adhesion to activated vascular endothelial cells and their migration into the vessel wall constitute the critical event in the initiation of atherosclerosis. This process is caused by the upregulation of adhesion molecules on the surface endothelial cells and an increased expression in the vascular wall of chemotactic factors to monocytes. Highly specific adhesive interactions between monocytes and endothelial cells are mediated by three main families of receptors: members of the immunoglobulin superfamily, selectins and integrins. Recent studies indicate that intercellular adhesion molecule 1 (ICAM-1) and vascular cell adhesion molecule 1 (VCAM-1), members of the immunoglobulin superfamily, are among the most common participants in monocyte attraction triggered by cytokines, homocystein, lipopolysaccharides and other stimuli. Recent studies demonstrated that carbamylated low-density lipoprotein (cLDL), a product of modification by urea-derived cyanate, seems to be the strongest candidate for the missing link (Ok *et al.* 2005). The most prominent pro-atherosclerotic activity of cLDL is the activation of adhesion molecules on the surface of endothelial cells that causes the attraction of monocytes (Apostolov *et al.* 2007b).

LDL CARBAMYLATION

Protein Carbamylation

There are exogenous and endogenous pathways that cause protein carbamylation. Both pathways facilitate carbamylation through the molecule of the cyanate or thiocyanate (Fig. 1). The exogenous pathway utilizes the molecules of cyanide or thiocyanate that are being oxidized by several peroxidases, hemoglobin or other proteins with oxidative properties (Qian *et al.* 1997). The only known endogenous pathway utilizes urea (Kraus and Kraus 2001). Urea spontaneously dissociates to cyanate and ammonia in aqueous solutions causing elevation of cyanate, which is more prominent in uremic patients (Nilsson *et al.* 1996). The active form of cyanate, isocyanic acid, reacts irreversibly with the N-terminal groups of amino acids by a process known as carbamylation or carbamoylation (Kraus and Kraus 2001). When a molecule of cyanate is used for carbamylation, another molecule of cyanate is formed to keep equilibrium between urea and cyanate.

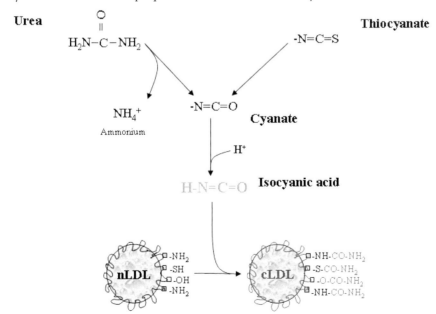

Fig. 1 Carbamylation of LDL. Urea and also thiocyanate of blood plasma are converted to cyanate, and then its isoform, isocyanic acid. The latter modifies ApoB protein in LDL particle to produce carbamylated LDL.

Unlike the exogenous pathway, which completely relies on the alimentary source of the pre-cyanate molecules, the endogenous pathway occurs permanently in the body and is exacerbated when renal function is impaired. The irreversible carbamylation forming epsilon-amino-carbamyl-lysine occurs at multiple lysine sites within a protein with accumulation over the life span of the protein (Kraus and Kraus 2001). The resulting *in vivo* carbamylation changes the structure of proteins and modifies (partly inactivates) the activity of enzymes, cofactors, hormones, and antibodies (Steinbrecher *et al.* 1984, Kraus and Kraus 2001). The best-studied carbamylated protein in patients with CKD is hemoglobin, which has been suggested to correlate with the severity and duration of the exposure to urea, and has been proposed as a measure of the degree of the uremic state (Hasuike *et al.* 2002). Other proteins that have been shown to be carbamylated include lens crystalline, actin and collagen. There are no specific inhibitors of carbamylation, but some studies showed that some decrease of carbamylation can be achieved using aspirin, ibuprofen or Bendazac.

Search for 'Uremic LDL'

Based mainly on studies of oxidized LDL (oxLDL), it is commonly accepted that endothelial cell injury by modified LDLs can initiate the atherosclerotic process. Higher predisposition of the uremic patients to atherosclerosis caused instant interest and search for the uremia-induced proatherogenic modified LDL. Oxidized LDL has been shown to be elevated in uremic patients (Futatsuyama *et al.* 2002, Van Tits *et al.* 2003), but LDL oxidation level was not quite associated with history of cardiovascular disease and its link with uremia remains questionable (Lonn 2001, Yusuf *et al.* 2000). There are also several other 'uremic toxins' that produce their effects through diverse actions on protein structure and function (Bailey and Mitch 1997). Many protein-derived 'uremic toxins' have been proposed, including urea, guanidines, aliphatic and aromatic amines, phenols, indole, aromatic hydroxyacids, oxalic acid, uric acid, and possibly other metabolites, all of which accumulate during renal failure and have various toxic affects. For a long time the role of urea in the clinical manifestations of uremia and in CKD-associated atherosclerosis has been underestimated.

A special type of LDL that appeared as a result of ApoB carbamylation in chronic renal failure patients was initially named 'uremic LDL' by Gonen and collaborators (Gonen *et al.* 1985). Kraus and Kraus (2001) described cLDL generation as a product of the chemical modification of native LDL (nLDL) by urea-derived isocyanate.

Carbamylated LDL: An Abundant LDL Isoform Both in Healthy Individuals and in Uremic Patients

Several chemical modifications of LDL have been reported since the 1970s: acetylated LDL (Basu *et al.* 1976), glycated LDL (Schmidt *et al.* 1995), oxLDL (Sawamura *et al.* 1997) and cLDL (Canal and Girard 1973, Horkko *et al.* 1992, Kraus and Kraus 2001). However, assays based on sandwich ELISA existed primarily for oxLDL and its isoform malondialdehyde-modified LDL (MDA-LDL) in studies by Holvoet *et al.*, Itabe *et al.*, Kohno *et al.*, Kotani *et al.*, Van Tits *et al.* (see references in Apostolov *et al.* 2005). MDA-LDL was detected in normal individuals at 1.9 ± 0.2 mg/L, 3.1 ± 1.6 mg/L, and 17.1 ± 50.2 mg/L. The oxLDL concentration was 10.8 ± 2.8 U/mL or 0.5 ± 0.3 U/µg LDL protein (1 U was 1 µg of mildly oxLDL).

Prior to our reports, tools for cLDL studies were very limited. For the purpose of measuring cLDL, we have developed sandwich ELISA assay. In this assay, anti-cLDL antibody does not cross-react with HDL, carbamylated high-density lipoprotein, very-low-density lipoprotein, native Apolipoprotein B, or seemingly any other human serum proteins (Apostolov *et al.* 2005). We showed that in addition to elevation of total plasma protein carbamylation in end-stage renal disease (ESRD) patients in comparison to a control group of healthy individuals (Ok *et al.* 2002), cLDL concentration is approximately tripled in ESRD patients (281.5 ± 46.9 µg/mL vs. 86.0 ± 29.7 µg/mL). The MDA-LDL was increased 2.5-5 times to 15.8 ± 15.0 µg/mL or 37 ± 2 µg/mL and oxLDL was increase eight-fold. Hence, the cLDL concentration in our study showed much higher values than oxLDL or MDA-LDL (Fig. 2). Therefore, currently cLDL may be considered to be the most abundant LDL isoform both in healthy individuals and in uremic patients on hemodialysis.

Association between cLDL and Atherosclerosis in Uremic Patients

To determine whether the concentration of cLDL was related to atherosclerosis in ESRD patients, serum cLDL and nLDL were quantified using sandwich ELISA in patients with normal common carotid artery intima-media thickness (N-IMT) and thickened common carotid artery intima-media thickness (T-IMT). These data showed that cLDL concentrations were significantly increased in T-IMT patients compared to N-IMT patients, which was not different from levels in healthy controls. These data suggested a potential link between carbamylation of LDL and atherosclerosis in hemodialysis patients.

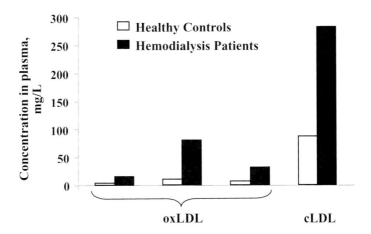

Fig. 2   Concentration of oxLDL and cLDL in blood plasma of healthy individuals and renal patients on hemodialysis. OxLDL and cLDL were measured in healthy individuals and renal patients on hemodialysis determined using sandwich ELISA in different studies: oxLDL (Holvoet *et al.* 1996, Kohno *et al.* 2000, Bosmans *et al.* 2001) and cLDL (Apostolov *et al.* 2005). Concentration of cLDL is several times higher in both patients and controls.

BIOLOGICAL PROPERTIES OF CARBAMYLATED LDL

cLDL Cytotoxicity and Binding by Endothelial Cells

Carbamylated LDL was shown to interact with cell surface receptors in human fibroblasts and to prevent the binding of native LDL in human fibroblasts (Weisgraber *et al.* 1978). LDL isolated from uremic patients as well as chemically carbamylated LDL had a slower clearance from plasma in rabbits than LDL from normal subjects or non-modified LDL (Horkko *et al.* 1992, 1994). Our studies showed that cLDL activates MAPK pathway and induces injury and dysfunction of endothelial cells *in vitro* (Ok *et al.* 2005, Apostolov *et al.* 2007a). The cLDL effects were dose-dependent and relevant to atherosclerosis: irreversible cell injury and apoptosis in endothelial cells, and vascular smooth muscle cell (VSMC) proliferation (Ok *et al.* 2005). Therefore, the cytotoxicity of cLDL is similar to the previously observed cytotoxicity of oxLDL (Galle *et al.* 1999) and acetylated LDL (Dart and Chin-Dusting 1999).

Our recent data suggest that cLDL is capable of binding by and internalization to endothelial cells (Apostolov *et al.* 2008). Additionally, cLDL may transmigrate through endothelium. It utilizes a unique spectrum of scavenger receptors, in which lectin-like oxLDL receptor (LOX-1), CD36, SREC-1 and SR-A1 receptors are essential for the pro-atherogenic effects of cLDL on human endothelial cells. These data are in agreement with other reports, which describe that scavenger receptors may upregulate both ICAM-1 and VCAM-1 expression in response to treatment with other modified LDLs (Chen *et al.* 2005, Inoue *et al.* 2005). Therefore, it seems plausible that cLDL-induced monocyte adhesion is mediated through scavenger receptors and subsequent ICAM-1/VCAM-1 overexpression.

Acceleration of Monocyte Adhesion to Endothelial Cells by cLDL

Our recent study showed that cLDL induces monocyte adhesion by endothelial cells *in vitro* (Apostolov *et al.* 2007b). To determine whether cLDL causes monocyte adhesion to endothelial cells, cLDL (200 µg/mL) was applied to endothelial cells for varying periods of time and fluorescently labeled U937 cells were allowed to adhere for 30 min. We found a significant increase of monocyte adhesion to the endothelial cells treated for 12 hr or longer with cLDL, while nLDL and the vehicle control did not cause any monocyte adhesion (Fig. 3). OxLDL that was previously described to induce monocyte adhesion (Li and Mehta 2000) was used as a positive control. A 6 hr or longer treatment with oxLDL (200 µg/mL) caused a four-fold increase of monocyte adhesion to the endothelial cells. It is interesting that once monocyte adhesion to oxLDL-treated endothelial cells reached the maximum, no further increase of adhesion was observed. As opposed to oxLDL, cLDL caused significant monocyte adhesion versus nLDL and vehicle control only after a 12 hr course, followed then by a further increase of monocyte adhesion over the level of oxLDL. At the end of the 24 hr course, the adhesion induced by cLDL was higher than that induced by oxLDL. In another experiment, freshly isolated human monocytes also had higher rate of adherence to endothelial cells pretreated with both cLDL and oxLDL (Fig. 4). cLDL- or oxLDL-activated endothelial cells also attracted monocytes under flow conditions, in laminar flow chambers. These results demonstrate that HCAECs treated with modified LDLs attracted more U937 cells or freshly isolated monocytes than vehicle- or nLDL-treated cells both in static adhesion experiments and in flow chambers.

cLDL Induces ICAM-1 and VCAM-1 Expression in Vascular Endothelial and Smooth Muscle Cells

The expression of adhesion molecules, ICAM-1, VCAM-1 and P-selectin, and a chemokine, MCP-1, which could potentially mediate cLDL-induced monocyte adhesion to endothelial cells, was recently studied (Apostolov *et al.* 2007b). HCAECs were treated with cLDL or nLDL for 24 hr, and the expressions of adhesion molecules were determined using cell ELISA. In this approach, the expression of MCP-1 and P-selectin was not induced by the treatment with cLDL in comparison to nLDL. However, the expression of ICAM-1 and VCAM-1 was significantly increased by cLDL. ICAM-1 was induced to a higher degree than VCAM-1. Contrary to cLDL, oxLDL did not affect ICAM-1 and VCAM-1

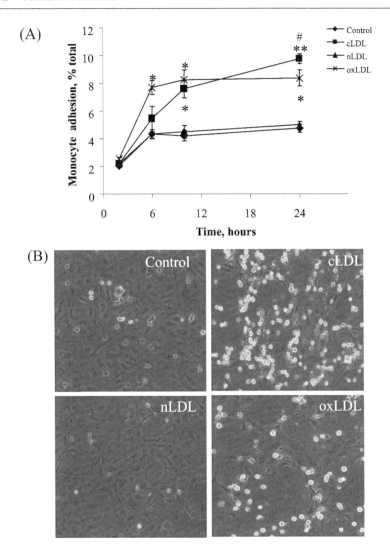

Fig. 3 Time course of the monocyte adhesion to endothelial cells treated with LDLs. (A) Cells were treated with 200 μg/mL LDL or vehicle control in 96-well plates and fluorescence was measured before and after fluorescence-labeled monocytes were allowed to adhere and non-adherent monocytes were washed out. The percentage of remaining fluorescence was calculated individually for each experimental well. Absolute data varied in the ranges of 622-685 and 12-58 units for total and remaining fluorescence measurement respectively. n = 4 per point, *P < 0.01 vs. either vehicle- or nLDL-treated cells, #P < 0.05 vs. oxLDL-treated cells at 24 hr. (B) Representative images. Endothelial cells are visualized with phase-contrast and labeled monocytes are detected using fluorescent microscopy. There is noticeable shrinkage and decreased density of HCAECs after cLDL or oxLDL treatment. Monocytes are adherent to remaining endothelial cells. Control cells were treated with vehicle or nLDL. Permission to publish obtained from Wolters Kluwer Health.

Color image of this figure appears in the color plate section at the end of the book.

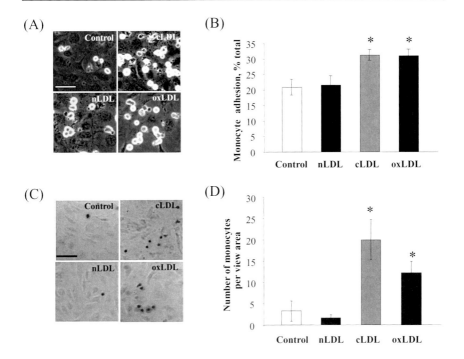

Fig. 4 Adhesion of freshly isolated primary human monocytes to endothelial cells treated with LDLs. Adherent monocytes visualized by microscopy (A) and quantified by the measurement of total fluorescence (B). Cells were treated with 200 µg/mL LDL or vehicle control in 6-well plates and fluorescence was measured before and after labeled monocytes were allowed to adhere and non-adherent monocytes were removed by washing. n = 3 per point, *P < 0.05 vs. either vehicle- or nLDL-treated cells. Scale bar, 25 µm. Adhesion of monocytes to HCAECs monolayer under flow conditions (C), and quantification of the total fluorescence (D). Endothelial cells were treated with 200 µg/mL LDL or vehicle control for 16 hr after and perfused with U937 monocytes as described in Materials and Methods. n = 3 per point, *P < 0.05 vs. either vehicle- or nLDL-treated cells. Scale bar, 25 µm. Permission to publish obtained from Wolters Kluwer Health.

expression while the expression of P-selectin and, to a lesser extent, MCP-1 molecules was increased. Native LDL did not cause significant change of either ICAM-1 or VCAM-1 expression.

The results of cell ELISA regarding ICAM-1 and VCAM-1 expressions after cLDL treatment were confirmed by immunocytochemistry: cLDL-treated HCAECs expressed significantly more ICAM-1 and VCAM-1 in comparison to the vehicle control or nLDL-treated cells. To determine whether or not the observed inductions of ICAM-1 and VCAM-1 are regulated at the transcription level, real-time PCR was performed with RNA extracted from the vehicle-, cLDL- (200 µg/mL) or nLDL-treated (200 µg/mL) endothelial cells. The results suggested that the transcription of both molecules was upregulated at 4 and 8 hr, and then decreased at later time points.

In another study, the VSMC were tested for the induction of the expression of ICAM-1 and VCAM-1 adhesion molecules by cLDL (Asci *et al.*). For this, human coronary artery VSMC were treated with cLDL or nLDL and expression of ICAM-1 and VCAM-1 was evaluated by cell ELISA. The data suggested that, unlike endothelial cells, cLDL-treated VSMC overexpress mostly VCAM-1 and to a lesser extent ICAM-1. However, because both adhesion molecules are overexpressed in vascular endothelial and smooth muscle cells, we may conclude that both ICAM-1 and VCAM-1 overexpression is likely to be a unique specific response from the vascular system to cLDL impact.

Role of ICAM-1 and VCAM-1 in cLDL-induced Monocyte Adhesion

Two approaches were applied to determine whether cLDL-induced ICAM-1 and/ or VCAM-1 overexpression cause the adhesion of monocytes to endothelial cells. In the first, the functions of ICAM-1 and VCAM-1 were abolished by antibodies, and in the second, the expression of the adhesion molecules was silenced using specific siRNA.

Because of the superficial location of adhesion molecules in endothelium, the antibodies to these proteins are widely used as a tool for determining the role of adhesion molecules. These studies showed that the inhibition of ICAM-1 caused a significant reduction of monocyte adhesion to endothelial cells, while the inhibition of VCAM-1 had only a minor and non-significant effect (Fig. 5). In the same experimental setting, simultaneous pretreatment of endothelial cells with both anti-ICAM-1 and anti-VCAM-1 antibodies caused the most significant inhibition of monocyte adhesion. Rabbit gamma-immunoglobulins, which were used as a negative control, did not have significant effect on monocyte adhesion regardless of the treatment.

The introduction of specific siRNAs resulted in the significant inhibition of ICAM-1 or VCAM-1 expression in modified LDL-treated cells as determined using real-time RT-PCR. After the application of siRNAs to HCAECs for 48 hr, the cells were treated for an additional 16 hr with vehicle, cLDL or nLDL and monocyte adhesion was measured. The data presented in Fig. 6 show that cLDL caused accelerated monocyte adhesion to endothelial cells, and it was significantly suppressed by anti-ICAM-1 siRNA. Anti-VCAM-1 siRNA had only a partial effect while simultaneously using anti-ICAM-1, while anti-VCAM-1 siRNAs caused the most prominent and significant suppression of monocyte adhesion. It did not reach the level of cells treated with the vehicle or nLDL, but it was consistent with the transfection efficiency observed in this experiment. Although nLDL did not accelerate monocyte adhesion, it was slightly suppressed by specific siRNAs.

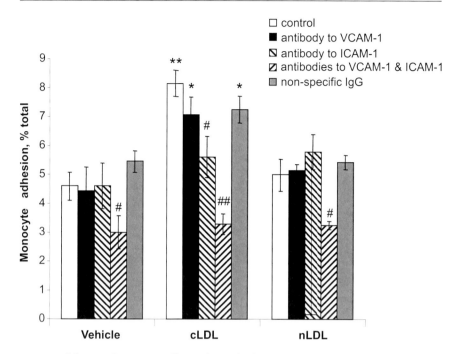

Fig. 5 Inhibition of monocyte adhesion by antibody to ICAM-1 or VCAM-1. HCAECs were treated with 200 µg/mL cLDL for 16 hr in 96-well plates. Control cells were treated with either vehicle or 200 µg/mL nLDL. Anti-ICAM-1, anti-VCAM-1 or both antibodies were added to HCAECs 2 hr prior to the application of labeled monocytes (final concentration of 10 ng/mL). Non-specific IgGs served as antibody treatment control. n = 3-4 per point, *P < 0.05, **P < 0.01 vs. vehicle control cells pretreated with the same antibody, #P < 0.05, ##P < 0.001 vs. no antibody control cells (white bars) subjected to the same treatment. Permission to publish obtained from Wolters Kluwer Health.

The control siRNA did not protect the endothelial cells from monocyte adhesion. These experiments provided evidence that ICAM-1 in cooperation with VCAM-1 is involved in monocyte adhesion by cLDL-activated human endothelial cells *in vitro*.

APPLICATIONS TO OTHER AREAS OF HEALTH AND DISEASE

Because urea is present in normal plasma, these observations may be applicable not only to renal patients but also to healthy individuals. While our data were obtained mainly using ESRD patients, the role of cLDL in atherosclerosis is certainly applicable to early stages of CKD, which includes more than 20 million adults in the United States and many more around the world.

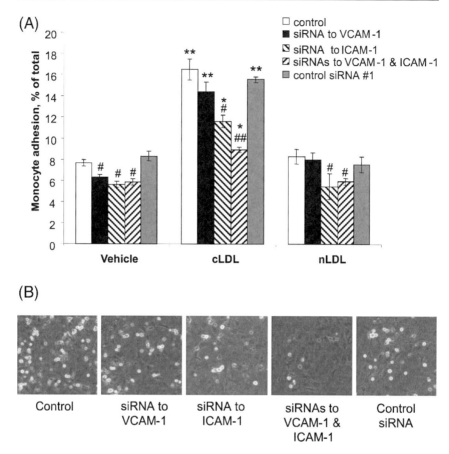

Fig. 6 Inhibition of monocyte adhesion by siRNA to ICAM-1 or VCAM-1. (A) HCAECs were transfected with anti-ICAM-1 or anti-VCAM-1 siRNA for 48 hr and then exposed with 200 μg/ mL cLDL for 16 hr. Control cells were treated with vehicle or 200 μg/mL nLDL. The monocyte adhesion was measured as described in the text. n = 3-4 per point, *P < 0.01, **P < 0.001 vs. vehicle control cells pretreated with the same siRNA, #P < 0.05, ##P < 0.01 vs. no siRNA control cells (white bars) subjected to the same treatment. (B) Representative images of cells treated with cLDL. Permission to publish obtained from Wolters Kluwer Health.

Color image of this figure appears in the color plate section at the end of the book.

CONCLUSION

Carbamylation is a non-enzymatic protein modification mainly by cyanate, a metabolite of urea, which is normally present in human plasma and is elevated in uremic patients. cLDL has several biological effects relevant to atherosclerosis. Activation of ICAM-1 and VCAM-1 adhesion molecules and attraction of monocytes to endothelial cells are some of the most prominent events. Other effects include the induction of injury in human coronary artery endothelial cells and stimulation of vascular smooth muscle cell proliferation *in vitro*. It has been well established that the risk of cardiovascular disease is increased several-fold in patients with CKD. The combination of *in vitro* studies demonstrating the ability of cLDL to induce monocyte adhesion and other biological effects relevant to atherosclerosis, along with elevated cLDL in hemodialysis patients, suggests an important role of LDL carbamylation in atherosclerosis in CKD patients. Future attempts to prevent LDL carbamylation and/or reduce the adhesion and other effects of cLDL may provide novel approaches for the prevention of atherosclerosis.

SUMMARY

- Carbamylated LDL (cLDL) is a product of spontaneous chemical modification of Apolipoprotein B in LDL by cyanate that is derived from urea.
- cLDL is present in blood plasma of healthy individuals and is elevated in uremic patients.
- cLDL concentration in renal patients plasmas correlate with early atherosclerosis measured by intima-media thickness.
- *In vitro*, cLDL induces proliferation and cell death of human coronary endothelial cells and vascular smooth muscle cells that is characteristic of atherosclerosis.
- *In vitro*, cLDL induces expression of adhesion molecules ICAM-1 and VCAM-1, thus promoting monocyte adhesion to endothelial cells.

Acknowledgements

This research was supported by a grant 1R21HL087405 from NIH/NHLBI (A.G.B.), AHA South Central Affiliate grant (E.O.A.), VA Merit Review Grants (A.G.B., S.V.S.), and an Arkansas Tobacco Settlement Award (E.O.A.).

Abbreviations

CKD	chronic kidney disease
cLDL	carbamylated LDL
ELISA	enzyme-linked immunosorbent assay
ESRD	end-stage renal disease
ICAM-1	intercellular adhesion molecule 1
IMT	intima-media thickness
HCAECs	human coronary artery endothelial cells
LDL	low-density lipoprotein

LOX-1	lectin-like oxLDL receptor
MAPK	mitogen-activated protein kinase
MCP-1	monocyte chemoattractant protein 1
MDA-LDL	malondialdehyde modified LDL
mRNA	matrix ribonucleic acid
nLDL	native LDL
N-IMT	normal intima-media thickness (of common carotid artery)
oxLDL	oxidized LDL
PBS	phosphate buffered saline
PCR	polymerase chain reaction
RT-PCR	reverse transcriptase polymerase chain reaction
siRNA	single-strand interfering RNA
SEM	standard error of mean
T-IMT	thickened intima-media thickness (of common carotid artery)
VCAM-1	vascular cell adhesion molecule 1
VSMC	vascular smooth muscle cells

Key Facts about Carbamylated LDL

1. It is a product of non-enzymatic modification of LDL by urea-derived cyanate or thiocyanate in blood plasma.
2. It is one of the most abundant modified LDL isoforms in human plasma.
3. It is present both in uremic patients and in healthy individuals.
4. Uremia caused by ESRD is associated with the elevation of plasma cLDL.
5. There are no inhibitors of LDL carbamylation.
6. Adhesion molecules ICAM-1 and VCAM-1 are induced by cLDL in endothelial and vascular smooth muscle cells *in vitro*.
7. It is absorbed from the bloodstream in minutes.
8. It induces endothelial and vascular smooth muscle cell proliferation and cell death.
9. Cytotoxicity of cLDL toward endothelial cells is compatible to that of oxidized LDL.

References

Apostolov, E.O. and S.V. Shah, E. Ok, and A.G. Basnakian. 2005. Quantification of carbamylated LDL in human sera by a new sandwich ELISA. Clin. Chem. 51: 719-728.

Apostolov, E.O. and A.G. Basnakian, X. Yin, E. Ok, and S.V. Shah. 2007a. Modified LDLs induce proliferation-mediated death of human vascular endothelial cells through MAPK pathway. Am. J. Physiol. Heart Circ. Physiol. 292: H1836-1846.

Apostolov, E.O. and S.V. Shah, E. Ok, and A.G. Basnakian. 2007b. Carbamylated low-density lipoprotein induces monocyte adhesion to endothelial cells through intercellular adhesion molecule-1 and vascular cell adhesion molecule-1. Arterioscler. Thromb. Vasc. Biol. 27: 826-832.

Apostolov, E.O. and S.V. Shah, K.D. Wagner, and A.G. Basnakian. 2008. Scavenger receptors participate in the effects of carbamylated LDL toward endothelial cells. *In* The Proceedings of the Arteriosclerosis, Thrombosis and Vascular Biology Annual Conference, Atlanta, GA. P. 98 (#450).

Asci, G. and A. Basci, S.V. Shah, A. Basnakian, H. Toz, M. Ozkahya, S. Duman, and E. Ok. 2008. Carbamylated low-density lipoprotein induces proliferation and increases adhesion molecule expression of human coronary artery smooth muscle cells. Nephrology (Carlton).

Bailey, J.L. and W.E. Mitch. 1997. The search for the uremic toxin: the case for metabolic acidosis. Wien Klin Wochenschr. 109: 7-12.

Basu, S.K. and J.L. Goldstein, G.W. Anderson, and M.S. Brown. 1976. Degradation of cationized low density lipoprotein and regulation of cholesterol metabolism in homozygous familial hypercholesterolemia fibroblasts. Proc. Nat. Acad. Sci. USA 73: 3178-3182.

Bosmans, J.L. and P. Holvoet, S.E. Dauwe, D.K. Ysebaert, T. Chapelle, A. Jurgens, V. Kovacic, E.A. Van Marck, M.E. De Broe, and G.A. Verpooten. 2001. Oxidative modification of low-density lipoproteins and the outcome of renal allografts at 1 1/2 years. Kidney Int. 59: 2346-2356.

Canal, J. and M.L. Girard. 1973. [Properties of heavy (Sf 4-6) carbamylated human betalipoproteins]. Ann. Biol. Clin. (Paris) 31: 97-99.

Chen, K. and J. Chen, Y. Liu, J. Xie, D. Li, T. Sawamura, P.L. Hermonat, and J.L. Mehta. 2005. Adhesion molecule expression in fibroblasts: alteration in fibroblast biology after transfection with LOX-1 plasmids. Hypertension 46: 622-627.

Dart, A.M. and J.P. Chin-Dusting. 1999. Lipids and the endothelium. Cardiovasc. Res. 43: 308-322.

Futatsuyama, M. and T. Oiwa, and Y. Komatsu. 2002. Correlation between oxidized low-density lipoprotein and other factors in patients on peritoneal dialysis. Adv. Perit. Dial. 18: 192-194.

Galle, J. and K. Heermeier, and C. Wanner. 1999. Atherogenic lipoproteins, oxidative stress, and cell death. Kidney Int. Suppl. 71: S62-65.

Gonen, B. and A.P. Goldberg, H.R. Harter, and G. Schonfeld. 1985. Abnormal cell-interactive properties of low-density lipoproteins isolated from patients with chronic renal failure. Metabolism 34: 10-14.

Hasuike, Y. and T. Nakanishi, K. Maeda, T. Tanaka, T. Inoue, and Y. Takamitsu. 2002. Carbamylated hemoglobin as a therapeutic marker in hemodialysis. Nephron 91: 228-234.

Holvoet, P. and J. Donck, M. Landeloos, E. Brouwers, K. Luijtens, J. Arnout, E. Lesaffre, Y. Vanrenterghem, and D. Collen. 1996. Correlation between oxidized low density lipoproteins and von Willebrand factor in chronic renal failure. Thromb. Haemost. 76: 663-669.

Horkko, S. and M.J. Savolainen, K. Kervinen, and Y.A. Kesaniemi. 1992. Carbamylation-induced alterations in low-density lipoprotein metabolism. Kidney Int. 41: 1175-1181.

Horkko, S. and K. Huttunen, K. Kervinen, and Y.A. Kesaniemi. 1994. Decreased clearance of uraemic and mildly carbamylated low-density lipoprotein. Eur. J. Clin. Invest. 24: 105-113.

Inoue, K. and Y. Arai, H. Kurihara, T. Kita, and T. Sawamura. 2005. Overexpression of lectin-like oxidized low-density lipoprotein receptor-1 induces intramyocardial vasculopathy in apolipoprotein E-null mice. Circ. Res. 97: 176-184.

Kohno, H. and N. Sueshige, K. Oguri, H. Izumidate, T. Masunari, M. Kawamura, H. Itabe, T. Takano, A. Hasegawa, and R. Nagai. 2000. Simple and practical sandwich-type enzyme immunoassay for human oxidatively modified low density lipoprotein using antioxidized phosphatidylcholine monoclonal antibody and antihuman apolipoprotein-B antibody. Clin. Biochem. 33: 243-253.

Kraus, L.M. and A.P. Kraus, Jr. 2001. Carbamoylation of amino acids and proteins in uremia. Kidney Int. Suppl. 78: S102-107.

Levey, A.S. and J. Coresh, E. Balk, A.T. Kausz, A. Levin, M.W. Steffes, R.J. Hogg, R.D. Perrone, J. Lau, and G. Eknoyan. 2003. National Kidney Foundation practice guidelines for chronic kidney disease: evaluation, classification, and stratification. Ann. Intern. Med. 139: 137-147.

Li, D. and J.L. Mehta. 2000. Antisense to LOX-1 inhibits oxidized LDL-mediated upregulation of monocyte chemoattractant protein-1 and monocyte adhesion to human coronary artery endothelial cells. Circulation 101: 2889-2895.

Lonn, E. 2001. Do antioxidant vitamins protect against atherosclerosis? The proof is still lacking. J. Am. Coll. Cardiol. 38: 1795-1798.

Nilsson, L. and P. Lundquist, B. Kagedal, and R. Larsson. 1996. Plasma cyanate concentrations in chronic renal failure. Clin. Chem. 42: 482-483.

Ok, E. and A.G. Basnakian, E.O. Apostolov, Y.M. Barri, and S.V. Shah. 2002. Elevation of both carbamylated LDL and autoantibody to carbamylated LDL in dialysis patients. J. Am. Soc. Nephrol. 13: 219A.

Ok, E. and A.G. Basnakian, E.O. Apostolov, Y.M. Barri, and S.V. Shah. 2005. Carbamylated low-density lipoprotein induces death of endothelial cells: a link to atherosclerosis in patients with kidney disease. Kidney Int. 68: 173-178.

Qian, M. and J.W. Eaton, and S.P. Wolff. 1997. Cyanate-mediated inhibition of neutrophil myeloperoxidase activity. Biochem. J. 326 (Pt 1): 159-166.

Sarnak, M.J. and A.S. Levey, A.C. Schoolwerth, J. Coresh, B. Culleton, L.L. Hamm, P.A. McCullough, B.L. Kasiske, E. Kelepouris, M.J. Klag, P. Parfrey, M. Pfeffer, L. Raij, D.J. Spinosa, and P.W. Wilson. 2003. Kidney disease as a risk factor for development of cardiovascular disease: a statement from the American Heart Association Councils on Kidney in Cardiovascular Disease, High Blood Pressure Research, Clinical Cardiology, and Epidemiology and Prevention. Circulation 108: 2154-2169.

Sawamura, T. and N. Kume, T. Aoyama, H. Moriwaki, H. Hoshikawa, Y. Aiba, T. Tanaka, S. Miwa, Y. Katsura, T. Kita, and T. Masaki. 1997. An endothelial receptor for oxidized low-density lipoprotein. Nature 386: 73-77.

Schmidt, A.M. and O. Hori, J.X. Chen, J.F. Li, J. Crandall, J. Zhang, R. Cao, S.D. Yan, J. Brett, and D. Stern. 1995. Advanced glycation endproducts interacting with their endothelial receptor induce expression of vascular cell adhesion molecule-1 (VCAM-1) in cultured human endothelial cells and in mice. A potential mechanism for the accelerated vasculopathy of diabetes. J. Clin. Invest. 96: 1395-1403.

Steinbrecher, U.P. and M. Fisher, J.L. Witztum, and L.K. Curtiss. 1984. Immunogenicity of homologous low density lipoprotein after methylation, ethylation, acetylation, or carbamylation: generation of antibodies specific for derivatized lysine. J. Lipid Res. 25: 1109-1116.

Van Tits, L. and J. De Graaf, H. Hak-Lemmers, S. Bredie, P. Demacker, P. Holvoet, and A. Stalenhoef. 2003. Increased levels of low-density lipoprotein oxidation in patients with familial hypercholesterolemia and in end-stage renal disease patients on hemodialysis. Lab. Invest. 83: 13-21.

Weisgraber, K.H. and T.L. Innerarity, and R.W. Mahley. 1978. Role of lysine residues of plasma lipoproteins in high affinity binding to cell surface receptors on human fibroblasts. J. Biol. Chem. 253: 9053-9062.

Yusuf, S. and G. Dagenais, J. Pogue, J. Bosch, and P. Sleight. 2000. Vitamin E supplementation and cardiovascular events in high-risk patients. The Heart Outcomes Prevention Evaluation Study Investigators. N. Engl. J. Med. 342: 154-160.

CHAPTER 26

Adhesion Molecules and Oxidized Low-density Lipoprotein

Masaaki Matsumoto

Department of Dermatology, Narita Memorial Hospital,
78 Shirakawa-cho, Toyohashi, Aichi 441-8021, Japan,
E-mail: matsumo@meiyokai.or.jp

ABSTRACT

Interactions between lipoproteins and vascular endothelium are considered to be a central component of the pathogenesis of atherosclerosis and cutaneous xanthomas. The binding of oxidized low-density lipoprotein to cell membrane receptors (including scavenger receptors) activates an intracellular signal transduction pathway that produces adhesion molecules on the surface of endothelial cells. Circulating monocytes adhere to these molecules and subsequently migrate across the cell membrane into the lesions, thus leading to the progression of the two diseases. Growing evidence indicates that the mechanisms of adhesion molecule expression vary depending on the oxidation process of low-density lipoprotein, the responsible molecules of oxidized low-density lipoprotein, and the organ specificity of the endothelial cells. This chapter summarizes the findings of recent studies involving the induction of adhesion molecule expression in vascular endothelial cells by oxidized low-density lipoprotein. A variety of drugs and nutrients have been used to regulate the adhesion molecule expression and degree of monocyte adhesion, and the mechanisms of these substances have also been elucidated in recent studies. Numerous signaling molecules have been implicated in the onset/progression of atherosclerosis and cutaneous xanthomas. Further investigations that clarify the mechanisms of adhesion molecule expression are expected to help

---

Key terms are defined at the end of the chapter.

establish more specific and effective treatment modalities for atherosclerosis and cutaneous xanthomas.

INTRODUCTION

The common histological feature of atherosclerosis and cutaneous xanthomas is the infiltration of the lesions by macrophage-derived lipid-laden foam cells. The migration of circulating monocytes into the lesions leads to the progression of the two diseases. Adhesion molecules on vascular endothelial cells play a critical role in leukocyte rolling, adhesion, and transmigration. E-selectin and P-selectin facilitate the rolling of leukocytes prior to firm adhesion. Monocyte adhesion to endothelial cells is mediated by the immunoglobulin superfamily, including intercellular adhesion molecule 1 (ICAM-1) and vascular cell adhesion molecule 1 (VCAM-1). An enlargement of atherosclerotic lesions is inhibited by either the inhibition or hypomorphic mutation of any of these four adhesion molecules.

In human atherosclerosis, E-selectin is expressed on endothelial cells in lipid-containing and fibrous plaques. P-selectin is detected in atherosclerotic plaques, but not in fibrous plaques. ICAM-1 expression is also increased in all subtypes of atherosclerotic lesions except for fibrous plaques. VCAM-1 is more prevalent in the intima of atherosclerotic plaques than in the non-atherosclerotic segments of coronary arteries. However, the E-selectin expression is more specific for the atherosclerotic intima than either ICAM-1 or VCAM-1 expression because E-selectin is not detected in any control segments.

E-selectin-positive endothelial cells are more prevalent in xanthoma lesions than in normal skin. In contrast, the ICAM-1 expression is less prevalent in xanthoma lesions than in normal skin. Almost all ICAM-1-positive endothelial cells in xanthomas co-express E-selectin, but there are also many endothelial cells in these lesions that only express E-selectin. Very few, if any, endothelial cells express VCAM-1 in either xanthoma lesions or normal skin.

Evidence from *in vitro* studies indicates that oxidation of low-density lipoprotein (LDL; Fig. 1; see also Key Facts about Low-density Lipoprotein) induces adhesion molecule expression on endothelial cells and enhances monocyte binding to the endothelial cell membrane (Fig. 2). A reasonable inference drawn from this evidence is that oxidized low-density lipoprotein (Ox-LDL) binds to endothelial receptors (including scavenger receptors) and activates an intracellular signal transduction pathway that induces the expression of adhesion molecules on the surface of endothelial cells (Fig. 3). Circulating monocytes adhere to the endothelial cells via the Ox-LDL-induced adhesion molecules and subsequently transmigrate into the lesions. The mechanisms of adhesion molecule expression vary depending on the process of LDL oxidation and the organ specificity of the endothelial cells. This chapter summarizes recent studies of the mechanisms of endothelial adhesion molecule expression induced by Ox-LDL.

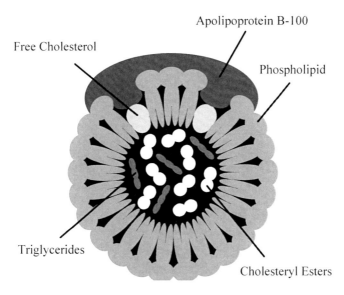

Fig. 1 Structure of low-density lipoprotein. Low-density lipoprotein is described as a spherical particle containing a hydrophobic core of cholesteryl esters and triglycerides, surrounded by an amphipathic monolayer of phospholipid and free cholesterol in which a single molecule of apolipoprotein B-100 is located.

THE MECHANISMS OF ENDOTHELIAL ADHESION MOLECULE EXPRESSION INDUCED BY Ox-LDL

LDL Oxidation

Circulating LDL transverses the subendothelial space of large arteries or extravasates into the dermis of the skin. It is theorized that the subendothelial space or dermis might be the primary site of LDL oxidation because some antioxidants provide potent protection against LDL oxidation in plasma. The existence of Ox-LDL in the circulation remains a controversial subject. Indeed, only a minor fraction of circulating LDL exhibits a high content of oxidized lipids, and human plasma demonstrates immunoreactivity of monoclonal antibodies against epitopes of Ox-LDL. In contrast to plasma, atherosclerotic lesions and xanthoma lesions contain substantial amounts of oxidized lipids. Despite this evidence, homogenates of the atherosclerotic lesions also contain ascorbate, uric acid, and α-tocopherol in quantities sufficient to effectively block LDL oxidation *in vitro*.

An important aspect of LDL oxidation is a lipid peroxidation chain reaction initiated and driven by free radicals that are generated by the endothelial cells, smooth muscle tissue, and migratory lymphocytes. In this process, the lipid peroxidation products fragment into reactive aldehydes such as malondialdehyde

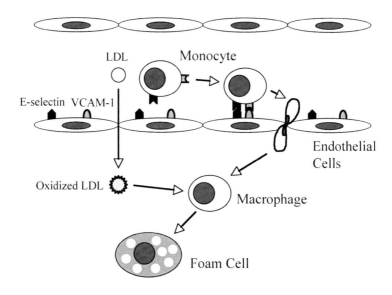

Fig. 2 Role of oxidized LDL in the pathogenesis of atherosclerosis and cutaneous xanthomas. Circulating LDL transverses the lesions of large arteries or skin, where the LDL is oxidized. The oxidized LDL induces adhesion molecule expression on endothelial cells. Circulating monocytes adhere to the endothelial cells via the adhesion molecules and subsequently transmigrate into the lesions, where the monocytes differentiate into macrophages. The macrophages incorporate the oxidized LDL and transform into foam cells. LDL, low-density lipoprotein; VCAM-1, vascular cell adhesion molecule 1.

and 4-hydroxynonenal. These aldehydes then react with amino phospholipids (such as phosphatidylethanolamine and phosphatidylserine) and the N^ε-amino groups of apolipoprotein B-100 lysine residues. LDL isolated from atherosclerotic plaques or xanthoma lesions possesses properties that resemble those of Ox-LDL formed *in vitro* (Ylä-Herttuala *et al.* 1989, Ikeda *et al.* 2006). Much of our knowledge on how LDL is converted to Ox-LDL originates from *in vitro* studies exposing LDL to metal ions or to cells cultured in transition metal-containing medium.

Oxidation by Chemicals

Copper ions are commonly used to oxidize LDL in prolonged periods of incubation. This method causes drastic alterations to the lipoprotein particle that, as a result, gains alternative functions. For example, the resulting Ox-LDL (referred to as highly oxidized Ox-LDL or Cu-LDL) is no longer able to effectively interact with

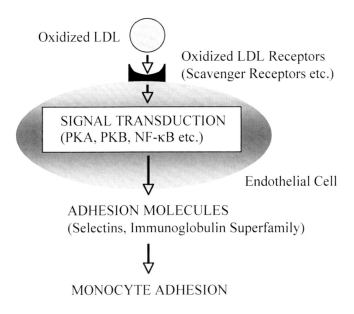

Oxidized LDL

Oxidized LDL Receptors
(Scavenger Receptors etc.)

SIGNAL TRANSDUCTION
(PKA, PKB, NF-κB etc.)

Endothelial Cell

ADHESION MOLECULES
(Selectins, Immunoglobulin Superfamily)

MONOCYTE ADHESION

Fig. 3 Mechanism of adhesion molecule expression in endothelial cells by oxidized LDL. Oxidized LDL binds to endothelial receptors (including scavenger receptors) and activates an intracellular signal transduction pathway that induces the expression of adhesion molecules on endothelial cells. LDL, low-density lipoprotein; PKA, protein kinase A; PKB, protein kinase B; NF-κB, nuclear factor-kappaB.

LDL receptors and instead is recognized only by scavenger receptors. Cu-LDL is obtained by incubating LDL with $CuSO_4$ in phosphate-buffered saline at 37°C. The concentration of LDL and $CuSO_4$ and the incubation time required for reaction vary among studies (Gebuhrer *et al.* 1995, Cominacini *et al.* 1997a, b, Takei *et al.* 2001, Matsumoto *et al.* 2003).

Minimally modified/oxidized LDL (MM-LDL) is LDL that is less drastically oxidized than Cu-LDL. In contrast to Cu-LDL, MM-LDL is recognized by LDL receptors but not scavenger receptors. MM-LDL is produced by enzyme modification using soybean lipoxygenase and phospholipase A_2 or by mild oxidation with iron. The levels of thiobarbituric acid-reactive substances are assayed to determine the lipid peroxide levels in the Ox-LDL samples. The level seen in MM-LDL is defined as less than 10 nmol malondialdehyde/mg LDL protein.

Oxidation by Cells and Tissues

LDL is oxidized *in vitro* by incubation with endothelial cells, smooth muscle cells, monocytes, macrophages, and fibroblasts. LDL oxidation is mediated by superoxide anions generated from monocytes and smooth muscle cells and by lipoxygenase from endothelial cells and macrophages. However, the precise mechanism of LDL oxidation is still controversial. Transition metals, such as Fe^{2+} and Cu^{2+}, are required for LDL oxidation by cells. Therefore, *in vitro* LDL oxidation experiments have usually been performed in Fe^{2+}-containing medium.

There are as yet only a few reports on LDL oxidation using tissue specimens. Gruel samples of human atherosclerotic lesions are capable of oxidizing LDL in the presence of transition metal ions. Ox-LDL is produced by incubation with the extracts from atherosclerotic plaques using centrifugation. Xanthoma lesion-oxidized LDL (X-LDL) is obtained by incubating human LDL with experimental rabbit xanthoma tissues in RPMI-1640 medium without Cu^{2+} and Fe^{2+}. The experimental xanthoma lesions are induced by intradermal injections of dextran sulfate solution at dorsal sites of rabbits fed a high cholesterol diet. The lipid compositions of the experimental rabbit xanthoma lesions are similar to those observed in human xanthoma lesions.

RECEPTORS THAT CAN BIND Ox-LDL

Ox-LDL binds to a diverse range of transmembrane proteins that are collectively termed scavenger receptors. These receptors are also capable of binding a wide variety of lipid- and lipoprotein-based ligands.

Scavenger Receptors

The scavenger receptor family can be broadly classified into eight subclasses (A-H) that bear little sequence homology to each other, but they do recognize common ligands. At least seven scavenger receptors have been detected in human vascular endothelial cells—scavenger receptor with C-type lectin (SRCL), scavenger receptor BI (SR-BI), CD36, lectin-like oxidized LDL receptor-1 (LOX-1), scavenger receptor expressed on endothelial cell-I (SREC-I), scavenger receptor for phosphatidylserine and oxidized lipoprotein (SR-PSOX), and fasciclin, epidermal growth factor (EGF)-like, laminin-type EGF-like and link domain-containing scavenger receptor 1 (FEEL-1; Fig. 4 and Table 1).

Other Receptors

Alternative receptors that bind Ox-LDL are described in this section. One of the earliest consequences of LDL oxidation is LDL aggregation, and the aggregated

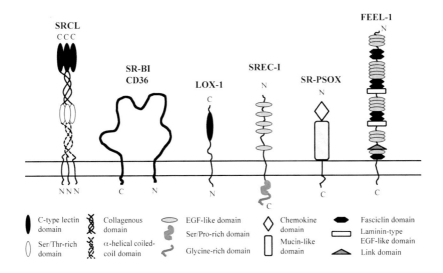

Fig. 4   Structure of scavenger receptors expressed in vascular endothelial cells. At least seven scavenger receptors have been detected as illustrated above. Specific domains are emphasized by the codes indicated in the figure. SRCL, scavenger receptor with C-type lectin; SR-BI, scavenger receptor BI; LOX-1, lectin-like oxidized LDL receptor 1; SREC-I, scavenger receptor expressed on endothelial cell I; SR-PSOX, scavenger receptor for phosphatidylserine and oxidized lipoprotein; FEEL-1, fasciclin, epidermal growth factor (EGF)-like, laminin-type EGF-like and link domain-containing scavenger receptor 1.

Table I   Key points of scavenger receptors expressed in human vascular endothelial cells

Scavenger receptors	Endothelial cells
SRCL/CL-P	Umbilical artery and vein
SR-BI	Umbilical vein
CD36	Dermal microvessels, especially small microvessels in the deep vascular plexus
LOX-1	Aortic, carotid, thoracic, and coronary arteries and veins
SREC-I	Umbilical vein
SR-PSOX	Umbilical vein
FEEL-1/stabilin-1/ CLEVER-1	Umbilical vein, dermal lymphatic vessels

The scavenger receptor family can be broadly classified into eight subclasses. At least seven scavenger receptors have been detected in human vascular endothelial cells as listed above. SRCL, scavenger receptor with C-type lectin. CL-P1, collectin placenta 1. SR-BI, scavenger receptor BI. LOX-1, lectin-like oxidized LDL receptor 1. SREC-I, scavenger receptor expressed on endothelial cell I. SR-PSOX, scavenger receptor for phosphatidylserine and oxidized lipoprotein. FEEL-1, fasciclin, epidermal growth factor (EGF)-like, laminin-type EGF-like and link domain-containing scavenger receptor 1. CLEVER-1, common lymphatic endothelial and vascular endothelial receptor 1.

LDL binds to LDL receptors and LDL receptor-related protein 1. As mentioned earlier, MM-LDL also binds to LDL receptors instead of scavenger receptors. Ox-LDL/immunoglobulin G complexes can be mediated via Fcγ receptors, and Ox-LDL bound by immunoglobulin M or C-reactive protein can in turn bind complement and undergo enhanced binding via complement receptors. Ox-LDL can bind to epidermal growth factor receptor and platelet-derived growth factor receptor. Ox-LDL can also bind to lysophosphatidic acid receptors.

Ox-LDL RECEPTOR STRUCTURES SUGGESTING SIGNAL TRANSDUCTION

The endothelial scavenger receptors CD36 and SREC may induce tyrosine kinase-related signaling in the presence of Ox-LDL stimulation. CD36 tightly associates with the Fyn, Lyn, and Yes tyrosine kinases of the Src family, and these kinases display an autophosphorylation activity *in vitro*. The carboxyl terminal domain of CD36 contains a sulfhydryl sequence ($CXCX_5K$) that may be involved in tyrosine kinase binding and may play a role in receptor-kinase-mediated signal transduction. SREC has a tyrosine residue (Tyr^{822}) fitting the consensus sequence (RXX-EXXY) for a site phosphorylated by tyrosine kinases, thus suggesting that this residue may play a role in some signal transduction processes. Ox-LDL also triggers the tyrosine phosphorylation of receptor tyrosine kinases including epidermal growth factor receptor and platelet-derived growth factor receptor.

Growth hormone-releasing peptides are a class of small synthetic peptides that are known to stimulate growth hormone release through the binding of a G-protein-coupled receptor. Hexarelin, a hexapeptide member of the growth hormone–releasing peptide family, binds to the amino acid residues 132-177 of CD36. The chemical cleavage of the hexarelin binding site results in the release of free ligands, and Met^{169} is the contact point for the ligands within the receptor-binding pocket. The binding domain on CD36 for Ox-LDL overlaps with that for hexarelin (the amino acid residues 155-183). In addition, Ox-LDL can bind to lysophosphatidic acid receptors, which are a type of G-protein-coupled receptor.

ADHESION MOLECULE EXPRESSION

This section shows a different pattern of adhesion molecule expression depending on the oxidation process of LDL, the responsible molecules of Ox-LDL, and the organ specificity of endothelial cells (Table 2).

Selectins

Cu-LDL treatment upregulates the expression of E- and P-selectins via LOX-1 on human coronary artery endothelial cells (HCAECs; Li *et al.* 2002). Both human

Table 2 Possible factors influencing the vascular endothelial adhesion molecule expression induced by oxidized low-density lipoprotein

1. LDL oxidation process

    Chemicals—copper ion, iron ion, lipoxygenase, phospholipase A2

    Cells—endothelial cells, smooth muscle cells, monocytes, macrophages, fibroblasts

    Tissues—atherosclerotic tissues, xanthoma tissues

2. Responsible molecules of Ox-LDL

    Lysophosphatidylcholine, PAPC, fatty acids

3. Organ specificity of endothelial cells

    Aorta, coronary artery, iliac artery, umbilical vein, glomerulus, dermal microvessels

The mechanisms of adhesion molecule expression vary depending on the oxidation process of LDL, responsible molecules of oxidized LDL, and the organ specificity of the endothelial cells. LDL, low-density lipoprotein. Ox-LDL, oxidized low-density lipoprotein. PAPC, 1-palmitoyl-2-arachidonoyl-*sn*-glycero-3-phosphorylcholine.

umbilical vein endothelial cell (HUVEC)-oxidized LDL and Cu-LDL made by incubation of LDL with 10 µM Cu^{2+} induce P-selectin, but not E-selectin, on HUVECs (Gebuhrer *et al.* 1995). LDL oxidized with 5 µM Cu^{2+} significantly induces E-selectin expression in HUVECs (Cominacini *et al.* 1997b). LDL oxidized with 1 µM Cu^{2+} and MM-LDL do not change E-selectin expression in HUVECs (Cominacini *et al.* 1997a). Oxidized lipoprotein (a) induces more P-selectin protein and mRNA expression than Ox-LDL in HUVECs (Zhao and Xu 2000). The treatment of human aortic endothelial cells (HAECs) with MM-LDL causes an increase in intracellular P-selectin protein that is associated with MM-LDL-induced cyclic adenosine monophosphate elevation (Vora *et al.* 1997). MM-LDL therefore causes little change in P-selectin cell surface expression. However, highly oxidized Ox-LDL, especially the oxidized fatty acids, has been shown to induce the surface expression of P-selectin (Vora *et al.* 1997). The effect of lysophosphatidylcholine on E-selectin expression is negligible in HAECs, human iliac artery endothelial cells, and HUVECs (Kume *et al.* 1992).

X-LDL upregulates E-selectin expression on human dermal microvascular endothelial cells (HDMECs; Matsumoto *et al.* 2003). The X-LDL-induced expression of E-selectin on HDMECs is mediated by the tyrosine kinase–related pathway, but not by the inhibitory G (G_i)-protein or protein kinase C (PKC) pathways (Matsumoto *et al.* 2003). Cu-LDL also induces E-selectin expression on HDMECs, and this expression is incompletely inhibited by pretreatment with pertussis toxin, a G_i-protein inhibitor (Matsumoto *et al.* 2003). This finding indicates that E-selectin expression might be mediated not only by the G_i-protein pathway but also by other pathways (except the tyrosine kinase-related pathway and the PKC pathway).

Immunoglobulin Superfamily

In HCAECs, Cu-LDL treatment upregulates the expression of ICAM-1 and VCAM-1 via LOX-1 (Li *et al.* 2002). In HUVECs, 5 µM Cu^{2+}-oxidized LDL significantly induces the expression of ICAM-1 and VCAM-1 (Cominacini *et al.* 1997b). Other investigators maintain that only LDL oxidized with 1 μM Cu^{2+}, but not LDL oxidized with 5 or 2.5 μM Cu^{2+}, will significantly induce the expression of these adhesion molecules (Cominacini *et al.* 1997a). Another study has demonstrated that highly oxidized Ox-LDL may not have the ability to stimulate ICAM-1 expression in HUVECs possibly due to LDL aggregation (Takei *et al.* 2001). HUVEC-oxidized LDL and MM-LDL cause a significant increase of ICAM-1 and VCAM-1 in HUVECs (Cominacini *et al.* 1997a).

Cu-LDL causes the PKC pathway to induce the ICAM-1 expression on HUVECs (Mason *et al.* 1997), but Cu-LDL does not induce PKC-mediated ICAM-1 expression on glomerular endothelial cells (Kamanna *et al.* 1999). Lysophosphatidylcholine induces ICAM-1 and VCAM-1 in human iliac artery endothelial cells, whereas only ICAM-1 is upregulated in HUVECs (Kume *et al.* 1992). The augmented expression of ICAM-1 by lysophosphatidylcholine is not mediated by the PKC pathway (Ochi *et al.* 1995). HDMECs stimulated with X-LDL show an enhanced expression of VCAM-1, but not ICAM-1 (Matsumoto *et al.* 2003). The VCAM-1 expression is mediated by a tyrosine kinase-related pathway, but not a G_i-protein or PKC pathway (Matsumoto *et al.* 2003). Cu-LDL induces the ICAM-1 and VCAM-1 expression on HDMECs (Matsumoto *et al.* 2003). The ICAM-1 expression is mediated by a G_i-protein pathway, but not a tyrosine kinase-related pathway or a PKC pathway. The Cu-LDL-induced VCAM-1 expression might be mediated by not only a G_i-protein pathway but also other pathways (except a tyrosine kinase-related pathway or a PKC pathway). Cu-LDL increases ICAM-1 expression on HDMECs depending on the degree of LDL oxidation. In contrast, X-LDL does not significantly upregulate the ICAM-1 expression even if X-LDL receives more extensive oxidation (longer incubation of LDL with xanthoma tissues).

MONOCYTE-BINDING TO ENDOTHELIAL CELLS

Ox-LDL increases the expression of P-selectin and ICAM-1 on HCAECs, and therefore enhances the adhesion of monocytes to the endothelial cells (Chen *et al.* 2003). The process of Ox-LDL–mediated monocyte adhesion to HCAECs is associated with phosphorylation at Tyr^{204} of ERK via LOX-1 (Li and Mehta 2000). MM-LDL induces monocyte binding to endothelial cells (Berliner *et al.* 1990). Oxidized 1-palmitoyl-2-arachidonoyl-*sn*-glycero-3-phosphorylcholine (Ox-PAPC), a component of MM-LDL, induces endothelial/monocyte interactions (Watson *et al.* 1997). 1-palmitoyl-2-(5-oxovaleroyl)-*sn*-glycero-3-phosphorylcholine and 1-palmitoyl-2-(5-hydroxy-8-oxooct-6-enoyl)-*sn*-glycero-3-phosphocholine in Ox-PAPC induce monocyte-binding to HAECs (Leitinger *et al.* 1997).

X-LDL induces monocyte adhesion to HDMECs through VCAM-1 and E-selectin, but not ICAM-1 (Matsumoto *et al.* 2003). Cu-LDL upregulates monocyte adhesion to HDMECs via ICAM-1, VCAM-1 and E-selectin (Matsumoto *et al.* 2003). Ox-LDL increases monocyte adhesion to EA.hy 926, an endothelial cell line, independently of ICAM-1, VCAM-1, and nuclear factor-kappaB (NF-κB) activation (Dwivedi *et al.* 2001). Chimeric sCD36-Ig protein blocked the Ox-LDL-induced adhesion of monocytes to endothelial cells through ICAM-1, thus indicating that this chimeric protein can effectively compete for the binding of Ox-LDL to membrane-expressed CD36 (Stewart and Nagarajan 2006). The toll-like receptor (TLR) 4/NF-κB signaling pathway plays a role in monocyte-endothelial adhesion induced by Ox-LDL (Yang *et al.* 2005). Anti-human TLR4 monoclonal antibody or transinfection with a functional mutant of TLR4 remarkably inhibit NF-κB activity and significantly reduce the degree of monocyte-endothelial adhesion, but anti-human TLR2 monoclonal antibody does not have a similar effect (Yang *et al.* 2005).

LDL oxidation can induce adhesion molecule expression on vascular endothelial cells and subsequent monocyte adhesion to the endothelial cells. These mechanisms are regulated by various intracellular signaling pathways in the cell membranes. A variety of drugs and nutrients have been used to regulate the adhesion molecule expression and the degree of monocyte adhesion, and the mechanisms of these substances have been elucidated by recent studies (Table 3). Further investigations that clarify the mechanisms of adhesion molecule expression will result in more specific and effective treatment modalities for atherosclerosis and cutaneous xanthomas.

Table 3 Key points of therapeutic targets of the molecules involved in the adhesion molecule expression induced by oxidized low-density lipoprotein

Drug	*Molecules*	*Adhesion molecules*	*References*
Simvastatin	NF-κB	E-selectin, P-selectin, ICAM-1, VCAM-1	Li *et al.* 2002
Atorvastatin	NF-κB	E-selectin, P-selectin, ICAM-1, VCAM-1	Li *et al.* 2002
SH-containing angiotensin converting enzyme inhibitor	NF-κB, ROS	E-selectin, ICAM-1, VCAM-1	Cominacini *et al.* 2002, Liu *et al.* 2002
Docosahexaenoic acid	PKB	P-selectin, ICAM-1	Chen *et al.* 2003
Eicosapentaenoic acid	PKB	P-selectin, ICAM-1	Chen *et al.* 2003
Diallyl disulfide	PKA, PKB	E-selectin, VCAM-1	Lei *et al.* 2008
Diallyl trisulfide	PKA, PKB	E-selectin, VCAM-1	Lei *et al.* 2008
Butyrate	NF-κB	ICAM-1, VCAM-1	Siennicka 2005
Propionate	NF-κB	ICAM-1, VCAM-1	Siennicka 2005

A variety of drugs and nutrients have been used to regulate the adhesion molecule expression, and the mechanisms of these substances have all been elucidated by recent studies. NF-κB, nuclear factor-kappaB. ROS, reactive oxygen species. PKA, protein kinase A. PKB, protein kinase B. ICAM-1, intercellular adhesion molecule 1. VCAM-1, vascular cell adhesion molecule 1.

APPLICATIONS TO OTHER AREAS OF HEALTH AND DISEASE

Although the pathological role of adhesion molecules in cardiovascular diseases remains uncertain, some studies have postulated that serum levels of soluble adhesion molecules can be useful risk predictors of cardiovascular events in healthy populations as well as in people in various stages of disease. In large prospective studies of healthy individuals, soluble ICAM-1 (but not soluble VCAM-1) appears to be consistently related to incidents involving coronary artery disease. Soluble VCAM-1 is considered to be the strongest predictor of future cardiovascular events in individuals with coronary artery disease, diabetes, or unstable angina. Soluble VCAM-1 is therefore a better marker of the extent and severity of atherosclerosis.

SUMMARY

- Interactions between lipoproteins and vascular endothelium are considered to be a central component of the pathogenesis of atherosclerosis and cutaneous xanthomas.
- The oxidation of low-density lipoprotein (LDL) is an important process in the pathogenesis and the oxidized LDL exists in the circulation and in the lesions.
- Oxidized low-density lipoprotein (Ox-LDL) binds to endothelial receptors (including scavenger receptors) and activates an intracellular signal transduction pathway that induces the expression of adhesion molecules on the surface of endothelial cells.
- Circulating monocytes adhere to the endothelial cells via the Ox-LDL-induced adhesion molecules and subsequently transmigrate into the lesions.
- The mechanisms of adhesion molecule expression vary depending on the process of LDL oxidation and the organ specificity of the endothelial cells.

Abbreviations

CLEVER-1	common lymphatic endothelial and vascular endothelial receptor 1
CL-P1	collectin placenta 1
Cu-LDL	low-density lipoprotein oxidized by copper-ion
FEEL-1	fasciclin, epidermal growth factor (EGF)-like, laminin-type EGF-like and link domain-containing scavenger receptor 1
G_i-protein	inhibitory G protein
HAECs	human aortic endothelial cells

HCAECs	human coronary artery endothelial cells
HDMECs	human dermal microvascular endothelial cells
HUVECs	human umbilical vein endothelial cells
ICAM-1	intercellular adhesion molecule 1
LDL	low-density lipoprotein
LOX-1	lectin-like oxidized LDL receptor 1
MM-LDL	minimally modified/oxidized low-density lipoprotein
NF-κB	nuclear factor-kappaB
Ox-LDL	oxidized low-density lipoprotein
Ox-PAPC	oxidized1-palmitoyl-2-arachidonoyl-*sn*-glycero-3-phosphorylcholine
PAPC	1-palmitoyl-2-arachidonoyl-*sn*-glycero-3-phosphorylcholine
PKA	protein kinase A
PKB	protein kinase B
PKC	protein kinase C
ROS	reactive oxygen species
SR-BI	scavenger receptor BI
SRCL	scavenger receptor with C-type lectin
SREC-I	scavenger receptor expressed on endothelial cell I
SR-PSOX	scavenger receptor for phosphatidylserine and oxidized lipoprotein
TLR	toll-like receptor
VCAM-1	vascular cell adhesion molecule 1
X-LDL	xanthoma lesion-oxidized low-density lipoprotein

Key Facts about Low-density Lipoprotein

The following are the basic facts about low-density lipoprotein that is a fraction of the total serum lipids. This lipoprotein is associated with an increased risk of atherosclerosis and cutaneous xanthomas.

1. A lipoprotein is a biochemical assembly that contains lipids and proteins.
2. Lipoproteins are classified by density: high-density lipoprotein, low-density lipoprotein, intermediate-density lipoprotein, very low-density lipoprotein, and chylomicrons.
3. Low-density lipoprotein consists of 50% cholesterol and 25% protein, and it carries cholesterol from the liver to peripheral tissues.
4. Low-density lipoprotein is associated with an increased risk of atherosclerosis and cutaneous xanthomas.

Definition of Terms

Atherosclerosis: A chronic inflammatory disease affecting the intima and media of large- and medium-sized arteries characteristically due to the accumulation of lipoproteins and macrophages. This inflammatory process leads to the formation of focal lesions (plaques), containing lipid and fibrous tissue. The plaques eventually rupture, causing the formation of a thrombus that will rapidly slow or stop blood flow, and thereby leading to death of the tissues fed by the artery.

Cutaneous xanthoma: A yellowish-orange, lipid-filled nodule or plaque in the skin, often occurring on an eyelid or over a joint. Histologically, the lesions show macrophage-derived lipid-laden foam cells. The formation of xanthomas may indicate an underlying disease, usually related to the abnormal metabolism of lipids, including cholesterol.

Endothelial cells: The cells lining the inner walls of blood vessels.

Low-density lipoprotein: A fraction of the total serum lipids, the so called 'bad' cholesterol. This lipoprotein is associated with an increased risk of atherosclerosis and cutaneous xanthomas.

Monocyte/macrophage: A macrophage is a highly phagocytic cell derived from circulating monocytes that occurs in the walls of blood vessels or in the connective tissue.

Scavenger receptors: A group of receptors that recognize modified low-density lipoprotein by oxidation or acetylation. This name is based on the function of cleaning (scavenging): scavenger receptors widely recognize and uptake macromolecules having a negative charge as well as modified low-density lipoprotein.

Signal transduction: Any process by which a cell converts one kind of signal or stimulus into another, most often involving ordered sequences of biochemical reactions inside the cell, which are carried out by enzymes, activated by second messengers resulting in what is thought of as a 'signal transduction pathway'. In many signal transduction processes, the number of proteins and other molecules participating in these events emanates from the initial stimulus and thus results in a 'signal cascade'.

References

Berliner, J.A. and M.C. Territo, A. Sevanian, S. Ramin, J.A. Kim, B. Bamshad, M. Esterson, and A.M. Fogelman. 1990. Minimally modified low density lipoprotein stimulates monocyte endothelial interactions. J. Clin. Invest. 85: 1260-1266.

Chen, H. and D. Li, J. Chen, G.J. Roberts, T. Saldeen, and J.L. Mehta. 2003. EPA and DHA attenuate ox-LDL-induced expression of adhesion molecules in human coronary artery endothelial cells via protein kinase B pathway. J. Mol. Cell. Cardiol. 35: 769-775.

Cominacini, L. and U. Garbin, A.F. Pasini, A. Davoli, M. Campagnola, G.B. Contessi, A.M. Pastorino, and V. Lo Cascio. 1997a. Antioxidants inhibit the expression of intercellular cell adhesion molecule-1 and vascular cell adhesion molecule-1 induced by oxidized LDL on human umbilical vein endothelial cells. Free Radic. Biol. Med. 22: 117-127.

Cominacini, L. and U. Garbin, A.F. Pasini, T. Paulon, A. Davoli, M. Campagnola, E. Marchi, A.M. Pastorino, G. Gaviraghi, and V. Lo Cascio. 1997b. Lacidipine inhibits the activation of the transcription factor NF-kappaB and the expression of adhesion molecules induced by pro-oxidant signals on endothelial cells. J. Hypertens. 15: 1633-1640.

Cominacini, L. and A. Pasini, U. Garbin, S. Evangelista, A.E. Crea, D. Tagliacozzi, C. Nava, A. Davoli, and V. Lo Cascio. 2002. Zofenopril inhibits the expression of adhesion molecules on endothelial cells by reducing reactive oxygen species. Am. J. Hypertens. 15: 891-895.

Dwivedi, A. and E.E. Anggard, and M.J. Carrier. 2001. Oxidized LDL-mediated monocyte adhesion to endothelial cells does not involve NF-kappaB. Biochem. Biophys. Res. Commun. 284: 239-244.

Gebuhrer, V. and J.F. Murphy, J.-C. Bordet, M.-P. Reck, and J.L. McGregor. 1995. Oxidized low-density lipoprotein induces the expression of P-selectin (GMP140/PADGEM/CD62) on human endothelial cells. Biochem. J. 306: 293-298.

Ikeda, M. and K. Nakajima, H. Nakajima, M. Matsumoto, M. Seike, and H. Kodama. 2006. Contribution of xanthoma tissue-derived LDL density substances in the transformation of macrophages to foam cells. J. Dermatol. Sci. 44: 161-168.

Kamanna, V.S. and R. Pai, H. Ha, M.A. Kirschenbaum, and D.D. Roh. 1999. Oxidized low-density lipoprotein stimulates monocyte adhesion to glomerular endothelial cells. Kidney Int. 55: 2192-2202.

Kume, N. and M.I. Cybulsky, and M.A. Gimbrone Jr. 1992. Lysophosphatidylcholine, a component of atherogenic lipoproteins, induces mononuclear leukocyte adhesion molecules in cultured human and rabbit arterial endothelial cells. J. Clin. Invest. 90: 1138-1144.

Lei, Y.P. and H.W. Chen, L.Y. Sheen, and C.K. Lii. 2008. Diallyl disulfide and diallyl trisulfide suppress oxidized LDL-induced vascular cell adhesion molecule and E-selectin expression through protein kinase A- and B-dependent signaling pathways. J. Nutr. 138: 996-1003.

Leitinger, N. and A.D. Watson, K.F. Faull, A.M. Fogelman, and J.A. Berliner. 1997. Monocyte binding to endothelial cells induced by oxidized phospholipids present in minimally oxidized low density lipoprotein is inhibited by a platelet activating factor receptor antagonist. Adv. Exp. Med. Biol. 433: 379-382.

Li, D. and J.L. Mehta. 2000. Antisense to LOX-1 inhibits oxidized LDL-mediated upregulation of monocyte chemoattractant protein-1 and monocyte adhesion to human coronary artery endothelial cells. Circulation 101: 2889-2895.

Li, D. and H. Chen, F. Romeo, T. Sawamura, T. Saldeen, and J.L. Mehta. 2002. Statins modulate oxidized low-density lipoprotein-mediated adhesion molecule expression in human coronary artery endothelial cells: role of LOX-1. J. Pharmacol. Exp. Ther. 302: 601-605.

Liu, G.X. and D.M. Ou, L.X. Li, L.X. Chen, H.L. Huang, D.F. Liao, and C.S. Tang. 2002. Probucol inhibits oxidized-low density lipoprotein-induced adhesion of monocytes to endothelial cells in vitro. Acta Pharmacol. Sin. 23: 516-522.

Mason, J.C. and H. Yarwood, K. Sugars, and D.O. Haskard. 1997. Human umbilical vein and dermal microvascular endothelial cells show heterogeneity in response to PKC activation. Am. J. Physiol. 273: C1233-1240.

Matsumoto, M. and M. Ikeda, M. Seike, and H. Kodama. 2003. Different mechanisms of adhesion molecule expression in human dermal microvascular endothelial cells by xanthoma tissue-mediated and copper-mediated oxidized low density lipoproteins. J. Dermatol. Sci. 32: 43-54.

Ochi, H. and N. Kume, E. Nishi, and T. Kita. 1995. Elevated levels of cAMP inhibit protein kinase C-independent mechanisms of endothelial platelet-derived growth factor-B chain and intercellular adhesion molecule-1 gene induction by lysophosphatidylcholine. Circ. Res. 77: 530-535.

Siennicka, A. 2005. The effect of short-chain fatty acids on expression of endothelial adhesion molecules stimulated by oxidatively modified LDL. Ann. Acad. Med. Stetin. 51: 117-126.

Stewart, B.W. and S. Nagarajan. 2006. Recombinant CD36 inhibits oxLDL-induced ICAM-1-dependent monocyte adhesion. Mol. Immunol. 43: 255-267.

Takei, A. and Y. Huang, and M.F. Lopes-Virella. 2001. Expression of adhesion molecules by human endothelial cells exposed to oxidized low density lipoprotein. Influences of degree of oxidation and location of oxidized LDL. Atherosclerosis 154: 79-86.

Vora, D.K. and Z.T. Fang, S.M. Liva, T.R. Tyner, F. Parhami, A.D. Watson, T.A. Drake, M.C. Territo, and J.A. Berliner. 1997. Induction of P-selectin by oxidized lipoproteins. Separate effects on synthesis and surface expression. Circ. Res. 80: 810-818.

Watson, A.D. and N. Leitinger, M. Navab, K.F. Faull, S. Horkko, J.L. Witztum, W. Palinski, D. Schwenke, R.G. Salomon, W. Sha, G. Subbanagounder, A.M. Fogelman, and J.A. Berliner. 1997. Structural identification by mass spectrometry of oxidized phospholipids in minimally oxidized low density lipoprotein that induce monocyte/endothelial interactions and evidence for their presence in vivo. J. Biol. Chem. 272: 13597-13607.

Ylä-Herttuala, S. and W. Palinski, M.E. Rosenfeld, S. Parthasarathy, T.E. Carew, S. Butler, J.L. Witztum, and D. Steinberg. 1989. Evidence for the presence of oxidatively modified low density lipoprotein in atherosclerotic lesions of rabbit and man. J. Clin. Invest. 84: 1086-1095.

Yang, Q.W. and L. Mou, F.L. Lv, J.Z. Wang, L. Wang, H.J. Zhou, and D. Gao. 2005. Role of Toll-like receptor 4/NF-kappaB pathway in monocyte-endothelial adhesion induced by low shear stress and ox-LDL. Biorheology 42: 225-236.

Zhao, S.P. and D.Y. Xu. 2000. Oxidized lipoprotein(a) enhanced the expression of P-selectin in cultured human umbilical vein endothelial cells. Thromb. Res. 100: 501-510.

CHAPTER 27

Adhesion Molecules in Atrial Fibrillation

Matthias Hammwöhner[1], Alicja Bukowska[2],
Rüdiger C. Braun-Dullaeus[3] and Andreas Goette [4,*]

Division of Cardiology
Otto-von-Guericke-University Hospital Magdeburg,
Leipzigerstr. 44, 39120, Magdeburg, Germany
E-mail: [1]matthias.hammwoehner@med.ovgu.de; [2]alicja.bukowska@med.vgu.de;
[3]r.braun-dullaeus@med.ovgu.de; [4]andreas.goette@med.ovgu.de

ABSTRACT

Atrial fibrillation (AF) is the most common cardiac arrhythmia. Its incidence rate and prevalence constantly rises with age. AF not only impairs quality of life, but also increases mortality due to its attributed stroke risk and related heart failure. AF is a major contributor to apoplectic stroke at all ages because of thrombus formation especially in the left atrial appendage with consecutive thromboembolism to brain arteries. Therefore, AF-related stroke contributes to about 30% of all strokes in elderly patients. Recent studies showed the importance of prothrombotic endocardial changes for the development of atrial thrombi. The initiating mechanism is the increased expression of adhesion molecules, which allows the endocardial recruitment of inflammatory cells and leukocyte-platelet conjugates to the atrial endothelium. This chapter summarizes AF-related cellular and molecular changes, focussing on the role that adhesion molecule expression

---

*Corresponding author
Key terms are defined at the end of the chapter.

plays in the context of AF and thrombus formation. It further discusses underlying mechanisms for altered adhesion molecule expression and possible treatment options that may influence these mechanisms.

INTRODUCTION

Atrial fibrillation (AF) is the most frequent sustained arrhythmia in clinical practice and is associated with a high risk of stroke, heart failure, and hospitalization (Wolf *et al.* 1991). The prevalence of AF constantly rises with advanced age (Fig. 1). Due to general ageing of our population and accumulation of predisposing conditions, the prevalence of AF will increase by at least 250% by the year 2050 (Go *et al.* 2001). Recent data even suggest that its prevalence is underestimated because of the frequent occurrence of clinically asymptomatic AF.

Thromboembolic stroke is the most common severe complication of AF, with a 5-fold increase in stroke incidence in nonvalvular and a 17-fold increase in valvular AF (Hammwöhner and Goette 2008). Risk factors for AF-induced thromboembolic events are age, diabetes mellitus, hypertension, heart failure,

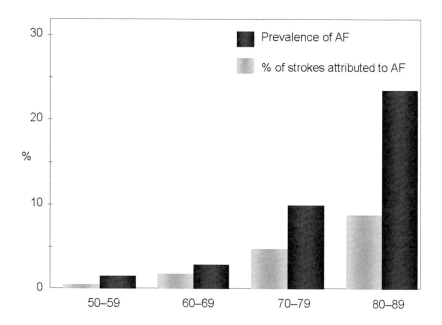

Fig. 1   Prevalence of atrial fibrillation (AF) and attributed stroke risk (Wolf *et al.* 1991).

and/or previous stroke (Hammwöhner *et al.* 2007a). Depending on these risk factors, the individual risk for thromboembolic events varies from about 2% to 18% per year.

Treatment of AF requires for optimized diagnostic and therapeutic management taking into consideration individual patient characteristics and determinants of clinical course and complications. AF therefore represents a significant health-care burden. Total annual costs for treatment of AF were estimated at US $6.65 billion in the United States in 2005.

PATHOPHYSIOLOGY OF ATRIAL FIBRILLATION

Cardiac tachyarrhythmia is known to induce significant electrophysiological and structural changes in cardiac tissue, which themselves may contribute to the persistence and aggravation of AF (Goette *et al.* 1996, Ausma *et al.* 1997). The overall pathophysiological mechanism for electrophysiological changes in atrial myocytes has been termed 'electrical remodeling'. Besides substantial shortening of the atrial action potential and alteration of other electrophysiological properties, AF also causes significant structural changes ('structural remodeling') in atrial tissue. Electrical and structural remodeling processes take place in atrial myocytes as well as in atrial endocardial cells (the interior surface layer of both upper heart chambers). In order to understand the impact of AF on adhesion molecule expression, one needs to understand these remodeling processes and their impact on downstream processes. Recent studies have provided initial insights into the molecular mechanisms involved in the development of cellular and subcellular changes. Altered intracellular calcium ion (Ca^{2+}) homeostasis and angiotensin II receptor type 1 (AT1R) activation have been identified as important remodeling factors contributing to cellular and cardiac hypertrophy, atrial extracellular matrix accumulation, and fibrosis in AF (Goette *et al.* 1996, Ausma *et al.* 1997). Cellular Ca^{2+}-overloading on the one hand and AT1R-induced nicotinamide adenine dinucleotide phosphate (NADPH) oxidase activity on the other hand have been demonstrated to cause excessive intracellular oxidative stress. Oxidative stress itself has also been identified to trigger atrial remodeling during AF ('positive feedback loop'). Gene expression profiling of atrial tissue samples from patients with SR and AF revealed a decreased expression of antioxidative genes, whereas expression of five reactive oxygen species (ROS)-producing genes was increased (Kim *et al.* 2005). At the molecular level, several AF-related alterations of atrial tissue are due to activation of different signal transduction systems. Recently, we were able to show the impact of the nuclear factor kappa B (NF-κB) pathway in the process of rapid pacing–induced oxidative stress (Bukowska *et al.* 2008). NF-κB in turn leads to upregulation of atrial adhesion molecule expression (Fig. 2).

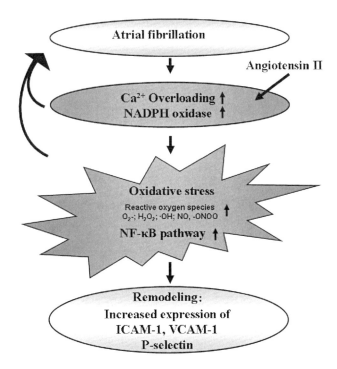

Fig. 2 Concept of atrial remodeling cascade. Atrial fibrillation (AF) induces angiotensin II-dependent changes in cellular calcium ion (Ca^{2+}) handling and nicotinamide adenine dinucleotide phosphate (NADPH) oxidase activity leading to oxidative stress. Accumulating reactive oxygen species (ROS) induce the nuclear factor kappa B (NF-κB) pathway. This in turn leads to upregulation of cellular adhesion molecule expression (ICAM-1, VCAM-1, P-selectin).

OXIDATIVE STRESS AND OXIDATIVE STRESS-DEPENDENT SIGNALING

Oxidative stress induced by ROS plays a pivotal role in AF. ROS are known to be physiologically generated in response to growth factors, cytokines, G-protein coupled receptor agonists or shear stress. They can also function as signaling molecules. Several enzymes including mitochondrial electron transport system, xanthine oxidase, the cytochrome p450, the NAPDH oxidase, uncoupled NO synthase (NOS) and myeloperoxidase have been reported to produce ROS.

Mihm *et al.* (2001), were the first to demonstrate extensive oxidative damage in atrial myocardium of patients with AF. This process was mainly driven by peroxynitrite. The same group also showed increased rates of protein carbonylation in fibrillating human tissue. Later it was experimentally shown by Carnes *et al.* (2001) that rapid atrial pacing led to decreased tissue levels of ascorbic acid, whereas protein nitration was increased. Carnes *et al.* (2001) showed that protein nitration among other processes accounts for electrical remodeling. Nitration and carbonylation of structural proteins especially alter Ca^{2+} handling of the endocardial cell. Abnormal calcium handling, however, not only changes electrophysiological properties, but also impairs mitochondrial function and thus myocardial energetics. In the cardiovascular system, voltage-activated Ca^{2+} channels are essential for the generation of normal cardiac rhythm, for induction of rhythm propagation through the atrioventricular node, and also for the contraction of the atrial and ventricular muscle. In diseased myocardium, calcium channels can contribute to abnormal impulse generation and cardiac arrhythmias. Under physiological state, mitochondria serve their well-known role in generation of adenosine triphosphate (ATP) by oxidative phosphorylation. Massive Ca^{2+} influx into the cytosol and subsequent Ca^{2+} sequestration by mitochondria result in marked alterations of mitochondrial morphology in atrial tissue of patients with AF (Schild *et al.* 2006). Recently, using two rapid pacing models of *in vitro* differentiated P19 cardiomyocytes and of human atrial tissue slices, we demonstrated that tachycardia is associated with mitochondrial swelling, and impairment of mitochondrial ATP production as evidenced by decreased endogenous respiration in combination with decreased ATP levels in intact cells (Schild *et al.* 2006). Inhibition of Ca^{2+} inward current with the calcium receptor antagonist verapamil protected against hypertrophic response and oxidative stress. Verapamil thus ameliorated morphological changes and dysfunction of mitochondria. Moreover, dysfunctional mitochondria seemed to contribute to signaling pathways aggravating oxidative stress and initiating the inflammatory status in atrial tissue. Data from our studies suggest that non-physiological Ca^{2+} entry is the primary trigger for mitochondrial dysfunction and oxidative stress within myocytes during tachycardia (Bukowska *et al.* 2008). The finding that calcium receptor antagonists exert a protective effect on oxidative stress is supported by other studies. They showed that mibefradil, which blocks L-type and T-type Ca^{2+} channels, and verapamil prevent the oxidation of cellular constituents and have cytoprotective effects.

Besides oxidative stress induced by changes in intracellular Ca^{2+} handling, Kim *et al.* (2005) have identified the NADPH oxidase to be the major source of ROS in the atrial system, whereas mitochondrial oxidase contribution is rather low (Schild *et al.* 2006). In AF, NADPH oxidase activity was shown to be increased (Dudley *et al.* 2005). Angiotensin II (Ang II) coupling of the AT1R was shown to stimulate the generation of ROS by induction of NADPH oxidase. AF is associated with a strong activation of the atrial renin-angiotensin-system (RAS), including induction of AngII and AngII-generating angiotensin-converting enzyme (ACE). Besides induction of NADPH oxidases, AngII binding to the G-protein-linked AT1R induces a cascade of phosphorylation that activates mitogen-activated protein kinases, leading to cell proliferation and cellular hypertrophy. The two major contributors to oxidative stress during AF are depicted in Fig. 3.

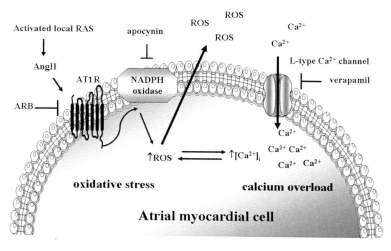

Fig. 3 Oxidative stress in AF. Altered intracellular calcium (Ca^{2+}) homeostasis and activation of AT1R have been identified as factors contributing to oxidative stress in atrial myocardium. Local activation of the renin angiotensin system (RAS) leads to activation of the angiotensin receptor (AT1R) via ligation of angiotensin II (AngII). This leads to increased activation of the nicotinamide adenine dinucleotide phosphate (NADPH) oxidase and consecutive overproduction of reactive oxygen species (ROS). Angiotensin receptor blockers (ARB) and NADPH oxidase inhibitor apocynin can decrease ROS production. Altered Ca^{2+} handling leads to intracellular Ca^{2+} overload enhancing oxidative stress. This process can be mitigated by calcium receptor antagonists, like verapamil.

Elevated ROS and Ca^{2+} levels are major activators of transcription factor NF-κB. Nuclear accumulation of NF-κB was observed in both *ex vivo* atrial tissue from patients with AF and in atrial tissue slices rapidly paced *in vitro*. As concluded from a supershift analysis we performed, the classical p50/p65 NF-κB components are involved in this setting (Bukowska *et al.* 2008). Typical target genes of NF-κB are pro-inflammatory cytokines such as interleukin-8 and tumor necrosis factor alpha (TNF-α), but also the endothelial adhesion molecules vascular cell adhesion molecule 1 (VCAM-1), intercellular adhesion molecule 1 (ICAM-1), and the endothelial lectin-like oxidized low-density lipoprotein receptor receptor (LOX-1). The LOX-1 promoter contains binding sites for numerous transcription factors, including NF-κB and AP-1. Recent studies demonstrated that AngII, by activating NF-κB, induces LOX-1 promoter activation in endothelial cells. Our studies confirmed these findings and also showed that LOX-1 induction could be prevented by the NADPH oxidase inhibitor apocynin and resveratrol, an inhibitor of NF-κB (Bukowska *et al.* 2008). Phosphorylation of IκB is a critical regulatory step in the activation of NF-κB. Release from IκB unmasks the nuclear localization signal of NF-κB and thus mediates its translocation into the cell nucleus. In our *in vitro* model we were able to show that rapid atrial pacing increased amounts of phospho-IκB-α substantially. Nuclear accumulation of NF-κB p50 was observed in atrial tissue from patients with AF and in paced atrial tissue slices. In line with the activation of NF-κB signaling, we observed an increase of pro-inflammatory and prothrombotic adhesion molecules such as ICAM-1 and VCAM-1, but also LOX-1 in atrial tissue from patients with AF and in *in vitro* paced atrial tissue slices (Bukowska *et al.* 2008) (Fig. 4). Interestingly, LOX-1 ligation itself was also shown to increase expression of adhesion molecules such as ICAM-1, VCAM-1, P-selectin, and E-selectin.

These findings strongly imply that oxidative stress in atrial tissue activates NF-κB signaling, which accounts for endocardial activation followed by an increased risk of atrial thrombus formation during AF. Tachycardia-associated factors (activated compounds of the RAS and Ca^{2+} signaling pathway) are involved in this response, which are directly or indirectly linked to oxidative stress. Accordingly, AT1R blockade, inhibition of L-type calcium channels, inhibition of NADPH oxidase, applications of antioxidants, and inhibition of NF-κB activation were all found to abolish or decrease the pacing-dependent changes in the atrial tissue (Figs. 3 and 4).

PROTHROMBOTIC ENDOCARDIAL REMODELING

Thrombus formation within the vascular system depends on pathological endothelial changes. The remodeling processes mentioned earlier affect myocardial cells and the atrial endothelium ('endocardial remodeling'). Endothelial damage and prothrombotic endothelial alteration are present in fibrillating atria. They are a prerequisite for the development of atrial clots, since atrial thrombi always start to grow from the atrial wall.

Virchow's triad defines circumstances under which thrombus formation is likely. It also applies to atrial thrombus formation during AF. Accordingly, thrombi develop in the presence of reduced blood velocities (*circulatory stasis*), when the activity of the clotting system is increased (*hypercoagulable state*), and in the presence of endothelial alterations (*endothelial injury*). While the loss of regular atrial contractions reduces blood flow velocities particularly in the atrial appendages, AF is also associated with an activation of the plasmatic clotting system and of platelets (Sohara *et al.* 1997, Goette *et al.* 2000a). However, until now, the contribution of endothelial alterations to atrial thrombogenesis has not yet been fully understood.

Numerous studies have shown that AF is associated with an inflammatory response measurable by systemic inflammatory markers (Conway *et al.* 2004). In turn, increased inflammatory markers and increased leukocyte-platelet interactions are predictors of atrial thrombus formation and thromboembolic stroke (Kawamura *et al.* 2006). An increased endothelial expression of adhesion

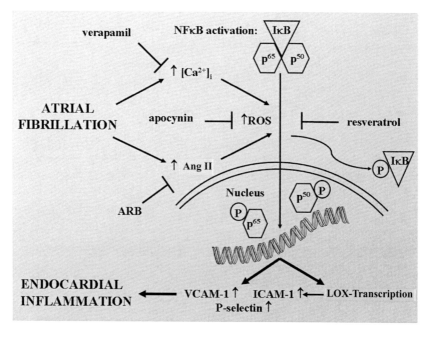

Fig. 4 NF-κB induced inflammation and adhesion molecule expression in AF. Atrial fibrillation (AF) leads to increased ROS-production followed by IκB phosphorylation. This unmasks the p50/p65 NF-κB components and their nuclear localization signal. After translocation to the cell nucleus they account for increased expression of adhesion molecules and lectin-like oxidized low-density lipoprotein receptor receptor (LOX-1), which itself is a stimulus for increased expression of the adhesion molecules ICAM-1, VCAM-1 and P-selectin. Angiotensin receptor blockers (ARB), calcium antagonist verapamil, NADPH oxidase inhibitor apocynin and NF-κB inhibitor resveratrol can reduce adhesion molecule production on various stages of the NF-κB signaling pathway.

molecules could be an important link between the initiation of pro-inflammatory and prothrombogenic mechanisms responsible for atrial thrombus formation (Gawaz *et al.* 1998, Hammwöhner *et al.* 2007b). As demonstrated in Fig. 5, AF-induced thromboemboli are very leukocyte rich, showing the close link between coagulation and inflammation in atrial thrombus formation.

In AF, interaction between inflammatory cells and platelets is increased by enhanced platelet P-selectin expression binding to its endothelial receptor P-selectin glycoprotein ligand 1 (PSGL-1) (Fig. 6) (Goette *et al.* 2000a). Kamiyama (1998) was also able to show increased P-selectin expression on the endothelial surface layer, which was accompanied by positively ICAM-1 stained adherent leukocytes in an AF-rabbit model. Endothelial adhesion molecule expression was also proven to be AngII dependent and declined after administration of an

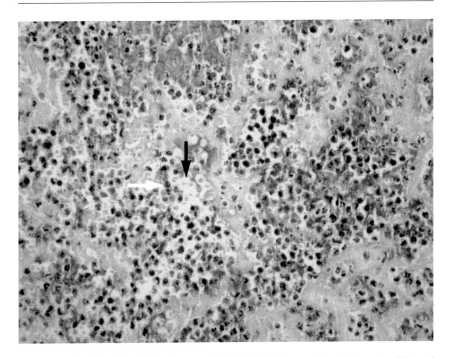

Fig. 5   Cardiac thromboembolus due to atrial fibrillation. Hematoxylin staining with additional myeloperoxidase antibody staining of a freshly embolized cell-rich cardiac thrombus in a patient with paroxysmal atrial fibrillation (modified after Hammwöhner *et al.* 2008). Note the abundant inflammatory cells (darkly stained myeloperoxidase, white arrow) mixed with platelets (small and lightly stained, black arrow) forming platelet-leukocyte aggregates embedded in a protein matrix of thrombin, fibrin and other proteins of the coagulation cascade. Original magnification level × 200. Microscope: Zeiss Axioskop 50; Camera: Nikon Coolpix 990; contrast enhanced for better differentiation of cell groups.

ACE inhibitor, as experiments in animal models showed. The granulocyte-platelet conjugates depicted in Figs. 5 and 6 are known to correlate with silent cerebral ischemia.

Interestingly, soluble ICAM-1 and VCAM-1 levels were also found to be elevated in patients with silent cerebral infarction (Kawamura *et al.* 2006). Of note, soluble adhesion molecule levels were found to correlate with a high predictive value with membrane-bound intramyocardial adhesion molecules (Noutsias *et al.* 2003). In line with these study results, we recently demonstrated that systemic plasma levels of ICAM-1 and VCAM-1 levels were increased in patients with AF, reaching the highest levels in patients with atrial thrombi. Soluble VCAM-1 levels were an independent predictor of atrial thrombi in multivariate analysis of variance (Hammwöhner *et al.* 2007b). Thus, AF-induced augmentation in endothelial

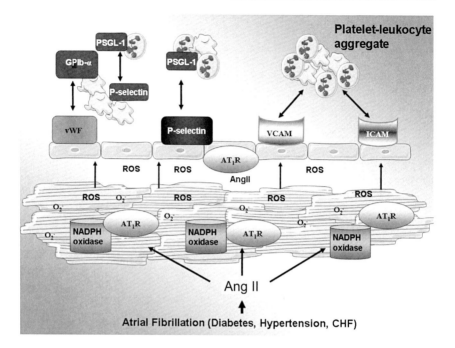

Fig. 6 Endocardial remodeling and thrombus formation in AF (modified after Hammwöhner and Goette 2008). Increased atrial expression of angiotensin II (Ang II) causes activation of the NADPH oxidase via activation of the angiotensin II type receptor (AT1R), which leads to generation of reactive oxygen species (ROS). Increased amounts of ROS induce via activation of NF-kB an increased expression of vascular cell adhesion molecule 1 (VCAM-1), intercellular adhesion molecule 1 (ICAM-1), and P-selectin on atrial endothelial cells. This causes activation and increased endocardial adhesion of leukocytes, which start to form aggregates with activated platelets. P-selectin is expressed on endothelial surfaces and in large amounts on platelets, binding to P-selectin glycoprotein ligand 1 (PSGL-1) on leukocytes. Von Willebrandt factor (vWF) on endothelial cells binds to glycoprotein receptor (GPIb-α) on platelets, intensifying the binding of leukocyte-platelet aggregates to the endothelial wall.

expression of adhesion molecules might be an early pathophysiological step in thrombogenesis by promoting adherence of leukocytes, monocytes, and platelets to the endocardial surface. Activated platelet-leukocyte aggregates in turn can further boost adhesion molecule expression by activating the NFκ-B pathway (Gawaz *et al.* 1998). In addition to increased adhesion molecule expression, AF leads to downregulation of myocardial nitric oxide synthase (eNOS) as shown by Cai *et al.* (2002) in an *in vivo* animal model. Loss of NO production in vascular endothelium further enhances the expression of adhesion molecules, thereby fostering the development of vascular thrombi. The pathophysiological significance of adhesion molecules in AF is further supported by the study of Kamiyama (1998) mentioned above.

Recently, in *ex vivo* experiments using organotypic human atrial tissue cultures and *in vivo* experiments, we found an increased atrial expression of VCAM-1 in patients with paroxysmal and persistent AF. Importantly, we were able to demonstrate that the increased expression of atrial adhesion molecules occurs within hours of rapid atrial pacing (Goette *et al.* 2008a). This may explain why there was no difference in atrial VCAM-1 expression in patients with persistent and paroxysmal AF. Of note, blockade of the AT1R by irbesartan or olmesartan reduced adhesion molecule expression in atrial tissue during rapid pacing. These findings support the assumption that therapy with AT1R blockers (ARB) may effect atrial thrombus formation. Nevertheless, the clinical benefit of ARB therapy in reducing thromboembolism in AF has still to be proven.

One important factor that may influence endocardial protein expression is left ventricular failure causing cardiac pressure and volume overload. In our recent study (Goette *et al.* 2008a), we were also able to demonstrate an association between left ventricular function and atrial VCAM expression using multivariable analysis. However, hemodynamic alterations were absent in the *in vitro* system or were constant in the *in vivo* experiments. This demonstrates that AF itself can induce alterations in VCAM expression. Nevertheless, heart failure is a known risk factor for stroke in patients with AF. One explanation for this finding might be the impact of left ventricular dysfunction on atrial expression of VCAM-1. Accordingly, concomitant diseases (e.g., heart failure, diabetes) and gender seem to influence atrial VCAM-1 levels during AF, which may also explain persistently elevated VCAM-1 levels in the systemic circulation after successful cardioversion of AF (Hammwöhner *et al.* 2007b). Nevertheless, the contribution of diabetes mellitus and heart failure per se on atrial VCAM-1 expression needs further evaluation.

Importantly, we found quantitative differences in VCAM-1 expression in the right and left atria as a response to AF. During rapid atrial pacing, adhesion molecule levels increased more significantly in the left atrial tissue than in the right atrium. This may help to explain why thrombus formation is typically observed in the left atrial appendage, whereas right atrial thrombus formation with consecutive pulmonary embolism is a rare event in patients with AF. For our recent large series of tissue micro-array analyses (Goette *et al.* 2008a), only right atrial tissue samples were available. However, the results of our animal experiments suggest that analyses of left atrial tissue samples might have revealed even more pronounced differences in endocardial VCAM expression.

VON WILLEBRANDT FACTOR

Of note, besides ICAM-1, VCAM-1, and P-selectin, no other adhesion molecules (integrins, cadherins, lymphocyte homing receptors, molecules of the immunoglobulin superfamily cell adhesion molecules (IgSF-CAMs), or other selectins) have been investigated or demonstrated to be influenced by AF. Von Willebrandt factor (vWF) is by definition not an adhesion molecule. Its function, however, resembles those of adhesion molecules as a binding target for platelet ligation and initiation of thrombus formation. Thus it is an interesting target to reduce the incidence of thromboembolism and stroke during AF. Under normal flow conditions no measurable binding of circulating vWF to platelet GPIbα occurs. However, under high fluid shear-stress, platelets are activated and vWF is conformationally changed. Platelets aggregate as a result of vWF binding to GpIbα (Fig. 5).

Conway *et al.* (2002) showed that clinical risk factors for stroke were closely related to vWF-levels in patients with AF. Furthermore, endocardial expression of vWF tends to be increased in AF. Moreover, atrial volume overload appears important for regulation of endocardial vWF expression. Thus, increased binding of platelets to vWF may predominantly occur in patients with atrial volume overload such as heart failure and, therefore, inhibition of vWF factor could influence atrial thrombogenesis in this specific clinical situation.

PLATELET ADHESION MOLECULES

Adenosine diphosphate (ADP) and thrombin are known potent stimuli to activate platelet aggregability and adhesion. Activation of the platelet ADP or thrombin receptor is followed by the appearance of alpha-granule proteins on the platelet cell surface. P-selectin is functionally the most important among them. Thus, P-selectin levels can be used to determine the stimulatory effect of ADP and thrombin on platelet function. P-selectin mediates platelet-neutrophil adhesion, which promotes intracellular signaling and enhances neutrophil superoxide anion release. We were able to demonstrate that platelet expression of P-selectin is significantly increased in a patient cohort with AF (Goette *et al.* 2000a). In addition, *ex vivo* platelet stimulation with ADP and thrombin receptor activating peptide (TRAP) resulted in increased amounts of platelet P-selectin in patients with AF as compared to patients in SR (Hammwöhner *et al.* 2007b). These data are supported by Minamino *et al.*, (1998), who demonstrated not only that rapid atrial pacing causes increased P-selectin expression, but also that platelet P-selectin expression is another predictor for silent cerebral infarction due to thromboembolism. Minamino *et al.* (1998), further demonstrated that inhibition of NO synthesis increases the expression of P-selectin. The same group had found reduced plasma NO levels in patients with AF before. Thus, decreased NO synthesis by damaged endothelial cells and/or platelets may be a possible trigger for the observed increased P-selectin levels during AF *in vivo*. However, according to the results of our studies the underlying pathophysiological mechanism causing an upregulation of P-selectin is not related to elevated atrial pressure or increased ventricular rate. Turbulent blood flow during AF, however, may be a potent

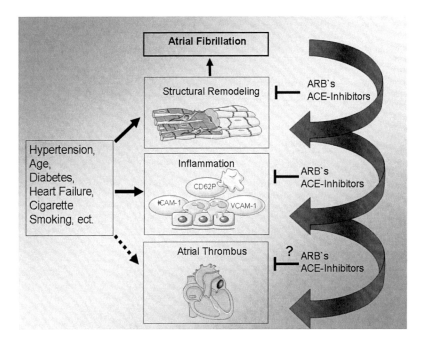

Fig. 7 Influence of ACE-Inhibitors and ARB on pathophysiology of AF. Atrial fibrillation (AF) causes structural remodeling and inflammation (Curved arrows on the right). Upregulation of adhesion molecules among other predisposing factors leads to thrombus formation. Angiotensin receptor blockers (ARB) and angiotensin converting enzyme (ACE) inhibitors positively influence remodeling and inflammation. Further studies are warranted to show a clinically significant impact of ACE-inhibitors and ARB on thrombus generation.

stimulus to cause frequent collisions of platelets and injury of endothelial cells leading to platelet activity and thrombus formation.

THERAPEUTIC IMPLICATIONS

The impact of ARB therapy on systemic adhesion molecule levels has been shown in patients with atherosclerosis and heart failure (Tsutamoto *et al.* 2000). Figure 7 summarizes the points of action of ARB therapy on AF-related processes and myocardial changes. In our recent AF-studies we were able to demonstrate the direct impact of ARB treatment on adhesion molecule expression. Our *ex vivo* experiments using organotypic human atrial tissue cultures and *in vivo* experiments showed that ARB treatment reduces adhesion molecule expression in atrial tissue during rapid pacing. Thus, rapid atrial stimulation influences prothrombotic endocardial remodeling via the AT1R. In support of this view we were able to demonstrate elevated plasma concentrations of AngII in response to rapid pacing *in vivo*. These findings are in line with studies that have already shown a significant impact of the RAS system on the occurrence of AF (Goette *et al.* 2000b). Previous experiments demonstrated that ACE inhibitors and ARBs influence short-term atrial electrical alterations, which we confirmed in our study. This effect, however, appears to decrease in the long term (Shinagawa *et al.* 2002). Besides the variable electrophysiological effects, ACE inhibitors and ARBs have been clearly demonstrated to reduce the amount of pro-arrhythmic structural alterations such as atrial fibrosis. The experimentally observed anti-arrhythmic effects of ACE inhibitors and ARBs are supported by various retrospective analyses of clinical trials (TRACE, ValHeFT, SOLVD, CHARM, LIFE).

Nevertheless, it still has to be proven whether the effect of ARBs observed in our short-term *in vitro* and *in vivo* models corresponds to a clinically relevant reduction of thromboembolic events in patients with AF. Interestingly, the very first clinical results from the LIFE study already point in that direction. In addition, further prospective clinical trials (ANTIPAF, ACTIVE-I, and CREATIV-AF (Goette *et al.* 2008b)) are being carried out to determine the impact of ARB therapy on the occurrence of thromboembolic stroke during AF. Thus, further experiments in chronic AF models are warranted to elucidate the impact of ARB in the long term on prothrombotic atrial alterations at the molecular level.

Recent analyses of the PROGRESS trial including more than 6,000 patients showed that, in addition to the traditional risk factors, systemic levels of VCAM-1 provide prognostic information about recurrent ischemic stroke (Campbell *et al.* 2006).

CONCLUSIONS

The present studies provide evidence that AF and pacing-induced atrial tachycardia increase the expression of adhesion molecules in the atrial endocardium within hours. In addition, AF-induced expression of VCAM-1 is higher in the left atrium than in the right atrium. The prothrombogenic process of endocardial remodeling can be substantially attenuated by ARB treatment, suggesting that angiotensin II has a significant pathophysiological role in endocardial remodeling. However, experiments in chronic AF models are warranted to determine the long-term efficacy of ARBs in this process.

SUMMARY

- AF and pacing-induced atrial tachycardia increase the concentration of soluble serum adhesion molecules.
- AF and pacing-induced atrial tachycardia also increase expression of membrane-bound adhesion molecules in the atrial endocardium and on platelets. These processes occur rapidly within hours.

- AF-induced expression of VCAM-1 is higher in the left atrium than in the right atrium.
- The prothrombogenic process of endocardial remodeling can be substantially attenuated by treatment with angiotensin receptor blockers.
- This suggests that angiotensin II has a significant pathophysiological role in endocardial remodeling and consecutive adhesion molecule expression.
- Experiments in chronic AF models are warranted to determine the long-term efficacy of ARBs in this process.

Abbreviations

ACE	angiotensin-converting enzyme
ADP	adenosine diphosphate
AF	atrial fibrillation
Ang II	angiotensin II
ARB	angiotensin receptor blockers
AT1R	angiotensin II receptor type 1
ATP	adenosine triphosphate
Ca^{2+}	calcium ion
eNOS	endothelial nitric oxide synthase
GPIbα	gylcoprotein receptor 1 alpha
ICAM-1	intercellular adhesion molecule 1
IgSF-CAMs	immunoglobulin superfamily cell adhesion molecules
LOX-1	endothelial lectin-like oxidized low-density lipoprotein receptor receptor
NADPH	nicotinamide adenine dinucleotide phosphate
NF-κB	nuclear factor kappa B
NO	nitric oxide
PSGL-1	P-selectin glycoprotein ligand 1
RAS	renin-angiotensin-system
ROS	reactive oxygen species
SR	sinus rhythm
TNFα	tumor necrosis factor alpha
TRAP	thrombin receptor activating peptide
VCAM-1	vascular cell adhesion molecule 1
vWF	von Willebrandt factor

Definition of Terms

Atrial fibrillation (AF): Abnormal and arrhythmic heart rhythm due to chaotic propagation of disorganized electrical impulses and electrical wavefronts throughout both atria. AF can occur intermittently, at intervals lasting from seconds to weeks ('paroxysmal'), or be continuous ('persistent').

Electrical remodeling: Changes in electrical properties of cells (progressive shortening of the atrial action potential) due to a state of disease, in this case AF.

Nonvalvular atrial fibrillation: AF that is not associated with a diseased heart valve ('lone' AF).

Oxidative stress: Accumulation of intracellular ROS leading to membrane damage, inflammation, damage or alteration of cell genes or (programmed) cell death ('apoptosis').

Reactive oxygen species (ROS): Ions or very small molecules that include oxygen ions, free radicals, and peroxides, both inorganic and organic. ROS are chemically highly reactive due to unpaired valence shell electrons.

Sinus rhythm (SR): Physiological heart rhythm that is generated in the 'sinus node' located in the right atrium (right upper heart chamber). Heart beats occur in regular intervals with a heart rate depending on physical activity.

Stroke: Rapidly developing and permanent loss of brain function(s) due to disturbance in the blood supply either by occlusion of brain arteries or hemorrhage with consecutive brain tissue compression.

Structural remodeling: Morphological and functional changes of cells and intercellular matrix due to a state of disease, in this case AF.

Tachyarrhythmia: Synonym for AF.

Thromboembolic stroke: Stroke caused by a blood clot (thrombus) that was generated elsewhere and wandered with the bloodstream (embolized), finally occluding a brain artery.

Valvular atrial fibrillation: AF that is associated with a diseased heart valve.

References

Ausma, J. and M. Wijffels, F. Thone, L. Wouters, M. Allessie, and M. Borgers. 1997. Structural changes of atrial myocardium due to sustained atrial fibrillation in the goat. Circulation 96: 3157-3163.

Bukowska, A. and L. Schild, G. Keilhoff, D. Hirte, M. Neumann, A. Gardemann, K.H. Neumann, F.W. Rohl, C. Huth, A. Goette, and U. Lendeckel. 2008. Mitochondrial dysfunction and redox signaling in atrial tachyarrhythmia. Exp. Biol. Med. (Maywood) 233: 558-574.

Cai, H. and Z. Li, A. Goette, F. Mera, C. Honeycutt, K. Feterik, J.N. Wilcox, S.C. Dudley, Jr., D.G. Harrison, and J.J. Langberg. 2002. Downregulation of endocardial nitric oxide synthase expression and nitric oxide production in atrial fibrillation: potential mechanisms for atrial thrombosis and stroke. Circulation 106: 2854-2858.

Campbell, D.J. and M. Woodward, J.P. Chalmers, S.A. Colman, A.J. Jenkins, B.E. Kemp, B.C. Neal, A. Patel, and S.W. MacMahon. 2006. Soluble vascular cell adhesion molecule 1 and N-terminal pro-B-type natriuretic peptide in predicting ischemic stroke in patients with cerebrovascular disease. Arch. Neurol. 63: 60-65.

Carnes, C.A. and M.K. Chung, T. Nakayama, H. Nakayama, R.S. Baliga, S. Piao, A. Kanderian, S. Pavia, R.L. Hamlin, P.M. McCarthy, J.A. Bauer, and D.R. Van Wagoner. 2001. Ascorbate attenuates atrial pacing-induced peroxynitrite formation and electrical remodeling and decreases the incidence of postoperative atrial fibrillation. Circ. Res. 89: E32-E38.

Conway, D.S. and P. Buggins, E. Hughes, and G.Y. Lip. 2004. Relationship of interleukin-6 and C-reactive protein to the prothrombotic state in chronic atrial fibrillation. J. Am. Coll. Cardiol. 43: 2075-2082.

Conway, D.S. and L.A. Pearce, B.S. Chin, R.G. Hart, and G.Y. Lip. 2002. Plasma von Willebrandt factor and soluble p-selectin as indices of endothelial damage and platelet activation in 1321 patients with nonvalvular atrial fibrillation: relationship to stroke risk factors. Circulation 106: 1962-1967.

Dudley, S.C., Jr. and N.E. Hoch, L.A. McCann, C. Honeycutt, L. Diamandopoulos, T. Fukai, D.G. Harrison, S.I. Dikalov, and J. Langberg. 2005. Atrial fibrillation increases production of superoxide by the left atrium and left atrial appendage: role of the NADPH and xanthine oxidases. Circulation 112: 1266-1273.

Gawaz, M. and F.J. Neumann, T. Dickfeld, W. Koch, K.L. Laugwitz, H. Adelsberger, K. Langenbrink, S. Page, D. Neumeier, A. Schomig, and K. Brand. 1998. Activated platelets induce monocyte chemotactic protein-1 secretion and surface expression of intercellular adhesion molecule-1 on endothelial cells. Circulation 98: 1164-1171.

Go, A.S. and E.M. Hylek, K.A. Phillips, Y. Chang, L.E. Henault, J.V. Selby, and D.E. Singer. 2001. Prevalence of diagnosed atrial fibrillation in adults: national implications for rhythm management and stroke prevention: the AnTicoagulation and Risk Factors in Atrial Fibrillation (ATRIA) Study. JAMA 285: 2370-2375.

Goette, A. and A. Bukowska, U. Lendeckel, M. Erxleben, M. Hammwöhner, D. Strugala, J. Pfeiffenberger, F.W. Rohl, C. Huth, M.P. Ebert, H.U. Klein, and C. Rocken. 2008a. Angiotensin II receptor blockade reduces tachycardia-induced atrial adhesion molecule expression. Circulation 117: 732-742.

Goette, A. and A. D'Alessandro, A. Bukowska, S. Kropf, C. Mewis, C. Stellbrink, J. Tebbenjohanns, C. Weiss, and U. Lendeckel. 2008b. Rationale for and design of the CREATIVE-AF trial: randomized, double-blind, placebo-controlled, crossover study of the effect of irbesartan on oxidative stress and adhesion molecules in patients with persistent atrial fibrillation. Clin. Drug Invest. 28: 565-572.

Goette, A. and C. Honeycutt, and J.J. Langberg. 1996. Electrical remodeling in atrial fibrillation. Time course and mechanisms. Circulation 94: 2968-2974.

Goette, A. and A. Ittenson, P. Hoffmanns, S. Reek, W. Hartung, H. Klein, S. Ansorge, and J.C. Geller. 2000a. Increased expression of P-selectin in patients with chronic atrial fibrillation. Pacing Clin. Electrophysiol. 23: 1872-1875.

Goette, A. and T. Staack, C. Rocken, M. Arndt, J.C. Geller, C. Huth, S. Ansorge, H.U. Klein, and U. Lendeckel. 2000b. Increased expression of extracellular signal-regulated kinase and angiotensin-converting enzyme in human atria during atrial fibrillation. J. Am. Coll. Cardiol. 35: 1669-1677.

Hammwöhner, M. and A. D'Alessandro, O. Wolfram, and A. Goette. 2007a. New pharmacologic approaches to prevent thromboembolism in patients with atrial fibrillation. Curr. Vasc. Pharmacol. 5: 211-219.

Hammwöhner, M. and A. Goette. 2008. Will Warfarin Soon Be Passe? New Approaches to Stroke Prevention in Atrial Fibrillation. J Cardiovasc. Pharmacol. 52: 18-27.

Hammwöhner, M. and A. Ittenson, J. Dierkes, A. Bukowska, H.U. Klein, U. Lendeckel, and A. Goette. 2007b. Platelet Expression of CD40/CD40 Ligand and Its Relation to Inflammatory Markers and Adhesion Molecules in Patients with Atrial Fibrillation. Exp. Biol. Med. (Maywood) 232: 581-589.

Hammwöhner, M. and J. Tautenhahn, D. Kuester, and A. Goette. 2008. Inadvertend catch of a factor VIII rich thrombus with a femoral sheath during cardiac catheterization. Int. J. Cardiol. 130: e39-e41.

Kamiyama, N. 1998. Expression of cell adhesion molecules and the appearance of adherent leukocytes on the left atrial endothelium with atrial fibrillation: rabbit experimental model. Jpn. Circ. J. 62: 837-843.

Kawamura, T. and T. Umemura, A. Kanai, M. Nagashima, N. Nakamura, T. Uno, M. Nakayama, T. Sano, Y. Hamada, J. Nakamura, and N. Hotta. 2006. Soluble adhesion molecules and C-reactive protein in the progression of silent cerebral infarction in patients with type 2 diabetes mellitus. Metabolism 55: 461-466.

Kim, Y.M. and T.J. Guzik, Y.H. Zhang, M.H. Zhang, H. Kattach, C. Ratnatunga, R. Pillai, K.M. Channon, and B. Casadei. 2005. A myocardial Nox2 containing NAD(P)H oxidase contributes to oxidative stress in human atrial fibrillation. Circ. Res. 97: 629-636.

Mihm, M.J. and F. Yu, C.A. Carnes, P.J. Reiser, P.M. McCarthy, D.R. Van Wagoner, and J.A. Bauer. 2001. Impaired myofibrillar energetics and oxidative injury during human atrial fibrillation. Circulation 104: 174-180.

Minamino, T. and M. Kitakaze, S. Sanada, H. Asanuama, T. Kurotobi, Y. Koretsune, M. Fukunami, T. Kuzuya, N. Hoki, and M. Hori. 1998. Increased expression of P-selectin on platelets is a risk factor for silent cerebral infarction in patients with atrial fibrillation: role of nitric oxide. Circulation 98: 1721-1727.

Noutsias, M. and C. Hohmann, M. Pauschinger, P.L. Schwimmbeck, K. Ostermann, U. Rode, M.H. Yacoub, U. Kuhl, and H.P. Schultheiss. 2003. sICAM-1 correlates with myocardial ICAM-1 expression in dilated cardiomyopathy. Int. J. Cardiol. 91: 153-161.

Schild, L. and A. Bukowska, A. Gardemann, P. Polczyk, G. Keilhoff, M. Tager, S.C. Dudley, H.U. Klein, A. Goette, and U. Lendeckel. 2006. Rapid pacing of embryoid bodies impairs mitochondrial ATP synthesis by a calcium-dependent mechanism—a model of in vitro differentiated cardiomyocytes to study molecular effects of tachycardia. Biochim. Biophys. Acta 1762: 608-615.

Shinagawa, K. and H. Mitamura, S. Ogawa, and S. Nattel. 2002. Effects of inhibiting Na(+)/H(+)-exchange or angiotensin converting enzyme on atrial tachycardia-induced remodeling. Cardiovasc. Res. 54: 438-446.

Sohara, H. and S. Amitani, M. Kurose, and K. Miyahara. 1997. Atrial fibrillation activates platelets and coagulation in a time-dependent manner: a study in patients with paroxysmal atrial fibrillation. J. Am. Coll. Cardiol. 29: 106-112.

Tsutamoto, T. and A. Wada, K. Maeda, N. Mabuchi, M. Hayashi, T. Tsutsui, M. Ohnishi, M. Sawaki, M. Fujii, T. Matsumoto, and M. Kinoshita. 2000. Angiotensin II type 1 receptor antagonist decreases plasma levels of tumor necrosis factor alpha, interleukin-6 and soluble adhesion molecules in patients with chronic heart failure. J Am. Coll. Cardiol. 35: 714-721.

Wolf, P.A. and R.D. Abbott, and W.B. Kannel. 1991. Atrial fibrillation as an independent risk factor for stroke: the Framingham Study. Stroke 22: 983-988.

CHAPTER 28

Apolipoproteins and Cell Adhesion Molecules

Chunyu Zheng[1], Frank M. Sacks[2] and Masanori Aikawa[3],[*]

Department of Nutrition, Harvard School of Public Health (C.Z., F.M.S.), and the Center for Excellence in Vascular Biology, Cardiovascular Division, Department of Medicine, Brigham and Women's Hospital, Harvard Medical School (M.A.).

[1]Department of Nutrition, Harvard School of Public Health, Building 2 Room 202B, 665 Huntington Avenue, Boston, MA 02115.
E-mail: czheng@hsph.harvard.edu

[2]Department of Nutrition, Harvard School of Public Health, Building 1 Room 201, 665 Huntington Avenue, Boston, MA 02115. E-mail: fsacks@hsph.harvard.edu

[3]Center for Excellence in Vascular Biology, Brigham and Women's Hospital, Harvard Medical School, 77 Avenue Louis Pasteur, NRB741J, Boston, MA 02115.
E-mail: maikawa@rics.bwh.harvard.edu

ABSTRACT

Cell adhesion molecules, by mediating the recruitment of circulating leukocytes to the blood vessel wall and their subsequent adherence and transendothelial migration, play critical roles in all stages of atherosclerotic lesion development. Various inflammatory stimuli induce expression of adhesion molecules. This chapter highlights the direct functions of apolipoproteins, surface constituents of plasma circulating lipoproteins, in monocyte activation and vascular endothelium dysfunction. Apolipoproteins regulate lipoprotein metabolism, and their dysregulation closely associates with hypercholesterolemia, hypertriglyceridemia and insulin resistance, dysmetabolic states that predispose patients at elevated

---

Corresponding author
Key terms are defined at the end of the chapter.

risk for coronary heart disease. Recent preclinical studies by our group and others have demonstrated that apolipoprotein C-III, alone or as a component of plasma apoB lipoproteins, promotes monocyte adhesion to vascular endothelial cells via activation of cellular adhesion molecules. Apolipoprotein C-III also induces the activation of nuclear factor κB, a pleiotropic transcription factor that regulates expression of many pro-inflammatory molecules including cytokines and their receptors. Besides apolipoprotein C-III, emerging evidence suggests that other apolipoproteins such as lipoprotein(a), apolipoprotein E, and apolipoprotein A-I also mediate monocyte adhesion to the vascular endothelium, in addition to their regulatory functions in lipoprotein metabolism. These exciting findings, therefore, provide a direct link between dyslipidemia and monocyte activation and endothelial cell dysfunction.

INTRODUCTION

Atherosclerosis is a chronic inflammatory process that involves the formation of plaques consisting of immune cells including macrophage-derived foam cells, vascular endothelial cells (EC), smooth muscle cells, platelets, extracellular matrix, and a lipid-rich core with extensive necrosis and fibrosis of surrounding tissues (Ross 1999, Libby 2002, Aikawa and Libby 2004,). One early phase of atherosclerosis involves the recruitment of inflammatory cells from the blood circulation and their transendothelial migration. This process involves cell adhesion molecules expressed on the vascular endothelium and on circulating leukocytes in response to inflammatory stimuli.

Much effort has explored the inciting factors for the pathogenesis of atherosclerosis. The lipid hypothesis has played a central role in our understanding of atherogenesis over the past fifty years. Elevated blood cholesterol level is one of the major risk factors for cardiovascular disease, and pre-clinical evidence has suggested that hypercholesterolemia induces the accumulation of oxidized low-density lipoproteins (oxLDL) in the arterial wall, promoting EC activation and leukocyte recruitment (Libby and Aikawa 2002). In the 1990s, clinical studies established that cholesterol lowering with HMG-CoA reductase inhibitors (statins) reduces the onset of acute thrombotic complications of coronary atherosclerosis (Libby and Aikawa 2002, Aikawa and Libby 2004). Our studies provided pre-clinical evidence that cholesterol lowering prevents acute coronary events by reducing oxidative stress and EC activation, thus limiting vascular inflammation (Aikawa *et al.* 2002) (Fig. 1). However, even with aggressive cholesterol lowering with statin therapy, the majority of patients experiencing clinical coronary heart disease remain unsaved from these events, driving research that explores mechanisms of vascular inflammation beyond elevated LDL cholesterol. A series of studies from our laboratory and others have identified that apolipoproteins, surface constituents of all lipoprotein particles, may independently promote atherogenesis.

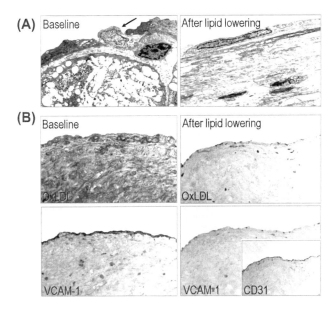

Fig. 1 Lipid lowering normalizes endothelial ultrastructure and reduces oxLDL accumulation and VCAM-1 expression in atherosclerotic rabbits. (A) Electron microscopy demonstrated that aortic EC in hypercholesterolemic rabbits fed an atherogenic diet for 4 mon (baseline, top left) showed a cuboidal structure typical of an 'activated' phenotype, whereas EC in rabbits after 16 mon lipid lowering (low, top right) had a more squamous morphology. A monocytic cell (arrowhead) appears to be entering the intima of baseline lesion. Original magnification × 3000. (B) OxLDL epitopes (MDA-lysine) accumulated in the aortic intima beneath EC immunoreactive for VCAM-1 in baseline rabbits fed the atherogenic diet for 4 mon (top left and bottom left), whereas oxLDL and VCAM-1 were barely detectable in the intima after dietary cholesterol lowering for 16 mon (top right and bottom right). CD3, an EC marker, indicated an intact monolayer (inset).
Color image of this figure appears in the color plate section at the end of the book.

This chapter reviews recent progress in this field, focusing on the direct actions of apolipoprotein C-III on monocyte activation and EC dysfunction.

APOLIPOPROTEIN C-III

Apolipoprotein C-III (apoC-III), an 8.8-kDa protein, resides on the surface of some very low-density lipoproteins (VLDL) and low-density lipoproteins and affects their metabolism. Plasma apoC-III levels correlate positively with plasma triglycerides over the entire spectrum from normo- to hypertriglyceridemia. Overexpression of apoC-III in mice causes hypertriglyceridemia, while apoC-III deficiency protects against it. ApoC-III regulates apoB lipoprotein metabolism by multiple mechanisms (Fig. 2). Most importantly, our results and others suggest that apoC-III impairs clearance of apoB lipoproteins from the circulation and

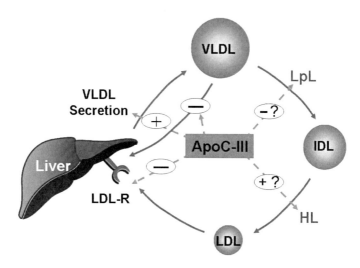

Fig. 2 Apolipoprotein C-III regulates plasma lipoprotein metabolism by multiple mechanisms. LpL, lipoprotein lipase; HL, hepatic lipase; LDL-R, LDL receptor.

stimulates hepatic VLDL assembly and secretion (Sundaram *et al.* 2007, Zheng *et al.* 2007).

Clinical evidence from us and others has demonstrated the importance of apoC-III as a predictor of cardiovascular disease outcomes (Lee *et al.* 2003, Sacks *et al.* 2000). The Cholesterol and Recurrent Events (CARE) trial demonstrated that plasma concentration of apoC-III in apoB lipoproteins (VLDL and LDL) is a strong, independent risk factor for recurrent coronary heart disease (Sacks *et al.* 2000) (Fig. 3A). In a substudy of the CARE trial, the concentration of LDL particles containing apoC-III was the strongest lipoprotein predictor of risk for recurrent cardiovascular events in type 2 diabetic patients (Lee *et al.* 2003) (Fig. 3B). Since the concentration of LDL containing apoC-III is much lower than the majority of plasma LDL that does not have apoC-III, and in view of the large risk associated with these relatively small increments of concentration of LDL with apoC-III, we hypothesized that apoC-III possesses direct atherogenic properties in addition to its deleterious effects on lipoprotein metabolism.

Following this thinking, we and Dr. Yoshida's group have systematically investigated direct pro-inflammatory and pro-atherogenic effects of apoC-III on vascular cells, demonstrating that this apolipoprotein, alone or as a component of plasma apoB lipoproteins, promotes EC and monocyte activation and their adhesion. Figure 4 summarizes our current understanding of these pleiotropic effects of apoC-III.

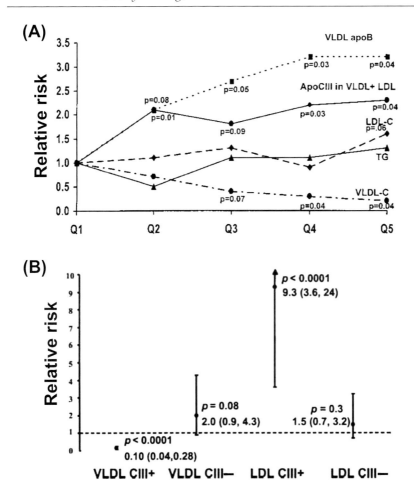

Fig. 3 ApoC-III as an independent risk factor for coronary heart disease. Multivariate analysis of apoC-III in VLDL and LDL (Panel A) and apoB lipoproteins with or without apoC-III (Panel B) to risk of recurrent coronary events in the CARE trial. (A) VLDL-apoB concentration, apoC-III concentration in VLDL+LDL, VLDL and LDL cholesterol concentrations in quintiles (Q1-Q5) were included together in multiple logistic regression model, along with HDL-cholesterol and plasma triglycerides. Other covariates were age, smoking, hypertension, and left ventricular ejection fraction. (B) Relative risk of apoB concentration in VLDL and LDL with (CIII+) or without apoC-III (CIII-) for 4th versus 1st quartile, were included together in multiple logistic regression model, along with treatment group (pravastatin or placebo), risk factors, LDL cholesterol, HDL cholesterol, and triglycerides. Reproduced from Sacks *et al.* (2000) (Panel A) and Lee *et al.* (2003) (Panel B), with permission.

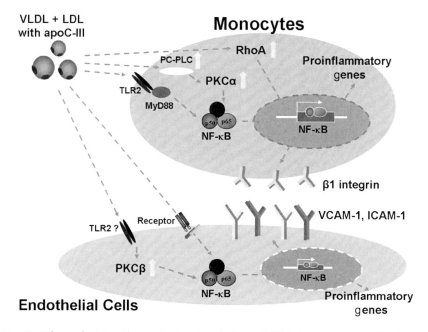

Fig. 4 Schema depicting the mechanisms by which apoC-III induces monocyte adhesion to vascular EC via activation of cellular adhesion molecules. ApoC-III, alone or as a component of circulating apoB lipoproteins, recognized by TLR-2, activates NF-κB through MyD88-dependent pathway, resulting in increased monocyte expression of β1-integrin (Kawakami *et al.* 2008b). PKCα activation, through PC-PLC, participates in apoC-III-induced β1-integrin activation via NF-κB (Kawakami *et al.* 2006a, 2007), and RhoA may also be involved (Kawakami *et al.* 2006a). In vascular endothelial cells, apoC-III induces VCAM-1 and ICAM-1 expression via PKCβ-mediated activation of NF-κB (Kawakami *et al.* 2006b). Induction of monocyte β1-integrin and endothelial cell VCAM-1 and ICAM-1 by apoC-III enhances monocyte adhesion under static and flow conditions. VCAM-1, vascular cell adhesion molecule 1; ICAM-1, intercellular adhesion molecule 1; NF- κB, nuclear factor κB; PKC, protein kinase C; TLR2, toll-like receptor 2; MyD88, myeloid differentiation primary response gene 88.

ApoC-III Induces Monocyte Adhesion to Vascular EC under Static and Flow Conditions

The adhesion of circulating monocytes to vascular endothelium contributes importantly to the inflammatory aspects of atherogenesis (Libby 2002, Libby and Aikawa 2002). Monocytes from hypercholesterolemic patients have increased expression of integrins and other adhesion molecules and show increased adhesion to EC *in vitro*. In this study we hypothesized that apoB lipoproteins with apoC-III could induce monocyte activation and subsequent adhesion to EC.

We isolated apoC-III containing VLDL and LDL lipoproteins from plasma of healthy volunteers by immunoaffinity chromatography and ultracentrifugation.

At physiological concentrations of apoB (50-100 μg/mL), VLDL containing apoC-III (CIII+) significantly increased the adhesion of THP-1 cells, a human monocytic cell line, to human saphenous vein EC (HSVEC) by more than 5-fold under static conditions, while LDL CIII+ significantly increased adhesion by more than 4-fold (Kawakami *et al.* 2006a). In addition, pretreatment of VLDL or LDL CIII+ with an anti-apoC-III antibody significantly reduced the activation of THP-1 cell adhesion, while anti-apoC-I, anti-apoC-II and anti-apoE antibodies had no effect, indicating that apoC-III in these lipoprotein fractions enhances THP-1 cell adhesion to vascular EC.

Similarly, VLDL CIII+, LDL CIII+, or apoC-III alone also induces the adhesion of THP-1 cells to vascular EC under flow conditions while VLDL CIII– or LDL CIII– did not affect THP-1 cell adhesion. Interestingly, THP-1 cells treated with apoC-III alone or in combination with VLDL or LDL demonstrated 90% to 95% of firm adhesions with minimal rolling interactions, suggesting integrin activation after treatment with apoC-III.

In addition to these deleterious effects on apoB lipoproteins, apoC-III inhibited the protective effects of HDL. In our experiments, HDL CIII– significantly reduced monocyte adhesion to EC by almost 30% in both static and flow condition. In contrast, HDL CIII+ did not reduce adhesion under either condition.

Apoc-III Induces Monocyte β1-integrin Activation

We then examined the activation of monocyte adhesion molecules by apoC-III (Kawakami *et al.* 2006a). Indeed, treatment with VLDL CIII+, LDL CIII+, or apoC-III alone increased the active forms of β1-integrin in THP-1 cells, an effect that was abolished by a binding-blocking β1-integrin antibody. In contrast, VLDL or LDL CIII– had no effect.

We further explored the molecular and cellular mechanisms involved in apoC-III-induced β1-integrin activation (Kawakami *et al.* 2007). We found that apoC-III activated RhoA, a ubiquitous intracellular GTPase. RhoA mediates cytoskeletal responses to extracellular signals and plays a crucial role in the migration and adhesion of monocytes by increasing the expression and/or binding affinity of cell surface integrins.

Protein kinase C (PKC) isoforms regulate monocyte adhesion in cooperation with or independently of RhoA. PKCα protein in the membrane fraction increased substantially after incubation with VLDL CIII+, LDL CIII+ or apoC-III alone. PKCβ was slightly activated, while little effect was observed on PKCδ and PKCζ isoforms. In contrast, VLDL or LDL CIII– had little effect on all PKC isoforms. In addition, a selective inhibitor of PKCα significantly inhibited THP-1 cell adhesion induced by VLDL CIII+, LDL CIII+, or apoC-III while anti-apoC-III antibody inhibited PKCα activation induced by apoC-III.

Phospholipases participate importantly in the activation of conventional PKC isoforms. We investigated their potential involvement in apoC-III-induced PKCα activation. ApoC-III treatment of THP-1 cells increased PC-PLC activity, an effect that was abolished by an anti-apoC-III antibody. A selective phosphatidylcholine-specific phospholipase C (PC-PLC) inhibitor inhibited apoC-III-induced PKCα activation in a concentration-dependent manner. In addition, an inhibitor to $G_{\alpha i}$ protein, pertussis toxin (PTX), inhibited apoC-III-induced PKCα activation.

Taken together, these results indicated that apoC-III-containing plasma apoB lipoproteins increase monocyte adhesion to vascular EC via PTX-sensitive G protein and PKCα-mediated β1-integrin activation.

Apoc-III Induces VCAM-1 and ICAM-1 Activation in Vascular EC

The induction of adhesion molecules in vascular EC is another initiating step in atherogenesis. Cell adhesion molecules of the immunoglobulin superfamily, vascular cell adhesion molecule 1 (VCAM-1) and intercellular adhesion molecule 1 (ICAM-1), mediate firm adhesion and transmigration across the vascular endothelium of activated circulating leukocytes. We therefore examined whether apoC-III regulates EC expression of VCAM-1 and ICAM-1 (Kawakami *et al.* 2006b).

Our results showed that pre-incubation of human saphenous vein EC with VLDL CIII+, or purified apoC-III, increased THP-1 cell adhesion. Incubating resting EC with apoC-III significantly increased VCAM-1 protein expression (5-fold), while moderately affecting ICAM-1 (1.4-fold). Function-blocking anti-VCAM-1 antibody abolished apoC-III-induced THP-1 cell adhesion to EC, whereas ICAM-1 antibody reduced adhesion by only 30%, indicating a predominant role of VCAM-1 in apoC-III-mediated EC activation and monocyte adhesion. We then examined apoC-III-induced PKC activation in vascular EC. Indeed, apoC-III treatment increased membrane-bound PKCβ in HSVEC, while minimally affecting PKCα activation. Treatment of vascular EC with a selective PKCβ inhibitor abolished the increased expression of VCAM-1 by apoC-III, indicating PKCβ's prominent role. Collectively, these results suggest that apoC-III triggers PKCβ activation in vascular EC, which leads to the induction of VCAM-1 expression and enhanced monocyte adhesion.

Apoc-III Induces PKC-mediated NF-κB Activation in Monocyte and Vascular EC

In vascular cells, PKC participates in several pro-inflammatory and pro-atherogenic pathways, including activation of NF-κB. NF-κB plays a pivotal role in the control of the inflammatory and the innate immune responses. We initially examined whether apoC-III affects NF-κB activation in monocytes (Kawakami *et al.* 2007). ApoC-III treatment of THP-1 cells induced degradation of Iκ-Bα in the cytosol and subsequent nuclear translocation of NF-κB p65, indicating NF-κB activation. A PKCα inhibitor attenuated apoC-III-induced NF-κB activation, although it did not affect baseline activity, indicating that NF-κB activation by apoC-III depends partly on PKCα. Pretreatment of THP-1 cells with the NF-κB inhibitor peptide SN50 reduced β1-integrin activation and THP-1 cell adhesion by apoC-III.

ApoC-III also induces NF-κB in vascular EC (Kawakami *et al.* 2006b). Incubation with apoC-III decreased cytosolic Iκ-Bα and induced NF-κB nuclear translocation in cultured HSVEC. NF-κB inhibitor peptide SN50 inhibited increased expression of VCAM-1 by apoC-III. NF-κB activation was mediated by PKCβ, as a PKCβ inhibitor substantially reduced NF-κB p65 nuclear translocation and degradation of cytosolic Iκ-Bα in response to apoC-III. These results identify NF-κB as the key molecular link between apoC-III-induced PKC activation and increased expression of adhesion molecule in monocytes and vascular EC.

TLR-2 Mediates ApoC-III-induced Monocyte Activation

Our results identified direct effects of apoC-III on EC activation and monocyte adhesion. However, the exact cellular mechanisms involved in these responses to apoC-III remained incompletely understood. Kawakami *et al.* recently identified toll-like receptor 2 (TLR2) as a molecular target of apoC-III that induced monocyte activation (Kawakami *et al.* 2008b).

TLR family receptors are principal sensors of the innate immune system and provide a molecular link between infection, inflammation and atherosclerosis development. In this study, Kawakami *et al.* observed that anti-TLR2 antibody treatment of THP-1 cells inhibited apoC-III-induced THP-1 cell adhesion in a concentration-dependent manner. Anti-TLR2 antibody also attenuated apoC-III-induced PKCα activation and NF-κB activation. MyD88, a TIR domain-containing adaptor, plays a crucial role in the induction of inflammatory cytokines triggered by most TLRs. ApoC-III stimulation of THP-1 cells induced the association of TLR2 with MyD88. These results suggested that apoC-III stimulation of THP-1 cells was mediated by a TLR2- and MyD88-dependent pathway.

Kawakami *et al.* further found that apoC-III-mediated NF-κB activation and β1-integrin expression were further enhanced in 293 cells transfected with human TLR2, effects that were inhibited by an anti-TLR2 antibody. These results indicate that TLR2 may function as the modulator of apoC-III-induced pro-inflammatory signal transduction. However, other receptors might also be involved as apoC-III was able to induce NF-κB activation and β1-integrin expression in 293 cells that did not express TLRs or CD14.

Apoc-III Impairs Insulin Stimulation of NO Production in Vascular EC

Apoc-III directly activates pro-inflammatory and atherogenic signaling in vascular EC through PKCβ. Because PKCβ inhibits insulin action in endothelial cells and apoC-III activates PKCβ in the same cells, Kawakami *et al.* then examined whether apoC-III affects insulin signaling in vascular EC (Fig. 5). The results suggested that apoC-III inhibited insulin-induced tyrosine phosphorylation of insulin receptor substrate 1 (IRS-1), decreasing phosphatidylinositol 3-kinase (PI3K)/Akt activation in human umbilical vein EC (HUVECs). These effects of apoC-III led to reduced endothelial nitric oxide synthase (eNOS) activation and NO release into the media. Apoc-III impaired insulin stimulation of NO production by vascular endothelium and induced endothelial dysfunction *in vivo*. This adverse effect of apoC-III was mediated by its activation of PKCβ, which inhibits the IRS-1/PI3K/Akt/eNOS pathway.

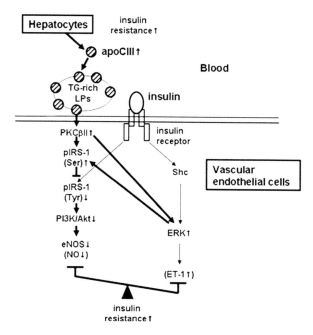

Fig. 5   Schema depicting the mechanisms by which apoC-III causes insulin resistance in vascular endothelial cells. Insulin resistance in hepatocytes increases apoC-III production and secretion into blood. Apoc-III in TG-rich lipoproteins in turn impairs insulin-induced NO production by inhibiting IRS-1function in EC. The inhibitory effects of apoC-III on NO production are likely to be mediated by the activation of PKCβ and partly by ERK, which induce activation of IRS-1.

Apoe

Apoe, a 34-kDa glycoprotein, circulates in plasma in association with various lipoproteins, including chylomicrons, VLDL, IDL and HDL. In addition to its well-known regulatory functions in lipid metabolism, apoE may also possess direct anti-atherosclerotic effects on vascular cells. For example, apoe-deficient mice exhibited elevated VCAM-1 and ICAM-1 levels in the vessel wall and aortic lesions (Nakashima *et al.* 1998), an effect that could be corrected by transgenic expression of low levels of apoE (Ma *et al.* 2008) or bone marrow transplantation from normal mice (Linton *et al.* 1995). Furthermore, reconstitution of apoE expression in the liver of apoe-deficient mice normalized the LPS-induced plasma protein levels of IL-12 p40, indicating that apoE may suppress the type I inflammatory response (Ali *et al.* 2005). ApoE also suppresses cytokine-induced VCAM-1 expression in human EC (Stannard *et al.* 2001). The suppression of endothelial activation by apoE most likely occurs via stimulation of eNOS; apoE increased levels of intracellular NO and its surrogate marker, cyclic guanosine monophosphate, while the eNOS inhibitor, ethyl-isothiourea, blocked its effect (Stannard *et al.* 2001). It is possible that apoE receptor 2 (apoER2), a member of the LDL receptor family, mediates apoE's activation of eNOS.

LIPOPROTEIN(a)

Plasma Lp(a) represents a major risk factor for premature coronary heart disease in patients with concomitant hypercholesterolemia. Lp(a) may promote atherosclerosis at least in part through induction of adhesion molecules in vascular EC. Lp(a) activates EC expression of VCAM-1 and E-selectin and ICAM-1 (Takami *et al.* 1998), and also induces secretion of monocyte chemotactic protein (Gosling *et al.* 1999). The exact mechanism behind Lp(a)-induced ICAM-1 activation remains incompletely understood, but it may account for TGF-β inhibition (Takami *et al.* 1998). Lp(a) also directly interacts with β2-integrin Mac-1, thereby promoting monocyte adhesion and transendothelial migration (Sotiriou *et al.* 2006). In addition, elevated Lp(a) levels are associate with impaired endothelium-dependent vasodilation in coronary arteries (Tsurumi *et al.* 1995), indicating a role of Lp(a) in inhibition of NO and endothelial dysfunction.

Apoa-I

Epidemiological studies have firmly established an inverse relationship between plasma HDL levels and risk of coronary heart disease. In addition to reverse cholesterol transport, a process of removing excess cholesterol from peripheral cells, HDL also exhibit direct antioxidant and anti-inflammatory properties on vascular cells. These pleiotropic atheroprotective functions of HDL include suppression of vascular adhesion molecule activation. However, the direct contribution of apoA-I, the main structural protein of HDL, in these actions remains obscure. HDL inhibit cytokine-induced expression of VCAM-1, ICAM-1 and E-selectin in vascular EC, but this effect strongly depends on the phospholipid composition of HDL as the inhibitory effects are not exerted by lipid-free apoA-I or apoA-II (Nofer *et al.* 2003). On the other hand, a recent study demonstrates that lipid-free apoA-I can inhibit expression of integrin CD11b on monocytes in a process mediated by ABC-A1 (Murphy *et al.* 2008). In addition, apoA-I induces TGF-β2, an anti-inflammatory modulator, in vascular EC (Norata *et al.* 2005). These data suggest an anti-inflammatory role of apoA-I in EC and monocytes.

CONCLUSIONS

These recent studies by our group and others suggest that apolipoproteins, the dominant protein moieties of circulating lipoproteins, have two-fold functions. First, apolipoproteins are crucial regulators of lipoprotein homeostasis, and their dysmetabolism underlies the pathogenesis of many lipid disorders. Second, apolipoproteins directly interact with vascular cells and exert either pro- or anti-atherogenic effects. This dual role makes apolipoproteins attractive targets for pharmacological intervention of lipid disorders and coronary heart disease. In particular, our results suggest that apoC-III, alone or as a component of triglyceride-rich lipoproteins, activates monocytes and vascular EC, enhancing adhesion. Taken together with apoC-III's established function in hypertriglyceridemia, lowering or modulating apoC-III may not only improve lipid profile but also prevent the development of inflamed atherosclerotic plaques, opening a new dimension in lipoprotein research.

SUMMARY

- Mechanisms responsible for atherosclerotic lesion development have largely focused on the accumulation and retention of LDL in the arterial intima and their subsequent oxidative modification.
- The oxidation leads to activation of the endothelium and, in particular, expression of cell adhesion molecules that mediate leukocyte adherence.
- Recent novel findings by us and others that apolipoproteins, protein constituents of circulating lipoproteins, induce adhesion molecules provide another direct link between dyslipidemia and atherogenesis.
- Among these lipoproteins, apoC-III and Lp(a) induce expression of adhesion molecules in monocytes and vascular EC, whereas apoE and apoA-I seem to suppress it.

- In particular, independent from its deleterious functions in lipid metabolism, apoC-III, alone or as a component of triglyceride-rich lipoproteins, activates pro-inflammatory and pro-atherogenic pathways in monocytes and endothelial cells, leading to endothelial dysfunction and enhanced monocyte adhesion (Fig. 6).

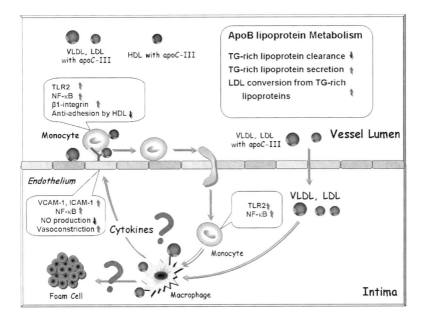

Fig. 6 The pleiotropic actions of apoC-III that may contribute to atherogenesis. ApoC-III is a major component of TG-rich lipoproteins (VLDL, chylomicrons) and HDL. ApoC-III delays TG-rich lipoprotein clearance and uptake by the liver and enhances VLDL synthesis and secretion. As such, apoC-III is a major modulator of TG-rich lipoprotein metabolism. ApoC-III is also a major constituent of HDL, and our results suggest HDL CIII+ lose the anti-adhesion function of HDL CIII−. Recent studies by us and others showed that apoC-III increases adhesion molecule expression and decreases NO production in endothelial cells and consequently results in endothelial dysfunction and enhances vasoconstriction. ApoC-III also promotes pro-inflammatory actions by activating NF-κB and inducing β1-integrin expression in monocytes that leads to enhanced adhesion to endothelial cells. ApoC-III may also induce cytokine expression in macrophages and promote foam-cell formation. These multiple apoC-III-mediated effects may independently contribute to enhance the risk of cardiovascular diseases. Reproduced from Kawakami *et al.* (2008a) with permission.

- Elevated plasma apoC-III is a common phenotype in patients with hypertriglyceridemia, insulin resistance or obesity, subjects who are also at increased risk for cardiovascular disease.

- Therefore, apoC-III can be a new therapeutic target to prevent coronary heart disease in patients with dysmetabolic states.

Abbreviations

apoB	apolipoprotein B
ApoC-III	apolipoprotein C-III
EC	endothelial cells
eNOS	endothelial nitric oxide synthase
HDL	high density lipoprotein
HSVEC	human saphenous vein endothelial cells
HUVEC	human umbilical vein endothelial cells
ICAM-1	intercellular adhesion molecule 1
LDL	low density lipoprotein
MyD88	myeloid differentiation primary response gene 88
NF-κB	nuclear factor κB
NO	nitric oxide
oxLDL	oxidized low density lipoproteins
PKC	protein kinase C
PTX	pertussis toxin
RLP	remnant lipoprotein
TLR	toll-like receptor
VCAM-1	vascular cell adhesion molecule 1
VLDL	very low density lipoprotein

Key Facts about Apolipoproteins

1. Apolipoproteins are protein constituents of lipoproteins, carriers of lipids in circulation.
    (a) Apolipoprotein(apo) B is the structural protein for apoB lipoproteins, including VLDL, IDL and LDL.
    (b) ApoA-I is the main structural protein for HDL. ApoA-I may possess direct anti-inflammatory and anti-atherogenic activities, at least in part through the inhibition of adhesion molecule expression.
2. ApoC-III is a small apolipoprotein present mainly on triglyceride-rich lipoproteins (VLDL, IDL) and HDL.
    (a) ApoC-III inhibits apoB lipoprotein clearance by interfering with apoB- or apoE-mediated receptor binding.

    (b) Recent findings from our laboratory and others demonstrate that apoC-III, alone or as a component of apoB lipoproteins, directly promotes the expression of adhesion molecules in vascular endothelial cells and monocytes, enhancing monocyte adhesion.

  3. ApoE is another small apolipoprotein commonly found on the surface of triglyceride-rich lipoproteins and HDL.

    (a) ApoE facilitates apoB lipoprotein clearance by binding with hepatic receptors.

    (b) There is limited evidence suggesting apoE may directly inhibit adhesion molecule expression in vascular cells.

Definition of Terms

Atherosclerosis: A chronic inflammatory disease affecting the intima and media of large- and medium-sized arteries. This inflammatory process leads to the formation of atherosclerotic plaques, containing lipid and fibrous tissue. The plaques rich in macrophages often rupture, causing the formation of a thrombus that will rapidly block blood flow, and thereby leading to ischemic death of the cardiac muscle fed by the artery.

Endothelial cells: The thin layer of cells that line the interior surface of blood vessels, forming an interface between circulating blood in the lumen and the rest of the vessel wall.

Lipoproteins: A biochemical assembly that contains both proteins and lipids. Examples include the high-density and low-density lipoproteins that enable fats (triglyceride, phospholipid, cholesterol ester, and free cholesterol) to be carried in the bloodstream. In addition to its lipid content, VLDL also carries protein components on its surface called apolipoproteins, such as apoC-III and apoE.

Monocytes: A type of white blood cell that in response to chemotactic signals can move quickly to sites of infection or inflammation in the tissues and divide/differentiate into macrophages and dendritic cells to elicit an immune response.

Signal transduction: Any process by which a cell converts one kind of signal or stimulus into another. Most processes of signal transduction involve ordered sequences of biochemical reactions inside the cell, which are carried out by enzymes, activated by second messengers, resulting in a signal transduction pathway.

References

Aikawa, M. and P. Libby. 2004. The vulnerable atherosclerotic plaque: pathogenesis and therapeutic approach. Cardiovasc. Pathol. 13: 125-138.

Aikawa, M. and S. Sugiyama, C.C. Hill, S.J. Voglic, E. Rabkin, Y. Fukumoto, F.J. Schoen, J.L. Witztum, and P. Libby. 2002. Lipid lowering reduces oxidative stress and endothelial cell activation in rabbit atheroma. Circulation 106: 1390-1396.

Ali, K. and M. Middleton, E. Pure, and D.J. Rader. 2005. Apolipoprotein E suppresses the type I inflammatory response in vivo. Circ. Res. 97: 922-927.

Gosling, J. and S. Slaymaker, L. Gu, S. Tseng, C.H. Zlot, S.G. Young, B.J. Rollins, and I.F. Charo. 1999. MCP-1 deficiency reduces susceptibility to atherosclerosis in mice that overexpress human apolipoprotein B. J. Clin. Invest. 103: 773-778.

Kawakami, A. and M. Aikawa, P. Libby, P. Alcaide, F.W. Luscinskas, and F.M. Sacks. 2006a. Apolipoprotein CIII in apolipoprotein B lipoproteins enhances the adhesion of human monocytic cells to endothelial cells. Circulation 113: 691-700.

Kawakami, A. and M. Aikawa, P. Alcaide, F.W. Luscinskas, P. Libby, and F.M. Sacks. 2006b. Apolipoprotein CIII induces expression of vascular cell adhesion molecule-1 in vascular endothelial cells and increases adhesion of monocytic cells. Circulation 114: 681-687.

Kawakami, A. and M. Aikawa, N. Nitta, M. Yoshida, P. Libby, and F.M. Sacks. 2007. Apolipoprotein CIII-induced THP-1 cell adhesion to endothelial cells involves pertussis toxin-sensitive G protein- and protein kinase C alpha-mediated nuclear factor-kappaB activation. Arterioscler. Thromb. Vasc. Biol. 27: 219-225.

Kawakami, A. and M. Osaka, M. Tani, H. Azuma, F.M. Sacks, K. Shimokado, and M. Yoshida. 2008a. Apolipoprotein CIII links hyperlipidemia with vascular endothelial cell dysfunction. Circulation 118: 731-742.

Kawakami, A. and M. Osaka, M. Aikawa, S. Uematsu, S. Akira, P. Libby, K. Shimokado, F.M. Sacks, and M. Yoshida. 2008b. Toll-like receptor 2 mediates apolipoprotein CIII-induced monocyte activation. Circ. Res. 103: 1402-1409.

Lee, S.J. and H. Campos, L.A. Moye, and F.M. Sacks. 2003. LDL containing apolipoprotein CIII is an independent risk factor for coronary events in diabetic patients. Arterioscler. Thromb. Vasc. Biol. 23: 853-858.

Libby, P. 2002. Inflammation in atherosclerosis. Nature 420: 868-874.

Libby, P. and M. Aikawa. 2002. Stabilization of atherosclerotic plaques: new mechanisms and clinical targets. Nat. Med. 8: 1257-1262.

Linton, M.F. and J.B. Atkinson, and S. Fazio. 1995. Prevention of atherosclerosis in apolipoprotein E-deficient mice by bone marrow transplantation. Science 267: 1034-1037.

Ma, Y. and C.C. Malbon, D.L. Williams, and F.E. Thorngate. 2008. Altered gene expression in early atherosclerosis is blocked by low level apolipoprotein E. PLoS ONE 3: e2503.

Murphy, A.J. and K.J. Woollard, A. Hoang, N. Mukhamedova, R.A. Stirzaker, S.P. McCormick, A.T. Remaley, D. Sviridov, and J. Chin-Dusting. 2008. High-density lipoprotein reduces the human monocyte inflammatory response. Arterioscler. Thromb. Vasc. Biol. 28: 2071-2077.

Nakashima, Y. and E.W. Raines, A.S. Plump, J.L. Breslow, and R. Ross. 1998. Upregulation of VCAM-1 and ICAM-1 at atherosclerosis-prone sites on the endothelium in the ApoE-deficient mouse. Arterioscler. Thromb. Vasc. Biol. 18: 842-851.

Nofer, J.R. and S. Geigenmuller, C. Gopfert, G. Assmann, E. Buddecke, and A. Schmidt. 2003. High density lipoprotein-associated lysosphingolipids reduce E-selectin expression in human endothelial cells. Biochem. Biophys. Res. Commun. 310: 98-103.

Norata, G.D. and E. Callegari, M. Marchesi, G. Chiesa, P. Eriksson, and A.L. Catapano. 2005. High-density lipoproteins induce transforming growth factor-beta2 expression in endothelial cells. Circulation 111: 2805-2811.

Ross, R. 1999. Atherosclerosis—an inflammatory disease. N. Engl. J. Med. 340: 115-126.

Sacks, F.M. and P. Alaupovic, L.A. Moye, T.G. Cole, B. Sussex, M.J. Stampfer, M.A. Pfeffer, and E. Braunwald. 2000. VLDL, apolipoproteins B, CIII, and E, and risk of recurrent coronary events in the Cholesterol and Recurrent Events (CARE) trial. Circulation 102: 1886-1892.

Sotiriou, S.N. *et al.* 2006. Lipoprotein(a) in atherosclerotic plaques recruits inflammatory cells through interaction with Mac-1 integrin. FASEB J. 20: 559-561.

Stannard, A.K. and D.R. Riddell, S.M. Sacre, A.D. Tagalakis, C. Langer, A. von Eckardstein, P. Cullen, T. Athanasopoulos, G. Dickson, and J.S. Owen. 2001. Cell-derived apolipoprotein E (ApoE) particles inhibit vascular cell adhesion molecule-1 (VCAM-1) expression in human endothelial cells. J. Biol. Chem. 276: 46011-46016.

Sundaram, M. and P. Links, M.B. Khalil, Y. Wang, S. Zhong, and Z. Yao. 2007. New insights into the roles of apolipoprotein C-III in stimulating the production of hepatic VLDL. Arterioscler. Thromb. Vasc. Biol. Annu. Conf. P142.

Takami, S. and S. Yamashita, S. Kihara, M. Ishigami, K. Takemura, N. Kume, T. Kita, and Y. Matsuzawa. 1998. Lipoprotein(a) enhances the expression of intercellular adhesion molecule-1 in cultured human umbilical vein endothelial cells. Circulation 97: 721-728.

Tsurumi, Y and H. Nagashima, K. Ichikawa, T. Sumiyoshi, and S. Hosoda. 1995. Influence of plasma lipoprotein (a) levels on coronary vasomotor response to acetylcholine. J. Am. Coll. Cardiol. 26: 1242-1250.

Zheng, C and C. Khoo, K. Ikewaki, and F.M. Sacks. 2007. Rapid turnover of apolipoprotein C-III-containing triglyceride-rich lipoproteins contributing to the formation of LDL subfractions. J. Lipid Res. 48: 1190-1203.

CHAPTER 29

Marine n-3 Polyunsaturated Fatty Acids and Cellular Adhesion Molecules in Healthy Subjects and in Patients with Ischaemic Heart Disease

Ole Eschen[1, *], Jeppe Hagstrup Christensen[1,2] and Erik Berg Schmidt[1]

[1]Department of Cardiology, Center for Cardiovascular Research, Aalborg Sygehus, Aarhus University Hospital, Aalborg, Denmark.
[2]Department of Nephrology, Aalborg Sygehus, Aarhus University Hospital, Aalborg, Denmark

Departmental contact: Ole Eschen, Center for Cardiovascular Research, Department of Cardiology, Aalborg Sygehus, Aarhus University Hospital, Hobrovej 16-22, 9100 Aalborg, Denmark. E-mail: oe@dadlnet.dk
Jeppe Hagstrup Christensen, E-mail: jhc@dadlnet.dk
Erik Berg Schmidt, E-mail: ebs@dadlnet.dk

ABSTRACT

Cellular adhesion molecules (CAMs) may play an important role in initiation and progression of atherosclerosis in both healthy subjects and in patients with ischaemic heart disease (IHD). Accumulating evidence supports a cardioprotective effect of n-3 polyunsaturated fatty acids (PUFA) of marine origin and the beneficial effect could in part be due to anti-inflammatory and anti-atherosclerotic properties of these fatty acids. *In vitro* and animal data suggest that n-3 PUFA are inversely associated with CAMs. However, human studies of the effects of n-3 PUFA on serum levels of soluble CAMs in both healthy subject and in patients with IHD are not entirely consistent. Most randomized controlled trials have been small and rather heterogeneous with respect to dose of n-3 PUFA and the period of

---

*Corresponding author
Key terms are defined at the end of the chapter.

intervention. Larger studies over extended periods are warranted for a conclusion on whether n-3 PUFA exerts beneficial effects with respect to vascular disease by modulation of CAMs.

INTRODUCTION

Over the past decades, increasing evidence has accumulated in support of an inverse association between fish consumption and mortality from ischaemic heart disease (IHD). Large observational cohort studies and nested case control studies have pointed at a reduction in cardiovascular mortality by fish consumption, but results are not entirely consistent (He *et al.* 2004). In large randomized controlled trials, interventions with n-3 PUFA have improved the outcome in the primary and secondary prevention of IHD (Lee *et al.* 2008), but not all trials have shown positive effects. IHD mainly results from atherosclerosis of the coronary arteries, and atherosclerosis is now considered, at least in part, an inflammatory disease of the arteries. There is evidence suggesting an anti-inflammatory effect of n-3 PUFA (Calder 2006). One of the earliest events in atherogenesis is recruitment of leucocytes/monocytes and their transendothelial migration and accumulation in the vessel wall. This process is predominantly mediated by CAMs that are expressed on the vascular endothelium, platelets and circulating leucocytes in response to a variety of inflammatory stimuli. The CAMs are transmembrane molecules, but a soluble part can be measured in serum (sCAMs) and may serve as markers of the inflammatory state of the endothelium (Blankenberg *et al.* 2003).

During inflammation a wide variety of cell-to-cell and cell-to-extracellular matrix interactions are important. CAMs are transmembrane proteins located on the cell surface that facilitate the transport of leucocytes through vascular endothelium into the intima, where they participate in the formation and growth of the atherosclerotic plaque. Expression of several CAMs has been reported on established atherosclerotic lesions in humans (Huo and Ley 2001). Resting endothelium does not support the adhesion of leucocytes under the shear stress in arterial blood flow. When the endothelium is activated through a variety of inflammatory mediators, the endothelial cells increase (or change) expression of several CAMs, which makes the endothelium sticky to leucocytes (Huo and Ley 2001, Blankenberg *et al.* 2003). First, the leucocyte rolls and slows down along the endothelial cell surface, mediated by the selectins. More firm adhesion of the leucocyte to the endothelium is mediated by ICAM-1 and VCAM-1. The final step is extravasation of the leucocyte.

Immunohistochemical analysis of human atherosclerotic plaques has shown a strong expression of P-selectin by the endothelium overlying atherosclerotic plaques characterized by inflammation. However, this is not found in the normal 'healthy' arterial endothelium or in endothelium overlying fibrous plaques. P-selectin is also involved in the adhesion of platelets to monocytes and neutrophils, playing a central role in neutrophils accumulation within thrombi. E-selectin is almost absent on resting inactive epithelium, but spotty expression occurs on the endothelium over fibrous and lipid-containing plaques.

ICAM-1 and VCAM-1 belong to the immunoglobulin superfamily, a large group of cell surface and soluble proteins involved in cell binding, adhesion and recognition. Both ICAM-1 and VCAM-1 are expressed on the vascular endothelium and by smooth muscle cells within the vascular wall. VCAM-1 expression is low in healthy arteries but is concentrated (2-4 times) in the vascular epithelium overlying the complicated atherosclerotic plaque. ICAM-1 is more abundantly expressed in the vessel tree, and upregulation over complicated atherosclerotic plaques is much less prominent (Duplaa *et al.* 1996). Moreover, fibroblast and several cells in the haematopoietic lineage may also express ICAM-1. The biological role of sCAMs is not clarified, but one study has suggested that sCAMs may inhibit monocyte adhesion to activated endothelium, thus indicating a negative feedback mechanism for monocyte adhesion (Abe *et al.* 1998).

CAMs AND ATHEROSCLEROSIS

The evidence that CAMs play a role in atherogenesis derives from both animal studies and *in vitro* studies. Genetically modified atherosclerotic animals lacking the ability to express CAMs have markedly reduced atherosclerosis (Huo and Ley 2001). CAMs are expressed within the first week after initiating an atherogenic diet for experimental animals prone to developing atherosclerosis, and expression of CAMs precedes the extravasation of monocytes in early atherosclerotic lesions (fatty streaks). At later stages in the evolution of the complicated atherosclerotic plaque in humans, CAMs might also recruit leucocytes through the neo-vasculature in the intima underlying the plaque. Thus, CAMs are critically involved in all stages of atherogenesis (Huo and Ley 2001, Blankenberg *et al.* 2003). Furthermore, sCAMs may be a good marker of atherosclerosis. We have previously reported a positive association between sCAMs and the extent of atherosclerosis in a patient referred for coronary angiography due to suspected angina pectoris (Eschen *et al.* 2005), as illustrated in Fig. 1. Studying endothelial cells *in vitro* has given insight into the mechanisms by which n-3 PUFA may exert anti-inflammatory and anti-atherosclerotic effects. In cell cultures, n-3 PUFA reduce monocyte adhesion to activated endothelial cells (De Caterina and Libby 1996, Sanderson and Calder 1998) and reduce expression of ICAM-1 and L-selectin on lymphocytes. DHA (De Caterina and Libby 1996) and EPA have been demonstrated to reduce expression of VCAM-1, ICAM-1 and E-selectin in cultures of human endothelial cells activated by inflammatory cytokines. The magnitude of this decrease in sCAMs paralleled the incorporation of DHA in the endothelium and a reduction in VCAM-1 mRNA (De Caterina and Libby 1996). EPA and DHA may also produce anti-inflammatory effects through direct actions on the intracellular signalling pathways regulating inflammatory gene expression. One of the key transcription factors involved in the expression of inflammatory genes is nuclear factor kappa B (NF-κB). Studies have shown that n-3 PUFA can downregulate the activity of NF-κB (Weber *et al.* 1995),

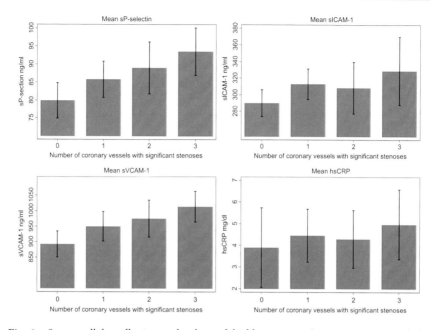

Fig. 1 Serum cellular adhesion molecules and highly sensitive C-reactive protein and the extent of coronary atherosclerosis among 291 patients referred for coronary angiography (mean with 95% CI). The extent of coronary atherosclerosis is quantified on the basis of the number of affected coronary vessels into 0, 1, 2, or 3-vessel disease. A significant stenosis is present if there is a reduction in the luminal diameter of more than 50%. The figure shows increasing levels of soluble cellular adhesion molecules with increasing number of affected vessels. (Authors' own data; details of study can be found in Eschen et al. 2005).

which may explain the negative association between n-3 PUFA and expression of CAMs.

Marine n-3 PUFA may also exert anti-inflammatory actions by modulation of production of pro-inflammatory and anti-inflammatory cytokines. After dietary intake, EPA and DHA are incorporated into the membranes of inflammatory cells and serve as substrates for the production of eicosanoids and cytokines that modulates the inflammatory response (Calder 2006). An increased intake of fish and a reduction in the intake of n-6 PUFA will result in a higher level of EPA and DHA and less arachidonic acid (AA) in leucocyte cell membranes (Schmidt *et al.* 1991). The inflammatory eicosanoids derived from EPA are much less potent than eicosanoids derived from AA. Thus, while AA is metabolized via the lipoxygenase pathway in leucocytes to leucotriene B4 with very powerful pro-inflammatory effects, EPA is metabolized to leucotriene B5 with an inflammatory activity of only 10% of leucotriene B4 (Schmidt *et al.* 1991). EPA and DHA may also decrease the production of inflammatory cytokines like tumour necrosis factor alpha, interleukin-1 and interleukin-6 (Calder 2006).

THE EFFECT OF MARINE n-3 PUFA ON CORONARY ATHEROSCLEROSIS

Several trials have been conducted to examine the effects of n-3 PUFA in patients with IHD (Arnesen 2001). The effects of n-3 PUFA on the course of coronary atherosclerosis in angiographically proven coronary artery disease have been investigated in two randomized controlled trial (RCT). In the Study on Prevention of Coronary Atherosclerosis with Marine Omega-3 Fatty Acids (SCIMO), a large daily dose of 6 g of n-3 PUFA was given for 3 mon followed by a daily dose of 3 g n-3 PUFA for 21 mon. Quantitative coronary angiographies were performed before and after supplementation, and a beneficial but modest effect on the course of coronary atherosclerosis was observed in patients allocated to n-3 PUFA compared to patients given placebo. Thus, more coronary lesions showed regression in the n-3 PUFA group, while no differences were seen in the number of lesions in which progression was observed between the two groups. An earlier but smaller RCT study using a similar design did not find any significant effect of 6.0 g of n-3 PUFA given for 28 mon.

Restenosis after coronary intervention may result from thrombosis, from intimal hyperplasia and in part from atherosclerosis. Prevention of restenosis after coronary interventions should be mentioned in this context as interventions with high doses of n-3 PUFA have been attempted to reduce restenosis after coronary artery bypass grafting (CABG) and percutaneous coronary intervention (PCI). One large trial investigated the effect of a high dose of n-3 PUFA on the occlusion of bypass grafts after CABG. Venous graft occlusion was reduced by 23% in the patients receiving n-3 PUFA compared to controls. Likewise, in the pre-stent era of PCI, the prevalence of restenoses after PCI was very high, and larger RCT trials have been performed and none have revealed any effect on restenosis of high doses of 5.1-8.0 g/day n-3 PUFA (Arnesen 2001). Studies in patients treated with coronary stents and n-3 PUFA have yet not been published.

N-3 PUFA AND PLAQUE STABILIZATION

n-3 PUFA may stabilize atherosclerotic plaques, making them less likely to rupture, as reported from a double-blind RCT in patients awaiting carotid endarterectomy (Thies *et al.* 2003). Patients were randomly allocated to receive either 1.4 g of EPA+DHA, 3.6 g alpha-linolenic acid or a control oil. Histological examination and fatty acids composition of the plaques removed at carotid endarterectomy were undertaken. In the EPA+DHA group, plaques were more likely to have a thick fibrous cap, less inflammation and fewer macrophages compared to the groups receiving the other interventions. However, the endothelial expression of ICAM-1 or VCAM-1 between intervention groups was not statistically different.

RELATION BETWEEN LEVELS OF MARINE n-3 PUFA AND sCAMs

In cross-sectional studies of healthy individuals (Eschen *et al.* 2004) and in patients with IHD (Eschen *et al.* 2005), we have reported only weak and inconsistent association between sCAMs and levels of n-3 PUFA in various cells or in adipose tissue. In contrast to our studies, a highly significant baseline inverse relation between EPA and sE-selectin, sICAM-1 and sVCAM-1, and DHA and sICAM-1 and sVCAM-1 was found in a subpopulation from the Diet and Omega 3 Intervention Trial on atherosclerosis (DOIT) (Yli-Jama *et al.* 2002).

EFFECTS OF SUPPLEMENTS WITH MARINE n-3 PUFA ON sCAMs

A limited number of studies have investigated the effect of marine n-3 PUFA supplements on sCAMs in healthy subjects or in patients at risk of IHD or with established IHD. These studies are summarized in Table 1. Most of these studies are small and they are rather heterogeneous. A number of factors may influence the effect of supplementations with n-3 PUFA on levels of sCAMs, including dose, age, and duration of supplementation, and the effect may differ between men and women. The presence of other risk factors such as dyslipidaemia, hypertension and DM may also influence the effect of n-3 PUFA on sCAMs. Overall, most RCT studies examining the effects of n-3 PUFA on levels of sCAMs have included a limited number of subjects, and no data exist from a larger RCT with n-3 PUFA.

Considering the lessons learned from ecological, epidemiological and large cohort studies, it is tempting to hypothesize that n-3 PUFA exert anti-inflammatory and anti-atherosclerotic effects in a dose-dependent way as populations with a very high intake of seafood have remarkably lower incidence of IHD (Newman *et al.* 1993). We have examined a possible dose-dependent effect of n-3 PUFA on sCAMs in healthy subjects. A high dose of 6.6 g of n-3 PUFA decreased sP-selectin, whereas no effects were seen on sICAM-1 and sVCAM-1. A moderate dose of 2.0 g of n-3 PUFA had no effect on levels of sCAMs. However, a possible dose-response relation between sCAMs and intake of n-3 PUFA is poorly defined, when evaluating the results from other studies. Recently, a report from a placebo-controlled randomized dose-response study of healthy subjects found no definite dose-response relation between increasing doses of n-3 PUFA and sCAMs (Cazzola *et al.* 2007), the highest dose being 4.9 g of n-3 PUFA/day. sE-selectin was significantly increased by the highest dose of n-3 PUFA among younger males, whereas sVCAM-1 tended to be reduced in both younger and older men receiving n-3 PUFA with no clear dose-response relationship. No change was observed in sICAM-1. The DOIT trial is the largest trial (n = 563) to provide insights in the association between sCAMs and n-3 PUFA. Recruited subjects had a long history of hyperlipidaemia, and they were randomized in a placebo-

Table 1 Clinical trial investigating the effect of supplementing marine n-3 PUFA on levels of soluble cellular adhesion molecules (sCAMs)

Reference	Design, population, intervention time	Intervention	sCAMs	Significant results (↓ or ↑)
Healthy subjects				
Eschen et al. 2004	RCT, double-blind N = 60 12 wk	Gr. 1: Placebo Gr. 2: 2.0 g EPA + DHA Gr. 3: 6.6 g EPA + DHA	sICAM-1 sP-selectin sVCAM-1	Gr. 3: sP-selectin ↓
Miles et al. 2001	RCT, double-blind (n = 16) males < 40 yr, (n = 12) subjects > 55yr 12 wk	Gr. 1: 1.2 g EPA + DHA Gr. 2: Placebo	sVCAM-1 sE-selectin sVCAM-1	Gr. 1: sE-selectin ↑ (males < 40 yr) sVCAM-1 ↓ (subjects > 55 yr)
Thies et al. 2001	RCT, double-blind (n = 46) age 55-70 yr 12 wk.	6 groups, different mix of FA: Gr. 5: 0.7 g DHA Gr. 6: 1.0 g EPA + DHA	sICAM-1 sE-selectin sVCAM-1	Gr. 6: sVCAM-1 ↓ sE-selectin ↓
Cazzola et al. 2007	RCT, double-blind (n = 93) males, 18-42 yr, (n = 62) males, 53-70 yr 12 wk	Gr. 1: Placebo Gr. 2: 1.35 g EPA + .27 g DHA Gr. 3: 2.7 g EPA + .54 g DHA Gr. 4: 4.05 g EPA + .81 g DHA	sICAM-1 sE-selectin sVCAM-1	Gr. 4: sE-selectin↑ (males, age 18-42 yr)
Patients at increased risk of IHD				
Hjerkinn et al. 2005	RCT, double-blind, factorial 2 × 2. (n = 563) long-standing dyslipidaemia. 3 yr	± 2.4 g EPA + DHA ± Dietary advice	sICAM-1 sE-selectin sVCAM-1	n-3 PUFA group and diet group: sICAM-1 ↓
Abe et al. 1998	A: RCT, double-blind.. (n = 39) hypertrigliceridaemia 6 wk B: Open label (n = 27) hypertrigliceridaemia > 6 mon	A: Gr.1: 4.0 g EPA + DHA Gr. 2: Placebo B: 4.0 g EPA + DHA	sICAM-1 sE-selectin sVCAM-1	A: Gr. 1: sE-selectin ↑ B: sE-selectin ↓, sICAM-1 ↓

contd...

contd...

Seljeflot et al. 1998	RCT, double-blind, factorial 2 × 2 (n = 41) male hyperlipidaemic smokers 6 wk	± 4.8 g EPA + DHA ± antioxidants	sICAM-1 sE-selectin sVCAM-1	n-3 PUFA group: E-selectin ↑, sVCAM-1 ↑
Patients with CAD				
Johansen et al. 1999	Open label (n = 54) a study extension for a subgroup of patients from a larger trial 4 wk	Gr. 1: continued 5.1 g EPA + DHA (= control) Gr. 2: No prior n-3 PUFA, now 5.1 g EPA + DHA.	sICAM-1 sE-selectin sVCAM-1	Gr. 2: sE-selectin ↑ sVCAM-1 ↑
Patients with a MI				
Grundt et al. 2003	RCT, double-blind. (n = 300), 4-6 d after a MI 12 mon	Gr. 1: 4.0 g EPA + DHA Gr. 2: Placebo	sICAM-1 sE-selectin	No differences

PUFA, polyunsaturated fatty acids; CAD, coronary artery disease; RCT, randomized controlled trial; FO, fish oil; Gr., group; FA, fatty acids; EPA, eicosapentaenoic acid; DHA, docosahexaenoic acid; MI, Myocardial infarction; IHD, ischaemic heart disease.

controlled 2x2 factorial design to receive ± dietary advice (including advice to ensure a high intake of seafood) and ± 2.4 g EPA + DHA from fish oil capsules (Hjerkinn *et al.* 2005). After 3 yr, a significant decrease in sICAM-1 was observed both in the group receiving dietary counselling and in the group given fish oil capsules, while no effect was seen on sE-selectin and sVCAM-1. The largest RCT in patients with IHD (n = 300) examined the effect of a high dose of 4 g of n-3 PUFA or corn oil in patients included within a week after an MI (Grundt *et al.* 2003). No effect on levels of sE-selectin or sICAM-1 was observed, despite the fact that a beneficial effect on serum lipoproteins was observed in subjects receiving n-3 PUFA. Furthermore, an increase in markers of lipid peroxidation was found in the n-3 PUFA group.

Compared to the low intake of EPA + DHA in most Western countries (Lee *et al.* 2008), relatively high doses of n-3 PUFA have been tested (0.7 g/d to 6.6 g/d) in RCT. Moderate doses of less than 2 g/d of n-3 PUFA have resulted in significant changes in some of the measured sCAM (Miles *et al.* 2001, Thies *et al.* 2001), or no effect at all (Cazzola *et al.* 2007). Decreased levels of sVCAM-1 with moderate doses have been reported mostly among healthy older subjects (Miles *et al.* 2001, Thies *et al.* 2001), whereas an increase in sE-selectin has been reported among healthy younger subjects (Miles *et al.* 2001). In contrast, studies using high doses of n-3 PUFA have tended to report increased levels sCAMs (Abe *et al.* 1998, Seljeflot *et al.* 1998, Johansen *et al.* 1999, Cazzola *et al.* 2007), except in trials using a high dose supplementation for extended time periods (Abe *et al.* 1998, Hjerkinn *et al.* 2005).

A concern when giving n-3 PUFA to humans is the risk of deleterious modifications of serum lipoproteins (Nestel 2000). n-3 PUFA are highly unsaturated and prone to oxidation. n-3 PUFA are also incorporated in lipoproteins and may increase LDL susceptibility to oxidation (Brude *et al.* 1997), and oxidized LDL is a known stimulus for expression of vascular CAMs. Increased lipid-peroxidation after supplementation with high doses of n-3 PUFA was found in the dose-response study of healthy subjects (Cazzola *et al.* 2007), in male hyperlipidaemic smokers (Seljeflot *et al.* 1998) and in 300 patients after an MI (Grundt *et al.* 2003). Interestingly, addition of antioxidants to n-3 PUFA may protect LDL from oxidation (Brude *et al.* 1997). However, conflicting results have been reported regarding whether or not EPA and DHA render LDL susceptible to oxidation.

A time-dependent effect of n-3 PUFA on levels of sCAMs could also be relevant. Most studies have examined the effect of supplementation of n-3 PUFA for 3-12 wk. This may be sufficient time to allow incorporation of marine n-3 PUFA in cellular membranes but insufficient to allow for long-term effects of n-3 PUFA on sCAMs. Beneficial effect of long-term supplementation (36 mon) was suggested by the DOIT trial (Hjerkinn *et al.* 2005). A sub-study from the DOIT trial (n = 171) provided further evidence in support of the need for long-term supplementation as no effect of 2.4 g/d of n-3 PUFA on sCAMs was evident after 18 mon of supplement (Berstad *et al.* 2003). Likewise, the high dose of n-3 PUFA given to patients with hypertriglyceridaemia increased sE-selectin after 6 wk, whereas prolonged treatment for more than 7 mon decreased both sE-selectin and sICAM-1 (Abe *et al.* 1998). In line with this, a plaque-stabilizing effect of n-3 PUFA was reported if supplementation of n-3 PUFA was given for more than 6 wk (Thies *et al.* 2003).

Levels of sCAMs may vary with age. Age has been reported to be a strong and independent predictor of increased levels of sVCAM-1 (Eschen *et al.* 2005). In Table 1, some studies point at a differential effect of supplements of n-3 PUFA on sCAMs in young and in older subjects indicating that older subjects may benefit the most from n-3 PUFA supplements (Miles *et al.* 2001, Thies *et al.* 2001, Cazzola *et al.* 2007). Age-related differences in handling of n-3 PUFA may in part explain these findings, and older subjects may incorporate EPA and DHA in serum phospholipids and mononuclear immune cells to a higher extent than younger subjects (Cazzola *et al.* 2007). However, age-related differences in the metabolism of n-3 PUFA may also be due to differences in body composition and to what extent young and old subjects may oxidize n-3 PUFA.

Effects of n-3 PUFA on sCAMs may also depend on gender. In a gender sub-analysis in our study of healthy subjects (Eschen *et al.* 2004), the decrease in sP-selectin was most pronounced in men. In women, a significant decrease in sICAM-1 in the 2.0 g group and a significant increase in sVCAM-1 in the 6.6 g n-3 PUFA group was observed. Compared to men, women had significantly lower levels of sVCAM-1 and sP-selectin at baseline in this study and lower levels of all sCAMs in yet another of our studies in healthy subjects (Eschen *et al.* 2008), but in a study of patients with IHD no significant differences were observed between men and women (Eschen *et al.* 2005). Further studies are needed to address this issue.

CONCLUSIONS

Epidemiological and clinical trials have provided evidence for a cardioprotective effect of marine n-3 PUFA. However, the mechanism by which n-3 PUFA offer protection is incompletely understood. Among the proposed beneficial effects are an anti-inflammatory and an antiatherosclerotic effect. Experimental animal and *in vitro* studies have demonstrated an inverse relation between expression of CAMs and n-3 PUFA.

However, studies in healthy humans or patients with IHD have not convincingly demonstrated a reduction in sCAMs after supplements of n-3 PUFA. The effects on serum levels of sCAMs seem to depend on the dose of n-3 PUFA and the presence of cardiovascular risk factors. Most of the studies have been small and relative high doses of n-3 PUFA have been used for short time periods. There are no reports on sCAMs from larger intervention studies with major cardiovascular endpoints, and such studies are warranted.

SUMMARY

- Epidemiological studies have reported an inverse relation between high intake of fish and risk of ischaemic heart disease.
- n-3 PUFA have been shown to be cardioprotective in primary and secondary intervention trials.
- Cellular adhesion molecules (CAMs) play a central role in the development of atherosclerotic disease.
- In vitro studies have shown an inverse association between n-3 PUFA and expression of CAMs.
- n-3 PUFA may also decrease the expression of pro-inflammatory genes, including transcriptional genes for CAMs.
- n-3 PUFA exert anti-inflammatory actions by modulation of production of anti-inflammatory and pro-inflammatory cytokines.
- Animal studies have inconsistently confirmed an anti-atherosclerotic effect of n-3 PUFA and a decreased expression of CAMs after dietary intervention with n-3 PUFA.
- Randomized, controlled trials of healthy subjects or patients with ischaemic heart disease so far have reported diverging effects of supplements of n-3 PUFA on serum levels of sCAMs.

Abbreviations

AA	arachidonic acid (20:4, n-6)
CABG	coronary artery bypass graft
CAMs	cellular adhesion molecules
DHA	docosahexaenoic acid (22:6, n-3)
DM	diabetes mellitus
EPA	eicosapentaenoic acid (20:5, n-3)
IHD	ischaemic heart disease
MI	myocardial infarction
PCI	percutaneous coronary intervention
PUFA	polyunsaturated fatty acids
RCT	randomized controlled trial
sCAM	soluble cellular adhesion molecule

Key Facts about n-3 Polyunsaturated Fatty Acids

1. Dietary fatty acids are carboxylic acids with chains of carbon atoms named according to the number of carbon atoms. They are divided according to the presence of double bonds between carbon atoms in saturated (no double bonds) and unsaturated fatty acids.

2. n-3 polyunsaturated fatty acids are an important class of dietary acids required for normal growth, development and optimal function of many organs.

3. n-3 polyunsaturated fatty acids cannot be synthesized de novo in mammals and are thus essential nutrients in the diet.

4. n-3 polyunsaturated fatty acids serve as substrate for the production of many cytokines play an important role in modulating the physical state of biological membranes and modify membrane protein activities.

5. The long-chained n-3 polyunsaturated fatty acids of marine origin (eicosapentaenoic acid (20:5, n-3) and docosahexaenoic acid (22:6, n-3)) especially have important regulatory functions in the inflammatory response in humans.

Definition of Terms

Atherosclerosis: A disease affecting arterial blood vessel, in which a chronic inflammatory response results in deposition of inflammatory cells and lipids in the vessel wall.

Carotid endarterectomy: A surgical procedure performed to remove atherosclerotic plaques in the carotid arteries.

Coronary angiography: A minimal invasive diagnostic procedure in which a contrast medium is injected into the coronary arteries in order to evaluate the blood supply of the myocardium.

Ischaemic heart disease: A disease of the myocardium due to reduced blood supply mainly due to atherosclerosis of the coronary arteries.

Restenosis: Recurrence of a significant reduction in the luminal diameter of a vessel after a prior intervention (dilation).

References

Abe, Y. and B. El Masri, K.T. Kimball, H. Pownall, C.F. Reilly, K. Osmundsen, C.W. Smith, and C.M. Ballantyne. 1998. Soluble cell adhesion molecules in hypertriglyceridemia and potential significance on monocyte adhesion. Arterioscler. Thromb. Vasc. Biol. 18: 723-731.

Arnesen, H. 2001. n-3 fatty acids and revascularization procedures. Lipids 36 Suppl. S103-S106.

Berstad, P. and I. Seljeflot, M.B. Veierod, E.M. Hjerkinn, H. Arnesen, and J.I. Pedersen. 2003. Supplementation with fish oil affects the association between very long-chain n-3 polyunsaturated fatty acids in serum non-esterified fatty acids and soluble vascular cell adhesion molecule-1. Clin. Sci. (Lond.) 105: 13-20.

Blankenberg, S. and S. Barbaux, and L. Tiret. 2003. Adhesion molecules and atherosclerosis. Atherosclerosis 170: 191-203.

Brude, I.R. and C.A. Drevon, I. Hjermann, I. Seljeflot, S. Lund-Katz, K. Saarem, B. Sandstad, K. Solvoll, B. Halvorsen, H. Arnesen, and M.S. Nenseter. 1997. Peroxidation of LDL from combined-hyperlipidemic male smokers supplied with omega-3 fatty acids and antioxidants. Arterioscler. Thromb. Vasc. Biol. 17: 2576-2588.

Calder, P.C. 2006. Polyunsaturated fatty acids and inflammation. Prostaglandins Leukot. Essent. Fatty Acids 75: 197-202.

Cazzola, R. and S. Russo-Volpe, E.A. Miles, D. Rees, T. Banerjee, C.E. Roynette, S.J. Wells, M. Goua, K.W. Wahle, P.C. Calder, and B. Cestaro. 2007. Age- and dose-dependent effects of an eicosapentaenoic acid-rich oil on cardiovascular risk factors in healthy male subjects. Atherosclerosis 193: 159-167.

De Caterina, R. and P. Libby. 1996. Control of endothelial leukocyte adhesion molecules by fatty acids. Lipids 31 Suppl. S57-S63.

Duplaa, C. and T. Couffinhal, L. Labat, C. Moreau, M.E. Petit-Jean, M.S. Doutre, J.M. Lamaziere, and J. Bonnet. 1996. Monocyte/macrophage recruitment and expression of endothelial adhesion proteins in human atherosclerotic lesions. Atherosclerosis 121: 253-266.

Eschen, O. and J.H. Christensen, R. De Caterina, and E.B. Schmidt. 2004. Soluble adhesion molecules in healthy subjects: a dose-response study using n-3 fatty acids. Nutr. Metab. Cardiovasc. Dis. 14: 180-185.

Eschen, O. and J.H. Christensen, E. Toft, and E.B. Schmidt. 2005. Soluble adhesion molecules and marine n-3 fatty acids in patients referred for coronary angiography. Atherosclerosis 180: 327-331.

Grundt, H. and D.W. Nilsen, M.A. Mansoor, O. Hetland, and A. Nordoy. 2003. Reduction in homocysteine by n-3 polyunsaturated fatty acids after 1 year in a randomised double-blind study following an acute myocardial infarction: no effect on endothelial adhesion properties. Pathophysiol. Haemost. Thromb. 33: 88-95.

He, K. and Y. Song, M.L. Daviglus, K. Liu, L. Van Horn, A.R. Dyer, and P. Greenland. 2004. Accumulated evidence on fish consumption and coronary heart disease mortality: a meta-analysis of cohort studies. Circulation 109: 2705-2711.

Hjerkinn, E.M. and I. Seljeflot, I. Ellingsen, P. Berstad, I. Hjermann, L. Sandvik, and H. Arnesen. 2005. Influence of long-term intervention with dietary counseling, long-chain n-3 fatty acid supplements, or both on circulating markers of endothelial activation in men with long-standing hyperlipidemia. Am. J. Clin. Nutr. 81: 583-589.

Huo, Y. and K. Ley. 2001. Adhesion molecules and atherogenesis. Acta Physiol. Scand. 173: 35-43.

Johansen, O. and I. Seljeflot, A.T. Hostmark, and H. Arnesen. 1999. The effect of supplementation with omega-3 fatty acids on soluble markers of endothelial function in patients with coronary heart disease. Arterioscler. Thromb. Vasc. Biol. 19: 1681-1686.

Lee, J.H. and J.H. O'Keefe, C.J. Lavie, R. Marchioli, and W.S. Harris. 2008. Omega-3 fatty acids for cardioprotection. Mayo Clin. Proc. 83: 324-332.

Miles, E.A. and F. Thies, F.A. Wallace, J.R. Powell, T.L. Hurst, E.A. Newsholme, and P.C. Calder. 2001. Influence of age and dietary fish oil on plasma soluble adhesion molecule concentrations. Clin. Sci. (Lond.). 100: 91-100.

Nestel, P.J. 2000. Fish oil and cardiovascular disease: lipids and arterial function. Am. J. Clin. Nutr. 71: 228S-231S.

Newman, W.P. and J.P. Middaugh, M.T. Propst, and D.R. Rogers. 1993. Atherosclerosis in Alaska Natives and non-natives. Lancet 341: 1056-1057.

Sanderson, P. and P.C. Calder. 1998. Dietary fish oil diminishes lymphocyte adhesion to macrophage and endothelial cell monolayers. Immunology 94: 79-87.

Schmidt, E.B. and J.O. Pedersen, K. Varming, E. Ernst, C. Jersild, N. Grunnet, and J. Dyerberg. 1991. n-3 fatty acids and leukocyte chemotaxis. Effects in hyperlipidemia and dose-response studies in healthy men. Arterioscler. Thromb. 11: 429-435.

Seljeflot, I. and H. Arnesen, I.R. Brude, M.S. Nenseter, C.A. Drevon, and I. Hjermann. 1998. Effects of omega-3 fatty acids and/or antioxidants on endothelial cell markers. Eur. J. Clin. Invest. 28: 629-635.

Thies, F. and J.M. Garry, P. Yaqoob, K. Rerkasem, J. Williams, C.P. Shearman, P.J. Gallagher, P.C. Calder, and R.F. Grimble. 2003. Association of n-3 polyunsaturated fatty acids with stability of atherosclerotic plaques: a randomised controlled trial. Lancet 361: 477-485.

Thies, F. and E.A. Miles, G. Nebe-von-Caron, J.R. Powell, T.L. Hurst, E.A. Newsholme, and P.C. Calder. 2001. Influence of dietary supplementation with long-chain n-3 or n-6 polyunsaturated fatty acids on blood inflammatory cell populations and functions and on plasma soluble adhesion molecules in healthy adults. Lipids 36: 1183-1193.

Weber, C. and W. Erl, A. Pietsch, U. Danesch, and P.C. Weber. 1995. Docosahexaenoic acid selectively attenuates induction of vascular cell adhesion molecule-1 and subsequent monocytic cell adhesion to human endothelial cells stimulated by tumor necrosis factor-alpha. Arterioscler. Thromb. Vasc. Biol. 15: 622-628.

Yli-Jama, P. and I. Seljeflot, H.E. Meyer, E.M. Hjerkinn, H. Arnesen, and J.I. Pedersen. 2002. Serum non-esterified very long-chain PUFA are associated with markers of endothelial dysfunction. Atherosclerosis 164: 275-281.

CHAPTER 30

Soluble Adhesion Molecules in Brain Infarction

Toshitaka Umemura[1, *], Takahiko Kawamura[2],
Toshimasa Sakakibara[3], Nigishi Hotta[4] and Gen Sobue[5]

[1]Departments of Neurology, Chubu Rosai Hospital,1-10-6 Komei, Minato-ku, Nagoya, 455-8530, Japan,E-mail: t.umemura@bg7.so-net.ne.jp

[2]Departments of Metabolism and Endocrine Internal Medicine, Chubu Rosai Hospital, 1-10-6 Komei, Minato-ku, Nagoya, 455-8530, JAPAN, E-mail: Kawamura.hsc@chubuh.rofuku.go.jp

[3]Departments of Neurology, Chubu Rosai Hospital,1-10-6 Komei, Minato-ku, Nagoya, 455-8530, Japan, E-mail: tsrfn@yahoo.co.jp

[4]Departments of Metabolism and Endocrine Internal Medicine, Chubu Rosai Hospital, 1-10-6 Komei, Minato-ku, Nagoya, 455-8530, Japan, E-mail: hotta@chubuh.rofuku.go.jp

[5]Departments of Neurology, Nagoya University Graduate School of Medicine,65 Tsurumai-cho, Showa-ku, Nagoya, 466-8550, Japan, E-mail: sobueg@med.nagoya-u.ac.jp

ABSTRACT

In brain ischemia, stimulation by various inflammatory cytokines increases the expression of adhesion molecules, which enhances interaction between leukocytes and vascular endothelial cells. Though this mechanism is considered important to brain infarction, in particular cerebral small-vessel disease (SVD), it has not yet been studied in detail. In this chapter, we primarily discuss the association between brain infarction and microcirculation dysfunction in cerebral small vessels at the acute and chronic phage.

---

*Corresponding author
Key terms are defined at the end of the chapter.

We review data concerning adhesion molecules and brain infarction, summarizing the results of important findings including those from animal studies for the acute phage and summarizing the findings of studies on the association between brain infarction and endothelial dysfunction in large and small vessels level for the chronic phage. Importance is being attached to the involvement of leukocytes at the capillary level as a cause of the no-reflow phenomenon after brain ischemia and reperfusion. In clinical trials on the use of anti-intercellular adhesion molecule 1 (ICAM-1) antibody in the acute phase, no significant effect has been observed and no progress has been made in the clinical application of anti-adhesion molecule therapy targeting leukocytes. At the chronic phage, there is thought to be repeated ischemia and reperfusion in microvessels in the area surrounding the ischemia focus, and it has been suggested that there is the association between endothelial dysfunction and the development and progression of SVD with lacunar infarction and white matter lesions. It is therefore necessary to study the usefulness of antiplatelet and antihypertensive agents having protection of the vessel endothelium.

Endothelial dysfunction in microvessels is thought to play an important role in the pathogenesis of SVD and its progression. There is therefore a need to study therapeutic strategies targeting vascular endothelial cells in more detail.

INTRODUCTION

Brain ischemia induces expression of inflammatory mediators in vascular endothelial cells, which causes adhesion molecules to be expressed on them, and enhances interaction between leukocytes and the endothelial cells. Frijns and Kappelle (2002) reviewed the association between leukocyte-endothelial cell and adhesion molecules in ischemic cerebrovascular disease. Although endothelial dysfunction may play an important role in the progression of brain ischemia in large and small vessels, the pathogenesis differs in vascular bed levels. In particular, the pathogenesis of brain infarction in small and microvessels is still unclear. It is necessary to target therapy on the vascular endothelium in order to prevent the development and progression of ischemic stroke. In this chapter, we primarily discuss the association between brain infarction and microcirculation dysfunction with small-vessel disease (SVD) at the acute and chronic phage.

ADHESION MOLECULES (AMs) IN ACUTE BRAIN INFARCTION

AM Expression Associated with Brain Ischemia and Reperfusion

In brain infarction, stimulation by various cytokines and endotoxins enhances AM expression, which increases coagulation activation, leukocyte accumulation and microvascular permeability, and gives rise to the no-reflow phenomenon locally in microvessels. As an important factor in the no-reflow phenomenon occurring after brain ischemia and reperfusion, the involvement of leukocytes at the capillary level has received attention (del Zoppo *et al.* 1991). The strong adhesion resulting from the action of the Cd11b/CD18 (Mac-1) of leukocytes and intercellular adhesion molecule 1 (ICAM-1) of vascular endothelial cells results in occlusion of microvessels and, following this, erythrocytes, platelets and fibrinogen cohere to form a thrombus. While ICAM-1 is only expressed to a small extent on monocytes and macrophages and vascular endothelial cells under normal conditions, its expression on vascular endothelial cells is enhanced by stimulation due to such inflammatory cytokines as interleukin 1 (IL-1) and tumor necrosis factor α (TNF-α) or endotoxins. In addition to vascular endothelial cells, ICAM-1 is also expressed on leukocytes and in this regard, marked ICAM-1 expression was observed in the autopsied brain within 3 d of an ischemic attack with marked infiltration of polymorphonuclear leukocytes (Lindsberg *et al.* 1996). ICAM-1 is principally an AM expressed on vascular endothelial cells; it is involved in the initial adhesion between leukocytes and vascular cells, as well as in the migration of leukocytes outside blood vessels, and plays important roles in the pathogenesis of inflammatory reactions, arteriosclerosis and ischemic heart disease. Further, in rat ischemia and reperfusion model of the middle cerebral artery in which ICAM-1 was studied immunohistologically, there was an increase in ICAM-1 expression in capillaries in the ischemic focus and its environs around 2 hr after reperfusion, which peaked at 46 hr and increased for about a week (Zang *et al.* 1995).

Vascular cell adhesion molecule 1 (VCAM-1) is another adhesion expressed on vascular endothelial cells and, for its expression, very late antigen 4 (VLA-4)— which is present in lymphocytes, eosinophils and monocytes—is used as a ligand. The expression of VCAM-1 is enhanced by arteriosclerosis, diabetes, smoking, and ischemic heart disease and it is an important marker associated with vascular lesions. It was reported that VCAM-1 is expressed after 2 hr when induced by such cytokines as IL-1 and TNFα and that expression is maintained for at least 72 hr (Osborn *et al.* 1989). Furthermore, the influence of P-selectin, a cell AM expressed by platelets and vascular endothelial cells on leukocytes, is important to brain microcirculation dysfunction in the brain ischemia and reperfusion state (Ishikawa *et al.* 2004). Vascular shear stress is also important in the progression of brain ischemia with a reduction in shear stress in capillaries enhancing ICAM-1 expression, which causes adhesion between platelets or leukocytes and vascular endothelial cells, producing accumulations of platelets and leukocytes that occlude capillaries.

Anti-ICAM-1 Therapy and Prospects for Medical Treatment

Several studies indicate that AMs may play an important role in the pathogenesis of acute ischemic stroke. Serum ICAM-1 levels were elevated in acute ischemic stroke within 24 hr, though there was no significant change in E-selectin levels (Shyu *et al.* 1997). Bitsch *et al.* (1998) repeatedly determined soluble AMs by ELISA in 38 patients over a period of 14 d after acute cerebral ischemia. In patients with completed stroke (n = 26), sICAM-1 peaked within 24 hr and sVCAM-1 reached a maximum after 5 d but not in patients with transient ischemic attack (n = 12) (Fig. 1). In a clinical trial, anti-ICAM-1 therapy (enlimomab) was not effective in human ischemic stroke (Enlimomab Acute Stroke Trial Investigators 2001) (Table 1). Although there are various possible reasons for the failure of enlimomab, ICAM-1 blockade may have caused the upregulation of other AMs or the activation of complement-mediated inflammation. While anti-ICAM-1 therapy targeting leukocytes has not advanced since this study, findings of various studies investigating the role of AMs in acute ischemic stroke were as follows. Castellanos *et al.* (2002) showed that high concentrations of ICAM-1 and TNF-α in blood were associated with early neurological deterioration and poor functional outcome in lacunar infarctions. Further, Wang *et al.* (2006) also reported that elevated serum ICAM-1 levels on admission were associated with neurological deterioration in ischemic stroke. Another study that investigated an association between changes in brachial flow-mediated vasodilation and ischemic stroke subtypes

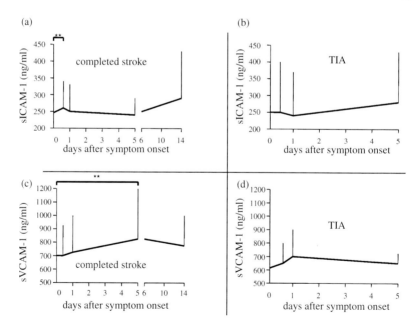

Fig. 1   Profiles of soluble adhesion molecule levels after completed stroke and TIA. Fluctuations in transient ischemic attack (TIA) patients were not statistically significant. Values are mean ± SD. **P < 0.05, corrected for multiple comparisons. (Adapted from Bitsch *et al.* 1998.)

found significantly impaired flow-mediated vasodilation vs the control only in lacunar infarction (Chen *et al.* 2006). The above studies indicate that awareness of the involvement of endothelial dysfunction is important to understanding the pathogenesis of SVD and, in the future, with respect to brain infarction subtypes at the acute phase, it will be necessary to conduct intervention studies with drugs having protection of the vessel endothelium such as phosphodiesterase inhibitors like dipyridamol and cilostazol and HMG-CoA inhibitors (statins).

ADHESION MOLECULES IN CHRONIC BRAIN INFARCTION

Carotid Atherosclerosis

To further clarify associations between chronic cerebral thrombosis and AMs, investigations should target both large and small blood vessels. However, few studies have been done on the latter. Regarding large blood vessels, in the past many studies have noted an association with carotid artery plaque or carotid artery stenosis, while an association between carotid artery stenosis and AMs is still controversial. DeGraba *et al.* (1998) reported that ICAM-1 expression was greater in symptomatic plaques than in asymptomatic plaques and that, for symptomatic plaques, ICAM-1 expression was more enhanced in the high-grade region than it was in the low-grade region. Also, the results of the Atherosclerosis Risk In Communities (ARIC) study indicated that the relationship between ICAM-1 and E-selectin and carotid artery atherosclerosis was independent of other traditional risk factors (Hwang *et al.* 1997). Another study showed that levels of sE-selectin and sP-selectin were significantly elevated in patients who had previously symptomatic carotid stenosis, though sICAM-1 and sVCAM-1 were not elevated. In contrast, it was also reported that symptomatic carotid disease was not associated with increased expression of AMs in the endothelium of advanced carotid plaques.

Cerebral Small-Vessel Disease (SVD)

At the chronic phage, brain ischemia and reperfusion is repeated in microvessels in the region of the ischemic focus and the inducing factor of endothelial dysfunction is likely to lead to development of small infarcts and progression of white matter lesions (WMLs). Fassbender *et al.* (1999) reported that serum concentrations of sE-selectin and sICAM-1 were significantly increased in patients with extracranial and intracranial large-vessel disease and patients with cerebral SVD. These results indicated that inflammatory endothelial activation and adhesion of leukocytes play similarly important roles in cerebral large- and small-vessel disease. Chronic hyperglycemia and obesity with insulin resistance have been seen to reduce eNOS activation and nitric oxide (NO) production. Also, in patients with diabetes mellitus, development of endothelial dysfunction can be associated with hyperglycemia, insulin resistance, and oxidative stress. In this regard, acute hyperglycemia in normal subjects was observed to induce an increase in plasma sICAM-1 but not plasma sVCAM-1 concentrations. L-arginine supplementation in hyperglycemic diabetic patients was found to normalize elevated sICAM-1 plasma levels. These results suggest that NO might have an inhibitory effect on the expression of some cellular AMs in human vascular endothelial cells (Marfella *et al.* 2000).

The following explains a possible basis of the close association between sICAM-1 and SVD in patients with diabetes mellitus. The activation of vascular endothelium by elevated blood glucose levels induces ICAM-1 expression and then ICAM-1 facilitates adhesion of leukocyte, especially neutrophils to the vascular endothelium, which causes small vessels in the brain to occlude, and leads to development of SVD such as silent brain infarction (SBI). In patients with diabetes mellitus, endothelial dysfunction can be worsened when oxidative stress enhances the blood-brain barrier (BBB) permeability, which may promote the development of microangiopathy such as SBI and WMLs. Oxidative stress increases tight junction permeability in vascular endothelium, leading to disruption of the BBB. Starr *et al.* (2003) reported that increased BBB permeability was detected in patients with type 2 diabetes on MRI. A recent study investigating an association between SBI incidence and AMs showed that high concentrations of sICAM-1 are significantly associated with SBI incidence and the relationship between endothelial dysfunction and presence of SBI may be stronger in diabetic patients than in non-diabetic subjects (Umemura *et al.* 2008) (Fig. 2). These results suggest that there is a greater possibility of endothelial dysfunction developing in diabetic patients with SVD than in non-diabetic subjects.

The presence of vascular risk factors such as hypertension, diabetes and smoking increases the production of reactive oxygen species in mitochondria, which can cause cell membrane damage and induce apoptosis. On the other hand, these risk factors decrease the production of NO, which is important to maintaining cerebral blood flow. This situation facilitates the progression of endothelial dysfunction. Previous study demonstrated that eNOS activity is reduced and NO production disrupted in SVD, and the results of more recent research have indicated the possibility that BBB dysfunction is an important mechanism in brain disorders due to cerebral small-vessel and microvessel disease (Wardlaw *et al.* 2008). These findings suggest that the systemic progression of endothelial dysfunction is an important reason for the pathogenesis in SVD.

In a study on 47 patients with isolated lacunar infarction, 63 patients with ischemic leukoaraiosis, and 50 controls, Hassan *et al.* (2003) graded the size of lacunar infarctions and severity of leukoaraiosis on CT or MRI to determine whether there were differences in markers of endothelial activation and damage between patient groups. These results showed that the mean levels of ICAM-1

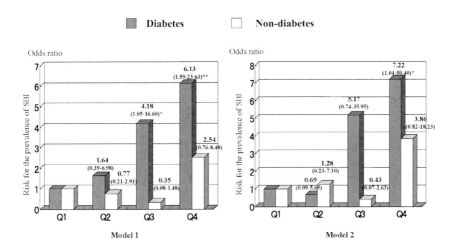

Fig. 2 Adjusted odds ratio for the association between sICAM-1 levels and SBI for diabetes and non-diabetes separately. Model 1 was adjusted for age and sex. Model 2 was additionally adjusted for systolic and diastolic blood pressure, current smoking, fasting blood glucose, serum creatinine, intima-media thickness, medication use: aspirin, HMG-CoA reductase inhibitors, angiotensin-converting enzyme inhibitors, calcium-channel blockers, and angiotensin receptor blockers. Cutoff for quartiles in diabetes: 1st, < 186.6 µg/L; 2nd, 186.6-232.7 µg/L; 3rd, 232.8-280.1 µg/L; 4th, > 280.1 µg/L. Cutoff for quartiles in non-diabetes: 1st, < 161.4 µg/L; 2nd, 161.4-195.6 µg/L; 3rd, 195.7-237.4 µg/L; 4th, > 237.4 µg/L, *P < 0.05, **P < 0.01 for comparison with first quartile.

and thrombomodulin were significantly higher in patients with isolated lacunar infarction and those with ischemic leukoaraiosis than in the controls. They also found that ICAM-1 levels were particularly high in patients with isolated lacunar infarction; and thrombomodulin levels were particularly high in patients with ischemic leukoaraiosis (Fig. 3). In contrast, tissue factor pathway inhibitor levels were higher only in the patients with isolated lacunar infarction, and there was no significant difference in tissue factor levels among the groups. These findings provide evidence of systemic endothelial activation in patients with lacunar infarction, and of different patterns of markers of endothelial dysfunction between patients with isolated lacunar infarction and those with ischemic leukoaraiosis. Furthermore a in 6 yr longitudinal study on WMLs, Markus *et al.* (2005) reported that sICAM-1 was a useful marker for monitoring disease progression.

Recently, several studies have been carried out on the prediction of ischemic stroke using inflammatory biomarkers. Tanne *et al.* (2002) indicated that elevated concentrations of sICAM-1 were associated with increased risk of ischemic stroke independent of other traditional cerebrovascular risk factors in a 8.2 yr prospective case control study. Further, a 3 yr longitudinal observational study on type 2 diabetic patients found that high sICAM-1 levels were an independent risk factor for SBI progression and endothelial dysfunction was a predictive factor for

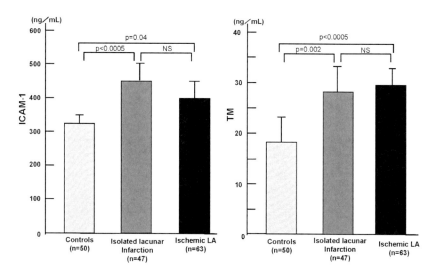

Fig. 3 Mean concentration of endothelial markers in isolated lacunar infarction, ischemic leukoaraiosis and controls. Levels expressed as geometric mean (95% CI), Overall P and within-group comparison (Scheffe's post hoc test); ICAM-1, intercellular adhesion molecule 1; TM, thrombomodulin; LA, leukoaraiosis; NS, not significant. (Modified from Hassan *et al.* 2003.)

SVD progression (Kawamura *et al.* 2006). In addition, high-sensitivity C-reactive protein (hs-CRP) has received attention in recent years as a new risk factor for stroke, which has value as predictor for future ischemic stroke. Recent studies have found that hs-CRP levels are elevated in patients with SBI, suggesting that the inflammatory mechanism is associated not only with carotid atherosclerotic disease but also with small-vessel or microvessel disease (Hoshi *et al.* 2005, Ishikawa *et al.* 2007). On the other hand, in the Rotterdam Scan Study (van Dijk *et al.* 2005), hs-CRP levels were observed to be significantly associated with WMLs but not with lacunar infarction, which suggests that the degree of involvement of inflammatory markers may differ between WMLs and lacunar infarction. Furthermore, the results of a recent study on type 2 diabetic patients indicated that elevated combination levels of ICAM-1 and hs-CRP were significantly increased subsequent symptomatic ischemic stroke (Kanai *et al.* 2008) (Fig. 4).

Prospects for Therapeutic Strategies

In order to prevent the incidence and recurrence of ischemic stroke, it is therefore necessary to make vascular endothelium the target of therapy. For antiplatelet agents, a secondary prevention in ischemic stroke has generally been established but a primary prevention has not yet been observed. As large-scale clinical trials have not yet been performed on the different types of ischemic stroke, it will be

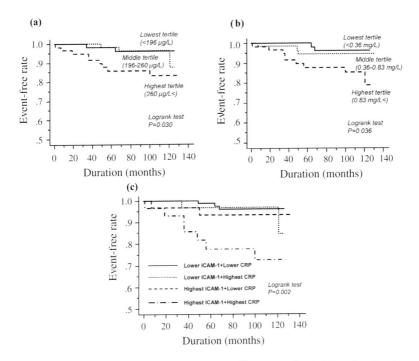

Fig. 4   Incidence of brain infarction related to levels of hs-CRP and sICAM-1 at baseline in type 2 diabetic patients. (a) Brain infarction event-free rates in relation to sICAM-1 levels. (b) Event-free rates in relation to hs-CRP levels by tertile. (c) Event-free rates in relation to both sICAM-1 and hs-CRP levels by combinations of highest tertile and lower tertile (lowest tertile + middle tertile).

necessary to look into this in the future. Aspirin, dipyridamol and clopidogrel are used globally as secondary prevention in ischemic stroke and while there is no doubt concerning their efficacy for atherothrombosis, it is unclear whether they are effective for lacunar infarctions or not. Current therapy to prevent the progression of SVD with SBI and WMLs, therefore, consists mainly of risk factor management. Cilostazol, a selective inhibitor of phosphodiesterase 3(PDE3), is only approved for intermittent claudication in the United States and United Kingdom, while it is approved under the national health insurance for the secondary prevention of noncardioembolic ischemic stroke in Japan, Korea, China and other Asian countries. Cilostazol has pleiotrophic effects which, besides inhibition of platelet aggregation, include vasodilation, improvement of vascular endothelium function, reduction of AM and inhibition of vascular smooth muscle cell proliferation. It has also been observed to prevent the progression of carotid intima-media thickness and the recurrence of lacunar infarction. In previous study, cilostazol significantly inhibited the expression of ICAM-1 and P-selectin induced by high glucose in cultured human endothelial cells (Omi *et al.* 2004) (Fig. 5). In chronic SVD with lacunar infarctions and WMLs, an increase in AMs has been observed (Hassan *et al.* 2003, Markus *et al.* 2005, Kawamura *et al.* 2006, Umemura *et al.* 2008) and in many cases coexists with microbleeds as observed in MRI T2*-weighted imaging, which has received attention in the last few years. In the CASISP study (Huang *et al.* 2008) published in Korea, compared to an aspirin group, a cilostazol group had significantly fewer hemorrhagic stroke events and in all subjects in the latter group who had experienced such events, microbleeds had been observed in imaging diagnosis conducted beforehand. In consideration of the above, cilostazol might be the drug of first choice in the treatment of SVD in view of its decreasing of AM and protection of the vessel endothelium in prevention of the incidence and progression of the disease and its low hemorrhagic risk. As statins and rennin-angiotensin system inhibitors have similar pleiotropic effects to cilostazol, they should be investigated to see if they would effective in suppressing the incidence and progression of SVD. In this regard, a study on the combined use of antiplatelet agents, angiotensin-converting enzyme inhibitors and statins prior to the incidence of acute SVD found that these agents significantly reduced severity once the disease had developed (Kumar *et al.* 2006).

In the disruption of the microcirculation that occurs in acute brain infarctions, there is an increase in AM expression that enhances interaction between leukocytes

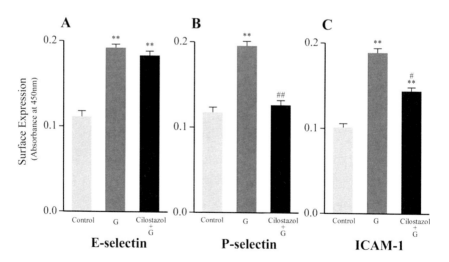

Fig. 5 Effects of cilostazol on surface expression on the endothelial adhesion molecules E-selectin, P-selectin, and ICAM-1 induced by high glucose. Cells were treated without (control) or with 27.8 mM glucose for 48 hr (G) in the presence or absence of 1 μM cilostazol. Values are expressed as means ± SEM. **P < 0.01 compared to the respective control, #P < 0.05, ##P < 0.01 compared to the respective cells treated with high glucose alone. (Adapted from Omi *et al.* 2004.)

and vascular endothelial cells. The involvement of leukocytes at the capillary level is particularly important and interaction with ICAM-1 reinforces adhesion, leading to occlusion of microvessels and the formation of microthrombosis. In these parts, there is also an increase in microvessel permeability and endothelial dysfunction develops from disruption of the BBB. These changes play an important role in the pathogenesis of SVD with lacunar infarctions and WMLs. It is necessary to target therapy on the vascular endothelium in order to prevent the development and progression of SVD and it is likely that combination therapy with drugs that protect the vessel endothelium, like cilostazol, rennin-angiotensin system inhibitors and statins, would be useful.

SUMMARY

- Brain ischemia induces expression of inflammatory mediators in vascular endothelial cells, which increases the expression of adhesion molecules and enhances interaction between leukocytes and the endothelial cells.
- As an important factor in the no-reflow phenomenon occurring after ischemia and reperfusion, the involvement of leukocytes at the capillary level has received attention.
- Endothelial dysfunction in microvessels is thought to play an important role in the pathogenesis of cerebral small-vessel disease and its progression.
- Vascular risk factors such as hypertension, diabetes, and smoking decrease the production of nitric oxide and facilitate the progression of endothelial dysfunction.
- It is necessary to target therapy on the vascular endothelium in order to prevent the development and progression of cerebral small-vessel disease and it is likely that combination therapy with drugs having protection of the vessel endothelium would be useful.

Abbreviations

AMs	adhesion molecules
BBB	blood brain barrier
hs-CRP	high-sensitivity C-reactive protein
MRI	magnetic resonance imaging
SBI	silent brain infarction
sICAM-1	soluble intercellular adhesion molecule 1
sVCAM-1	soluble vascular cell adhesion molecule 1
SVD	small-vessel disease
TNF-α	tumor necrosis factor α
WMLs	white matter lesions

Key Facts about Brain Infarction

1. Brain infarction can be classified by the mechanism of ischemia: hemodynamic, thromboembolic and the pathology of the vascular lesion: atherosclerotic, lacunar, cardioembolic, or undetermined.

2. The mechanism for stroke in atherothrombus is initially contributed to perfusion failure distal to the site of severe stenosis or occlusion of the large vessel.

3. Lacunar infarction and cerebral white matter lesions are caused by small-vessel disease.

4. Although the pathophysiology of cerebral small-vessel disease is still unclear, aging and hypertension are considered the main risk factors.

5. In elderly people, atrial fibrillation and ischemic heart disease are the most common causes of cardioembolic stroke.

6. Approved treatment for acute brain infarction is limited to intravenous thrombolysis with recombinant tissue plasminogen activator within 3 hr after CT-based exclusion of intracerebral hemorrhage or to aspirin within 48 hr.

7. The benefit of antiplatelet therapy for secondary stroke prevention in patients with a previous ischemic stroke or TIA is well established. However, there is no evidence that antiplatelet agents reduce the risk of primary stroke in the general population.

Definitions of Key Terms

Carotid plaque: Carotid plaques usually develop at the bifurcation of the common carotid artery and echolucent plaques detected by ultrasound are associated with increased risk for cerebrovascular events.

Cerebral large-vessel disease: Ischemic stroke due to large-vessel disease may be caused by atheroma on the extracranial and intracranial arterial stenosis.

Cerebral small-vessel disease: Although the pathophysiology of cerebral SVD is still unclear, aging and hypertension are considered the main risk factors, small vessel occlusion is associated with thickening or hyaline deposition of small perforating arterioles.

Silent brain infarction: Asymptomatic lesions detected in the basal ganglia, thalamus, brain stem, or subcortical white matter on brain MRI, these lesions are commonly lacunar infarctions.

White matter lesions: Periventricular or subcortical ischemic lesions, which are pathologically neuronal loss, ischemic demyelination, and gliosis.

Table I Results of anti-ICAM-1 therapy with enlimomab in acute ischemic stroke

Treatment	Patients, no Dead	Patients, no Alive	NIH Stroke Scale score 0	1-5	6-10	11-15	16-20	>20	p Value* Wilcoxon	p Value* Proportional odds
Day 0										
Placebo	0	308	0.0	4.9	28.6	26.6	15.9	24.0		
Enlimomab	0	315	0.0	5.4	25.7	28.3	17.8	22.9	0.79	NA
Day 5										
Placebo	14	294	8.8	28.2	15.6	15.6	11.6	20.1		
Enlimomab	23	292	3.8	24.0	18.8	18.2	13.0	22.3	0.03	0.02
Day 30										
Placebo	38	270	17.4	36.7	15.9	11.5	7.8	10.7		
Enlimomab	52	263	14.4	33.1	18.3	14.1	10.6	9.5	0.17	0.27
Day 90										
Placebo	50	258	22.9	40.3	17.4	7.0	6.2	6.2		
Enlimomab	70	245	18.4	38.8	19.6	11.8	5.7	5.7	0.17	0.34

The table indicates NIH Stroke Scale as percentage of surviving patients. Enlimomab has a negative effect on the NIHSS score at 5 d. *p Values compare enlimomab with placebo in patients surviving to a given study day; proportional odds likelihood ratio tests adjusted for age and baseline NIH score. (Adapted from Enlimomab Acute Stroke Trial Investigators, 2001.)

References

Bitsch, A. and W. Klene, L. Murtada, H. Prange, and P. Rieckmann. 1998. A longitudinal prospective study of soluble adhesion molecules in acute stroke. Stroke 29: 2129-2135.

Castellanos, M. and J. Castillo, M. García, R. Leira, J. Serena, A. Chamorro, and A. Dávalos. 2002. Inflammation-mediated damage in progressing lacunar infarctions. A potential therapeutic target. Stroke 33: 982-987.

Chen, P.L. and P.Y. Wang, W.H. Sheu, Y.T. Chen, Y.P. Ho, H.H. Hu, and H.Y. Hsu. 2006. Changes of brachial flow-mediated vasodilation in different ischemic stroke subtypes. Neurology 67: 1056-1058.

DeGraba, T.J. and A.L. Sirén, L. Penix, R.M. McCarron, R. Hargraves, S. Sood, K.D. Pettigrew, and J.M. Hallenbeck. 1998. Increased endothelial expression of intercellular adhesion molecule-1 in symptomatic versus asymptomatic human carotid atherosclerotic plaque. Stroke 29: 1405-1410.

del Zoppo, G.J. and G.W. Schmid-Schönbein, E. Mori, B.R. Copeland, and C.M. Chang. 1991. Polymorphonuclear leukocytes occlude capillaries following middle cerebral artery occlusion and reperfusion in baboons. Stroke 22: 1276-1283.

Enlimomab Acute Stroke Trial Investigators. 2001. Use of anti-ICAM-1 therapy in ischemic stroke: Results of the Enlimomab Acute Stroke Trial. Neurology 57: 1428-1434.

Fassbender, K. and T. Bertsch, O. Mielke, F. Mühlhauser, and M. Hennerici. 1999. Adhesion molecules in cerebrovascular disease. Evidence for an inflammatory endothelial activation in cerebral large- and small-vessel disease. Stroke 30: 1647-1650.

Frijns, C.J.M. and K.J. Kappelle. 2002. Inflammatory cell adhesion molecules in ischemic cerebrovascular disease. Stroke 33: 2115-2122.

Hassan, A. and B.J. Hunt, M. O'Sullivan, K. Parmar, J.M. Bamford, D. Briley, M.M. Brown, D.J. Thomas, and H.S. Markus. 2003. Markers of endothelial dysfunction in lacunar infarction and ischemic leukoaraiosis. Brain 126: 424-432.

Hoshi, T. and K. Kitagawa, H. Yamagami, S. Furukado, H. Hougaku, and M. Hori. 2005. Relations of serum high-sensitivity C-reactive protein and interleukin-6 levels with silent brain infarction. Stroke 36: 768-772.

Huang, Y. and Y. Cheng, J. Wu, Y. Li, E. Xu, Z. Hong, Z. Li, W. Zhang, M. Ding, X. Gao, D. Fan, J. Zeng, K. Wong, C. Lu, J. Xiao, and C. Yao. 2008. Cilostazol as an alternative to aspirin after ischaemic stroke: a randomized, double-blind, pilot study. Lancet Neurol. 7: 494-499.

Hwang, S.J. and C.M. Ballantyne, A.R. Sharrett, L.C. Smith, C.E. Davis, A.M. Gotto, and E. Boerwinkle. 1997. Circulating adhesion molecules VCAM-1, ICAM-1, and E-selectin in carotid atherosclerosis and incident coronary heart disease cases. The Atherosclerosis Risk In Communities (ARIC) Study. Circulation 96: 4219-4225.

Ishikawa, J. and Y. Tamura, S. Hoshide, K. Eguchi, S. Ishikawa, K. Shimada, and K. Kario. 2007. Low-grade inflammation is a risk factor for clinical stroke events in addition to silent cerebral infarcts in Japanese older hypertensives. The Jichi Medical School ABPM Study, Wave 1. Stroke 38: 911-917.

Ishikawa, M. and D. Cooper, T.V. Arumugam, J.H. Zhang, A. Nanda, and D.N. Granger. 2004. Platelet-leukocyte-endothelial cell interactions after middle cerebral artery occlusion and reperfusion. J. Cereb. Blood Flow Metab. 24: 907-915.

Kanai, A. and T. Kawamura, T. Umemura, M. Nagashima, N. Nakamura, M. Nakayama, T. Sano, E. Nakashima, Y. Hamada, J. Nakamura, and N. Hotta. 2008. Association between future events of brain infarction and soluble levels of intercellular adhesion molecule-1 and C-reactive protein in patients with type 2 diabetes mellitus. Diabet. Res. Clin. Pract. 82: 157-164.

Kawamura, T. and T. Umemura, A. Kanai, M. Nagashima, N. Nakamura, T. Uno, M. Nakayama, T. Sano, Y. Hamada, J. Nakamura, and N. Hotta. 2006. Soluble adhesion molecules and C-reactive protein in the progression of silent cerebral infarction in patients with type 2 diabetes mellitus. Metabolism 55: 461-466.

Kumar, S. and S. Savitz, G. Schlaug, L. Caplan, and M. Selim. 2006. Antiplatelets, ACE inhibitors, and statins combination reduces stroke severity and tissue at risk. Neurology 66: 1153-1158.

Lindsberg, P.J. and O. Carpén, A. Paetau, M.L. Karjalainen-Lindsberg, and M. Kaste. 1996. Endothelial ICAM-1 expression associated with inflammatory cell response in human ischemic stroke. Circulation 94: 939-945.

Markus, H.S. and B. Hunt, K. Palmer, C. Enzinger, H. Schmidt, and R. Schmidt. 2005. Markers of endothelial and hemostatic activation and progression of cerebral white matter hyperintensities. Longitudinal Results of the Austrian Stroke Prevention Study. Stroke 36: 1410-1414.

Marfella, R. and K. Esposito, R. Giunta, G. Coppola, L.D. Angelis, B. Farzati, G. Paolisso, and D. Giugliano. 2000. Circulating adhesion molecules in humans. Role of hyperglycemia and hyperinsulinemia. Circulation 101: 2247-2251.

Omi, H. and N. Okayama, M. Shimizu, T. Fukutomi, A. Nakamura, K. Imaeda, M. Okouchi, and M. Itoh. 2004. Cilostazol inhibits high glucose-mediated endothelial-neutrophil adhesion by decreasing adhesion molecule expression via NO production. Micovasc. Res. 68: 119-125.

Osborn, L. and C. Hession, R. Tizard, C. Vassallo, S. Luhowskyj, G. Chi-Rosso, and R. Lobb. 1989. Direct expression cloning of vascular cell adhesion molecule 1, a cytokine-induced endothelial protein that binds to lymphocytes. Cell 59:1203-1211.

Shyu, K.G. and H. Chang, and C.C. Lin. 1997. Serum levels of intercellular adhesion molecule-1 and E-selectin in patients with acute ischaemic stroke. J. Neurol. 244: 90-93.

Starr, J.M. and J.M. Wardlaw, K. Ferguson, A. MacLullich, I.J. Deary, and I. Marshall. 2003. Increased blood-brain barrier permeability in type II diabetes demonstrated by gadolinium magnetic resonance imaging. J. Neurol. Neurosurg. Psychiatry 74: 70-76.

Tanne, D. and M. Haim, V. Boyko, U. Goldbourt, R. Reshef, S. Matetzky, Y. Adler, Y.A. Mekori, and S. Behar. 2002. Soluble intercellular adhesion molecule-1 and risk of future ischemic stroke. A nested case-control study from the bezafibrate infarction prevention (BIP) study cohort. Stroke 33: 2182-2186.

Umemura, T. and T. Kawamura, T. Sakakibara, A. Kanai, T. Sano, N. Hotta, and G. Sobue. 2008. Association of soluble adhesion molecule and C-reactive protein levels with silent brain infarction in patients with and without type 2 diabetes. Curr. Neurovasc. Res. 5: 106-111.

van Dijk, E.J. and N.D. Prins, S.E. Vermeer, H.A. Vrooman, A. Hofman, P.J. Koudstaal, and M.M.B. Breteler. 2005. C-reactive protein and cerebral small-vessel disease: The Rotterdam Scan Study. Circulation 112: 900-905.

Wang, J.Y. and D.H. Zhou, J. Li, M. Zhang, J. Deng, C. Gao, J. Li, Y. Lian, and M. Chen. 2006. Association of soluble intercellular adhesion molecule 1 with neurological deterioration of ischemic stroke: The Chongqing Stroke Study. Cerebrovasc. Dis. 21: 67-73.

Wardlaw, J.M. and A. Farrall, P.A. Armitage, T. Carpenter, F. Chappell, F. Doubal, D. Chowdhury, V. Cvoro, and M.S. Dennis. 2008. Changes in background blood-brain barrier integrity between lacunar and cortical ischemic stroke subtypes. Stroke 39: 1327-1332.

Zang, R.L. and M. Chopp, C. Zaloga, Z.G. Zhang, N. Jiang, S.C. Gautam, W.X. Tang, W. Tsang, D.C. Anderson, and A.M. Manning. 1995. The temporal profiles of ICAM-1 protein and mRNA expression after transient MCA occlusion in the rat. Brain Res. 682: 182-188.

CHAPTER 31

Adhesion Molecules in Dementia

Sabina Janciauskiene[1], Sun Yong-Xin[2] and Jia Jianping[3]
[1]Lund University, Department of Medicine, University Hospital Malmö, 20502
Malmö, Sweden, E-mail: Sabina.Janciauskiene@med.lu.se
[2]Department of Neurology, Xuan Wu Hospital of Capital Medical University,
Beijing, P. R. 100053, China, E-mail: dailynewsun@yahoo.com
[3]Department of Neurology, Xuan Wu Hospital of Capital Medical University,
Beijing, P. R. 100053, China, E-mail: jiajp@vip.sina.com

ABSTRACT

During the past decade it has become evident that immunological, inflammatory and vascular processes play an important role in the etiology and pathogenesis of various neuro-degenerative diseases. Alzheimer's disease (AD) is a complex and genetically heterogeneous disease that is the most common form of dementia and affects up to 15 million individuals worldwide. The presenting pathology of AD includes extacellular neuritic plaques composed of beta-amyloid peptide (Aβ) and intracellular neurofibrillary tangles composed of hyperphosphorylated tau, with neuronal loss in specific brain regions. A large body of evidence suggests that some form(s) of the polymorphic Aβ are neurotoxic and induce neuronal death, tau hyperphosphorylation and neuronal death. However, the mechanisms underlying these pathological changes are still largely unknown. The early stages of symptomatic AD are characterized by memory impairment and subtle behavioral changes that are associated with changes in synaptic function. The loss of synapses strongly correlates with cognitive decline in AD and is now thought to result from the interactions of toxic forms of Aβ peptide with molecules that are essential for neuronal integrity and synaptic connections. A combination of cell culture and animal studies has recently shown that adhesion molecules play important roles in synapse initiation, maturation, and function. Functional studies of individual adhesion molecules have begun to provide information on their role in synapse assembly and synaptic plasticity. In this chapter, we review the roles of different

---

Key terms are defined at the end of the chapter.

families of adhesion molecules, including the immunoglobulins, integrins, cadherins and selectins, in normal brain and in dementia, particularly Alzheimer's disease.

INTRODUCTION

Almost 2% of the population of western industrialized countries is affected by Alzheimer's disease (AD). As AD advances, symptoms include confusion, irritability and aggression, language breakdown, and long-term memory loss. Individual prognosis is difficult to assess, as the duration of the disease varies. AD develops for an indeterminate period of time before becoming fully apparent, and it can progress undiagnosed for years. The mean life expectancy following diagnosis is approximately seven years. AD is characterized by loss of neurons and synapses in the cerebral cortex and certain subcortical regions. However, the pathogenetic processes leading to this are essentially unknown. Intracellular trafficking and proteolytic processing of amyloid precursor protein (APP) have been the focus of numerous investigations over the past two decades. APP is the precursor of the amyloid beta-protein (Aβ), the 38-43–amino acid residue peptide that is at the heart of the amyloid cascade hypothesis of AD. Today it is suggested that mild cognitive impairment in early AD may be due to synaptic dysfunction caused by the accumulation of non-fibrillar, oligomeric Aβ, long before widespread synaptic loss and neurodegeneration occurs.

Memory impairment is a process associated with alterations in neuronal plasticity, synapses formation and stabilization. The synapse is the point of communication where virtually all important brain activity emerges, including interactions between one neuron and a neighboring neuron, muscle cell or gland cell. Synapses require many classes of cell-adhesion proteins for recognition of pre- and post-synaptic sides, specification of neurotransmitter type, structural cohesion, signaling, and many other properties. Therefore, dysregulated production of adhesion molecules may be one important event that leads to cognitive decline and dementia.

In the past 10 years, many of the molecules that are responsible for general adhesive interactions between a growth cone and its environment, and others that help in choosing which surfaces axons grow on, have been identified. Many of the molecules involved in adhesion are glycoproteins and belong to the immunoglobulin superfamily (their adhesion properties are Ca^{2+}-independent), the cadherins (their adhesion properties are Ca^{2+}-dependent) and the integrins. These mediate interactions between the cell-surface and the extracellular matrix.

THE IMMUNOGLOBULIN SUPERFAMILY IN DEMENTIA

The superfamily of immunoglobulin molecules contains extracellular immunoglobulin domains that mediate intercellular adhesion. These molecules are upregulated in expression during the immune response and are important for cell activation and migration (Table 1). Neuronal cell adhesion molecules of the immunoglobulin superfamily (IgCAMs) play crucial roles in the formation of neural circuits, including cell migration, axonal and dendritic targeting and synapse formation. Furthermore, in perinatal and adult life, neuronal IgCAMs are required for the formation and maintenance of specialized axonal membrane domains, synaptic plasticity and neurogenesis. An excellent monograph on neuronal immunoglobulin superfamily proteins has summarized most of the earlier work on Ig superfamily proteins in the nervous system (Sonderegger 1998). Here, we present an update of the developments in the field in recent years.

Table 1   Cell adhesion molecules of the immunoglobulin superfamily (IgCAMs) in the brain

Molecule	Ligands	Distribution
AMOG	–	Glial
CNTN1	NOTCH1	Neural; neuromuscular junctions
L1 (CD171)	Axonin	Neural
MAG	MAG	Myelin
MOG	–	Myelin; Oligodendrocytes Cytoplasmic membranes
NCAM-1 (CD56)	NCAM-1 via polysialic acid modulated by sialytranferase X, Polysialytransferase	Neural cells
NrCAM	Ig superfamily	Neural
NrCAM-2		Neural
OBCAM	Opioids; acidic lipids	Brain
Po protein	Po	Myelin
PMP-22	PMP-22	Myelin
Necl1 (SynCAM3; CADM3)	Necl1, Necl2, Necl4	Peripheral nervous system, myelinated axons; dorsal root ganglion; central nervous system neurons
Necl2 (SynCAM1; CADM1)	Ig superfamily Necls and Nectins	Pre- and post-synaptic dorsal root ganglion neurons
Necl3 (SynCAM2; CADM2)		Myelinated axons
Necl4 (SynCAM4; CADM4)	Necl1 Ankyrin-G	Myelinating Schwann cells, dorsal root ganglion neurons
Neurofascin		Neural; synapse
MDGA1		Neural

This table shows the important neuronal cell adhesion molecules of the immunoglobulin superfamily (IgCAMs) that play a crucial role in the formation of neural circuits at different levels: cell migration, axonal and dendritic targeting as well as synapse formation. AMOG, adhesion molecule on glial; CNTN1, contactin-1; NOTCH1, Notch homolog 1, a human gene encoding a single-pass transmembrane receptor; L1(CD171), L1 cell adhesion molecule with important functions in the development of the nervous system; MAG, myelin-associated glycoprotein; MOG, myelin-oligodendrocyte glycoprotein; NCAM-1 (CD56), neural cell adhesion molecule 1; NCAM-2, neural cell adhesion molecule 2; NrCAM, neuron-glia-related cell adhesion molecule; NrCAM-2, neuron-glia-related cell adhesion molecule 2; OBCAM, opioid-binding cell adhesion molecule; PMP-22, peripheral myelin protein; Necl1, nectin-like molecule 1 (plays a role in the formation of synapses, axon bundles and myelinated axons); Necl2, nectin-like molecule 2; Necl3, nectin-like molecules 3; Necl4, nectin-like molecule 4; MDGA1, MAM domain-containing glycosylphosphatidylinositol anchor protein 1.

Recent evidence suggests that members of the immunoglobulin superfamily are involved in functional neuronal plasticity, such as that associated with learning, and that both short- and long-term memory require adhesion molecules for synapse formation. For example, neural cell adhesion molecule, L1 (CD171), netrin-1, and DCC (deleted in colorectal cancer) are widely expressed IgCAMs during development, and mediate homophilic binding between neighboring cells such as neurons, astrocytes, and muscle cells and heterophilic interactions between cells and extracellular matrix components. Several other members of this superfamily, SynCAM, Sdk-1 and Sdk have been discovered that appear to have important functions in the earliest stages of the formation and organization of synaptic contacts (Fig. 1).

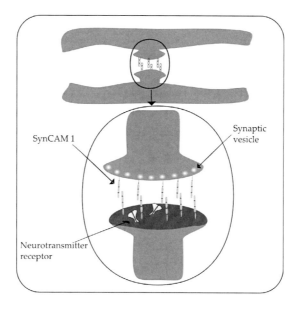

Fig. 1 Role of cell adhesion molecules in synapse formation. Synapse formation requires the precise alignment and attachment of pre-synaptic and post-synaptic cells. Hemophilic cell adhesion molecules have now been found to have a role in these processes on both sides of the synaptic cleft. SynCAM, an example of one of several synaptically localized adhesion molecules, remains at the synapse to hold the pre-synaptic and post-synaptic terminals together. The pre-synaptic terminal is filled with synaptic vesicles (white spheres indicated by arrow), and neurotransmitter receptors are recruited to the post-synaptic membrane (Philip Washbourne *et al.* 2004).

Another class of Ig proteins that has been implicated in synapse formation comprises the nectins, a family of four members that show a striking similarity in domain organization to SynCAM. Nectin-1 and nectin-3 are positioned in an asymmetric manner in a specialized form of synaptic junctions in adult hippocampal neurons, the puncta adherentia junctions, which are thought to be mechanical anchoring sites that stabilize neurotransmitter release sites and which also contain homophilic cadherin complexes. Interestingly, the disruption of nectin interactions at the developing synapse *in vitro* was found to result in a significant decrease in synapse size followed by a possibly compensatory increase in the overall number of synapses (Mizoguchi *et al.* 2002).

Circulating adhesion molecules have been found in the sera of healthy individuals, and increased levels have been described in dementia-related neuro-degeneration and aging. These observations point to an abnormal processing and/or shedding of IgCAMs, which may reflect changes in adhesion molecule–related cell interactions. There is also some evidence that circulating adhesion molecules interfere with various cell-cell interactions and may act as signal transducers, but their full biological roles remain unclear.

Concentrations of soluble L1 and NCAM were found to be significantly increased in the cerebrospinal fluid (CSF) of AD patients compared to a normal control group. Proteolytic fragments of L1, but not NCAM, were also elevated in patients with vascular dementia and dementia of mixed type (Strekalova *et al.* 2006). Elevated levels of soluble forms of intercellular adhesion molecule 1 (sICAM-1, or CD54), vascular cellular adhesion molecule 1 (VCAM-1, or CD106), and the CD40 ligand (CD40L, or CD154) have been reported in patients with various brain diseases, such as atherosclerosis, ischemic stroke, cerebrovascular disease and AD (Fassbender 1999, Rentzos 2004).

ICAM-1 can be upregulated by several cytokines such as interferon-γ, interleukin-1 and tumor necrosis factor α, and in central nervous system (CNS) pathologies ICAM-1 is aberrantly expressed by cerebral endothelial cells and astrocytes. *In vitro*, the Alzheimer's peptides (Aβ) have also been shown to induce expression of ICAM-1 in cultured vascular endothelial cells and to induce leukocyte adherence and endothelial and smooth muscle cell activation. Experimental studies also show that Aβ-induced microglia activation and tau phosphorylation are dependent on CD40/CD40L interaction, and that the inhibition of CD40L results in decreased amyloid production and microglia activation in a transgenic animal model of AD (Tan *et al.* 1999). Thus, higher levels of sICAM-1 and sCD40 in AD patients correlate with Aβ pathology. This is further supported by post-mortem studies of AD patients that revealed upregulated CD40 in blood vessel walls (Togo *et al.* 2000) and localization of ICAM-1 in senile plaques (Akiyama *et al.* 1993).

While ICAM-1 is expressed on a broad variety of cells, the expression of VCAM-1, another member of the immunoglobulin superfamily, is more restricted to endothelial cells, and VCAM-1 is assumed to be involved in lymphocyte, monocyte, and eosinophil but not neutrophil adhesion to activated endothelium. After firm adhesion, leukocytes proceed to transendothelial migration through adhesion mediated by platelet-endothelial cell adhesion molecule 1 (PECAM-1), which is expressed by endothelial cells, platelets, and various leukocytes. PECAM-1 is also a major constituent of the endothelial cell intercellular junction, which is capable of mediating homophilic as well as heterophilic interactions. A soluble form of VCAM-1 as well as PECAM have been identified in AD, but their biological significance remains to be elucidated.

The question whether circulating levels of various adhesion molecules are associated with a risk for dementia remains to be answered. For instance, The Rotterdam Study has shown that levels of the soluble forms of ICAM-1 and VCAM-1 are not associated with dementia risk (Engelhart *et al.* 2004). However, from recent studies it appears that cognitively intact elderly subjects with CSF levels of sICAM-1 above the median (> 893 ng/L) have a regional cerebral blood flow pattern similar to that seen in AD patients, i.e., increased frontal and fronto-temporal and reduced temporo-parietal region flow (Janciauskiene 2008). Moreover, a significant correlation was found between CSF levels of sCD40 and poor cognitive performance in elderly controls without cognitive symptoms (Buchhave *et al.* 2009). Together, these findings indicate that some soluble adhesion molecules may be elevated in the very early stages of preclinical AD.

THE INTEGRIN SUPERFAMILY MEMBERS IN DEMENTIA

Integrins constitute a superfamily of membrane proteins that allow individual cells to communicate with their environment. Through their interactions with the extracellular matrix, external stimuli are transmitted internally through the activation of various signaling cascades, thus allowing cells to adapt to environmental changes. Integrins are expressed in all cells throughout the brain and influence long-term potentiation (LTP) in hippocampal neurons, regulate neurite outgrowth induced by growth factors, and regulate survival and death signals. Integrins are expressed widely at dendritic spines and synapses in the brain and are known to be involved in the regulation of synaptic transmission and synaptic plasticity (Table 2).

Consistent with this, synaptic plasticity was found to be strongly reduced in Drosophila mutants in which a gene encoding synaptic integrins is deleted (Rohrbough *et al.* 2000). Moreover, peptides containing the arginine-glycine-aspartate (RGD) binding sequence, which is recognized by a large number of integrins present in neurons, have been found to cause an inhibition of LTP (Bahr *et al.* 1997) and mutant mice lacking integrin-associated protein, which is associated with integrin and functions as a cell adhesion molecule, are also deficient in LTP (Chang *et al.* 1999). There is also an alteration of synaptic plasticity in mutant mice lacking laminin α2, a glycoprotein abundant in basement membrane (Anderson *et al.* 2005).

Table 2 Members of the integrin superfamily in the brain

Molecule	Ligand	Distribution
α1β1	Laminin, collagen, tenascin	Glial, perineurium, Schwann, endothelial cells
α2β1	Laminin, collagen	Endothelium, astrocytes, Schwann cells
α3β1	Laminin, collagen, fibronectin	Endothelium, epithelium, astrocytes
α4β1	α4β1, α4β7, fibronectin, VCAM-1, thrombospondin-1	Endothelial, neural-crest-derived cells
α5β1	Fibronectin, murine L1	Epithelium, endothelium, astrocytes
α6β1	Laminin	Epithelium, endothelium, T-cells, glia
α8β1 (CD-/CD29; VLA-8)	Fibronectin, vitronectin, tenascin	Epithelium, neurons, oligodendroglia
αvβ1	Vitronectin, fibronectin, collagen, von Willebrandt factor, fibrinogen	Oligodendroglia
αL β2 (LFA-1α; CD11a)	ICAM-1, ICAM-2, ICAM-3	Macrophages, T-cells, microglia
αM β2 (CD11b)	ICAM-1, Factor X, iC3b, fibrinogen	Macrophages, microglia
αX β2 (CD11c)	iC3b, fibrinogen	Dendritic cells, microglia
αv β3 (CD51/CD61)	Fibronectin, osteopontin, von Willebrandt`s factor, PECAM-1, vitronectin, fibrinogen, human L1, thrombospondin, collagen	Glia, Schwann cells, endothelium
α 6 β4	Laminin	Endothelium, epithelium
αv β6	fibronectin, fibrinogen	Epithelium, oligodendroglia
αv β8	Vitronectin	Schwann cells, endothelium oligodendroglia, brain synapses

This table shows the important members of the brain integrin superfamily of cell-surface adhesion molecules consisting of eight different β-chains and 18 different α-chains that assemble as heterodimers to mediate cell-cell and cell-matrix interactions. VCAM-1, vascular cell adhesion molecule 1; ICAM, Inter-cellular adhesion molecule; PECAM-1, platelet/endothelial cell adhesion molecule 1; iC3b, the proteolytically inactive product of the complement cleavage fragment C3b. The table is based on data presented on the website http://neuromuscular.wustl.edu/lab/adhesion.htm

Neuronal dystrophy is associated with synaptic loss in culture and in the AD brain, and is a unique pathological feature of AD. The ability of the neuron to respond dynamically to extracellular cues is reminiscent of plasticity mechanisms.

In this regard, maladaptive neuronal plasticity may play a major role in AD. Several studies have shown that Aβ binds to integrins and activates the focal adhesion proteins paxillin and focal adhesion kinase, which are downstream of integrin receptors, suggesting that focal adhesion signaling cascades might be involved in Aβ-induced neuronal dystrophy and cell death. Therefore, characterization of the molecular pathway(s) by which Aβ induces neuronal dystrophy may suggest therapies to block maladaptive plasticity to preserve neuronal function and synaptic integrity.

Recent studies report that integrins bind both the soluble and fibrillar forms of Aβ, mediating Aβ signal transmission from extracellular sites into the cell and ultimately to the nucleus. Thus, integrins probably play a role in AD pathology as potential neuronal and glial receptors mediating Aβ-induced cytotoxicity. Aβ can form a meshwork that resembles an extracellular matrix, and cells are known to attach to both soluble and fibrillar Aβ-coated plates in an integrin-dependent manner. Evidence was provided by Wright *et al.* (2007) that integrins mediate attachment of Aβ to the cell surface: Aβ deposition on the cell surface was inhibited by specific adhesion-blocking antibodies for αv and β1 integrins. Integrins containing the αv subunit are widespread throughout the brain, including the hippocampus. The finding that the αv integrin-specific antibody was an effective antagonist of Aβ both *in vivo* and *in vitro* indicates that the prevention of LTP inhibition *in vivo* is due to a direct action in the hippocampus. Such evidence supports the potential therapeutic usefulness of antibodies to target αv integrins within the brain or small molecule antagonists of αv integrins. The compound SM256, when injected systemically, prevented the inhibition of LTP by centrally applied Aβ (Wang *et al.* 2008). This proof of concept provides encouragement for the search for potent and selective integrin subtype antagonists.

Fibrillar Aβ has been shown to induce apoptotic cascades in neurites and synapses. Thus, aberrant focal adhesion activation by Aβ may lead to the initiation of localized apoptotic cascades normally involved in adaptive plasticity in both neurites and synapses. Brain regions with the highest plasticity are most vulnerable in AD, suggesting that under pathological conditions such as presenilin mutations or Aβ deposition, neuronal plasticity may result in neuronal dysfunction.

Integrins can also interact with other membrane-bound proteins, including CD147, which is a transmembrane glycoprotein and a member of the immunoglobulin superfamily (IgSF) of receptors that associate with two integrin isoforms, α3β1 and α6β1, at points of cell-cell contact. CD147 has been observed to be a component of the γ secretase complex, which is an important regulator of the production of Aβ peptides. Results from experimental models have shown that the deletion of cellular CD147 from the γ secretase complex by RNA interference results in an increase in Aβ peptide production.

Interestingly, neurological abnormalities observed in CD147-null mice, including spatial learning and memory deficits, are similar to those observed in mouse models of AD. Although the mechanism by which CD147 attenuates the production of amyloid β peptides is unknown, further studies of CD147 function within the γ secretase complex may promote the development of new AD therapy. α6β1 integrins associated with CD147 bind amyloid β peptides and initiate a signal transduction cascade in microglial cells (Bamberger *et al.* 2003), indicating the potential for another level of regulation influenced by CD147 expression in AD.

Thus, a pathological relationship among Aβ deposition, neuronal dystrophy, and synaptic loss may lead to cognitive impairment in AD. The molecular mechanism by which neuronal dystrophy occurs is unclear. Neuronal dystrophy and cell death induced by fibrillar Aβ take place over different time intervals and at different Aβ concentrations, pointing to the possibility that the two events are mediated by two distinct molecular mechanisms. Extracellular signals producing alterations in the cytoskeleton are often transduced through adhesion proteins. In this regard, Aβ could promote dystrophy by aberrantly activating signal transduction cascades leading to cytoskeletal changes.

THE CADHERIN SUPERFAMILY MEMBERS IN DEMENTIA

Cadherins constitute a superfamily comprising more than 100 members in vertebrates, grouped into subfamilies that are designated as classic cadherins, desmosomal cadherins, protocadherins, Flamingo/CELSRs and focal adhesion targeting domain. With a few exceptions, cadherins are transmembrane proteins. In cadherin-based adherens junctions, the extracellular domains of transmembrane cadherins promote cell-cell adhesion by engaging in Ca^{2+}-dependent homophilic interactions, while the cytoplasmic domains are linked to the actin cytoskeleton via α- and β-catenins. Classical cadherins, including epithelial (E)- and neural (N)-cadherins, are major cell-cell adhesion receptors involved in axon sorting and in the regulation of physiological functions of the brain, such as long-term potentiation in the hippocampus. Both E- and N-cadherins have been found at the synapse and are reported to play key roles in synaptic structure and function.

Recent findings point out the important relationship between cadherins and presenilin-1 (PS1) in the brain. PS1 has been the subject of intensive study in relation to AD, since it has been shown that PS1 mutations are linked to familial Alzheimer's disease (FAD). PS1 is a member of the high molecular weight complex of γ-secretase, which generates the carboxyl end of β-amyloid peptide (gamma-cleavage) and leads to the accumulation of the Aβ peptide in amyloid plaques in AD-affected brains. Recent evidence suggests that upon formation of cell-cell contacts, presenilin colocalizes with cadherins and stabilizes the cadherin-based adhesion complex. PS1 forms complexes with E-cadherin and N-cadherin and it has been localized at synaptic sites. Thus, PS1 is suggested to have two distinct activities: (1) Under conditions promoting cell-cell adhesion, incorporation of PS1 into the E-cadherin-catenin complex may result in the stabilization of cell-cell links. (2) In contrast, under conditions of cell-cell dissociation or apoptosis, PS1 may promote disassembly of adherence junctions, thus facilitating cell separation. Having both activities in the same polypeptide may represent an efficient and quick way for the formation and dissolution of cell-cell contacts.

The association of PS1 with cadherins leads to the epsilon-cleavage [(ε-)-cleavage] of E- and N-cadherin by PS1 and the formation of an intracellular peptide (E-cad/C-terminal fragment 2 (CTF2) and N-cad/CTF2, respectively). Recent findings raise the exciting possibility that ε-cleavage of N-cadherin regulates the expression of genes implicated in memory and that decreased cleaved N-cadherin in FAD associated with PS1 mutations might contribute to the impaired memory observed in AD patients. These findings also suggest an alternative explanation for FAD that is distinct from the widely accepted 'amyloid hypothesis': dysfunction in transcription regulatory mechanisms.

The protocadherins (Pcdh) represent the largest subgroup of the cadherin superfamily. Protocadherin-γ (Pcdhγ) proteins are expressed throughout the nervous system and most are already present during embryogenesis. They are enriched at synapses and have been proposed to be involved in synapse formation, specification and maintenance. The loss of all Pcdhγ proteins results in a remarkable loss of specific subpopulations of spinal neurons, which might account for the early lethality of mice in which this gene cluster is knocked out. Despite the widespread expression of Pcdhγ proteins in the embryonic nervous system and their homology to classical cadherins, alterations in axonal growth, adhesion and migration were not observed in Pcdhγ mutants. Intriguingly, the loss of spinal neurons observed in Pcdhγ-deficient mice is restricted to interneurons, which undergo apoptosis late in embryogenesis. As yet, it is unclear whether Pcdhγ proteins directly activate anti-apoptotic pathways or whether the loss of spinal interneurons occurs indirectly as a consequence of secondary events caused by the ablation of the entire Pcdhγ gene cluster. Nevertheless, these observations indicate that single or several members of the γ-family of protocadherins are required for the survival of spinal interneurons.

To address the role of Pcdhγ proteins in synapse formation and maturation, Pcdhγ-deficient mice in which early cell death was circumvented were bred, either by intercrossing them with Bax-deficient mice or by generating a hypomorphic Pcdhγ allele, which leads to reduced apoptosis of interneurons in the spinal cord. Detailed analysis of these two mouse mutants revealed that although cell death was successfully prevented, both excitatory and inhibitory synapses in the spinal cord were severely altered in number and strength (Junghans *et al.* 2005). Reduced size in puncta labeling of excitatory synapse terminals together with reduced current amplitudes measured *in vitro* suggest that Pcdhγ proteins play an important role in the strengthening and maturation of synapses. It seems reasonable to speculate that these functions of Pcdhγ proteins at synapses might contribute to the loss of interneurons observed in Pcdhγ-deficient mice.

Most neurological disorders are characterized by extensive axonal degeneration or cell loss. Therefore, there is a clear need to develop strategies to reverse the lost functions.

SELECTIN SUPERFAMILY MEMBERS IN DEMENTIA

Increased lymphocyte trafficking across the blood-brain barrier is a prominent and early event in inflammatory and immune-mediated CNS diseases (Fig. 2). The function of adhesion molecules that control the entry of leukocytes into the brain has not been fully elucidated. Although the roles of ICAM-1 and VCAM-1 have been well documented, the expression and role of selectins is still a matter of controversy.

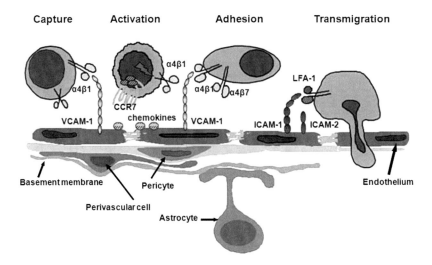

Fig. 2 Lymphocyte trafficking across the blood-brain barrier (BBB). In healthy individuals, lymphocyte traffic into the central nervous system (CNS) is very low and tightly controlled by the highly specialized BBB. Interaction of circulating leukocytes with the endothelium of the blood–spinal cord barrier and BBB therefore is a critical step in the pathogenesis of diseases of the CNS, such as AD. Leukocyte-endothelial interactions are mediated by adhesion molecules and chemokines and their respective chemokine receptors. Brain endothelial cells constantly express VCAM-1 and to a lower degree ICAM-1, and $\alpha 4\beta 1$ interaction with VCAM-1 mediates early steps of lymphocyte-BBB interaction. An interaction between endothelial ICAM-1 and ICAM-2 and LFA-1 is required for the cell adhesion and migration across the endothelial vessel wall of CNS microvessels (for review see Engelhardt 2006). VCAM-1, vascular cell adhesion molecule 1; ICAM-1 and -2, inter-cellular adhesion molecule 1 and 2; LFA-1, lymphocyte function-associated antigen 1; CCR7, chemokine (C-C motif) receptor 7; $\alpha 4\beta 1$, integrin that binds to the vascular cell adhesion molecule; $\alpha 4\beta 7$, integrin expressed on the surface of B and T lymphocytes, which plays an essential role in lymphocyte trafficking.

The selectins are a small family of lectin-like adhesion receptors comprising three members, L-, E-, and P- selectin. Selectins mediate heterotypic cell-cell interactions through calcium-dependent recognition of sialyated glycans. The best-defined physiological role of selectins is in leukocyte adherence to endothelial cells and platelets during inflammatory processes. The expression and function of selectins is tightly regulated so as to come into play only when leukocytes need to stick to the vessel wall as part of normal immune system cellular trafficking or during inflammation. E-selectin is synthesized and expressed on endothelial cells in response to inflammatory cytokines such as tumor necrosis factor α or interleukin-1. L-selectin is expressed constitutively on leukocytes, but its presentation at the cell surface may be regulated. P-selectin (CD62P) is a membrane glycoprotein that is rapidly mobilized to the surface of activated platelets and endothelial cells, where it mediates leukocyte-platelet and leukocyte-vascular endothelial cell adhesion. Moreover, P-selectin expression on activated platelets appears to be important for the formation of large, stable platelet aggregates and for the amplification of the leukocyte recruitment process and may prime monocytes for tissue factor and cytokine upregulation.

It is known that CD62E (E-selectin) and CD62P (P-selectin) are upregulated as part of the host response to injury or disease, where they play a key role in the initial tether-roll phase of the homing of leukocytes to sites of inflammation. The brain also utilizes the CD62 proteins, and they consequently offer an ideal, to date underexploited, biomarker for brain disease diagnosis.

Human studies have shown increased selectin levels in ventricular CSF from children with severe traumatic brain injury (Glasgow coma score < 8) (Whalen *et al.* 1998) and in patients with relapsing-remitting multiple sclerosis (Duran *et al.* 1999). Increased plasma levels of E-selectin have been shown to be associated with AD (Zuliani *et al.* 2008). Selectins are also thought to contribute to tissue injury in stroke. In multiple murine models of stroke, the use of selectin ligands to block selectin function has reduced infarct size. For example, van Kasteren and collaborators (van Kasteren *et al.* 2009) have demonstrated the application of carbohydrate-functionalized nanoparticles to the direct detection of endothelial markers E-/P-selectin (CD62E/CD62P) in acute inflammation. These first examples of glyconanoparticles visible by MRI display multiple copies of the natural complex of glycan ligand with selectins. Their resulting sensitivity and binding selectivity has made possible sensitive detection of disease in mammals with beneficial implications for treatment of an expanding patient population suffering from neurological disease.

Inflammatory mechanisms and immune activation play a role not only in neuro-degenerative conditions such as AD and vascular dementia, but also in age-associated cognitive decline. For instance, Yaffe *et al.* (2003) followed 3,031 well-functioning subjects for 2 yr and reported that a high baseline level of serum markers of inflammation, most notably C-reactive protein and P-selectin, was associated with poor cognitive performance and greater risk of cognitive decline over the follow-up period. Similar associations between elevated baseline C-reactive protein levels and greater cognitive decline have been reported in patients followed for 5 to 6 yr (Tilvis *et al.* 2004).

The trafficking of lymphocytes among blood and lymphoid and non-lymphoid tissue depends on the expression of specific adhesion molecules, including the membrane glycoprotein L-selectin (CD62L). L-selectin is constitutively expressed on all classes of leukocytes, including naïve T lymphocytes, which utilize L-selectin to migrate through lymph nodes and other secondary lymphoid tissue. L-selectin mediates leukocyte rolling on vascular endothelium at sites of inflammation and lymphocyte migration to peripheral lymph nodes. L-selectin is rapidly shed from the cell surface after leukocyte activation by a proteolytic mechanism that cleaves the receptor in an extracellular region proximal to the membrane. This process may allow rapid leukocyte detachment from the endothelial surface before entry into tissues.

The transition from a naïve to an activated memory T cell is generally accompanied by a shedding of L-selectin. CD8+ memory T cells, for example, acquire a distinct profile of other adhesion molecules that permits them to preferentially migrate through the tissue in which they were activated. A deficiency in L-selectin has been associated with alterations in both lymphocyte migration and reduced immune responsiveness (Steeber *et al.* 1996). Patients with AD who were classified as vulnerable showed reduced levels of circulating $CD4^+$ and $CD8^+$ and $CD62L^-$ T lymphocytes, possibly as a result of increased sympathomedullary activation (Mills *et al.* 1999). In general, these findings seem consistent with previous studies demonstrating functional immune deficits in elderly patients with AD and suggest the identity of specific lymphocyte subsets related to this phenomenon.

Internalized amyloid precursor protein either recycles to the plasma membrane, where α-secretase resides, or moves to acidic compartment(s) for β-secretase exposure. While the trans-Golgi network contains β-secretase activity, recent examination of the subcellular distribution of this protease, called β-amyloid cleaving enzyme, has led to the suggestion that β-secretase activity might also reside at the plasma membrane and in endosomes. To examine the role of endocytic compartments in β-secretase processing of amyloid precursor protein, the wild-type and endosomal sorting mutant P-selectin cytoplasmic domains were used to control movement of amyloid precursor protein through endosomes. Amyloid precursor protein/P-selectin, which is sorted from early to late endosomes, undergoes significantly less α-secretase cleavage, and more β-secretase cleavage, than amyloid precursor protein/P-selectin768A, a mutant that recycles more efficiently to the cell surface. These results demonstrate that endosomal sorting influences the relative exposure of the amyloid precursor protein/P-selectin chimeras to α- and β-secretase activities and suggest that, because delivery to late endocytic compartments favors β-secretase processing of amyloid precursor protein, it is likely that there is limited β-secretase activity in early endosomes and at the cell surface. It is proposed that the trans-Golgi network may be involved in both secretory and endocytic generation of Aβ protein (Daugherty and Green 2001).

SUMMARY

- Immunological, inflammatory and vascular processes play an important role in the etiology or pathogenesis of various neuro-degenerative diseases.
- Alzheimer's disease (AD) is the most common form of dementia and affects up to 15 million individuals worldwide.
- The pathology of AD includes extacellular neuritic plaques composed of β-amyloid peptide (Aβ) and intracellular neurofibrillary tangles composed of hyperphosphorylated protein tau.
- Neuronal death and loss of functional connections (synapses) between neurons, or between neurons and other types of cells in the cerebral cortex and hippocampus is a prominent feature of AD that is correlated with cognitive impairment.
- Adhesion molecules are suggested to play an important role in synapse assembly and synaptic plasticity.
- Cell-cell adhesions are important for brain morphology and highly coordinated brain functions such as memory and learning.
- It is necessary to learn about the mechanisms that govern the synthesis and degradation of cell adhesion molecules and their subsynaptic distribution and signaling in order to understand how these events contribute to the development of neuro-degenerative diseases such as AD.

Key Facts about Dementia and Adhesion Molecules in the Brain

1. The human brain is a highly complex organ and neurons are the basic units that make up the brain and nervous system.
2. Detailed delineation of the functional anatomy of different brain circuits, and discrimination of the unique computational activities performed by their component structures, has not yet been achieved in humans.
3. The progressive decline in cognitive function beyond what might be expected from normal aging is called dementia (from Latin *de-* 'apart, away' + *mens* 'mind').

4. Alzheimer's disease (AD) is among the leading causes of dementia in the elderly; it is suggested that up to 70% of dementia cases are due to AD.

5. AD is incurable and its diagnosis can only be confirmed on autopsy, by the presence of amyloid plaque, neurofibrillary tangles, neuronal and synaptic loss and brain atrophy in specific areas.

6. AD is characterized by defects in the remodeling of synapses (where nerve cells communicate), neuronal loss or dysfunction with implications for memory and other functions, and changes in the responses of non-neuronal cells (such as glial cells) involved in neuron survival and brain plasticity.

7. AD pathology involves molecules critical for the regulation of synaptic function, growth-inducing and growth-associated molecules, cytoskeletal proteins, synaptic molecules and adhesion molecules, among others.

8. A combination of cell culture and animal studies has recently shown that adhesion molecules play important roles in synapse initiation, maturation, and function.

9. Functional studies of individual adhesion molecules have begun to provide information on their role in synapse assembly and synaptic plasticity.

Definition of Terms

Adherens junctions: Junctions characterized by the presence of transmembrane, calcium-dependent cadherins that link to the actin cytoskeleton via catenins, and visible by electron microscopy as a focal subsurface change in density on plasma membranes.

Alzheimer's disease: The most common form of dementia. This incurable, degenerative, and terminal disease was first described by the German psychiatrist Alois Alzheimer in 1906 and was named after him.

Amyloid peptide (Aβ): A peptide of 39-43 amino acids, the main constituent of amyloid plaques in the brains of patients with Alzheimer's disease. Similar plaques appear in some variants of Lewy body dementia and in inclusion body myositis, a muscle disease. Aβ also forms aggregates coating cerebral blood vessels in cerebral amyloid angiopathy.

Amyloid precursor protein (APP): A transmembrane glycoprotein of undetermined function. The cleavage of AAP by the β and γ secretases generates Aβ protein.

Cerebral blood flow: A blood supply to the brain per unit time.

Cognitive decline: A decline in memory and cognitive (thinking) function, considered to be a normal consequence of aging. Assessment of cognitive function among older adults requires specialized training and refined psychometric tools.

Focal adhesions: The specific types of large macromolecular assemblies through which both mechanical force and regulatory signals are transmitted.

Long-term potentiation: The long-lasting improvement in communication between two neurons that results from stimulating them simultaneously. Since neurons communicate via chemical synapses, and because memories are believed to be stored within these synapses, LTP is considered one of the major cellular mechanisms that underly learning and memory.

Presenilins: A family of related transmembrane proteins that function as a part of the gamma-secretase protease complex. Vertebrates have two presenilin genes, called PSEN1 (located on chromosome 14 in humans), which encodes presenilin 1 (PS-1), and PSEN2 (on chromosome 1 in humans), which codes for presenilin 2 (PS-2). Mutations in the presenilin proteins are known to cause early onset alzheimer's disease through mechanisms which are still being elucidated.

Synaptic transmission: Also called neurotransmission, an electric movement within synapses caused by the propagation of nerve impulses.

Abbreviations

AD	Alzheimer's disease
AMOG	adhesion molecule on glial
APP	amyloid precursor protein
Aβ	beta-amyloid peptide
CNS	central nervous system
CNTN1	contactin-1
CSF	cerebrospinal fluid
CTF2	C-terminal fragment 2
DCC	deleted in colorectal cancer
FAD	familial Alzheimer's disease
IgCAMs	cell adhesion molecules of the immunoglobulin superfamily
IgSF	immunoglobulin superfamily
LTP	long-term potentiation
MAG	myelin-associated glycoprotein
MDGA1	MAM domain-containing glycosylphosphatidylinositol anchor
MOG	myelin-oligodendrocyte glycoprotein
NCAM-1	neural cell adhesion molecule 1
NCAM-2	neural cell adhesion molecule 2
Necl1	nectin-like molecule 1
Necl2	nectin-like molecule 2
Necl3	nectin-like molecule 3
Necl4	nectin-like molecule 4
NOTCH1	Notch homolog 1
NrCAM	neuron-glia-related cell adhesion molecule
NrCAM-2	neuron-glia-related cell adhesion molecule 2
OBCAM	opioid-binding cell adhesion molecule
Pcdh	protocadherins

Pcdhγ	Protocadherin-γ
PECAM-1	platelet-endothelial cell adhesion molecule 1
PMP-22	peripheral myelin protein
PS1	presenilin 1
RGD	arginine-glycine-aspartate
sICAM-1	soluble forms of intercellular adhesion molecule 1
VCAM-1	vascular cell adhesion molecule 1

References

Akiyama, H. and T. Kawamata, T. Yamada, I. Tooyama, T. Ishii, and P.L. McGeer. 1993. Expression of intercellular adhesion molecule (ICAM)-1 by a subset of astrocytes in Alzheimer's disease and some other degenerative neurological disorders. Acta. Neuropathol. 85: 628-634.

Anderson, J.L. and S.I. Head, and J.W. Morley. 2005. Synaptic plasticity in the dy2J mouse model of laminin alpha2-deficient congenital muscular dystrophy. Brain Res. 1042: 23-28.

Bahr, B.A. and U. Staubli, P. Xiao, D. Chun, Z.X. Ji, E.T. Esteban, and G. Lynch. 1997. Arg-Gly-Asp-Ser-selective adhesion and the stabilization of long-term potentiation: pharmacological studies and the characterization of a candidate matrix receptor. J. Neurosci. 17: 1320-1329.

Bamberger, M.E. and M.E. Harris, D.R. McDonald, J. Husemann, and G.E. Landreith. 2003. A cell surface receptor complex for fibrillar beta-amyloid mediates microglial activation. J. Neurosci. 23: 2665-2674.

Buchhave, P. and S. Janciauskiene, H. Zetterberg, K. Blennow, L. Minthon, and O. Hansson. 2009. Elevated plasma levels of soluble CD40 in incipient Alzheimer's disease. Neurosci. Lett. 450: 56-59.

Chang, H.P. and F.P. Lindberg, H.L. Wang, A.M. Huang, and E.H. Lee. 1999. Impaired memory retention and decreased long-term potentiation in integrin-associated protein-deficient mice. Learn Mem. 6: 448-457.

Daugherty, B.L. and S.A. Green. 2001. Endosomal sorting of amyloid precursor protein-P-selectin chimeras influences secretase processing. Traffic 2: 908-916.

Durán, I. and E.M. Martínez-Cáceres, J. Río, N. Barberà, M.E. Marzo, and X. Montalban. 1999. Immunological profile of patients with primary progressive multiple sclerosis. Expression of adhesion molecules. Brain 122: 2297-2307.

Engelhart, M.J. and M.I. Geerlings, J. Meijer, A. Kiliaan, A. Ruitenber, J.C. van Swieten, T. Stijnen, A. Hofman, J.C. Witteman, and M.M. Breteler. 2004. Inflammatory proteins in plasma and the risk of dementia: the Rotterdam study. Arch. Neurol. 61: 668-672.

Engelhardt, B. 2006. Molecular mechanisms involved in T cell migration across the blood-brain barrier. J. Neural. Transm. 113: 477-485.

Fassbender, K. and T. Bertsch, O. Mielke, F. Mühlhauser, and M. Hennerici. 1999. Adhesion molecules in cerebrovascular diseases. Evidence for an inflammatory endothelial activation in cerebral large- and small-vessel disease. Stroke 30: 1647-1650.

Janciauskiene, S.M. and C. Erikson, and S. Warkentin. 2009. A link between sICAM-1, ACE and parietal blood flow in the aging brain. Neurobiol. Aging 30: 1504-1511.

Junghans, D. and I.G. Haas, and R. Kemler. 2005. Mammalian cadherins and protocadherins: about cell death, synapses and processing. Curr. Opin. Cell Biol. 17: 446-452.

Mills, P.J. and J. Rehman, M.G. Ziegler, S.M. Carter, J.E. Dimsdale, and A.S. Maisel. 1999. Nonselective beta blockade attenuates the recruitment of CD62L(-)T lymphocytes following exercise. Eur. J. Appl. Physiol. Occup. Physiol. 79: 531-534.

Mizogushi, A. and H. Nakanishi, K. Kimura, K. Matsubara, K. Ozaki-Kuroda, T. Katata, T. Honda, Y. Kiyohara, K. Heo, M. Higashi, T. Tsutsumi, S. Sonoda, C. Ide, and Y. Takai. 2002. Nectin: an adhesion molecule involved in formation of synapses. J. Cell. Biol. 156: 555-565.

Rentzos, M. and M. Michalopoulou, C. Nikolaou, C. Cambouri, A. Rombos, A. Dimitrakopoulos, E. Kapaki, and D. Vassilopoulos. 2004. Serum levels of soluble intercellular adhesion molecule-1 and soluble endothelial leukocyte adhesion molecule-1 in Alzheimer's disease. J. Geriatr. Psychiatry Neurol. 17: 225-231.

Rohrbough, J. and M.S. Grotewiel, R.L. Davis, and K. Broadie. 2000. Integrin-mediated regulation of synaptic morphology, transmission, and plasticity. J. Neurosci. 20: 6868-6878.

Sonderegger, P. and S. Kunz, C. Rader, A. Buchstaller, P. Berger, L. Vogt, S.V. Kozlov, U. Ziegler, B. Kunz, D. Fitzli, and E.T. Stoeckli. 1998. Discrete clusters of axonin-1 and NgCAM at neuronal contact sites: facts and speculations on the regulation of axonal fasciculation. Prog. Brain Res. 117: 93-104.

Steeber, D.A. and N.E. Green, S. Sato, and T.F. Tedder. 1996. Humoral immune responses in L-selectin-deficient mice. J. Immunol. 157: 4899–4907.

Strekalova, H. and C. Buhmann, R. Kleene, C. Eggers, J. Saffell, J. Hemperty, C. Weiller, T. Müller-Thomsen, and M. Schachner. 2006. Elevated levels of neural recognition molecule L1 in the cerebrospinal fluid of patients with Alzheimer disease and other dementia syndromes. Neurobiol. Aging 27: 1-9.

Tan, J. and T. Town, D. Paris, T. Mori, Z. Suo, F. Crawford, M.P. Mattson, R.A. Flavell, and M. Mullan. 1999. Microglial activation resulting from CD40-CD40L interaction after beta-amyloid stimulation. Science 286: 2352-2355.

Tilvis, R.S. and M.H. Kahonen-Vare, J. Jolkkonen, J. Valvanne, K.H. Pitkala, and T.E. Strandberg. 2004. Predictors of cognitive decline and mortality of aged people over a 10-year period. J. Gerontol. A. Biol. Sci. Med. Sci. 59: 268-274.

Togo, T. and H. Akiyama, H. Kondo, K. Ikeda, M. Kato, E. Iseki, and K. Kosaka. 2000. Expression of CD40 in the brain of Alzheimer's disease and other neurological diseases. Brain Res. 885: 117-121.

van Kasteren, S.I. and S.J. Campbell, S. Serres, D.C. Anthony, N.R. Sibson, and B.G. Davis. 2009. Glyconanoparticles allow pre-symptomatic in vivo imaging of brain disease. Proc. Nat. Acad. Sci. USA 106: 18-23.

Wang, Q. and I. Klyubin, S. Wright, I. Griswold-Prenner, M.J. Rowan, and R. Anwyl. 2008. Alpha v integrins mediate beta-amyloid induced inhibition of long-term potentiation. Neurobiol. Aging 29: 1485-1493.

Washbourne, P. and A. Dityatev, P. Scheiffele, T. Biederer, J.A. Weiner, K.S. Christopherson, and A. El-Husseini. 2004. Cell adhesion molecules in synapse formation. J. Neurosci. 24: 9244-9249.

Whalen, M.J. and T.M. Carlos, P.M. Kochanek, S.R. Wisniewski, M.J. Bell, J.A. Carcillo, R.S. Clark, S.T. DeKosky, and P.D. Adelson. 1998. Soluble adhesion molecules in CSF are increased in children with severe head injury. J. Neurotrauma 15: 777-787.

Wright, S. and N.L. Malinin, K.A. Powell, T. Yednock, R.E. Rydel, and I. Griswold-Prenner. 2007. Alpha2beta1 and alphaVbeta1 integrin signaling pathways mediate amyloid-beta-induced neurotoxicity. Neurobiol. Aging 28: 226-237.

Yaffe, K. and K. Lindquist, B.W. Penninx, E.M. Simonsick, M. Pahor, S. Kritchevsky, L. Launer, L. Kuller, S. Rubin, and T. Harris. 2003. Inflammatory markers and cognition in well-functioning African-American and white elders. Neurology 61: 76-80.

Zuliani, G. and M. Cavalieri, M. Galvani, A. Passaro, M.R. Munari, C. Bosi, A. Zurlo, and R. Fellin. 2008. Markers of endothelial dysfunction in older subjects with late onset *alzheimer's* disease or vascular dementia. J. Neurol. Sci. 272: 164-170.

Index

About the Editor

Victor R. Preedy BSc, PhD, DSc, FIBio, FRCPath, FRSPH is Professor of Nutritional Biochemistry, King's College London and Professor of Clinical Biochemistry, Kings College Hospital. He is also Director of the Genomics Centre, King's College London.

Professor Preedy graduated in 1974 with an Honors Degree in Biology and Physiology with Pharmacology. He gained his University of London PhD in 1981 when he was based at the Hospital for Tropical Disease and The London School of Hygiene and Tropical Medicine. In 1992, he received his Membership of the Royal College of Pathologists and in 1993 he gained his second doctoral degree, i.e. DSc, for his outstanding contribution to protein metabolism in health and disease.

Professor Preedy was elected as a Fellow to the Institute of Biology in 1995 and to the Royal College of Pathologists in 2000. Since then he has been elected as a Fellow to the Royal Society for the Promotion of Health (2004) and The Royal Institute of Public Health (2004). In 2009, Professor Preedy became a Fellow of the Royal Society for Public Health.

In his career Professor Preedy has carried out research at the National Heart Hospital (part of Imperial College London) and the MRC Centre at Northwick Park Hospital. He has collaborated with research groups in Finland, Japan, Australia, USA and Germany. He is a leading expert on the pathology of disease and has lectured nationally and internationally. He has published over 570 articles, which includes over 165 peer-reviewed manuscripts based on original research, 90 reviews and 20 books.

Color Plate Section

Chapter 1

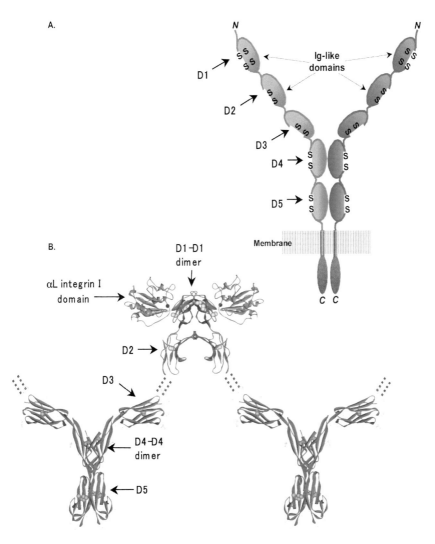

Fig. 2 Dimerization structure of intercellular adhesion molecule (ICAM-1). (A) Schematic of *cis*-dimerization of two ICAM-1 molecules (orange and blue), interactions of Ig-like domains four and five (D4, D5) are shown. (B) Ribbon diagram of two ICAM-1 dimers. The interaction of Ig-like domains D1 and D2 of one molecule from each dimer with its neighbor forms the binding site for the αA domain of αL integrin (green) (PDB code: 1MQ8) (Shimaoka *et al.* 2003). An approximate alignment of the D1-D2 structures with the dimer of D3-D5 (PDB code: 1P53) (Yang *et al.* 2004) is shown. Yellow sulfur molecules indicate the position of disulfide bonds. Divalent cations (Mg^{++}) are in pink. Figure adapted from Yang *et al.* (2004).

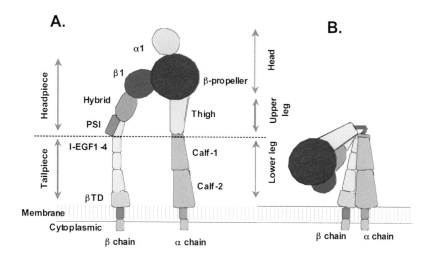

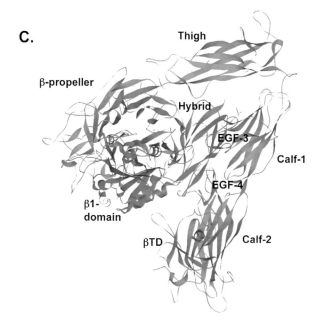

Fig. 3 Integrin organization. (A) Schematic of activated αXβ2-integrin. (B) The integrin in an inactive conformation. (A) and (B) were adapted from Luo *et al.* (2007). (C) Ribbon diagram showing the interaction between the extracellular segments of αV (blue) and β3 (orange) subunits (PDB code: 1JV2) (Xiong *et al.* 2001).

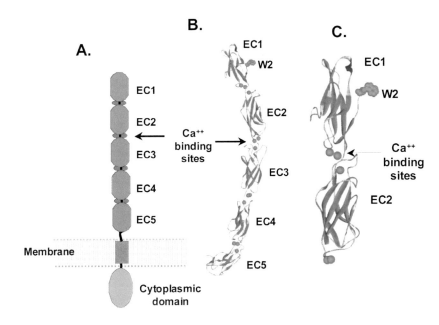

Fig. 5 Type 1 cadherin. (A) Schematic of E-cadherin, a classical cadherin with five extracellular cadherin domains (EC1-EC5) (blue) stabilized by Ca^{2+} at domain interfaces (green). (B) Ribbon diagram of the C-cadherin ectodomain (blue) with bound calcium molecules (green) from *Xenopus laevis* (PDB code:1L3W) (Boggon *et al.* 2002). (C) Ribbon diagram (represented according to secondary structure) of human E-cadherin domains EC1 and EC2 showing the Ig-like fold, the conserved tryptophan residue (W2) (orange) and the bound Ca^{2+} (PDB code: 2O72) (Parisini *et al.* 2007).

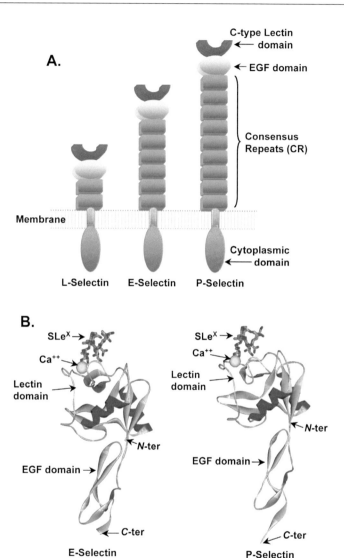

Fig. 7 The selectin family. (A) Schematics of L-selectin, E-selectin and P-selectin. (B) Structure of the C-type lectin and the EGF domains of E- and P-selectins (PDB codes: 1G1T and 1G1R) (Somers *et al.* 2000) depicted as ribbon diagrams represented according to secondary structure. Structural similarities are evident and, if optimally overlaid, bound Ca^{2+} ions are superimposed (Somers *et al.* 2000).

Chapter 2

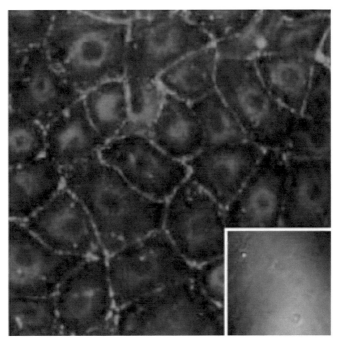

Fig. 1 Occludins tightly attaching adjacent cells. Occludins are special adhesion molecules in tight junctions between cells. Reprinted from Kaneda *et al.* (2006), with permission.

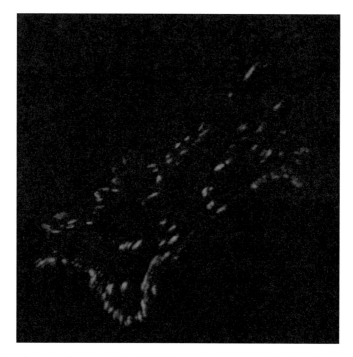

Fig. 2   Focal contacts demonstrated by TIRF. Focal contacts are specialized attachments between integrins and actin filaments that allow cells to pull on the substratum to which they are attached. This figure was taken using Total Internal Reflection Fluorescence (TIRF) imaging of a smooth muscle cell in culture using the antibody p-Tyr-FITC, Sigma #F3145 (clone #PT66). Courtesy of Soon-Mi Lim and Andreea Trache, Texas A&M Health Science Center College of Medicine.

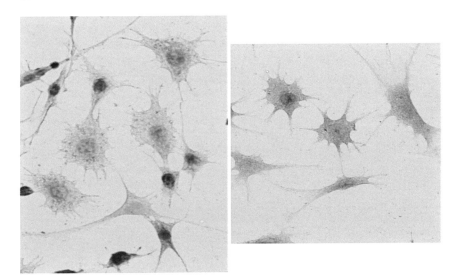

Fig. 3 Immunohistochemistry of p120 in MLO-Y4 osteocytes using a 3,3′-diaminobenzidine marker dye. Immunohistochemistry using monoclonal anti-mouse antibodies against p120 in MLO-Y4 cells. Left, a control with no primary antibody. Right, with a primary antibody and 3, 3′-diaminobenzidine marker dye. Original = 400 × (MLO-Y4 cells were kindly provided by Dr. Lynda Bonewald, Dept. Oral Biology, U. Missouri at KC School of Dentistry).

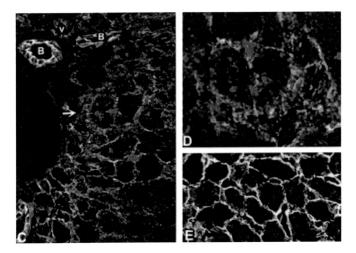

Fig. 4 Co-localizing of different proteins in the same tissue or cell. Double immunofluorescence staining for E-cadherin (red) and β-catenin (green) as seen with a laser scanning confocal microscope. Reprinted from Schmetz *et al.* (2001), with permission.

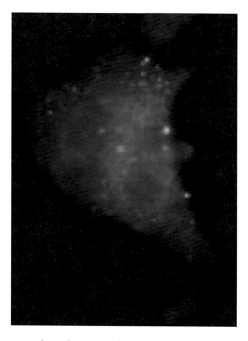

Fig. 5 Fluorescent immunohistochemistry of p120 in MLO-Y4 osteocytes using fluorescence microscopy. Fluorescent immunohistochemistry for labeled antibodies against mouse-p120 in MLO-Y4 cells. The primary antibody was mouse anti-p120 (BD Bioscience) with an Alexa Fluor 488 goat anti-mouse secondary antibody (Invitrogen). Yellow dots of labeled p120 are found scattered evenly throughout the cytoplasm of the MLO-Y4 cell.

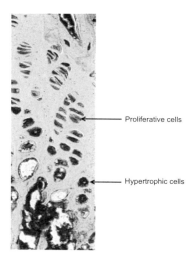

Fig. 7 Tibial growth plate demonstrating E-cadherin in various stages of development. E-cadherin seen in different stages of development in the tibial growth plate of a rat. Notice the nuclear localization in the zone of proliferation and cytoplasmic localization in the zone of hypertrophy.

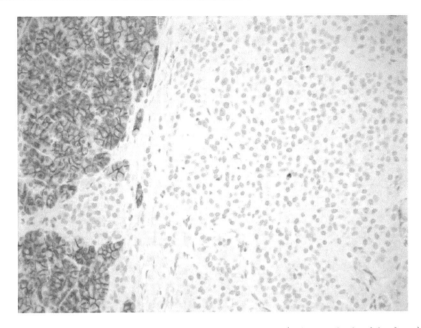

Fig. 8 'Cadherin switch'. E-cadherin is negative in the tumor (right two-thirds of the figure), whereas adjacent normal pancreatic parenchyma is heavily stained (left third of the figure). Reprinted from Chetty *et al.* (2008), with permission. © American Society for Clinical Pathology.

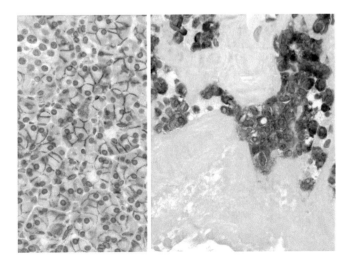

Fig. 9 Translocation of p120 catenin from the juxtamembrane position to cytoplasmic. Left, normal pancreatic parenchyma showing linear, crisp membrane staining for p120 in contrast to, right, the intense cytoplasmic positivity in a solid pseudopapillary tumor. Reprinted from Chetty *et al.* (2008), with permission. © 2008 American Society for Clinical Pathology.

Control   Ischemia

E-Cadherin

N-Cadherin

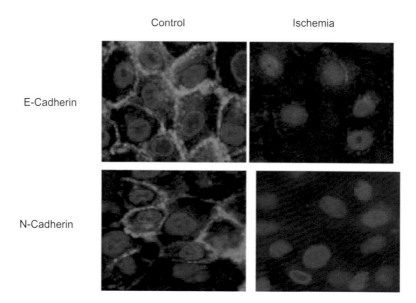

Fig. 10 Impact of ischemia on cadherin/catenin localization in pSM2 cells. pSM2 cells made ischemic by a mineral oil overlay model, demonstrating loss of E- and N-cadherin. Reprinted from Covington *et al.* (2006), with permission.

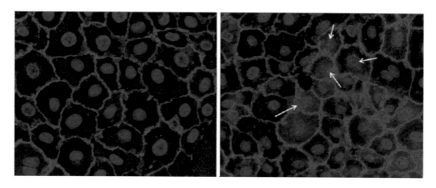

Fig. 11 Peri-nuclear translocation of β-catenin. Reperfusion (right) in a simulated ischemia model (left) in normal rat kidney causes translocation of β-catenin (red) to a perinuclear location (nucleus in blue).

Chapter 24

 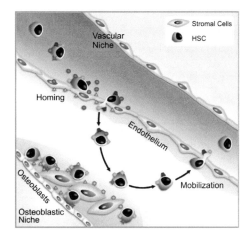

Fig. 1 Homing and mobilization in the hematopoietic stem cells niches. Hematopoietic stem cells express a number of adhesion molecules on their surface that are important to both homing and mobilization. Key molecules involved in these processes are selectins, integrins, CD44 and their ligands. Chemokines are important in guiding the homing of stem cells from the periphery into the bone marrow and ultimately into the stem cell niche.

Chapter 25

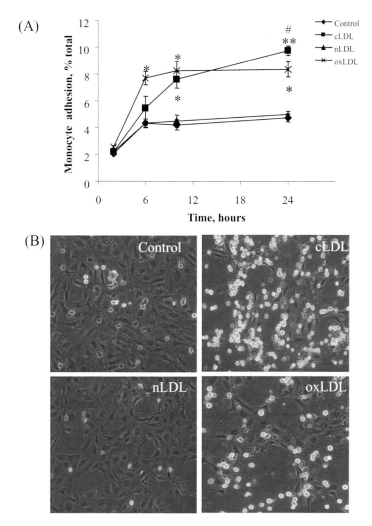

Fig. 3 Time course of the monocyte adhesion to endothelial cells treated with LDLs. (A) Cells were treated with 200 μg/mL LDL or vehicle control in 96-well plates and fluorescence was measured before and after fluorescence-labeled monocytes were allowed to adhere and non-adherent monocytes were washed out. The percentage of remaining fluorescence was calculated individually for each experimental well. Absolute data varied in the ranges of 622-685 and 12-58 units for total and remaining fluorescence measurement respectively. n = 4 per point, *P < 0.01 vs. either vehicle- or nLDL-treated cells, #P < 0.05 vs. oxLDL-treated cells at 24 hr. (B) Representative images. Endothelial cells are visualized with phase-contrast and labeled monocytes are detected using fluorescent microscopy. There is noticeable shrinkage and decreased density of HCAECs after cLDL or oxLDL treatment. Monocytes are adherent to remaining endothelial cells. Control cells were treated with vehicle or nLDL. Permission to publish obtained from Wolters Kluwer Health.

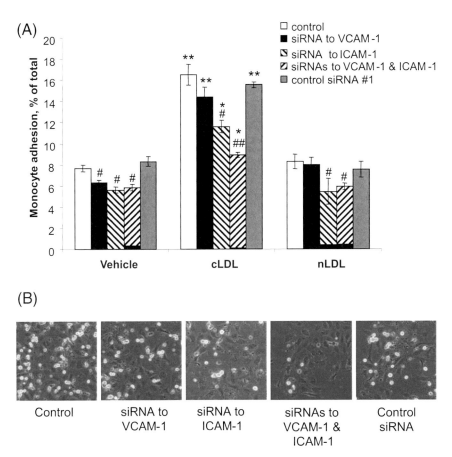

Fig. 6 Inhibition of monocyte adhesion by siRNA to ICAM-1 or VCAM-1. (A) HCAECs were transfected with anti-ICAM-1 or anti-VCAM-1 siRNA for 48 hr and then exposed with 200 μg/mL cLDL for 16 hr. Control cells were treated with vehicle or 200 μg/mL nLDL. The monocyte adhesion was measured as described in the text. n = 3-4 per point, *P < 0.01, **P < 0.001 vs. vehicle control cells pretreated with the same siRNA, #P < 0.05, ##P < 0.01 vs. no siRNA control cells (white bars) subjected to the same treatment. (B) Representative images of cells treated with cLDL. Permission to publish obtained from Wolters Kluwer Health.

Chapter 28

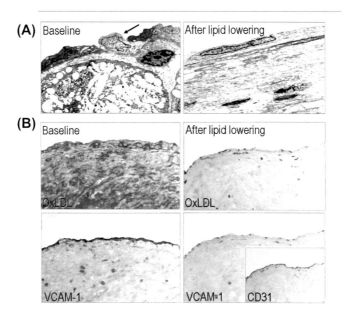

Fig. 1 Lipid lowering normalizes endothelial ultrastructure and reduces oxLDL accumulation and VCAM-1 expression in atherosclerotic rabbits. (A) Electron microscopy demonstrated that aortic EC in hypercholesterolemic rabbits fed an atherogenic diet for 4 mon (baseline, top left) showed a cuboidal structure typical of an 'activated' phenotype, whereas EC in rabbits after 16 mon lipid lowering (low, top right) had a more squamous morphology. A monocytic cell (arrowhead) appears to be entering the intima of baseline lesion. Original magnification × 3000. (B) OxLDL epitopes (MDA-lysine) accumulated in the aortic intima beneath EC immunoreactive for VCAM-1 in baseline rabbits fed the atherogenic diet for 4 mon (top left and bottom left), whereas oxLDL and VCAM-1 were barely detectable in the intima after dietary cholesterol lowering for 16 mon (top right and bottom right). CD3, an EC marker, indicated an intact monolayer (inset).